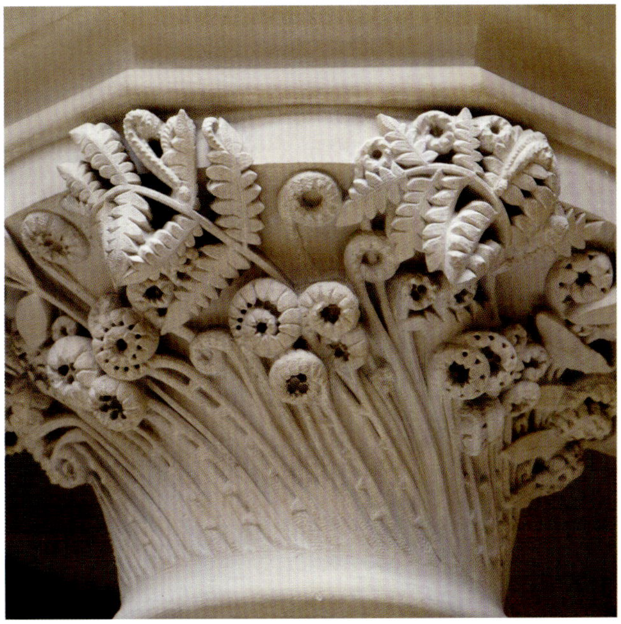

THE PRE-RAPHAELITES AND SCIENCE

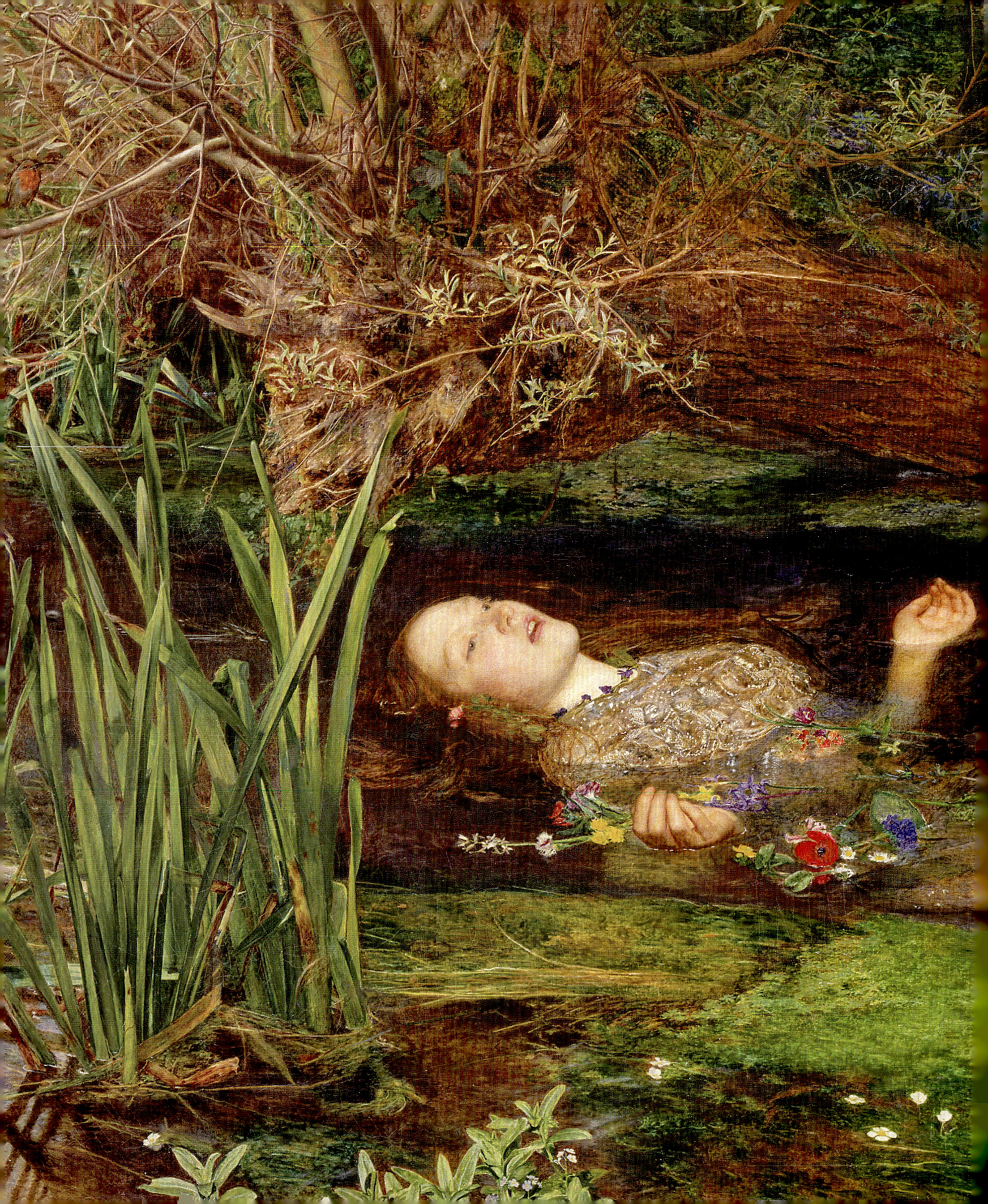

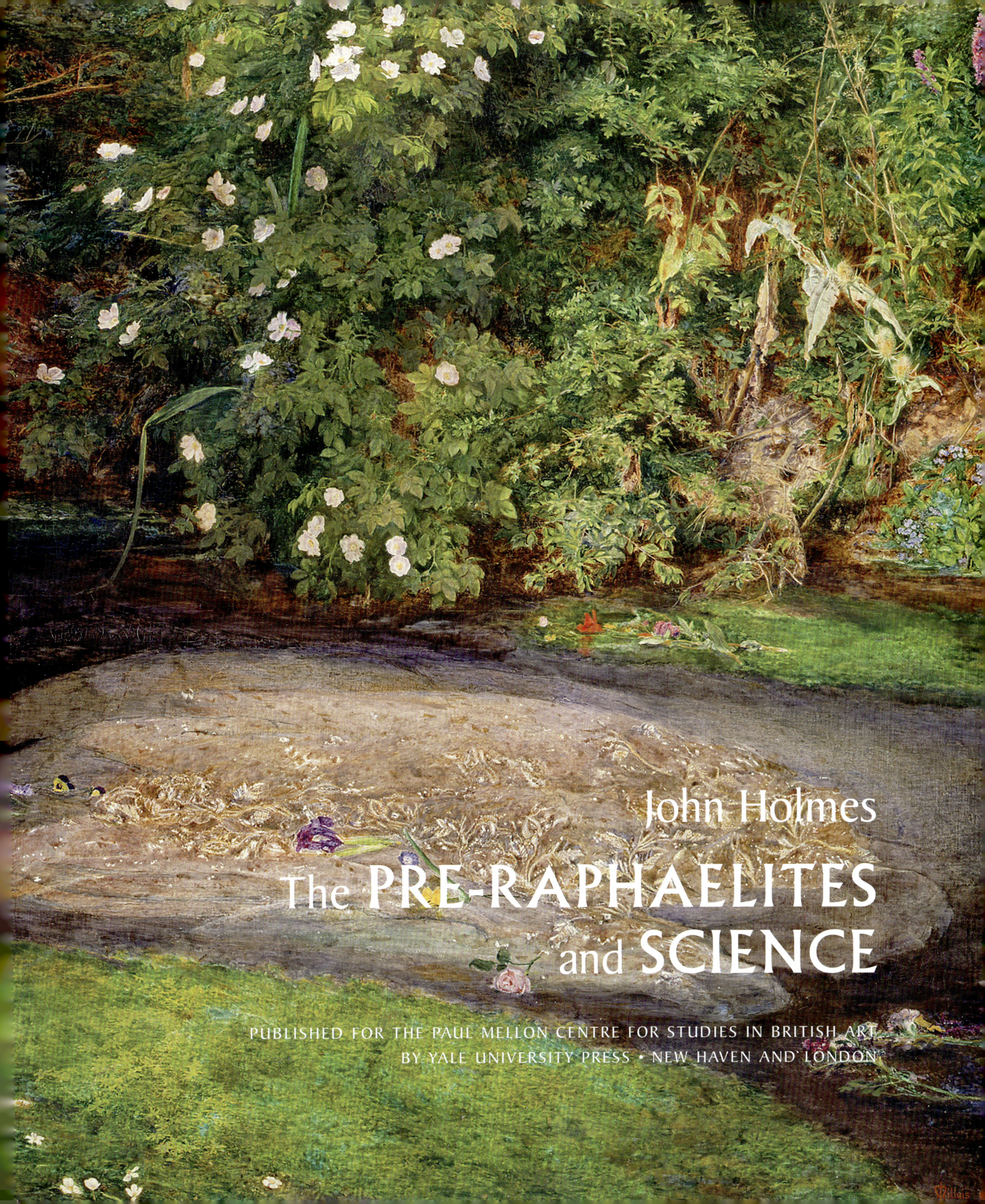

John Holmes

The PRE-RAPHAELITES and SCIENCE

PUBLISHED FOR THE PAUL MELLON CENTRE FOR STUDIES IN BRITISH ART
BY YALE UNIVERSITY PRESS • NEW HAVEN AND LONDON

ISBN 978-0300-232066 HB
Library of Congress Control Number: 2017954552

Designed by Emily Lees
Printed in China

Image on p. i James O'Shea, golden leather ferns, 1860 (see fig. 75).
Frontispiece John Everett Millais, *Ophelia*, 1851–2 (detail of fig. 7).

MIX
Paper from responsible sources
FSC® C008047
FSC
www.fsc.org

CONTENTS

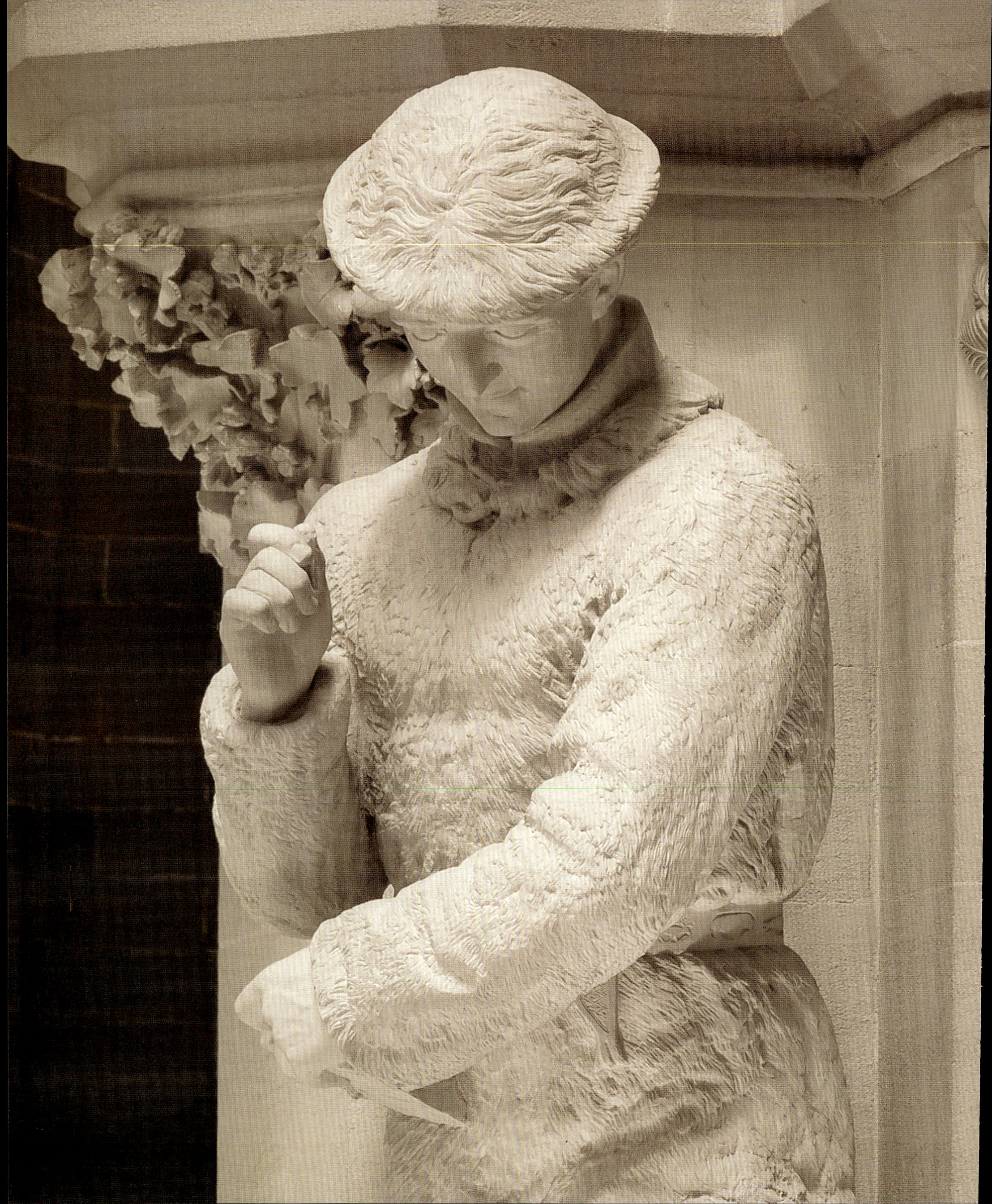

ACKNOWLEDGEMENTS

This book began as an experiment to see what happens to Pre-Raphaelite painting and poetry if you take seriously the Pre-Raphaelites' own claims that their art is modelled on science. For several years I had taught the Pre-Raphaelite periodical *The Germ*, where these claims were first made, on an MA course at the University of Reading, but an invitation from the Pre-Raphaelite Society to speak in September 2011 gave me the impetus to begin thinking seriously about this project. To my surprise, the relationship between Pre-Raphaelite art and science turned out to be a rich seam. A research fellowship from the Arts and Humanities Research Council, granted under their 'Science in Culture' theme, gave me the opportunity to research it in depth from October 2012 to June 2013, while in 2014 a term's study leave from the University of Reading gave me the chance to finish the first draft of the book. During my AHRC fellowship I worked closely with the Natural History Museum, the Oxford University Museum of Natural History, the Manchester Art Gallery and the Manchester Museum. Since July 2015, an Insight Development Grant from the Canadian Social Sciences and Humanities Research Council on 'Building the Book of Nature: The Poetics of the Natural History Museum' has supported further research on this project, extending its range to include museums in Ireland and Canada. In September 2015, I joined the University of Birmingham, which has again been very supportive of my project, both morally and financially, in particular contributing to the cost of the permissions for reproducing the images included within the book. My first thanks, then, go to the two research councils, the AHRC and SSHRC, and my two employers, the University of Reading and the University of Birmingham, for funding and otherwise

Facing page Detail of fig. 87.

supporting my research. Next I would like to thank the four partner museums, which provided me with all the support and encouragement I could have wanted during my fellowship, and to the Oxford University Museum and the Natural History Museum, in particular, for their continued enthusiasm for working together on further collaborations which have followed from this project. Third, I would like to thank Yale University Press and the Paul Mellon Centre for Studies in British Art for publishing and so generously supporting my work.

Since the project started in 2011, I have given public talks on my research at the Royal Society, the Natural History Museum and the Italian Cultural Institute in London; at the Oxford University Museum and Modern Art Oxford; at the Museum of English Rural Life in Reading; and for the Pre-Raphaelite Society in Birmingham. I have given keynote lectures at conferences at the universities of Northumbria, London and Reading; at the Birmingham Museum and Art Gallery (co-organised with Birmingham City University); and at a symposium at Mount Allison University in New Brunswick, funded by the Royal Society of Canada. I have spoken on my work at the Life Sciences Research Seminar at the Natural History Museum, the Ruskin Research Seminar at Lancaster University and the English Research Seminars at Durham and Birmingham. I have taught my research as a guest lecturer at the Courtauld Institute in London and lead a discussion of it as work in progress at the North American Victorian Studies Association in Banff, Alberta. I have presented further papers on aspects of the research towards this book at conferences organised by the British Society for Literature and Science in Cambridge, Oxford, Cardiff, Surrey, Liverpool and Bristol; the British Association for Victorian Studies in Sheffield and Kent; the 24th International Congress of History of Science, Technology and Medicine in Manchester; the Science in Public network in Kent; and the North American Victorian Studies Association in Madison, Wisconsin, and Venice (with the British Association for Victorian Studies and the Australasian Victorian Studies Association); and at a conference on Pre-Raphaelitism: Past, Present and Future, organised jointly by Oxford University and Oxford Brookes University. I would like to thank all these learned societies, museums, galleries and universities, and the organisers of and participants at all of these conferences and seminars, for providing me with invaluable opportunities to test and refine my ideas, and for all the comments and feedback which have helped to shape this book.

Continuing the institutional thread, my research has included archival work at the Natural History Museum, the Royal Institute of British Architects, the Anthropological Institute, the Oxford University Museum, the Bodleian Library, the Cadbury Research Library, the John Rylands Library, the Ruskin Library and Research Centre at Lancaster University, Trinity College Dublin and the Irish Architectural Archive, as well as more routine research using the Bodleian Library and the Radcliffe Science Library in Oxford and the university libraries at Reading and Birmingham. I would like to thank all the archivists and librarians who have facilitated this research. Although I have not visited it directly, I would like to thank too the Huntington Library in San Marino, California, and its staff for helping me to access Walter Deverell's poem 'The Garden' and allowing me to reproduce it. As well as the Manchester Art Gallery, my work on Pre-Raphaelite painting, sculpture and stained glass has taken me to Tate Britain and the Guildhall Art Gallery in London; the Ashmolean Museum in Oxford; the Birmingham Museum and Art Gallery and the Barber Institute at the University of Birmingham; the Walker Art Gallery, the Lady Lever Art Gallery and Sudley House in Liverpool; Manchester Town Hall; the Laing Art Gallery in Newcastle upon Tyne; and the Fitzwilliam Museum, the Herbarium, Trinity College and Peterhouse in Cambridge. I was extremely fortunate that the large retrospective exhibition 'The Pre-Raphaelites: Victorian Avant-Garde' opened at the Tate in London just as I was beginning my research fellowship, which gave me a unique opportunity to see the full range of the Pre-Raphaelites' art, and many of their most famous and significant works, in a single exhibition. Thank you to the curators of all these galleries for the access I have had to these collections – all of them delightfully still free to the public – and of the Tate exhibition in particular for giving me complimentary access to this definitive show. In addition to the museums in London, Oxford and Manchester, I have visited the Geological Museum at Trinity College Dublin and the Royal Ontario Museum in Toronto in studying the architectural origins and legacy of the Pre-Raphaelite natural history museum. Again, I would like to thank both the museums for their support and assistance with this project.

Moving on from institutions to individuals, there are so many people I owe a debt to for their help and encouragement that it is hard to know where to begin. Paul Smith, the Director of the Oxford University Museum of Natural History, and his team have been unstintingly generous and enthusiastic in their support of my research on the museum. Alongside Paul himself, I would like to thank Scott Billings, Mark Dickerson, Kate Diston, Chris Jarvis, Jim Kennedy, Darren Mann, Rachel Parle, Wendy Shepherd, Ellie Smith and Janet Stott (and to apologise to anyone else who has slipped off the list!). I would like to pay tribute to Scott particularly, who took most of the finest photographs of the museum included in this book – thanks again to Paul, Scott and the museum for their generosity in allowing me to use so many excellent images free of charge. It has been a real pleasure to work with the museum, as well as on it, and it has been immensely rewarding too to have the chance to revive the legacy of Pre-Raphaelitism at the museum through further collaborations, particularly during the 'Visions of Nature' year in 2016. Thanks too to the three poets-in-residence who contributed to 'Visions of Nature': John Barnie, Steven Matthews and Kelley Swain. I enjoyed very much sharing our responses to and interpretations of this beautiful and continually fascinating museum.

At the Natural History Museum in London, the person I worked most closely with was Julie Harvey, then Director of the Centre for Arts and Humanities Research. It was Julie's keenness for me to work with CAHR on this project that led me to work as closely as I did on the Natural History Museum and to recognise it as such an important example of how Pre-Raphaelitism came to shape the culture and communication of Victorian science. Julie was an exemplary host during my time as a visiting research fellow at the museum, providing me with opportunities to tour the building, from the basement to the rooftops, and always doing the utmost to ensure that I had access to the people, places and documents that I needed. Her kindness is unfailing, and I appreciate it enormously. More recently, Alex Burch, the Head of Visitor Experience, Learning and Outreach at the museum, and Justin Morris, the Director of Public Engagement, have been likewise encouraging. Julie, Alex and Justin have between them provided me with several opportunities to work with the museum, most rewardingly on the ongoing project to redisplay the collections and to landscape the grounds. I have especially enjoyed discussing the museum with the architects Kim Wilkie and Níall McLaughlin and their partners. Other people at the museum I would like to thank for their support include Mark Heslington, Zara Naghizadeh, Hellen Pethers, Helen Sturge. Thanks too to Isabel Martin and Kate Whittington for all their assistance with gathering images from the museum, and to the museum itself for its generosity in allowing me to draw on the existing image library free of charge.

I have met with the same generosity and encouragement across all the museums, galleries and libraries I have visited for my research. I am very grateful to all the people who have gone out of their way to guide me through their collections, exhibitions and buildings. Those whose time and knowledge have made a material difference to this book include Rebecca Milner from the Manchester Art Gallery, Alison Smith from the Tate, Stephen Wildman from the Ruskin Library and Research Centre at the University of Lancaster, Christine Bartram at the Herbarium in Cambridge, Patrick Wyse Jackson from the Geology Museum at Trinity College, Dublin, and Mark Engstrom from the Royal Ontario Museum. Although I have yet to have the chance to visit it in person, I have been given a virtual guided tour of the Naturhistorische Museum in Vienna by Stefanie Jovanović-Kruspel, who has very kindly shared her research on this museum with me. Further thanks to Stefanie and to John van Wye, Kim Wilkie, the Manchester Art Gallery, the National Portrait Gallery, Peterhouse, the Ruskin Library and Research Centre and the owners of various individual works in private hands for their kind permission to reproduce images free of charge.

As well as many of the people I have already thanked, Gowan Dawson and Michael Ruse have each read parts of my book in manuscript and given me very helpful advice which I have not hesitated to take, to the improvement of the book as a whole. I have benefited from indispensable advice from Patrick Parrinder, who corrected my entomology in my reading of William Michael Rossetti's 'Sheer Waste'; Karen Sayer, who sent me off to read up on the temperaments and behaviour of sheep to substantiate my interpretation of Holman Hunt's *Our English Coasts*; and Humphrey Masten, who saved me from my ignorance of sheep farming in the same painting. Other colleagues whose suggestions along the way have been valuable to me include Caroline Arscott, Barrie Bullen, Verity Burke, Katie Faulkner, Rosie Ibbotson and Paola Spinozzi, as well as numerous people I have forgotten to mention but who will, hopefully, mostly find themselves covered in one of the groupings above. Michael Wollaston and Serena Trowbridge of the Pre-Raphaelite Society have been very helpful too, especially in providing me with a series of opportunities to discuss my work with enthusiasts as well as fellow scholars.

There are three colleagues whom I would like to single out for my heartiest thanks. The first is Simon Dentith, who was my head of department at Reading when I began work on this project. Simon died in the autumn of 2014. The last of the many favours he did for me was to read sections of the first draft. His belief in the project, in spite of what he at first took to be its counter-intuitive approach to the Pre-Raphaelites ('The Pre-Raphaelites…and science?'), meant an enormous amount to me. I will always be grateful for his support and his unwavering kindness towards me. The second is Janine Rogers. Janine and I had been working in parallel on natural history museum architecture for some time before we decided to work together. It is through Janine that I have had the opportunity to extend my work beyond the UK, to explore the prehistory of the Pre-Raphaelite natural history museum in Dublin and its legacy in Canada. As well as leading our SSHRC project on museums, Janine has provided me with opportunities to present my research on the Pre-Raphaelites and science in Canada, and with a consistent sounding-board for (and at times a wise check on) my interpretations of architecture. Working together has been a treat throughout, and an enriching experience, I hope, for both of us. The third Anna Greutzner-Robins. Anna and I worked closely together for several years at the University of Reading, co-ordinating the joint undergraduate degree in History of Art and English. Anna's respect for my views on visual art helped me to credit them myself. It was Anna who encouraged me to send this book to Yale University Press. This takes me on to thanking all the people at Yale and the Paul Mellon Centre who have helped to see this book into production: my original commissioning editor, Gillian Malpass; the Director of Studies at the Paul Mellon Centre, Mark Hallett; my editor, Emily Lees; my copy editor, Nancy Marten and my proofreader, Jacquie Meredith. I would also particularly like to thank the anonymous readers who reviewed my book manuscript. One reader in particular sent me a twelve-page report, full to the brim with helpful suggestions. The book would have been more uneven, less cogent and less developed without this advice. I enjoyed incorporating it and reframing the book accordingly, and I am very grateful to have had such conscientious and encouraging mentoring. The errors and infelicities which no doubt remain are, of course, all my own work.

My final thanks, as always, go to my family – my wife Jo, my daughters Amy and Hannah, and my parents Margaret and Martin. Thanks to you all for bearing with me. Researching and writing this book has deepened my appreciation of Victorian art, enriched my experience of architecture, and added a new dimension to a forty-year-old love of natural history museums. It makes me very happy that I have been able to share these enthusiasms with you, even if I have not necessarily carried you all the way in all aspects of them. I will particularly treasure the memory of visiting the Pre-Raphaelite exhibition at the Tate with Amy, who made a very appreciative, good-natured and committed gallery-going companion, in spite of being only eight at the time, and of our family trip to Canada under the auspices of the SSHRC project, which gave us some grand experiences of nature itself, as well as of several natural history museums.

The last word must be for my father, who died in August 2017 while this book was in production. That I was able to spend time with him in his last weeks, and to join with my mother and brothers in helping him and one another through the process, felt a real privilege. This book is a small portion of all that would not be had it not been for him.

Small portions of the research towards this book have appeared in different forms in the following essays: 'Dante Gabriel Rossetti and Science' in *Review of the Pre-Raphaelite Society*,

21.3 (2013); 'Building a Vision of Nature: Owen, Waterhouse and the Design of the Museum' in *Evolve*, 17 (Autumn 2013); 'Poetry on Pre-Raphaelite Principles: Science, Nature and Knowledge in William Michael Rossetti's "Fancies at Leisure" and "Mrs. Holmes Grey"' in *Victorian Poetry*, 53.1 (2015); 'Pre-Raphaelitism, Science and the Arts in *The Germ*' in *Victorian Literature and Culture*, 43.4 (2015); and 'Monkey Business: The Victorian Natural History Museum, Evolution, and the Medieval Manuscript', co-authored with Janine Rogers, in *Romanticism and Victorianism on the Net* (forthcoming). Two other related essays on aspects of the Pre-Raphaelite engagement with science which are not discussed in detail here but which grew out of this project are 'Ruskin's Windows at the Oxford Museum' in *Ruskin Review and Bulletin*, 9.2 (2013), and 'Algernon Swinburne, Anthropologist' in *Journal of Literature and Science*, 9.1 (2016).

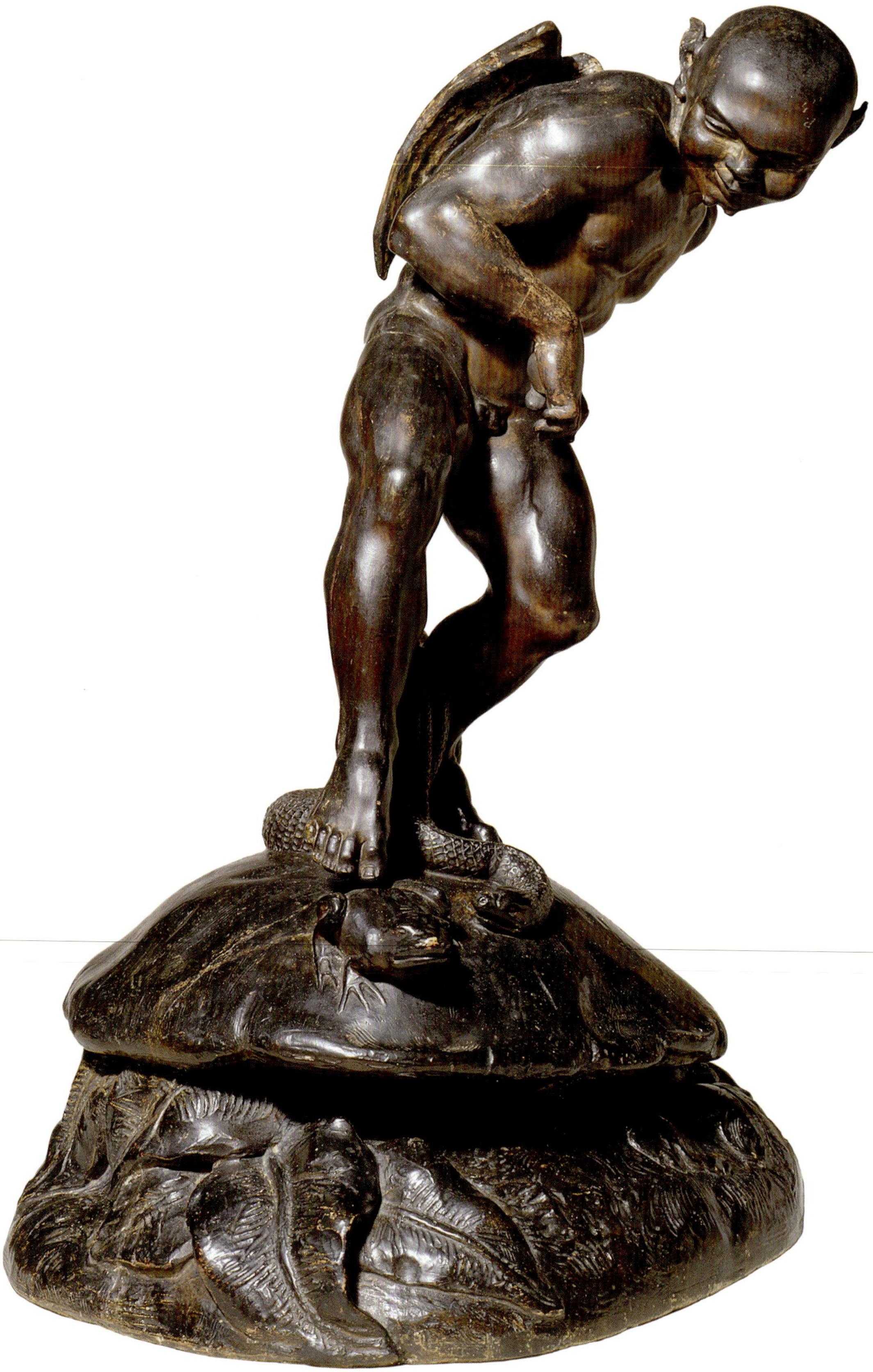

INTRODUCTION

The Pre-Raphaelites and Victorian Science

In March 1883 John Ruskin began a series of lectures on 'Recent English Art' as the Slade Professor of Fine Art at Oxford. In his first lecture, he revisited the paintings of two of his old friends, Dante Gabriel Rossetti, who had died the previous year, and William Holman Hunt. Thirty-five years earlier, when they were students at the Royal Academy, Hunt and Rossetti had joined four fellow students – John Everett Millais, Thomas Woolner, James Collinson and Frederic George Stephens – along with Rossetti's younger brother William, to form the Pre-Raphaelite Brotherhood. Founded as a protest against the Academy's hierarchical structures and conventional aesthetics, the Brotherhood would spearhead a revolution in Victorian art, with Ruskin himself as their most vocal champion. In his lecture, Ruskin noted how his friends had been committed to representing the truth of their subjects in 'sternly materialistic' terms in their paintings.[1] Returning to this theme in his second lecture, he declared: 'Truth is the vital power of the entire school, – Truth its armour, – Truth its war-word.' This truth, Ruskin insisted, was aligned to the truth of science. While the 'grotesque and wild forms of imagination' characteristic of Pre-Raphaelite painting might 'seem to be the reaction of a desperate fancy, and a terrified faith, against the incisive scepticism of recent science', they were, he declared, 'a part of that science itself'.[2]

Ruskin's analysis is as arresting today as he intended it to be to his original audience. Pre-Raphaelite art is more popular now that it has been at any point since the end of the nineteenth century, but much of its appeal lies in a richly imaginative medievalism and a mystical and erotic symbolism which seem at first glance utterly at odds with science. The Brotherhood's own name for themselves points to the

Facing page See fig. 5.

fourteenth and fifteenth centuries, not to the scientific nineteenth century, as their inspiration and the key to their aesthetic. Yet Ruskin was not alone in identifying Pre-Raphaelite art with science. In an article on contemporary English painting published in the *Fortnightly Review* in 1867, Sidney Colvin commented that 'the scientific spirit, coupled with the disgust of earnest men at academic pretensions and their reaction from academic principles, constituted the very essence of prae-Raphaelitism.'[3] Like Ruskin, Colvin recognised that the Pre-Raphaelite project was firmly and deliberately grounded in the idea of science. Their accounts of the movement chime with what the Pre-Raphaelites themselves wrote about their own aims as they launched their campaign for a new kind of art.

Since the late 1990s, critics and curators have increasingly recognised the importance of the Pre-Raphaelites' engagement with science. In *The Art of the Pre-Raphaelites*, published at the turn of the current century, Elizabeth Prettejohn identifies 'the burgeoning Victorian interest in the sciences' as among Pre-Raphaelite art's 'most important contemporary contexts.'[4] Several critics have identified the remorseless detail of early Pre-Raphaelite painting, and indeed poetry, as analogous to science. As Tim Barringer puts it, 'The attention which hard-edged Pre-Raphaelite naturalism of the 1850s paid to observing the individual object encapsulates this questioning, empirical spirit.'[5] Prettejohn and Barringer recognise too that this analogy was self-conscious. Both cite a passage from Stephens's essay on 'The Purpose and Tendency of Early Italian Art' as their key primary source. This passage, in which Stephens calls for the arts to imitate the methods of science, has been something of a touchstone for critics writing about Pre-Raphaelitism and science.[6] (I quote it myself at the beginning of Chapter 1 below.) As Prettejohn remarks, Stephens's 'description of scientific method matches the procedures of Pre-Raphaelite technique, at least in a general sense'. At the same time, 'it links this procedure to a strong faith in progress', allowing art to 'share in the spirit of progressive discovery attributed to science in the Victorian period and still powerful in scientific discourses today'.[7] Through its invocation of science, and in spite of its apparently backward-looking title, Stephens's essay makes precisely the kind of avant-garde claims which Prettejohn and other recent critics see as characterising the Pre-Raphaelite movement. In Barringer's words, 'It is here that the modernity of Pre-Raphaelitism can most clearly be recognised.'[8]

Prettejohn called for further research into this area back in 2000.[9] In spite of this, and of the general acceptance of the relevance of science to Pre-Raphaelite art, the relationship between the two has received only patchy attention to date. The revival of interest in Pre-Raphaelite landscape painting, crowned by a major exhibition in London, Berlin and Madrid in 2004–5, has led to new research into how these paintings engage with geology and natural history, which I build on in Chapter 2.[10] But beyond this particular subgenre of Pre-Raphaelite painting, coverage remains piecemeal.[11] This book is the first exploration of the Pre-Raphaelites' engagement with science across their work. My aim is to demonstrate how far the original Pre-Raphaelite project was indeed shaped by the ideal of an art modelled on science. In the process, I will re-examine Pre-Raphaelite art theory and painting, but also poetry, sculpture and architecture. I will show too how several Victorian scientists themselves embraced Pre-Raphaelite aesthetics, working with artists on the decoration of new museums of natural history, and making their own interventions into debates on how science could inform modern art and vice versa. And I will explore how later Pre-Raphaelite art and poetry, which increasingly abandoned science as a model, nevertheless played a significant role within the scientific culture of Victorian England. Overall, this study will reveal how the most influential movement in Victorian art was intimately bound up with the development of Victorian science in a relationship that was creative and reciprocal, if at times tense and critical.

Before making a case for the importance of science to the Pre-Raphaelites, it is worth pausing to set the terms of reference. What does 'science' cover, and who precisely counts as one of 'the Pre-Raphaelites'? To start with the second of these terms, the Pre-Raphaelite movement was the creation of a circle of young men who came together in the late 1840s. The Pre-Raphaelite Brotherhood itself was founded on the initiative of three young artists. William Holman Hunt and John Everett Millais had been friends for some time. Dante Gabriel Rossetti met them in 1848 at the Royal

Academy Schools. When they decided to form a brotherhood that autumn, Rossetti asked the painter James Collinson and the sculptor Thomas Woolner to join them, along with his own brother William Michael Rossetti, while Hunt roped in his pupil Frederic George Stephens. These seven comprised the P.R.B., but the P.R.B. itself was only one of a number of coteries that this circle of artists formed that year. In July the Rossetti brothers, Hunt and Collinson had formed a literary society to discuss one another's poetry.[12] Since the spring, the practising artists among the P.R.B. had all been involved in the Cyclographic Society, a somewhat wider group of art students who met to appraise one another's drawings and designs.[13]

The Cyclographic was founded by Rossetti's friend Walter Deverell, among others. Deverell was one of several artists closely involved with the Pre-Raphaelite Brotherhood in spite of never formally being members. Both Deverell and Millais's friend Charles Collins were touted as prospective replacements for Collinson when he quit the P.R.B. on religious grounds in May 1850.[14] Two somewhat older though still young artists – the painter Ford Madox Brown and the sculptor John Lucas Tupper – became integral members of the movement. Brown, who had taught Rossetti, was one of the Brotherhood's closest associates, 'virtually a PRB', as Hunt remarked to William Rossetti in 1852.[15] Tupper is barely remembered today, but he was an active member of the original Pre-Raphaelite circle, pioneering sculpture on Pre-Raphaelite principles alongside Woolner.[16] His diaries from the early 1840s were published in instalments in the American art journal *The Crayon* in the mid-1850s. As George Landow has observed, these diaries and Tupper's accompanying reminiscences include 'many important anticipations of Ruskin and Pre-Raphaelitism'.[17] Tupper appears to have been elevating quattrocento painters above Raphael, critiquing the hidebound and misleading conventionality of the Royal Academy Schools, and probing the possibilities of a mutually beneficial relationship between science and the arts, nearly a decade ahead of the Pre-Raphaelites themselves.[18] When they met, he became something of a mentor to them. On Tupper's death in 1879, Hunt wrote to Edward Lear that 'in many respects he was my superior' and 'I valued him accordingly.'[19]

Histories and exhibitions of Pre-Raphaelite art have tended to concentrate on Hunt, Millais and Rossetti, together with Brown, as the major artists of the movement. In one sense this is fair. They were the most gifted and imaginative painters among the original Pre-Raphaelites, and they had the longest and most productive careers. But this hierarchy is false to the democratic spirit of the Pre-Raphaelites themselves, especially in their early years. None of them apart from Brown had any public reputation. As young men of promise rather than fully realised achievements, they encouraged and supported one another. Similarly, although Pre-Raphaelite painting has a far wider public today than Pre-Raphaelite poetry or sculpture, the Pre-Raphaelites themselves did not rank the arts in this order or any other. The *P.R.B. Journal*, which William Rossetti kept as secretary to the Brotherhood, confirms this. According to Rossetti, the 'object' of the *Journal* was 'to be a record, from day to day, of the proceedings of *all* the Members, so far as these were of a professional or semi-professional character' – professional, that is, as artists.[20] The earliest surviving entry in the *Journal*, dated 15th May 1849, records Millais at work on a poem entitled 'Castle-moat' while William Rossetti works on a picture called 'Maud's Madness', an illustration to Coventry Patmore's poem 'The Woodman's Daughter'.[21] These works were abandoned – though Millais later took up Patmore's poem as a pictorial subject, as we shall see in Chapter 2 – but the mention of them in the *Journal* shows that both were attempted as part of the creative project of Pre-Raphaelitism. Indeed, all seven P.R.B.s wrote poetry, as did Deverell, Brown and Tupper, just as they all created works of visual art. Similarly, when Tupper's bas-relief illustrating Chaucer's 'Merchant's Second Tale' was rejected by the Royal Academy in May 1851, William Rossetti wrote that it 'carried out to the full' in sculpture the 'P.R.B. principle of uncompromising truth to what is before you'.[22] Rossetti later recalled that, in keeping the *Journal*, he 'used to make it a rule . . . to mention only the Members of the Brotherhood, along with some few other persons who came into close relation with them, their purposes, and their work'.[23] Insofar as this was a rule at all, it was one defined by its exceptions. That Tupper was not a P.R.B. did not exclude his contribution to the Pre-Raphaelite project from

No. 1. *(Price One Shilling.)* JANUARY, 1850.

With an Etching by W. HOLMAN HUNT.

The Germ:

Thoughts towards Nature

In Poetry, Literature, and Art.

When whoso merely hath a little thought
    Will plainly think the thought which is in him,—
    Not imaging another's bright or dim,
Not mangling with new words what others taught;
When whoso speaks, from having either sought
    Or only found,—will speak, not just to skim
    A shallow surface with words made and trim,
But in that very speech the matter brought:
Be not too keen to cry—"So this is all!—
    A thing I might myself have thought as well,
But would not say it, for it was not worth!"
    Ask: "Is this truth?" For is it still to tell
That, be the theme a point or the whole earth,
Truth is a circle, perfect, great or small?

London:

AYLOTT & JONES, 8, PATERNOSTER ROW.

G. F. Tupper, Printer, Clement's Lane, Lombard Street.

1 Frontispiece to *The Germ*, 1850.

the *Journal*. Deverell, Collins and Brown too all feature on equal terms with the P.R.B.s proper.

After forming the Pre-Raphaelite Brotherhood in 1848 and beginning the *P.R.B. Journal* in 1849, the Pre-Raphaelite circle launched a manifesto magazine in January 1850. Commercially, it was a failure, folding after only four issues. But those four issues are critical to understanding the Pre-Raphaelite project. Brown, Tupper, Deverell and all seven P.R.B.s contributed to the production of *The Germ* (fig. 1), as it was known for its first two issues, or *Art and Poetry*, as it became for its last two.[24] Like the Pre-Raphaelite movement as a whole, *The Germ* was very much a collective endeavour. Hunt, Collinson, Brown and Deverell designed etchings, as did Millais, though the magazine folded before his could be printed. Woolner, Brown, Tupper, Deverell and the Rossetti brothers wrote poetry. Essays on art theory were contributed by Tupper, Brown and Stephens, while William Rossetti wrote poetry reviews and edited the magazine as a whole, and Dante Gabriel Rossetti drafted its advertisement or brief.

The Pre-Raphaelites' aesthetic agenda was set out in the critical prose of *The Germ* and was exemplified in its poetry as much as in their early paintings. It was grounded in an aspiration to create an art which, unlike the derivative paintings of the Royal Academy, could lay claim to truth on scientific terms. As Hunt recalled in one of a series of retrospective articles he wrote for the *Contemporary Review* in 1886: 'We frequently talked over scientific and historical matters, for my previous reading and experience had led me to love them and to regard them as of the greatest poetic and pictorial importance for modern art; for then, as now, I concluded that the appeal we made could be strengthened by using the instruments of the age which human intellect had discovered.'[25] As Barringer notes, 'there never was, in fact, a single identifiable Pre-Raphaelite style,' which goes for poetry as well as painting and sculpture.[26] But there were shared principles. One of the most central, articulated by Stephens in his article on early Italian art, was a commitment to creating art through practices modelled on those of science. Exactly what that commitment entailed for the Pre-Raphaelites and what kinds of art it produced is the subject of this book.

Within the pages of *The Germ*, as in the *P.R.B. Journal*, all contributions to the Pre-Raphaelite project bear equal weight. This same principle governs my study of early Pre-Raphaelitism in this book (with one key exception – the poetry of Christina Rossetti – which I will come to shortly). I discuss paintings by Hunt, Millais, Rossetti and Brown, but I also consider how Collinson contributed to the Pre-Raphaelite project, along with George Boyce and John and Rosa Brett, three of the landscape painters who came into the Pre-Raphaelites' orbit in the 1850s. In sculpture, Woolner and Tupper are joined by Alexander Munro, who was a close associate of Dante Gabriel Rossetti from the late 1840s, and H. H. Armstead, who was a friend of Hunt's in the 1850s. In poetry, I include poems by Rossetti and, latterly, William Morris and Algernon Charles Swinburne, but I pay equal attention to poems by Rossetti's brother, Tupper, Deverell, Collinson and Woolner. My aim is to recover how these different artworks conduct the engagement with science proposed in Pre-Raphaelite art theory, so I draw heavily on the essays of Stephens, Tupper and William Rossetti (figs 2, 3, 4) as the leading theorists of the movement, at least in print.

Beyond the original Pre-Raphaelite circle, the major theorist of the movement was John Ruskin. As there is some disagreement over Ruskin's role within Pre-Raphaelitism, it may be worth spelling out how I see his relationship to the Pre-Raphaelites themselves. Ruskin defended the Pre-Raphaelites in two letters to *The Times*, published in May 1851.[27] Soon after, he became a friend and patron to Millais, Rossetti and several other Pre-Raphaelite artists in turn. Beforehand none of them had known him personally, and he had not been directly involved in their initial campaign to reinvigorate art. The influence of his early writings on Pre-Raphaelitism has long been taken for granted, but as Marcia Werner has shown it is a mistake to take Ruskin's treatise *Modern Painters*, the first two volumes of which appeared in 1843 and 1846, as a manual to Pre-Raphaelite art theory.[28] The testimony of the Pre-Raphaelites themselves suggests that probably only Hunt had read Ruskin before he weighed in on their behalf. In a letter drafted in 1880, Hunt wrote to Ruskin that 'all that the Preraphaelite brotherhood had of Ruskinism' came from his reading of *Modern*

2 John Everett Millais, *Frederic George Stephens*, 1853. National Portrait Gallery, London.

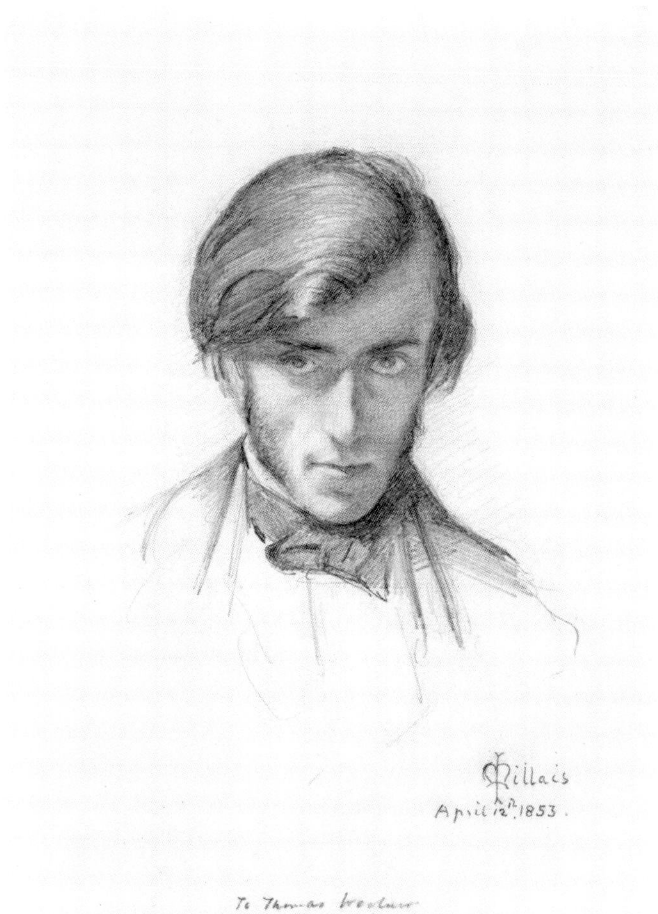

3 Anonymous photograph of John Lucas Tupper, mid-1850s. National Portrait Gallery, London.

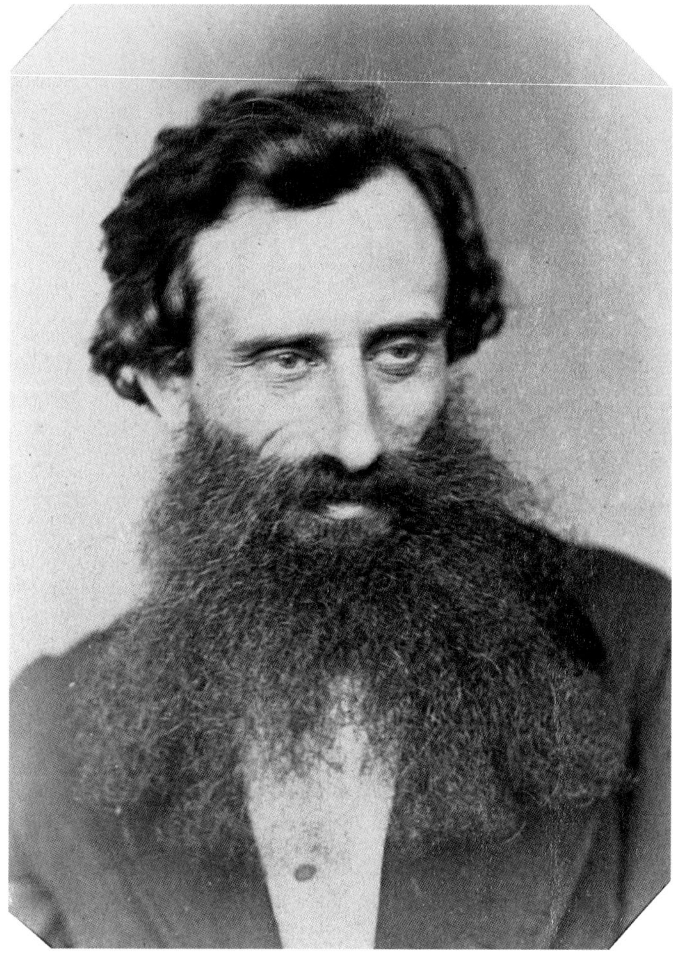

Painters.[29] In his *Memoir* of his brother, William Rossetti confirms: 'Hunt had read and laid to heart in 1847 the first volume of *Modern Painters*,' but he remarks that, as far as he can recollect, 'I do not think any other P.R.B. (with the possible exception of Collinson) had, up to 1848 or later, read him at all.'[30] In his *Life and Letters* of his father, John Guille Millais remarks likewise that, while Hunt read *Modern Painters* 'with enthusiasm, as partially embodying his own preconceived ideal of art', Millais refused to read it as he already had his own ideas which he was determined to carry out without being distracted from them by someone else's teachings.[31] Amy Woolner's *Life in Letters* of her father suggests that Woolner too did not read *Modern Painters* until 1854.[32] For his own part, although Ruskin seems initially to imply in his 1851 pamphlet *Pre-Raphaelitism* that the Pre-Raphaelite artists were following his lead, carrying out his advice in *Modern Painters* 'to the very letter', he affirms later on in the same essay that Millais and Hunt 'have conceived for themselves a totally independent and sincere method of study'.[33]

Weighing up the evidence, Ruskin is best seen not as the Pre-Raphaelites' guru but rather as an enthusiastic fellow traveller who understood, shared and latterly helped to articulate many of their aims. Ruskin's role did prove to be crucial to the Pre-Raphaelites' engagement with science, however. It was principally through his personal connections with scientists and architects that Pre-Raphaelitism moved decisively beyond the limits of the art world into the cultural sphere of science itself. Pre-Raphaelitism was adopted by

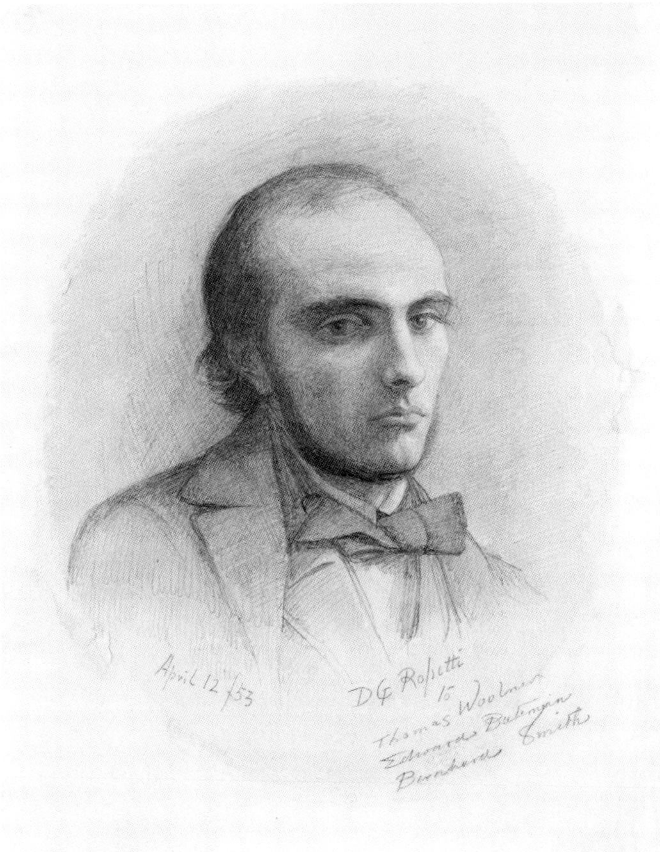

4 Dante Gabriel Rossetti, *William Michael Rossetti*, 1853. National Portrait Gallery, London.

Ruskin's friend the physician Henry Acland and his colleague the geologist John Phillips as the model for the decorative carvings at the new Oxford University Museum in the late 1850s. Benjamin Woodward, who built the museum, worked closely with the Pre-Raphaelites themselves. His master masons, the O'Shea brothers, followed Pre-Raphaelite principles as Acland stipulated – principles that were in keeping with their own practice as sculptors. At much the same time, the architect George Edmund Street identified his own Gothic Revival architecture, heavily influenced by Ruskin's architectural theories, as a form of Pre-Raphaelitism. To appreciate how Pre-Raphaelitism came to be the visual language of science at Oxford and, in a looser form, in the collaboration of Richard Owen and Alfred Waterhouse at the Natural History Museum in London, we need to extend the definition of who counts as a Pre-Raphaelite artist to encompass these architects and decorative sculptors, and even to a degree the scientists themselves.

This study thus foregrounds a number of Pre-Raphaelite artists, poets and theorists who tend to be under-represented in other accounts of the movement, and brings architects, sculptors and scientists into the fold. At the same time, my particular focus on the Pre-Raphaelites' engagement with science inevitably marginalises other well-known artists often identified with them. It is remarkable how pervasive the references to science as a model for the arts are in early Pre-Raphaelite art theory, and how many early Pre-Raphaelite paintings, poems and sculptures reflect this ideal. But this does not apply to the art of the second generation of Pre-Raphaelites led by Dante Gabriel Rossetti, along with Morris, Swinburne and Edward Burne-Jones, from the late 1850s onwards. Of the original Pre-Raphaelites, Rossetti had always been the most ambivalent in his attitude to science, and in this later phase of his art he abandons entirely the principle that art should model itself on science. This is not to say that second-generation Pre-Raphaelitism has no place in the cultural landscape of Victorian science. As I will show, late Pre-Raphaelite poetry in particular played a surprisingly important if largely unintended role in giving form and meaning to the world-view of Victorian scientific naturalism. But second-generation Pre-Raphaelite painting did not engage closely with science and did not set out to model itself on it, so Burne-Jones, for example, plays only an incidental role in the story told in this book.

For this same reason, the women artists and poets associated with the Pre-Raphaelites do not have the representation here which they would otherwise deserve. In the 1980s and 1990s Jan Marsh and Pamela Gerrish Nunn pioneered a new critical project to recover paintings by women from the Pre-Raphaelite circle.[34] In *The Art of the Pre-Raphaelites*, Prettejohn makes a powerful case for rethinking the place of women within Pre-Raphaelitism not 'merely to add some women to the Pre-Raphaelite canon' but by 'writing a wholly new, and different, story about Pre-Raphaelitism – a story in which the activities of women are no longer incidental, but necessary to the plot'.[35] This book signally fails to achieve even the more modest of these ambitions, but it does so for what I hope are defensible reasons. From the mid-1850s onwards, Elizabeth Siddal, Joanna Boyce, Marie Spartali, Emma Sandys, Lucy Madox Brown and Evelyn de Morgan would each make a significant contribution to the development and expression of later forms of Pre-Raphaelitism. As the visual art of the second-generation Pre-Raphaelites was only very loosely connected to science, however, their works fall outside the purview of this study.

In the late 1840s and early 1850s, when the Pre-Raphaelite project was conceived on scientific lines, its social and creative milieu was almost entirely masculine. (So, of course, was the world of professional science, as represented by universities, museums and – for the most part – the scientific press.) As William Rossetti later commented, 'Considering that the Præraphaelite Brothers were all very young men, it is a little remarkable that only two women can be named as rather closely associated with their early proceedings.'[36] One of these was Siddal. At this stage, however, she was a model and not yet an artist in her own right, and she is not mentioned in the *P.R.B. Journal* as contributing to the movement's 'professional' activities. The other was the Rossetti brothers' younger sister, the poet Christina Rossetti. According to William, Christina saw Hunt, Woolner and Stephens often, as well as her brothers, and she was briefly engaged to Collinson.[37]

As the author of several poems printed in *The Germ*, Christina Rossetti does feature in the *P.R.B. Journal*. On these grounds, she has the same right to be taken as a Pre-Raphaelite as, say, Deverell or Brown. Yet her early poetry cuts against the grain of the Pre-Raphaelite project, as it was conceived by the P.R.B. and their male associates, in very fundamental ways. Rather than attending to concrete details, Christina Rossetti's poems in *The Germ* shun the material world which science works to comprehend. Instead they inhabit worlds of dreams, allegories and Old Testament prophecy.[38] Where the poems by the male Pre-Raphaelites in *The Germ* stress the precise observation of details, she calculatedly evades any such precision. Her speakers exist without any context; their sorrows are given no explanation. Her strategy is at once to pose a riddle which her readers are left to solve as best they can, and to empty her poems of specifics so that the reader can experience their moods as universal precisely because they are not dependent on set circumstances.

Like Siddal's poetry and paintings, Christina Rossetti's poetry has a closer aesthetic affinity with aspects of what would emerge as second-generation Pre-Raphaelitism than with the Pre-Raphaelite project as it is mapped out and tried out by the other contributors to *The Germ*. Yet there remains another stark difference, besides the use or avoidance of detail, between her later poetry and that of her brother or Morris or Swinburne. As we will see in Chapter 7, their refusal to accept the consolations of Christian faith was central to the resonance that their poetry held for men like John Morley or Walter Pater, for whom scientific naturalism had given the lie to Christianity's promise of immortality and its trust in a benevolent world order. Christina Rossetti, by contrast, remained intensely committed to her religious beliefs. In the first six volumes of devotional prose which she wrote from the 1870s to the 1890s, she prayed: 'Suffer us not to be seduced from the faith by miscalled reason, or apparent facts of science, or the wit and learning of misbelievers, or subtilties of Satan.'[39] Science for Christina Rossetti existed under sufferance. She was happy to accept it unless and until it challenged the precepts of Christianity. If it did so, it was the Devil's work. Other members of the Pre-Raphaelite circle, including Tupper and Ruskin, contested the science of the 1860s and 1870s from a Christian standpoint, but their criticisms were predicated on the prior science of natural theology. Christina Rossetti had no need of natural theology as faith itself could overrule the 'apparent facts of science'. There is a book to be written about her engagement with science, but it is not this one, as she did not stand with either generation of Pre-Raphaelites on this issue.

It was the circle of young male artists who congregated around Hunt, Millais and Rossetti in the late 1840s and the early 1850s who launched the Pre-Raphaelite project to revitalise the arts, in large part by following the model of the sciences as they saw them. Most of these young men were also poets, and many of them wrote insightfully about their own and one another's practice in the arts. There were, of course, other components to the Pre-Raphaelite revolution, not least the Romantic turn to medieval models and sources for art and poetry, in search of a freshness that later, staler art had lost. Again, if this book largely ignores this aspect of Pre-Raphaelitism, it is not to deny its part in the story, but to trace the alternative thread that connects Pre-Raphaelitism to science. In the process I hope to reveal that science was more fundamental to and pervasive within the original Pre-Raphaelite project than has been recognised before; that through its scientific procedures Pre-Raphaelitism makes contributions to our understanding of the natural world and human interactions within it that can stand alongside those of Victorian science; and that Victorian scientists themselves recognised its capacity and integrity, taking up the Pre-Raphaelite project to express their own world-views.

The Pre-Raphaelites cite science as a model for their art. But what science did they actually know? Only Tupper had a specifically scientific training, as an anatomy student and, from 1849, an anatomical illustrator at Guy's Hospital in London.[40] By his own account, Hunt's early apprenticeship as an artist included making 'geological and astronomical diagrams'. Beyond that, he was largely self-educated, reading widely in his father's library, including books on mathematics and geometry as well as other sciences. He only mentions one book by name, however: the anonymous and controversial evolutionist account of the universe, *Vestiges of the Natural History of Creation* (1844), which was a publishing sensation in the mid-1840s, and which Hunt himself soon repudiated, blaming it alongside Voltaire, Byron and Shelley for converting him briefly to materialism.[41] William Rossetti enjoyed reading 'easy books on natural history' as a child, and remembered owning a book which included a biography of the eminent chemist Humphry Davy.[42] According to their children, Millais and Woolner were both keen amateur naturalists.[43]

Aside from these very slight biographical details, little documentary evidence survives to identify specific sources for the Pre-Raphaelite artists' knowledge or conception of science. Even their correspondence and archives reveal very little in the way of a detailed knowledge of science. Tupper, Hunt and perhaps Stephens aside, their interest in science seems to have been real but not assiduous, and their knowledge more casual than systematic. Yet their aspiration to ground their art in science was reiterated many times and was, as I hope to show, genuine. Given this, 'science' in this book is inevitably a capacious term. It encompasses those sciences that bear a direct relationship to the subject matter of Pre-Raphaelite painting – geology and natural history in the case of landscape, psychology in the case of historical paintings, portraits and poems. But it also includes sciences alluded to by the Pre-Raphaelites themselves and their associates that bear a less direct relationship to the themes of their work as artists and writers. Chemistry, anatomy and geometry are all important points of reference in *The Germ*, for example. All three bear a practical relation to the Pre-Raphaelites' artistic training: chemistry through the make-up of pigments, anatomy through the life class, geometry through the patterns of composition laid down as models by the Royal Academy and rejected as strictures by the Pre-Raphaelites themselves. Together these sciences take on a collective significance in the Pre-Raphaelites' writings on art, not as precise disciplines or bodies of knowledge, but as ideals of what art ought – or ought not – to be. Indeed, the thrust of my argument in this book is that it was the idea of 'science' that mattered most to the Pre-Raphaelites, refracted through a number of very different sciences, more than the findings of those sciences themselves. On these same terms, the findings the Pre-Raphaelites reach through their own investigative art are themselves, if not exactly a part of Victorian science, as Ruskin would have it, then certainly contributions to understanding that exist in parallel with it and that ask to be judged by the same standards.

We may not know exactly what the Pre-Raphaelites knew about science, but we can be sure of its significance and visibility within the wider culture of the 1840s and 1850s when they were pursuing their project to revitalise the arts. The glamour that science had for the Pre-Raphaelites was in part a result of their particular moment. They were born in the 1820s, just as a scientific and technological revolution was beginning to unfold. In his books *Visions of Science* and his earlier study of *Vestiges*, *Victorian Sensation*, the historian James Secord has shown how the new medium of industrial printing helped to energise and popularise scientific debate in the 1830s and 1840s.[44] Whether or not they read them themselves, the books that Secord writes about in *Visions of Science* – Charles Lyell's *Principles of Geology* (1830–33) or George Combe's exposition of phrenology, *The Constitution of Man* (1828), or John Herschel's primer in the philosophy of science, *A Preliminary Discourse on the Study of Natural*

Philosophy (1831) – would have been well-known points of reference for the Pre-Raphaelites and their contemporaries. As the scientific press started rolling in earnest in the 1830s, so scientific societies took off. The British Association for the Advancement of Science was founded in 1831 as an annual forum for the discussion of new scientific findings and theories. By contrast with the elite Royal Society, the BAAS was open to a wide public interested in joining in with the scientific spirit of the age. This spirit found expression both in the rapid growth of scientific knowledge, especially in geology and chemistry, and in the creation of new technologies, including railways and steamships, which were transforming the Victorians' world.

The Pre-Raphaelites looked to science because they were born into a self-consciously modern, scientific age. Far from rejecting this, first-generation Pre-Raphaelitism embraced it, in its methods if not in its subjects or its industrial aesthetics. At the same time, their conception of science had much deeper roots. As we shall see in Chapter 1, the Pre-Raphaelites shared with early Victorian scientists a commitment to empirical observation. Both groups identified this with the philosophy expounded by Sir Francis Bacon (see fig. 86) in the early seventeenth century. In principle, the Baconian inductive method begins with observations, then moves to generalisations where the observations allow it, then tests those generalisations through experiments and further observations to establish natural laws. It is disputable how far science has ever in fact followed this method, or indeed whether it even could, but both the early Victorian scientists and the Pre-Raphaelites presented science – and thus their own practice – as rigorously Baconian. In this, Pre-Raphaelitism was one manifestation of a widespread mid-Victorian aspiration to establishing truth through unprejudiced and scrupulous observation. Ruskin articulated this as truth to nature in *Modern Painters*. Matthew Arnold encapsulated it when he famously defined the aim of all criticism as 'to see the object as in itself it really is'.[45] The Pre-Raphaelites put this principle into practice in their art itself, in painting, poetry and sculpture, including through direct collaborations with scientists.

A second important aspect of the early Victorian conception of science which had a history going back to the seventeenth century was natural theology. The premise of natural theology is that nature is God's creation, as the Bible is his word. By studying the world through science, we can come to an understanding of the principles of its design, and so better understand God's purposes and our place within them. Natural theology was not coterminous with science in the mid-nineteenth century, but it had a significant hold on it, especially at the Anglican universities – Oxford, Cambridge and, from 1829, King's College London – where William Paley's elegantly argued textbook, *Natural Theology*, was the foundation of the science curriculum. It was further underscored in the 1830s by the publication of the Bridgewater Treatises, a series of eight volumes 'On the Power Wisdom and Goodness of God as Manifested in the Creation'.[46] Funded by a bequest from the Earl of Bridgewater, the treatises were written by prominent establishment men of science, among them the dons John Kidd and William Buckland from Oxford, William Whewell from Cambridge and Charles Bell from King's. They covered the full range of the sciences from geology, natural history and anatomy to chemistry, physics and astronomy, arguing that each of them provided evidence of God's benevolent design in nature, and affirming the compatibility of science as a whole with revelation.

The authority of the Church of England and the universities in the 1830s and the 1840s was such that science tended to be framed by a discourse of natural theology, even where the details of the science itself were theologically neutral or irrelevant. In his study of natural theology in the nineteenth century, Peter Addinall discusses the Bridgewater Treatises as representative of conservative natural theology, but he points out that Herschel, Lyell, Robert Chambers (the author of *Vestiges*) and others couched their own science as more liberal versions of the same discourse.[47] The scientists to whom the Pre-Raphaelites became personally and professionally closest in the 1850s and who embraced the Pre-Raphaelite project themselves – Acland (see fig. 58) and Phillips at Oxford, Richard Owen (see fig. 133) at the British Museum – were committed to natural theology in this more liberal, less literal, form. So were Tupper, Hunt and Ruskin. Natural theology underlies Stephens's justification for the Pre-Raphaelite project too. It even shapes William Rossetti's thinking, in spite of his having been an avowed free-thinker from the age of fourteen.[48]

With its avowal of Baconianism and natural theology, early Victorian science no longer looks very modern today. To recognise the Pre-Raphaelite project as scientific, therefore, requires us to suspend some of our more recent assumptions about science. The Pre-Raphaelites are not concerned to propose or test hypotheses, for example. Nor are they much interested in those Victorian scientific theories – evolution and thermodynamics – that most preoccupy scholars today. Not that they were untouched by the controversy around evolution in particular. From 1860, Charles Darwin's theory of natural selection became a rallying point for an increasing number of scientists who felt that the presumptions of natural theology – that nature was the creation of a benevolent God, and that humanity had a unique place at the apex of this creation – were hobbling science. In part because of its roots in the science of Jean-Baptiste Lamarck, which was associated with the French Revolution, evolutionism had always had a whiff of materialism about it. The polite consensus that science should be framed as natural theology had begun to fray in the 1850s, but it was in the 1860s that power and prestige within science began to slip away from the natural theologians towards a younger generation who characterised themselves as scientific naturalists.

Natural theology was not eliminated from science, but it was on the back foot from the 1860s onwards. By the 1870s, one of the leading scientific naturalists, John Tyndall, could declare: 'We claim, and we shall wrest from theology, the entire domain of cosmological theory.'[49] The Pre-Raphaelites were caught up in this struggle for dominance – arguably one of the most fundamental paradigm shifts ever to take place within British science. Scientists on both sides used the principles and practices of Pre-Raphaelitism to affirm their own conceptions of the natural world and the scientific enterprise. Hunt and Tupper stuck with the old natural theology, in the full knowledge that science was changing around them. By contrast, Rossetti, Morris and Swinburne found themselves in demand with the scientific naturalists and their supporters, who found in their poetry a scepticism and objectivity that they themselves shared. As the original Pre-Raphaelite project had been modelled on the science of early Victorian England, so the second-generation Pre-Raphaelites became the first major poets of a new world-view born of scientific naturalism.

The story of the Pre-Raphaelite engagement with science falls into two unequal halves. The first five chapters of this book discuss the Pre-Raphaelite project as it was conceived in the 1840s and pursued through the 1850s, when Bacon was still seen as the founder of the scientific method and when natural theology was still the dominant scientific world-view. In cultural history, specific dates serve as indicative markers for wider trends. In these terms, the publication of *On the Origin of Species* in 1859 marks not the sudden start of a new epoch but a changing of the tide. Natural theology began to recede in the face of scientific naturalism. While for many this was simply a matter of method, for others the methodological materialism of scientific naturalism revealed a deeper materialism which belied the Church's promises of immortality and a benevolent creation. As this rift opened up within science, so the Pre-Raphaelite movement too divided. Hunt and Tupper remained staunch in their commitment to both natural theology and the ideal of scientific art, while Rossetti and his circle turned away from science to aestheticism and symbolism. The four chapters that make up the second part of this book explore the roles that the different Pre-Raphaelite arts played in this new, more divided scientific culture.

These different contexts necessitate different approaches to the arts themselves. In the first part of the book, I consider painting, poetry and – in Chapter 5 – sculpture and architecture alongside one another, as contributions to the same project. My study of the early Pre-Raphaelite movement in these chapters, though far from exhaustive, especially as far as landscape painting is concerned, aims to be comprehensive. In Part 2 I take Pre-Raphaelite poetry, architecture and painting in turn, considering each within a separate chapter, because each played a different part in the scientific culture of later Victorian England. It is not possible to offer a comprehensive account of the relationship between Pre-Raphaelitism and science after 1860. This is partly because deep rifts had opened up in both of them, but also because there was no longer the same sustained engagement with science in the Pre-Raphaelite arts. As a result, individual responses to science in particular paintings or poems do not necessarily cohere into a distinctive movement that can be set apart from other late Victorian engagements with science in the arts. There are, however, substantial Pre-Raphaelite works which, when brought together, reveal the function that the major Pre-Raphaelite arts played in shaping perceptions of and responses to science in the late Victorian period.

After this introduction, I begin by explaining in detail what the Pre-Raphaelites meant by science and how they put into practice their programme for scientific art. Chapter 1 starts by examining the Pre-Raphaelites' art theory as they expounded it in *The Germ*, *The Spectator* and elsewhere. The terms of these early essays mirror how Victorian scientists themselves, including Herschel, wrote about their own practice. In this chapter, I trace how the precepts of Pre-Raphaelite art theory are carried out in experimental paintings and poems such as Millais's *Ophelia* and Deverell's 'The Garden'. To explore this in more detail, I contrast the engagement with science as subject matter in William Dyce's painting *Pegwell Bay* with the rigorous observation of movement in space and time in Millais's portrait of Ruskin and Dante Gabriel Rossetti's poems from his trip to Paris and Belgium with Holman Hunt in 1849. *Pegwell Bay* is typically cited as the Pre-Raphaelite painting which references science most directly, yet, as I show, Dyce's approach to science is very different from that of the close Pre-Raphaelite circle in the late 1840s and early 1850s. As well as stressing the potential for art to emulate the observational precision of science, Stephens argued that, by adopting scientific standards, the arts could further their own moral aims. The last section of this chapter shows how William Rossetti's poem 'Sheer Waste', published in *The Germ*, and Hunt's paintings *The Awakening Conscience* and *The Hireling Shepherd* bear out this claim.

As I explain in the first chapter, William Rossetti characterised Pre-Raphaelite art as independent investigation into areas of knowledge previously interpreted through received wisdom. In the next two chapters I offer readings of how Pre-Raphaelite paintings and poems work to investigate for themselves two of the major fields of Victorian scientific endeavour. Chapter 2 concerns natural history. Setting several early Pre-Raphaelite poems alongside Ruskin's prose and paintings by Rosa and John Brett, I argue that the Pre-Raphaelites' minutely discriminating but spatially indiscriminate attention to detail in landscapes and particularly their commemorating of living creatures that might be thought negligible or even grotesque – daisies but also thistles, flies, toadstools – is an act of valuing. In paintings such as Millais's *Woodman's Daughter* and Hunt's *Our English Coasts*, we can see how the value they ascribe to the lives of animals and plants leads them to recognise the ecological relations between ourselves and other creatures, well before ecology was named as a science along Darwinian lines. This ecological view of nature is predicated on a sympathy which reaches beyond our own kind. In his moving poem 'The Child Jesus', Collinson equates this sympathy with Christ's willingness to take on the suffering of not only mankind but other animals. In their suburban and rural landscape paintings, Ford Madox Brown and George Boyce suggest in more secular terms how our moral and psychological well-being depends on maintaining our own ecological grounding even in a modern Victorian world.

In Chapter 3 I move from landscape painting to history painting and from ecology to psychology. The Pre-Raphaelites were painting at the inception of the medical study of psychology, just as it was beginning to cohere into a science. With several competing discourses all claiming to be sciences of the mind, the Pre-Raphaelites were wary and sceptical. Where their landscape paintings complement and extend mid-Victorian natural history, their history paintings implicitly critique mid-Victorian psychology. Instead of the formulaic systems offered by physiognomy and phrenology, which purported to be able to determine mental character on the basis of the physical form of the face or the head, the Pre-Raphaelites returned to first principles, investigating for themselves how mood and temperament might be expressed not in fixed forms but in the tensions and inflections within an individual's body. Their method for studying this embodied psychology is spelled out in essays by Stephens, Brown and Tupper. It can be seen in practice in the way in which facial expressions and bodily postures become more subtly revealing across four series of Pre-Raphaelite pictures: the drawings and finished paintings for Millais's *Isabella* and Hunt's *Valentine Rescuing Sylvia from Proteus*; the successive paintings and poems in which Rossetti tries to conceive of the psychological impact of the incarnation on the Virgin Mary; and Collinson's paired paintings *Answering the Emigrant's Letter* and *The Emigration Scheme*. While these paintings explore psychology as it is expressed, typically unconsciously, in the bodies of others as we see them from without, Rossetti uses narrative poetry to take us directly into the mind of a man whose own bodily impulses lead him to a perverse fascination with a young girl's growing body in his brilliant and disturbing dramatic monologue, 'A Last Confession'. Finally, to close this chapter, I show how William Rossetti's experimental narrative poem 'Mrs. Holmes Grey' critiques the science of medical psychology even as it invokes science in its method and form.

Chapter 4 traces how the Pre-Raphaelite programme to model the arts on science shaped their religious art. Although there were fears that science might lead to materialism in the 1840s and 1850s, the thesis that science and religion were in conflict was a largely rhetorical product of the campaign for scientific naturalism from the 1860s onwards. Among liberal Christians in the 1850s, there was every reason to believe that science could help to reach towards a fuller and truer understanding of the foundations of faith. This premise, allied to new trends in biblical scholarship in the mid-nineteenth century, became the basis of Pre-Raphaelite religious painting. William Rossetti outlined its principles in an important essay on 'The Externals of Sacred Art' in the late 1850s. After considering both the success and the limits of Millais's daring experiment in creating a new form of sacred art in *Christ in the House of His Parents*, I move on to an extended discussion of Hunt's paintings from his first two trips to the Middle East in the 1850s and 1860s. Hunt visited the Holy Land, as he saw it, to paint scenes from Christ's life in situ, not merely out of a compulsion towards accuracy, but out of the conviction that, in order to grasp the full significance of the incarnation, it needed to be understood in its specific place and time. Hunt's paintings *The Scapegoat*, *The Finding of the Saviour in the Temple* and *The Shadow of Death* draw on and conduct research into archaeology, physical geography, natural history and ethnography to build up as close an approximation as they can to the reality of Christ's life and times.

Two events in 1860 marked the culmination of the Pre-Raphaelite project to create art on scientific principles. The first was the exhibition in London of *The Finding of the Saviour in the Temple*. The second was the opening of the new Oxford University Museum of Natural History. The scientists who commissioned the museum embraced Pre-Raphaelitism. Together with the architects, decorative artists and the Pre-Raphaelites themselves, they transformed it from an art conducted on scientific principles into an art that could express their own scientific world-view. In Chapter 5 I explore this remarkable collaboration between artists and scientists which led to the creation of what is arguably the greatest single work of Pre-Raphaelite art. The museum itself is the best example too of architecture as a representational art form on Pre-Raphaelite principles. It is both a model and an illustration of nature as understood by natural theology. At the same time, it is a bold affirmation of the modernity of science, an experiment to combine Gothic engineering with modern materials. The contributions made by Pre-Raphaelite artists to its design – both in drawings and in verse – help us to tease out its rich symbolic meanings. The Pre-Raphaelite project is fulfilled in the museum through two separate sculptural schema: the O'Sheas' vibrant natural history carvings on Pre-Raphaelite principles and the Pre-Raphaelites' own portrait statues of scientists. The one embodies an ecological vision of nature, the other a vision of science as the work of imagination. Both add new elements which sit alongside and at times jar against the natural theology epitomised in the museum's overall form.

The ecological vision of the O'Sheas' carvings at Oxford chime with the Darwinian view of nature that would come to eclipse the old natural theology. The relationship between the Pre-Raphaelites and Victorian science was transformed by the rise of scientific naturalism and the rift that opened up within Pre-Raphaelitism itself. Chapter 6 considers this transformation. In the first half of the chapter I look at how the surviving advocates of the original Pre-Raphaelite principles of scrupulous accuracy and independent investigation in art saw them as bearing on the question of whether science reveals God's design in nature or whether instead theological considerations of this kind should be rigorously excluded from science. For Tupper, Pre-Raphaelite practice in looking and drawing sustained natural theology, yet the same practice was advocated by Brett and by the astronomer Norman Lockyer in *Nature*, a new magazine in the vanguard of scientific naturalism which would become the premier journal of modern science. This same divide can be traced in the reception of Hunt's late painting *The Triumph of the Innocents*, which depicts the souls of the babies murdered on Herod's orders as literally present in company with Jesus on the flight into Egypt. For Ruskin, this painting gave material expression to a profound truth; for Stephens, it was a ridiculous misjudgement to conflate the spiritual and symbolic with the physical world.

If the aesthetic principles of the original Pre-Raphaelite project were being invoked by both sides in the struggle between natural theology and scientific naturalism, the second generation of Pre-Raphaelites had abandoned any aspiration to create art on the model of science. In the second half of Chapter 6 I explain how changes in Rossetti's art in particular over the 1850s and into the 1860s embody this move away from science towards aestheticism and symbolism. Yet the divorce between science and second-generation Pre-Raphaelitism was not absolute. In his *Studies in the History of the Renaissance*, the leading theorist of aestheticism, Walter Pater, made a powerful case for the value of art in a world in which life was finite and immortality redundant. In Chapter 7 I trace how Pre-Raphaelite poetry came to play an important role in confronting the materialist implications of scientific naturalism and in sustaining meaning in a world in which the reassuring promises of Christianity no longer held good. My two principle test cases here are the Pre-Raphaelite poems published in the influential liberal-to-radical periodical the *Fortnightly Review* in the late 1860s and early 1870s, when the editor John Morley was championing scientific naturalism, and William Morris's enormous and enormously popular poem, *The Earthly Paradise*, which was well received by readers sympathetic to science. Through close readings of these poems alongside respectively the articles on science published in the *Fortnightly* and reviews of *The Earthly Paradise*, we can see how poems that seem on the face of it neglectful of or even hostile towards science came to be aligned to scientific naturalism. Rossetti's own scepticism towards science was expressed in the last poem he contributed to the *Fortnightly*, his late design *The Question* and his very last poem which he composed on his deathbed to accompany it. Together these form a triple work of art, but even in their critique of science they fulfil the Pre-Raphaelite commitment to investigate received opinions, reaffirming scientific naturalism's conviction that no truths are sacred and no science is final.

Chapter 8 concentrates on a single case study to explore how Pre-Raphaelite architecture became an instrument of scientific controversy in the wake of Darwinism. Richard Owen was England's leading comparative anatomist and superintendent of the natural history collections at the British Museum. He was also a generous friend of several of the Pre-Raphaelites, including Hunt, Millais and Tupper, and a determined opponent of Darwin and the scientific naturalists. When Owen came to collaborate with the architect Alfred Waterhouse – who had his own Pre-Raphaelite connections – in designing the British Museum's new Natural History Museum in South Kensington, they adopted more loosely the Pre-Raphaelite principles that Acland, Phillips and Woodward had followed in Oxford. As at Oxford, the aim was to reinforce natural theology by setting its precepts in stone. Yet, in their forms, materials and decorations, the two buildings offer subtly but significantly different models of creation. The Natural History Museum differs from its predecessor too in being an act of resistance. When the Oxford museum was built, natural theology was still in the ascendant. When the London museum was built, it was under attack. For most of the scientific community, Owen's science was obsolete well before his museum was opened. Though it was praised in art magazines and newspapers, it was critiqued in much of the scientific press, including *Nature*. Even as Owen and Waterhouse adopted Pre-Raphaelite principles in building their museum, so *Nature* critiqued it on the same grounds. Both held that the museum should offer a true image of natural history as a subject and a science, but they disagreed radically over that truth itself.

In Chapter 9 I return to the art form that first attracted scientists such as Owen and Acland to Pre-Raphaelitism. In the 1880s the three surviving painters from the original Pre-Raphaelite circle revisited their own earlier preoccupation with science in paintings of scientists. These paintings reveal a new and acute awareness of the history of science as a process that is not straightforwardly progressive but that also includes the loss of earlier certainties and convictions, and the eclipse of one generation by the next. Hunt's portrait of the elderly Owen and Millais's painting *The Ornithologist; or, The Ruling Passion*, inspired by his meeting with the dying naturalist John Gould, both capture the collapse of natural theology as the natural theologians themselves aged and died. This is especially apparent when these late paintings are contrasted with the younger painter John Collier's portraits of the leading scientific naturalists. Where Hunt and Millais revisit natural theology, Brown returns to the Baconian model of science as the patient accumulation of individual observations in his historical paintings of the astronomer William Crabtree and the chemist John Dalton for the series of murals he painted in the new Manchester Town Hall. Again, even as he reaffirms the ideal of science held up by the Pre-Raphaelites forty years before, his paintings intimate that that ideal is to be found in the past and no longer maps on to current science.

These late paintings by Hunt, Millais and Brown bring the Pre-Raphaelite project full circle, at once embodying its original principles and recognising their contingent origins in what was by this time an outdated conception of science. But while both science and Pre-Raphaelite art itself had moved on, the Pre-Raphaelite project had achieved its ends. It had brought a new rigour and intellectual independence to the arts. As the historian of nineteenth-century British painting Luke Herrmann has remarked, the Pre-Raphaelites' influence was so decisive that 'the story of British painting of the nineteenth century has to be told in two halves: the "pre-Pre-Raphaelite" and the "post-Pre-Raphaelite".'[50] Pre-Raphaelitism had also provided science itself with a new architectural language and a standard by which to judge its own visual observations. Altogether, the Pre-Raphaelite paintings, poems, sculptures and buildings discussed in this book bear witness to one of the most fertile engagements with science by art, and to some of the richest collaborations between scientists and artists, of the nineteenth century.

Following pages Detail of fig. 30.

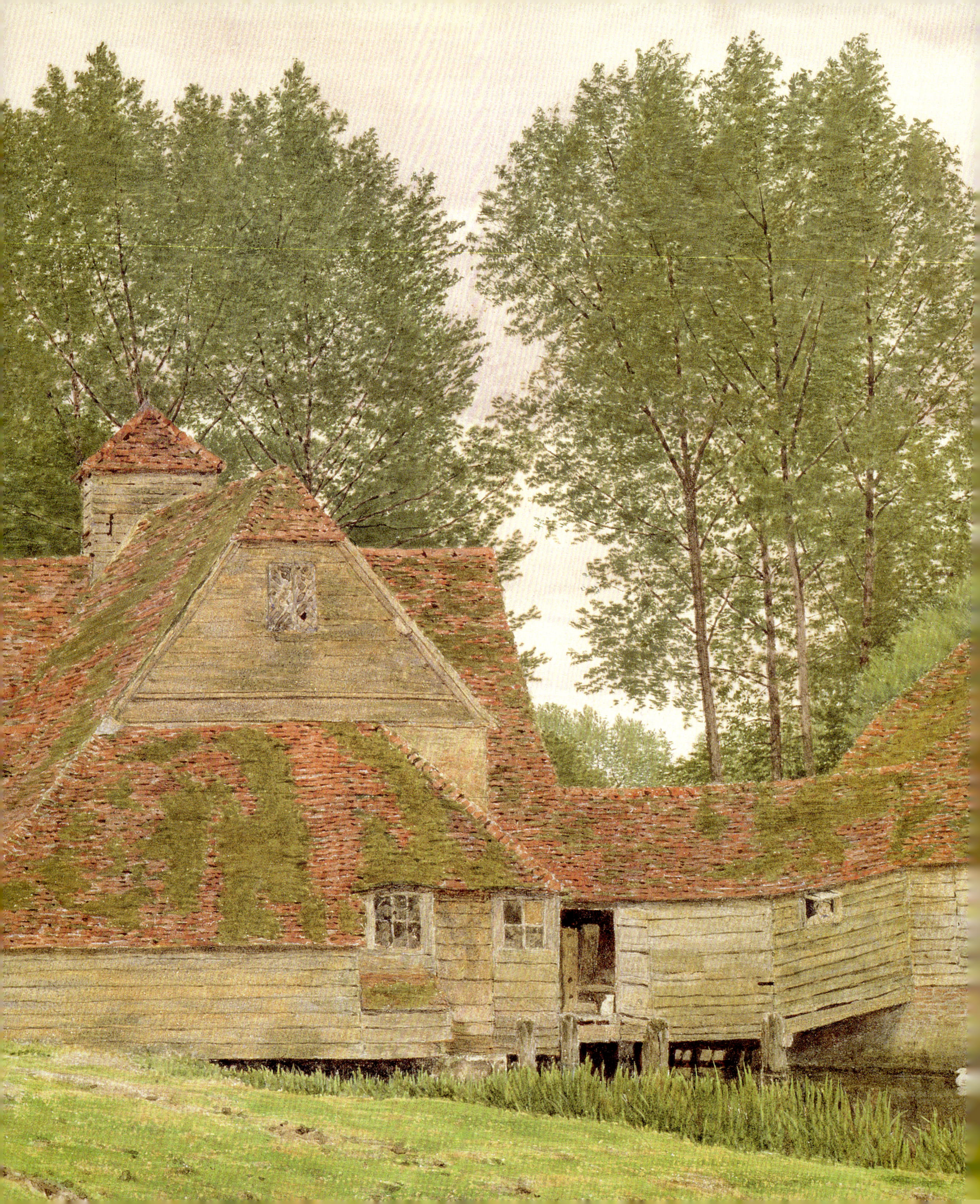

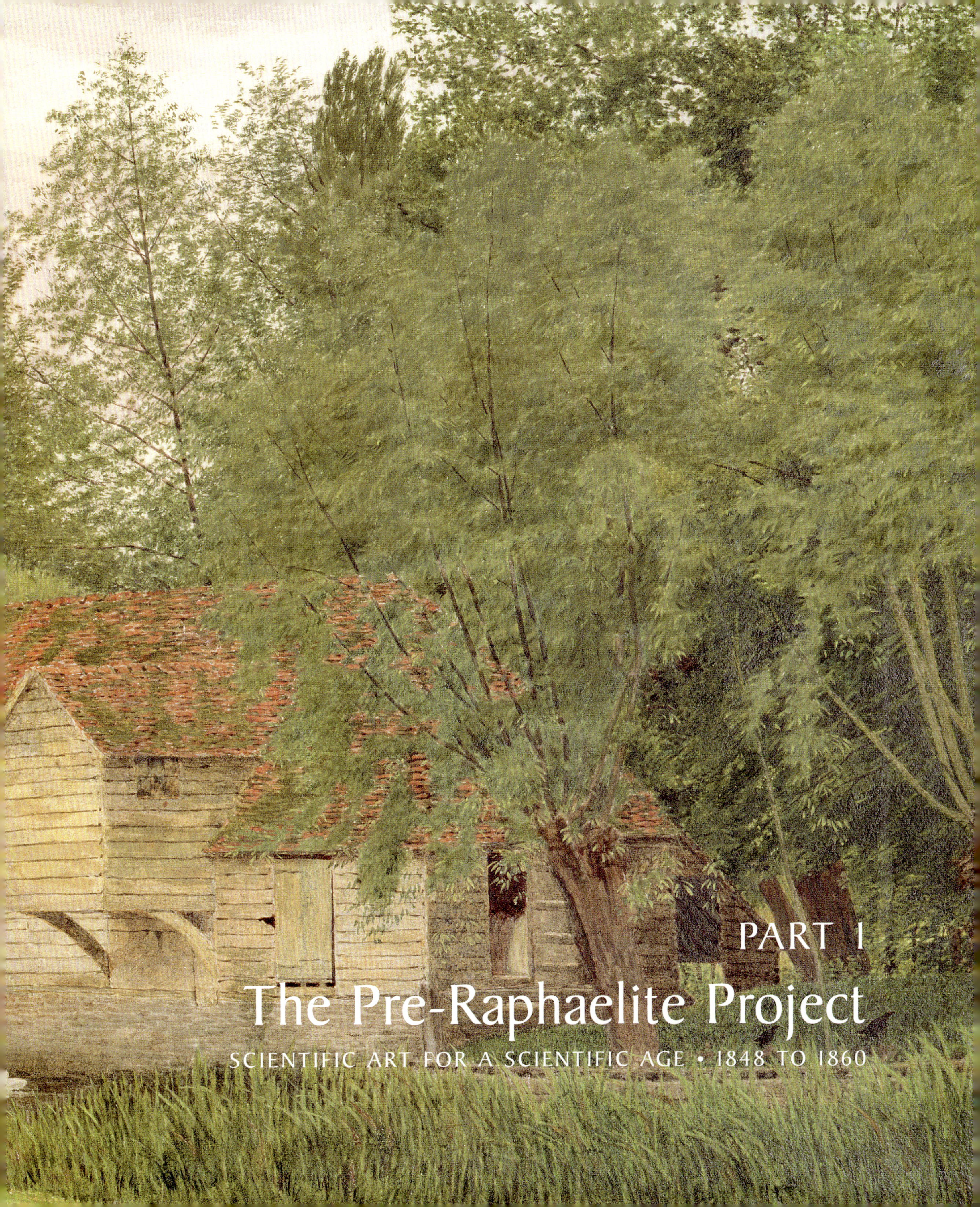

PART I

The Pre-Raphaelite Project

SCIENTIFIC ART FOR A SCIENTIFIC AGE • 1848 TO 1860

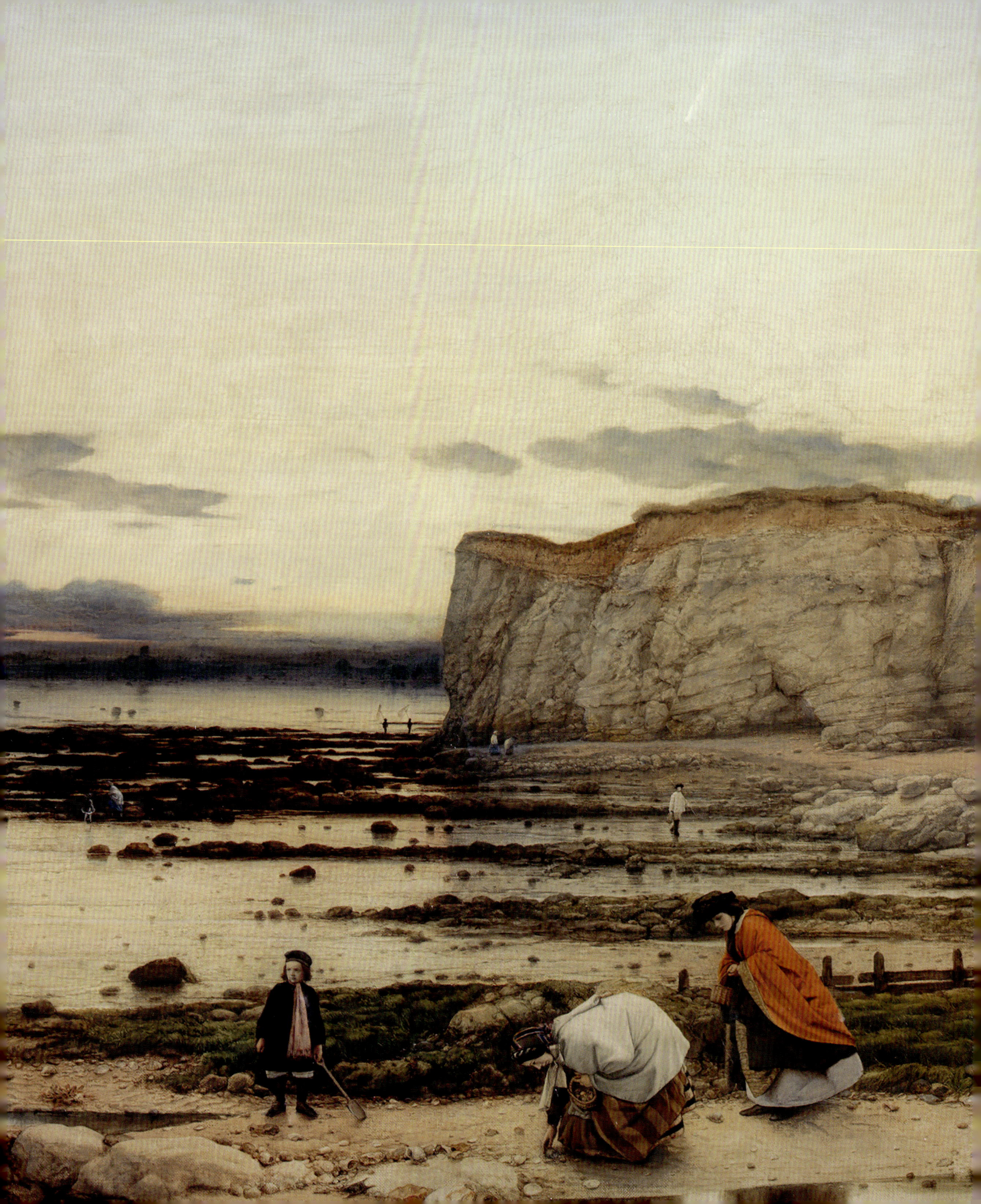

1 · The Search after Truth

ART ON THE MODEL OF SCIENCE

In January 1850 the Pre-Raphaelites and their associates launched a new magazine entitled *The Germ: Thoughts towards Nature in Poetry, Literature, and Art*. *The Germ* provided them with a forum in which to articulate their vision of the future of art and to debate it with one another. It is the closest thing to a single Pre-Raphaelite manifesto, and has been rightly seen as a precursor to the little magazines launched by later avant-garde art movements in the late nineteenth and early twentieth centuries.[1] For an art manifesto, *The Germ* is laced with references to science. There are explicit discussions of the relationship between science and art in each of its four issues. These essays make reference to geology, chemistry, physiology and other modern sciences, as well as the technological advances made possible by science. Even when science is not invoked directly, the critical vocabulary of *The Germ* aligns the Pre-Raphaelite project to it, with a recurrent emphasis on facts and the observation of nature.[2] The magazine's very title, for all that it is printed in a Gothic typeface, makes a 'biological and botanical reference', as Stephanie Grilli has remarked.[3]

The best-known reference to science in *The Germ* comes in Frederic George Stephens's essay 'The Purpose and Tendency of Early Italian Art', published in February 1850 in the magazine's second issue. Here Stephens identifies current science as just as important a model for modern art as the medieval paintings which are the ostensible subject of his essay and from which the Pre-Raphaelite Brotherhood took its name. Stephens writes:

> The sciences have become almost exact within the present century. Geology and chemistry are almost re-instituted. The first has been nearly created; the second expanded

Facing page Detail of fig. 11.

so widely that it now searches and measures the creation. And how has this been done but by bringing greater knowledge to bear upon a wider range of experiment; by being precise in the search after truth? If this adherence to fact, to experiment and not theory, – to begin at the beginning and not fly to the end, – has added so much to the knowledge of man in science; why may it not greatly assist the moral purposes of the Arts?[4]

As Stephens's questions imply, the idea of science itself was more important to the Pre-Raphaelites than any particular scientific fact or theory. Gowan Dawson and Bernard Lightman have pointed out that 'science' was still a flexible term in the early Victorian period, retaining wider connotations of 'knowledge acquired by study, or trained skill' as well as 'experimental method and the investigation of the natural world'.[5] This earlier sense of the term was not only broader than our own; it could even be diametrically opposed to it. In his popular *Treasury of Knowledge*, reprinted and expanded throughout the 1830s, the knowledge-monger Samuel Maunder defined 'science' as 'knowledge, art attained by precepts', and proceeded to list the seven liberal arts as examples.[6] When the landscape painter John Constable opened his series of lectures on 'The History of Landscape Painting' at the Royal Institution in June 1836, he spoke of painting as '*scientific* as well as *poetic*' simply because it was a 'regularly taught profession'.[7] But the Royal Institution was London's premier institution for disseminating scientific research and new technologies, headed up by Michael Faraday, and when Constable closed the last lecture in the same series by declaring, 'Painting is a science, and should be pursued as an inquiry into the laws of nature,' he used 'science' in its modern sense.[8] Given this slippage between different definitions, to see how Pre-Raphaelite paintings and poetry might have been 'scientific' on their own terms, we first need to establish first what 'science' itself meant to the Pre-Raphaelites.

PRE-RAPHAELITISM AND THE IDEA OF SCIENCE

For the Pre-Raphaelites, 'science' meant first of all an ethic of strict honesty in observation – what Stephens calls 'adherence to fact'. The design of a picture had to be 'subordinate to the truth of nature', as Ford Madox Brown puts it in the first of what was to have been a series of essays for *The Germ* 'On the Mechanism of a Historical Picture'.[9] Individual details too must be meticulously accurate, judged by the standards of science itself. In another essay in *The Germ*, 'Modern Giants', Stephens writes that it is fatuous for people to be taught to admire paintings containing 'rocks that make geologists wonder, their angles are so impossible, their fractures are so new'. 'Letting observation sleep,' he asks, 'what can you know of nature?'[10] John Ruskin, who had insisted on this kind of 'scientific representation' in *Modern Painters*, would extend this argument in his own Pre-Raphaelite manifesto, the pamphlet *Pre-Raphaelitism*, published in August 1851.[11] One of the duties of the artist, he suggests, is to record faithfully and accurately his impressions of the natural world. These '*seen facts*' can convey 'perfect knowledge' and even 'aid the advance of the sciences' themselves.[12]

One early Pre-Raphaelite work of art did indeed make a contribution to science in this way. When modelling a lively plaster statuette of *Puck* (fig. 5) from *A Midsummer Night's Dream* in the mid-1840s, Woolner realised that the tradition of representing satyrs and other supernatural creatures with pointy ears derived from an anatomical feature which he had observed in both humans and monkeys: a small, blunt protrusion on the fold of the outer ear as it curves over to form the shell. In his statue, he intensified this feature to give Puck the pointy ears that have become ubiquitous in pictures and models of pixies and elves. Twenty years later, Woolner shared his observations with Charles Darwin, who was sitting for a portrait bust (see fig. 98). Darwin, who was building up the case for human evolution that he would present to the public in *The Descent of Man* in 1871, was delighted. When Woolner sent him a drawing of the human ear (fig. 6) as an illustration, Darwin wrote back, 'My dear Mr. Woolner, Very many thanks for the drawing: it does excellently. – The "Woolnerian tip" is worth anything to me.'[13] According to Darwin, Woolner 'perceived the full significance' of his discovery, implying that the sculptor realised independently that the 'Woolnerian tip' was evidence for the common ancestry of humans and monkeys.[14]

Darwin's citation of Woolner is a rare direct reference to something integral to the Pre-Raphaelites' conception of

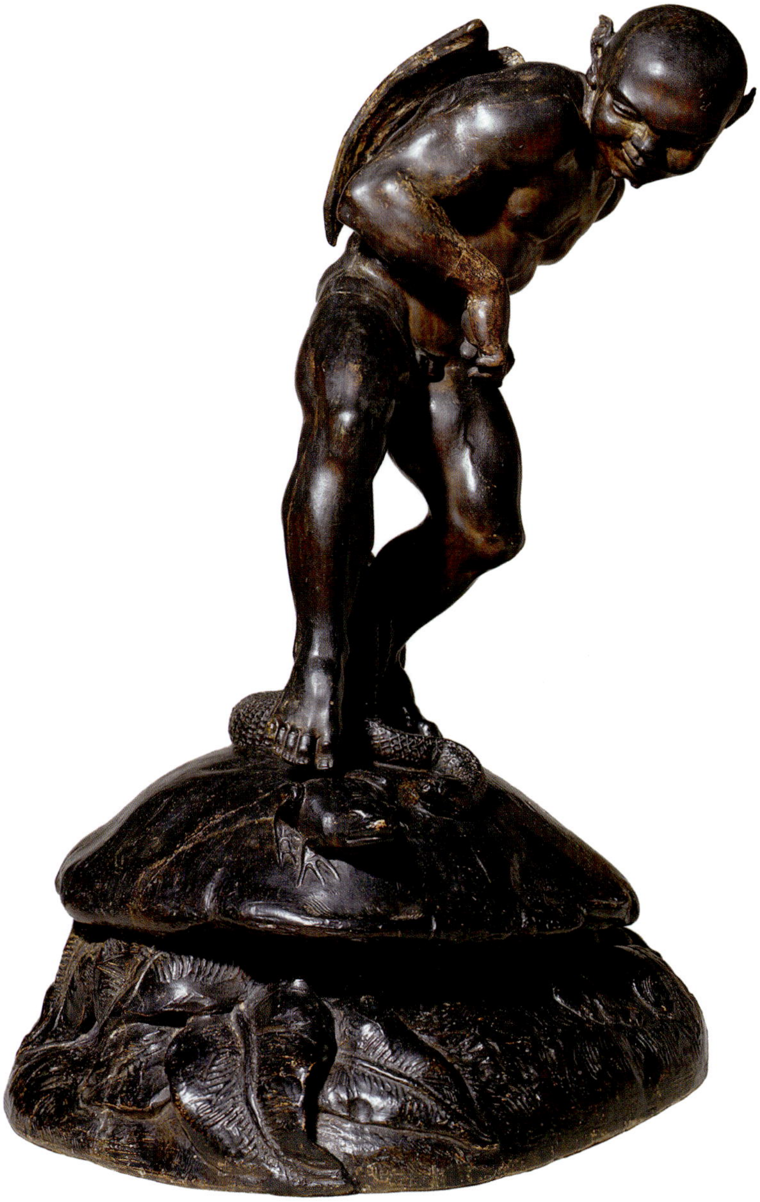

5 Thomas Woolner, *Puck*, 1845–7. Tate Gallery, London.

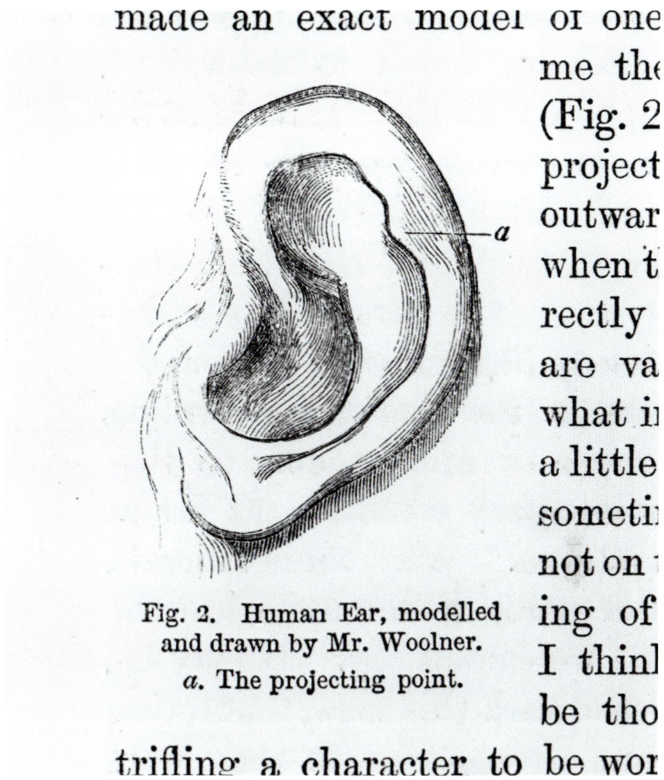

made an exact model of one
me the
(Fig. 2
project
outwar
when t
rectly
are va
what i
a little
someti
not on
ing of
I thinl
be tho
trifling a character to be wor

Fig. 2. Human Ear, modelled and drawn by Mr. Woolner.
a. The projecting point.

6 Charles Darwin, *The Descent of Man, and Selection in Relation to Sex*, 1871, volume 1, page 22, including Thomas Woolner's drawing of a human ear.

their art but which remains largely implicit within the art itself. The Pre-Raphaelites saw accurate observation as an artist's means to make independent discoveries which had the same claim to truth as science. William Rossetti set out this principle in another manifesto essay, published in *The Spectator*, five weeks after Ruskin's pamphlet. For Rossetti, merely recording the details of natural forms for their own sake is 'a low branch of art', while the 'conventional rules of painting are, and must ever be, matter of opinion: they are not fact.' By contrast with both documentary and conventional painters, he explains, the Pre-Raphaelites are dedicated to 'investigation for themselves on all points which have hitherto been settled by example or unproved precept, and unflinching avowal of the result of such investigation'. For the Pre-Raphaelites, science sets the standard for independent investigation that does not defer to the authority of conventional wisdom – in Rossetti's words, 'individual mind and views working from absolute data of fact'.[15] By following this procedure boldly and rigorously, the sciences had made rapid and cumulative progress since the turn of the nineteenth century. In Stephens's words, knowledge had become at once 'wider' in its range and more 'exact' in its details. Through imitating science's investigative approach, art too could advance towards a broader and more precise understanding.

For Stephens, it is chemistry above all – 'the modern science' as he calls it in 'Modern Giants' – that exemplifies this progress.[16] Charles Daubeny, Professor of Chemistry at Oxford, opened his *Introduction to the Atomic Theory*, published in 1831, by announcing that 'a few microscopic chemists' had 'done more towards the elucidation of one of the most abstruse questions on which the human mind can be engaged, than was effected by the profoundest intellects of the ages that preceded them'.[17] The same year, the astronomer John Herschel – son of William Herschel, who discovered Uranus and remapped the stars of the northern hemisphere in collaboration with his sister Caroline – wrote in his *Preliminary Discourse on the Study of Natural Philosophy* that the 'transformations of chemistry . . . are opening up to us every day sources of wealth and convenience of which former ages had no idea, and which have been pure gifts of science to man.'[18] In 1848, the year the Pre-Raphaelite Brotherhood was formed, the eminent physician Henry Holland impressed upon the readers of the *Quarterly Review* quite how rapid progress in this science had been, to the extent that even the works that had 'the greatest reputation' in the field only thirty years ago 'are now utterly out of date and useless'.[19]

By the mid-century, when Stephens was writing, chemistry had established itself as the paramount example of a science that had risen in fifty years from virtually nothing to the heights of both abstract and useful knowledge. Such rapid progress was a key element in the Pre-Raphaelites' conception of science. By pursuing their own independent investigations of nature, including human nature, they hoped to imitate in the arts the progress of sciences such as chemistry, transforming painting and sculpture from static arts, perpetuating old conventions, into dynamic arts capable of making advances in knowledge, in their own practices and in the impact they could have upon their viewers.

In another essay for *The Germ*, 'The Subject in Art', the sculptor John Lucas Tupper offers a scientific parable for this ideal of progress in art. Tupper draws an analogy between critics who judge a new work by the standards of an old one and geometers comparing 'two triangles in Euclid':

> the critics, certainly guilty of the most unpardonable blindness, blundered up to the masses of 'High Art,' left by antiquity, saying, 'there let us fix our observatory,' and here came out perspective glass, and callipers and compasses; and here they made squares and triangles, and circles, and ellipses, for, said they, 'this is "High Art," and this hath certain proportions;' then in the logic of their hearts, they continued, 'all these proportions we know by admeasurement, whatsoever hath these is "High Art," whatsoever hath not, is "Low Art."'[20]

Tupper's purblind geometers are the equivalent of the Royal Academy, insisting that modern art should ape the pre-existing models of classical sculpture and Raphaelesque painting. His analogy is not arbitrary. One of the conventional rules of painting taught in the Academy which the Pre-Raphaelites repeatedly condemned was precisely the insistence on pyramidal groupings of figures.[21] They were not alone in taking this stance. Later in 1850, the pioneer of photography and populariser of science Robert Hunt too used his column in *The Art-Journal*, 'On the Applications of Science to the Fine and Useful Arts', in which he explored how new materials and technologies were advancing art, to take the Scottish painter and decorative artist David Ramsay Hay to task for insisting that art students should be taught 'the set formula, by which the human form may be bound within geometrical lines' as the key to beauty in figure painting.[22] As a foil to such geometrical art critics and teachers, Tupper conjures up 'a then embryon-electrician', who could have explained to the Academicians that their 'high art marbles possessed an electric influence, which, acting in the brain of the observer, would awake in him emotions of so exalted a character, that he forthwith, inevitably nodding at them, must utter the tremendous syllables "High Art"'. In so doing, Tupper suggests, he would 'have done something more to the purpose than all the critics and the compasses'.[23]

For Tupper, different sciences have a different standing depending on how far they have kept pace with changes in knowledge and its practical applications. In his exuberant and unruly satire on the art establishment, the geometers stand for superseded science, not because their geometry is false in itself, but because they fail to realise that it no longer accounts for the phenomena in question. It was not until much later in the century that the non-Euclidean geometry pioneered on the Continent by Nikolai Lobachevsky and

János Bólyai made significant headway in Britain, and Tupper would not have had privileged access to these developments. But, as Alice Jenkins has shown, the place of Euclid's *Elements* as the cornerstone of geometry had been being debated and contested in Britain from the 1820s onwards.[24] Tupper's embryon-electrician embodies this move to displace Euclid with a more modern conception of science. In his portentousness, the 'embryon-electrician' can seem as foolish as the geometers themselves. Yet Tupper's language holds clues as to the difference between them. Tupper mocks the pretension of the phrase 'High Art', but his essay is nonetheless an attempt to define it, and this serio-comic parable is a key moment in that definition. The geometers are pedants. They speak in outdated English with undue, legalistic precision, using 'hath' for 'has' and 'admeasurement' where 'measurement' would do. The electrician's rhetoric may be overblown, but his thinking is dynamic. Where they look for static analogies ('forsooth'), he identifies active causes and effects ('forthwith').

Tupper worked as an anatomical illustrator at Guy's Hospital in London. Through his training, he would almost certainly have known Neil Arnott's *Elements of Physics*, published in several editions from 1827 onwards. The title of Arnott's popular textbook signals its ambition to supplant Euclid's *Elements*, while its contents include an extensive account of anatomy, described largely in mechanical terms.[25] Tupper may well also have been familiar with *Electricity: Its Nature, Operation, and Importance in the Phenomena of the Universe* by William Leithead, the secretary to the London Electrical Society. In the second part of his book, published in 1837, Leithead coined the word 'electrophysiology' to define a science of the human body understood in terms of electrical influences and impulses.[26] Tupper knew too the work of the ophthalmic surgeon Alfred Smee, whose *Elements of Electro-Biology* had come out as recently as 1849.[27] These books and others like them are mirrored in Tupper's allegory of the history of art criticism. Tupper proposes that modern science would attribute the quality of 'High Art' not to set forms which could be copied, but to its 'electric influence'. The marble statues surviving from antiquity exert such an influence, but other art in other forms can exert the same or an equivalent influence equally well. Tupper sums up his argument at the beginning of the second instalment of his essay, insisting that 'all objects so exciting mental activity and emotion in the highest degree, may afford subjects for High Art.'[28] To deny this is to cling to an outdated and increasingly irrelevant science; to embrace it is to open the door to new experiments in sculpture and painting. These in turn can lead to progress in the arts, extending their range and deepening their insights.

In his *Preliminary Discourse*, Herschel sought to delineate the philosophy of science that underpinned this progress. Herschel's book was the defining statement of early Victorian philosophy of science, but as James Secord has pointed out, unlike most substantial philosophical essays, it was published not as an expensive folio but rather as a small, six-shilling volume in Dionysius Lardner's *Cabinet Cyclopedia*. Fifteen thousand copies were printed and sold within three years of publication, and Secord estimates that overall tens of thousands of people read Herschel's book over the middle decades of the nineteenth century.[29] It is not certain that the Pre-Raphaelites were among them, but it is not unlikely that Hunt, for instance, might have encountered the *Preliminary Discourse* in the course of the autodidactic reading in his father's library which he later saw as foundational to his views on art. Secord argues that one of the main functions of Herschel's book was to develop a 'scientific frame of mind' and, in Herschel's own words, the 'habit of strict investigation' in its readers.[30] Whether or not the Pre-Raphaelites read Herschel's book itself, they affirmed precisely this commitment in their essays and put it into practice in their art.

Herschel distinguishes two sources of natural knowledge. The first, 'observation', he defines as 'noticing facts as they occur, without any attempt to influence the frequency of their occurrence, or to vary the circumstances under which they occur'. In effect, this describes the practices of natural history. The second, 'experiment', entails 'putting in action causes and agents over which we have control, and purposely varying their combinations, and noticing what effects take place'.[31] In 'The Purpose and Tendency of Early Italian Art', Stephens draws out the parallel between the Pre-Raphaelites' experimental art and experiments in science implicit in Tupper's essay. This might seem little more than semantics, a way of borrowing the prestige of science for a wholly

7 John Everett Millais, *Ophelia*, 1851–2.
Tate Gallery, London.

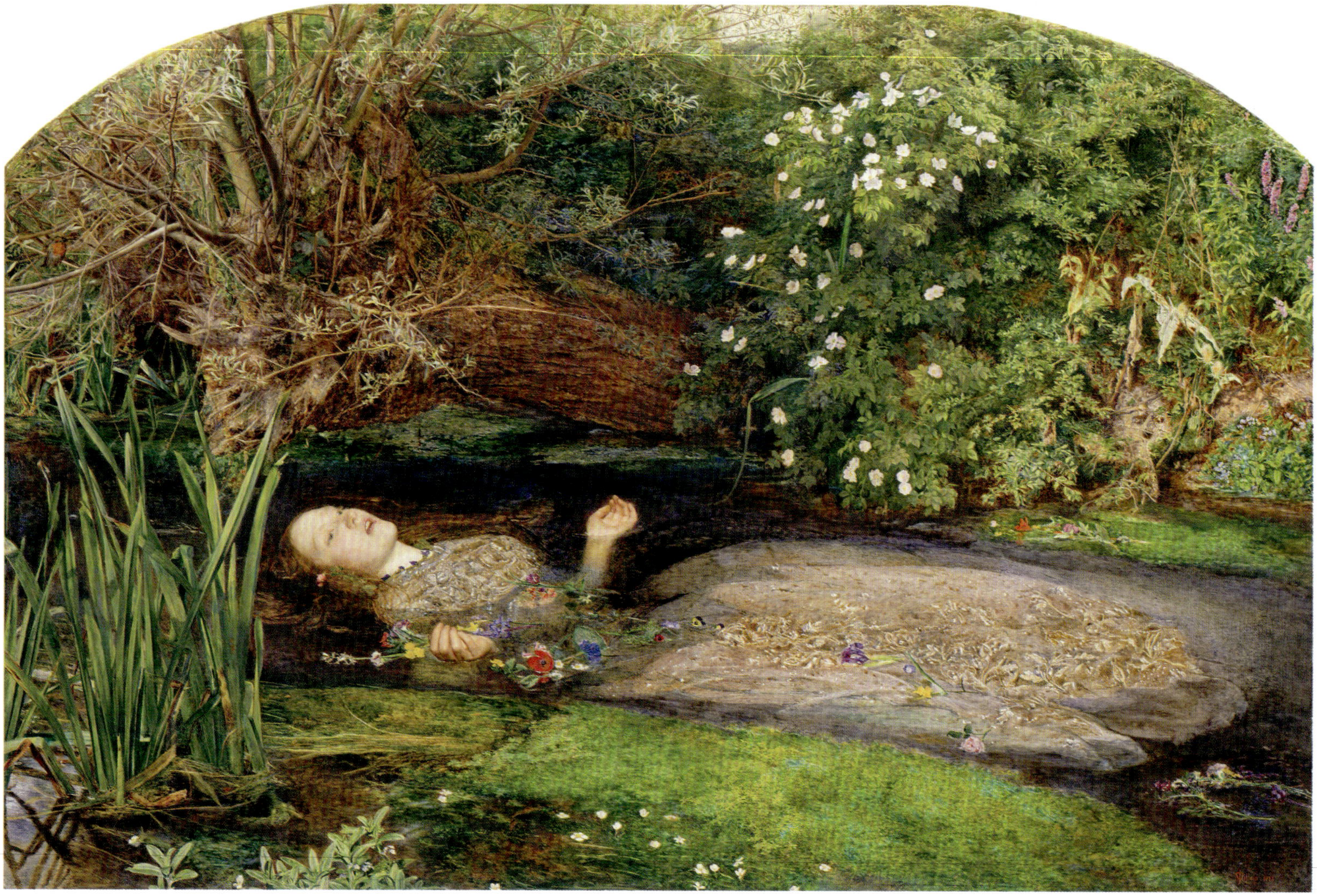

different concept of experimentation. In fact, the scientific experiment as defined by Herschel is another of the main elements of the Pre-Raphaelite conception of scientific art, alongside accuracy of observation, independent investigation, the refusal to defer to authority, and progress.

Stephens's analogy between the experimental procedures of science and the new practices that the Pre-Raphaelites introduced into painting bears a surprising amount of strain. The remarkable precision of the details of the landscape in John Everett Millais's *Ophelia* (fig. 7), for example, align it not only with the mid-Victorian enthusiasm for natural history, as Timothy Hilton, Lynn Merrill, Jason Rosenfeld and Alison Smith have all argued, but also with the scrupulous recording of numerical values in the variables and results of a scientific experiment.[32] In painting *Ophelia*, Millais sought to replicate as far as possible the conditions of Ophelia's suicide in *Hamlet*. The first stage in establishing these controlled conditions was to paint on location in the open air. Millais spent five months in 1851 painting a riverbank near Ewell in Surrey as the background for *Ophelia*. This was one of a series of paintings worked on by the Pre-Raphaelites in this way. Their practice again recalls both the naturalist making

8 Ford Madox Brown, *The Pretty Baa-Lambs*, 1851–9. Birmingham Museums and Art Gallery.

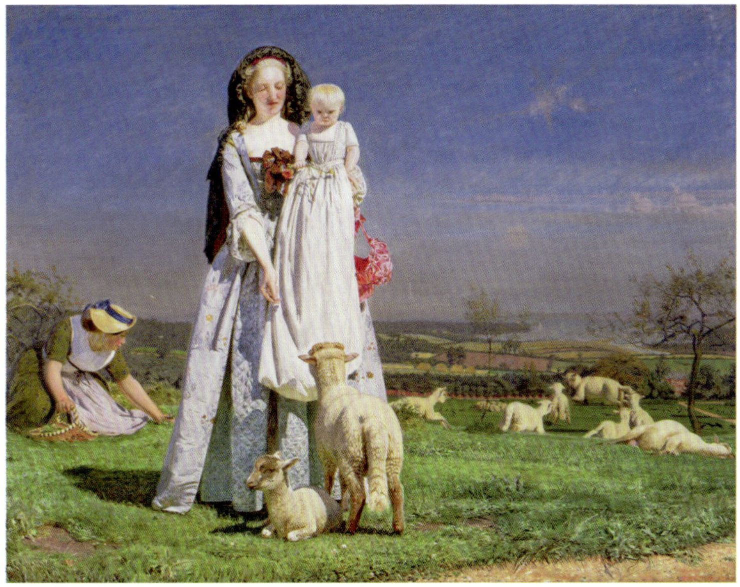

observations in situ and the experimental scientist determining the precise conditions required for an experiment.

At the same time, the paintings themselves comprise a series of experiments to test the viability of painting meticulously finished works outdoors. William Holman Hunt begun this collective experiment in 1848 by painting the background to *Rienzi Vowing to Obtain Justice for the Death of his Younger Brother* on Hampstead Heath. The following summer Hunt painted the setting for *A Converted British Family Sheltering a Christian Priest from the Persecution of the Druids* in the Lea Marshes in Essex, while Millais painted that for *Ferdinand Lured by Ariel* at Shotover Park a couple of miles up the London Road from Oxford. The next year, Millais returned to Oxford to paint the background to *The Woodman's Daughter* in Wytham Woods, while Hunt and Dante Gabriel Rossetti visited Sevenoaks in Kent, painting the backgrounds for *Valentine Rescuing Sylvia from Proteus* and a painting that Rossetti would eventually complete twenty years later as *The Bower Meadow*. In 1851 Hunt accompanied Millais to Ewell. While his friend worked on *Ophelia*, Hunt painted the landscape for *The Hireling Shepherd* (see fig. 16).[33]

For Stephens, this last painting typified Hunt's 'thoughtful power and deep scientific observation'. 'It is an absolute fact', he wrote in December 1856 in one of a series of articles entitled 'The Two Pre-Raphaelitisms' for the American magazine *The Crayon*, 'that never in this world before had any one attempted to paint the broad, bright effect of sunlight day: what were *called* sunlights were frequent enough, but no result of scientific observation of natural truth and learned execution had ever been laid before the world.'[34] Four years later in his pamphlet *William Holman Hunt and his Works*, Stephens gave a scientific lineage for Hunt's meticulous observation of the colour of shadows in full sunlight. The phenomena recorded by Hunt – the relationship between the colour of shadows and the colour of the surfaces on which they are cast – had been 'hinted at by Leonardo da Vinci, in one of his wonderful world-guesses, was partly explained by Newton, and fully developed by Davy and Brewster'.[35] But in spite of these scientific antecedents, Hunt 'discovered the fact for himself without reference to books' through 'his own observations', much as Woolner did the relationship between humans and monkeys while sculpting *Puck*.[36] In this Hunt resembles Leonardo, an exemplary elder contemporary of Raphael, the only painter given two stars on the list of 'Immortals' drawn up by Hunt and Rossetti in 1848.[37] But where even Leonardo only manages a guess, in spite of his reputation as a proto-scientist, through his experiment in painting in the open air Hunt succeeds in making an independent discovery that corroborates the scientific discoveries of the famous chemist Humphry Davy and the expert on optics David Brewster.[38] The results of Woolner's observations remained private until he shared them with Darwin, as *Puck* represents a supernatural figure, so any hints it may give as to the actual nature of human beings are obscured. Hunt's discoveries, on the other hand, were set out directly at the Royal Academy in 1852, beginning a process of re-education through which, in Stephens's words, the public might be taught to look 'at pictures with the eyes by which they look at nature', not 'at nature with the eyes by which they judged of pictures'.[39]

While Hunt and Millais were painting *The Hireling Shepherd* and *Ophelia*, Brown was extending the Pre-Raphaelite experiment further in *The Pretty Baa-Lambs* by painting the figures as well as the landscape outdoors (fig. 8). Stephens compares Brown to Hunt for his 'scientific knowledge of Art' in 'The

9 John Everett Millais, *Study of the Head of Elizabeth Siddal for 'Ophelia'*, 1852. Birmingham Museums and Art Gallery.

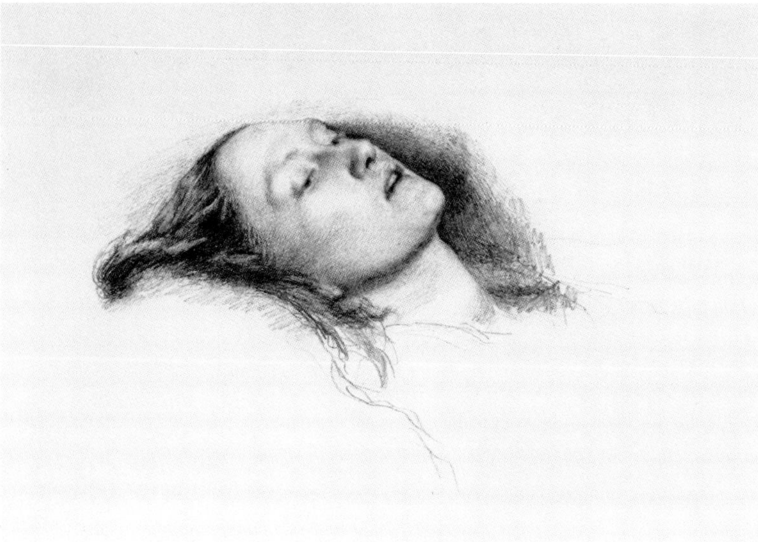

Two Pre-Raphaelitisms'.[40] Yet the notorious story of Millais's painting Elizabeth Siddal for hours on end in an increasingly chilly bath is no less a testimony to the Pre-Raphaelites' commitment to art as science than Brown's asking his wife Emma to pose in their garden in the full sun day after day. Millais's decision to paint the model for Ophelia in water was not an example of a Quixotic insistence on literalism. Rather, it was the second stage in his attempt to recreate as closely as possible the situation of the subject itself – to put in action 'causes and agents over which we have control', to borrow Herschel's words. To depict a woman on the verge of drowning, Millais simulates the physical conditions as best he can without the dangers or impracticability of attempting to paint her in the river itself. In effect, he replicates the scenario in laboratory conditions. Millais's control of these conditions was less effective than it might have been, as Siddal caught a chill, apparently from exposure to the cold water. Nevertheless, by posing her in water he was able to capture with far more precision than he could by any other means the appearance of her waterlogged dress, her floating hair, her submerged wrists.

A comparison of the finished painting with a study of Siddal's head done in preparation for it shows how the angle of her neck changed from the sketch, where her head is supported by a cushion (fig. 9), to the painting, where it appears at least to be floating.[41] In the painting, Ophelia is sinking from the feet first, her chest, face and hands alone remaining above the water. Her waist is only just visible in the darkness of the river while her skirt below it balloons up to the surface. These closely observed details of how a thin body like Siddal's would float and sink are completed by the colour of her face, the pale skin turning at once blue and pink with the cold. The differences between Siddal's expression in the study and Ophelia's in the painting can be seen to be the result of Millais's botched and yet as it turned out unwittingly successful experiment. In the study, she seems abstracted but also relaxed, in spite of the slightly awkward and uncomfortable turn of her neck. Time appears to be irrelevant to her state of mind, as though she were out of time entirely. In the painting, by contrast, she seems not so much thoughtful as numbed, as if she has acquiesced long before to this moment. Time here is crucial, yet also arrested. While the river flows on, as the prevailing sway of the water plants towards the left of the canvas reveals, Ophelia herself is witnessed on the verge of sinking. Like Siddal in the bath, she is held static by the painting at the instant when she ceases to be part of the flow of the river itself.

In *The Art of the Pre-Raphaelites*, Elizabeth Prettejohn observes that the Pre-Raphaelites' procedure in paintings such as *Ophelia* 'is like that of a research scientist who documents each experimental observation scrupulously, without regard to whether it confirms an overall hypothesis'.[42] Perhaps the most peculiar aspect of the Pre-Raphaelites' conception of science to modern eyes – certainly to modern scientists – is how unimportant hypotheses are within it. Stephens characterises science as making its discoveries through 'experiment and not theory'. This is out of keeping with today's science, both in terms of public perceptions, which tend to emphasise overarching theories rather than particular facts – the theory of evolution, the theory of relativity, chaos theory, string theory and so on – and in terms of the dominant methodology, which assumes that a prior hypothesis is needed before an experiment can be designed to test it. In practical terms, much early Victorian science shared this methodology, yet British scientists in particular understood and wrote about their own enterprise in ways that were very close to Stephens's account of it.

10 Frontispiece to John Herschel, *A Preliminary Discourse on the Study of Natural Philosophy*, 1831.

The thinker who encapsulated scientific rigour for the early Victorians was the Jacobean statesman, essayist and empirical philosopher Sir Francis Bacon (see fig. 86). In his philosophical writings, Bacon expounded what is known as the inductive method. He proposed that by doggedly gathering observations it was possible to infer patterns. These could in turn be experimentally tested to confirm scientific causes and ultimately natural laws. As Richard Yeo notes, appeals to Bacon were central to the rhetoric of Victorian science, being 'commonly associated with attacks upon hypotheses and theoretical conjecture, and with a celebration of factual observation'.[43] Herschel in particular underscored the centrality of Bacon's ideas for Victorian conceptions of scientific method.[44] He put Bacon's portrait on the frontispiece of his *Preliminary Discourse* (fig. 10) and insisted on his primacy as the founder of natural philosophy as a discipline.[45] Herschel's championing of Bacon did not amount to a naive rejection of theorisation, while other mid-Victorian philosophers of science, most notably William Whewell, went further in qualifying and ultimately repudiating the inductive method. Even so, the frequency with which early Victorian scientists invoked Bacon suggests that their loyalty to him was not mere rhetoric, and that he played an important role in defining their self-image and their conception of how scientific discovery worked.

These appeals to Bacon are characteristic of the two sciences singled out by Stephens – chemistry and geology – well into the mid-century. Holland pays tribute to Bacon in his 1848 essay on 'Modern Chemistry' for the *Quarterly Review*, while Daubeny repeats his judgement from his textbook on atomic chemistry that the achievements of modern chemistry are a 'triumph for the Baconian school of philosophy' in a second edition, published in 1850.[46] As Secord has shown, the founding work of modern British geology – Charles Lyell's *Principles of Geology*, first published in the early 1830s – is characterised by the 'vogue for "Baconian" induction and empiricist theories of perception'.[47] Other prominent geologists followed where Lyell led. In 1852 Hugh Edwin Strickland, the Deputy Reader in Geology at Oxford, wrote that it was through the 'sound principles' of 'a strictly Baconian induction of facts' that modern geologists had discovered 'Laws of Nature'.[48] The following

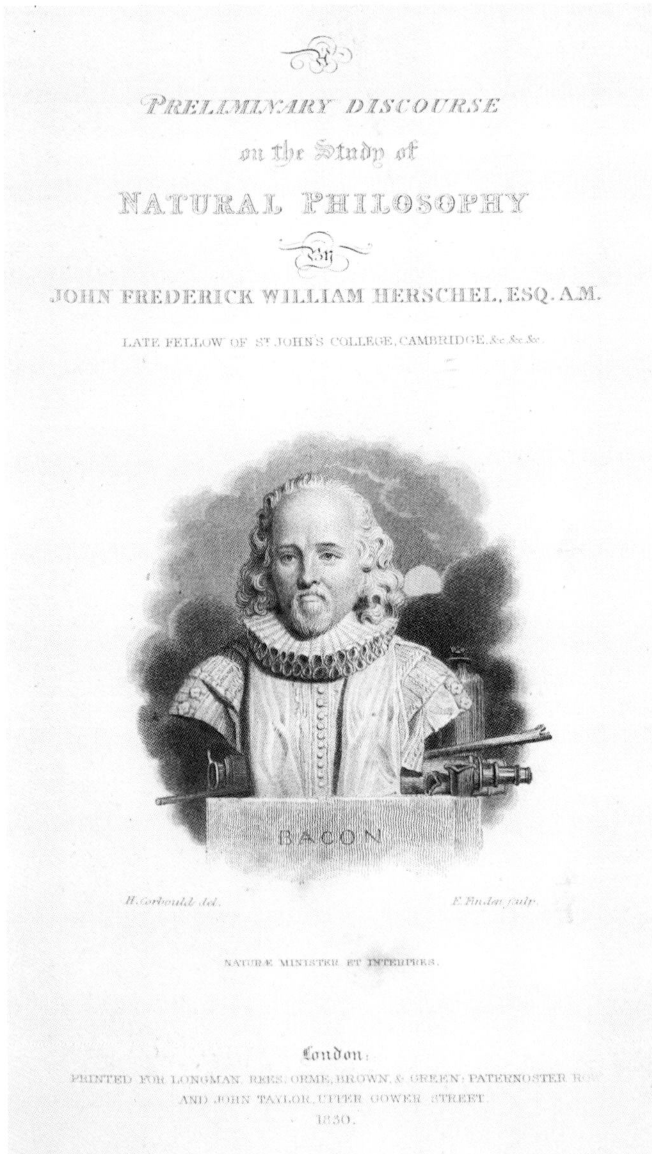

year Strickland was killed by a train while practising his Baconian geology in a railway cutting. He was succeeded by John Phillips, formerly Professor of Geology at King's College London and Trinity College Dublin. In each of the first four editions of his popular *Guide to Geology*, published between 1834 and 1854, Phillips too insisted that 'so long as men adhere to the method of philosophy taught by Bacon, geology can never again become a speculative science, never again be an arena for discussing delusive hypothesis and unsubstantial conjecture.'[49]

For these British men of science, Bacon's emphasis on fact and experiment gave their work a rigour lacking from the freer, less well-supported theorisation of some of their French and German colleagues. The Pre-Raphaelites sought the same rigour in their art, and they voiced their own loyalty to Bacon surprisingly forcefully. For Hunt and Rossetti, he was one of the Immortals. Decades later Hunt would preface the second chapter of his memoirs, where he begins his training at the Royal Academy and first meets Millais, with an epigraph from Bacon counselling his readers 'to accept of nothing but examined and tried'.[50] Woolner wrote that Bacon's prose 'sheds truth as the sun does light'.[51] For Stephens, he was nothing less than 'the greatest of all Englishmen'.[52] Like the scientists, the Pre-Raphaelites mark themselves out as Baconians.[53] In a major study of Baconian science and the nineteenth-century imagination, Jonathan Smith observes that, increasingly, '"Baconianism" was seen as too sterile, too mechanical, and too impersonal to capture the artistic quality of the scientist doing science.'[54] In a striking contrast, for the most ambitious and imaginative artists of the mid-Victorian period, Baconianism was precisely what was needed to inculcate a proper scientific attitude in the artist doing art.

As Smith observes, by the beginning of the nineteenth century Bacon and Newton had been firmly canonised as the two heroes of the English scientific revolution.[55] It is striking to see Newton paired with Bacon on the Pre-Raphaelites' list of Immortals (see also fig. 144). Newton's notoriously misleading motto 'Hypotheses non fingo' – I make no hypotheses – chimes with Stephens's emphasis on fact and experiment over theory. The fact that he and Bacon are the only men on the list who were principally remembered as scientists – there are no living scientists named, though there are several living writers – confirms the impression that the Pre-Raphaelites were drawn less to particular scientific discoveries, let alone recent ones, than to the idea of science itself.[56]

Newton features in a short poem written by the Pre-Raphaelite painter Walter Deverell a year or two before he died of kidney disease at the age of twenty-seven in 1854. It is preserved in an unpublished manuscript edited by William Rossetti in the Huntington Library in California. 'The Garden' elegantly encapsulates the Pre-Raphaelite ideal of an art modelled on science:

Except the voice of birds, there is no sound.
  The fruit-trees rustle not: the sultry whirr
Of insects, it is true, is all around,
  Swimming about me with a swoony stir.
The sparrows peck the peaches. Branches green,
  In most admired confusion interlaced,
Look down upon the walks – a shady screen
Of languid coolness; emerald and bright,
  With golden patches of the Sun's warm eye.
Flecked only by the faintest cirrhi white,
  A blank pale blueness reigns throughout the sky.
It might be thus that Isaac Newton paced,
  And saw the apple drop upon the mould.
That apple more than Adam's sure was graced –
  Telling the laws of Nature's Goddess old,
  A revelation vast and manifold.[57]

Deverell's poem may not seem at first glance to be especially remarkable or successful. The verse is mostly the studied, regular iambic pentameter that might be expected from a not particularly original or practised poet. The language and mood, with its sultriness, swooniness and languor, is rather obviously imitated from Keats, as Eben Bass – one of the very few critics to have discussed the poem – has remarked.[58] On looking at it more closely, however, Deverell's poem reveals surprisingly sophisticated and complex artistry. Its opening lines set up the expectation that this is going to be a straightforward sonnet. The rhyme scheme opens with the regular *abab* pattern of Shakespeare's sonnets. Generically, the poem seems to fit with the more recent Romantic model of sonnets that cast the speaker not as a lover but alone, placing him or her in a particular physical spot evoked through the verse, and meditating on his or her communion with nature. A garden is not, of course, a natural location. The peaches here are planted, the walks laid out for the owner's or visitor's pleasure. Yet Deverell's concentration, like Millais's in *Ophelia*, is focussed primarily on the natural phenomena he can observe within his immediate surroundings – on birds and insects, trees, clouds. The fact that he is concentrating is borne out by the sheer number of details he records over what turns out to be eleven lines before the volta or turn in the sonnet, which usually occurs after only eight lines in the Petrarchan model preferred by the Romantics. It is as if there is too much here to fit into the constraints of a conventional sonnet, too many sounds to distinguish, too many colours to differentiate. His concentration is borne out too by the concern with accuracy suggested in the meteorological register he slips into to name the clouds in line 10, and in the phrase 'it is true' in line 3, which might seem redundant, a mere filler to stretch the line and complete the metre, but which signals rather that he is taking care to correct his initial, inaccurate impression that birdsong is the only sound he can hear.

The density of observation in Deverell's poem forces it to overspill the limits of a typical sonnet. But he also breaks with convention in the rhyme scheme. By rights, the eighth line should rhyme with the sixth, but instead this rhyme is held over and a new quatrain embarked on. This is signalled in the manuscript by the break in the usual pattern of indentation at this line. Tellingly, the line that is not completed with a rhyme describes the branches of the trees 'In most admired confusion interlaced'. Deverell's departure from the conventional rhyme scheme results in precisely such interlacing. Like the garden itself, the poem's form imposes order on observed nature, but not completely. It is only with the intervention of Newton, the archetypal scientist, that the rhyme is completed and order is restored. The last five lines of the poem work like the sextet in a Petrachan sonnet, overlaying a new level of interpretation on to the observations made in the octet, or in this case over the first eleven lines. As the rhyme from the sixth line is held over to this five-line sextet, so that sixth line becomes in effect the point of departure for these last five lines. Through Newton's scientific gaze, the 'confusion', 'admired' as it is, is transformed into a 'revelation' of the laws of nature – a revelation, moreover, that reverses the Fall of Man. Knowledge is no longer a curse, but a 'vast and manifold' good, as science reveals the truth about the natural world. Where his poetic master Keats spurned Newton, Deverell venerates him, setting him up as the culmination of the same project that the poet is more modestly undertaking himself. The familiar legend that Newton saw an apple fall and so discovered gravity is an emblem in the poem of what attentive observation can achieve. Through Newton, the poem gestures towards the possibility that poetry itself might be grounded in the same close observation, while the vivid visual details of the poem hint at the intense clarity of Hunt's and Millais's paintings.

ART AS SCIENCE VERSUS SCIENCE IN ART

According to eighteenth- and early nineteenth-century histories of science, Newton brought Bacon's scientific revolution to fruition. Through their paintings, poems and sculptures, the Pre-Raphaelites in turn look to bring about their own 'revelation' in line with and as true as that of science itself. Following the example of science, their art would be meticulous in its observations and independent in its investigations. It would be experimental: not casually but in a sustained and systematic way. Through this self-imposed rigour, the Pre-Raphaelites themselves sought to carry the arts forward into new territory and restore their power and relevance in a modern, scientific age.

In *The Germ* science is a recurrent frame of reference, in poetry as well as prose. Where Stephens invokes geology in his essays, Thomas Woolner refers to geological change as 'time's cycles' in his poem 'Emblems'.[59] Across the page, Dante Gabriel Rossetti too hears 'the lapse of time' in the sound of the eroding sea in 'From the Cliffs: Noon'.[60] Where Tupper makes play with electricity in 'The Subject in Art', Rossetti weaves the astrophysical concept of 'ether' into his cosmology in 'The Blessed Damozel'.[61] The fullest and most specific references to science are in Tupper's comic poems 'An Incident in the Siege of Troy, seen from a modern Observatory' and 'Smoke'. These appeared in the enigmatic series 'Papers of "The M.S. Society"' in the third and fourth issues of *The Germ*, by now retitled *Art and Poetry*. Tupper's poems are rapidly associating jumbles of puns, jokes and satire on contemporary science, taking in anatomy and algebra, geometry and geology, astronomy and modern technology. 'An Incident in the Siege of Troy', in particular, includes jokes about the use of letters in anatomical illustration, with a pun on the anatomist Jones Quain's 'note on the Tendo Achillis' from his book *The Muscles of the Human Body*, published in 1836; an association between transcendental anatomy and 'the *dead sell* Physiology', clinching a reference to Robert Knox, the evolutionist and transcendental anatomist who had been caught up in the Burke and Hare scandal in Edinburgh twenty years before, but who was still lecturing and publishing in London on a range of subjects, including against the use of anatomy in teaching art; and a scene in which the Greek prophet and turncoat Helenus is imagined:

> In a coffee-house, an upper room,
> Where none but hungry devils come,
> Wherein 'tis said, with animation
> He read 'Vestiges of Creation.'

Like the allusions to Quain and Knox, this reference to Robert Chambers's anonymous, immensely popular and even notorious book promoting evolutionary theory, *Vestiges of the Natural History of Creation*, published in 1844, is strikingly current. Helenus reads it 'with animation', but the gossip about him for reading it is animated too. The poem's closing reference is more recent still, as the sixteen Special Constables entrusted with the defence of 'Priam's keep':

> . . . smoked their pipes in silence deep
> Till there was such a fog – any
> Attempt to discover the priest in the smother
> Had bothered old Airy and Adams and t'other
> And – every son of an *English* mother.

Tupper's jibe here is at the astronomers George Biddell Airy, John Couch Adams and James Challis. In 1846 they had failed to locate the planet Neptune ahead of their French and German competitors, in spite of Adams having predicted its existence mathematically. Finding Helenus in the fug of pipe smoke, Tupper jokes, is as difficult as it is for an English astronomer to find a planet he knows to be there. A modern observatory, it turns out, is not a place where you can get a clear view of anything.[62]

Tupper's comic play with science is in keeping with the challenge set to readers of *The Germ* by Stephens in 'Modern Giants' when he bates us: 'You revel in Geology: but in chemistry, the modern science, possessing thousands of powers as great as any used yet, you see no glory.'[63] If the jokes in Tupper's poems are beyond us, it is because we are not as up-to-date with science as we should be. On the other hand, if we find them funny, then we are alive to the place of science within modern, Victorian society. For the most part, this does not require us to have specialist knowledge. Quain excepted, all the scientists, books and discoveries that Tupper riffs on here were either newsworthy or part of a general education. Nor does it demand a po-faced reverence for science. Instead Tupper is playful. His poems remind us that this irreverence in the face of established authority is crucial to the independence of the Pre-Raphaelite project, as it is to science itself. At the same time, in placing science at the centre of his comedy, Tupper paradoxically reaffirms the Pre-Raphaelites' more serious concern with the place of science in artistic theory and practice.

The Germ was a crucial forum within which the Pre-Raphaelites could define their artistic project. Looking beyond this early magazine, however, direct references to science in their art are rare. Science is not a major theme within their poetry or painting. Instead it is a method, a guiding principle. This can be seen in the contrast between a painting such as *Ophelia* and William Dyce's *Pegwell Bay, Kent* (fig. 11) – the painting most often discussed as an example of Pre-Raphaelite art's engagement with science. Dyce was a much older contemporary of the Pre-Raphaelites whose work and ideas overlap with their own. Born in 1806, he fell in with the Nazarene group of painters in Rome in the 1820s. In their reaction against neoclassicism and return to the models of early Italian art, the Nazarenes anticipate one aspect of the Pre-Raphaelite project. Nonetheless, William Rossetti repeatedly and advisedly rejected the common assumption that these two movements were aligned, dismissing 'the revivalism and pietism of that modern German art which has sometimes been termed Præraphaelite, but which is more properly Raphaelite without the backbone of Raphael'.[64] Certainly Dyce's early paintings, such as the *Madonna and Child* in Tate Britain in London, look much more directly imitative of Nazarene practices and quattrocento art than anything painted by the Pre-Raphaelites themselves. Later, in the 1840s, he would anticipate further elements of the Pre-Raphaelites' project, including the revival of Arthurian

11 William Dyce, *Pegwell Bay, Kent – a Recollection of October 5th, 1858*, c.1858–60. Tate Gallery, London.

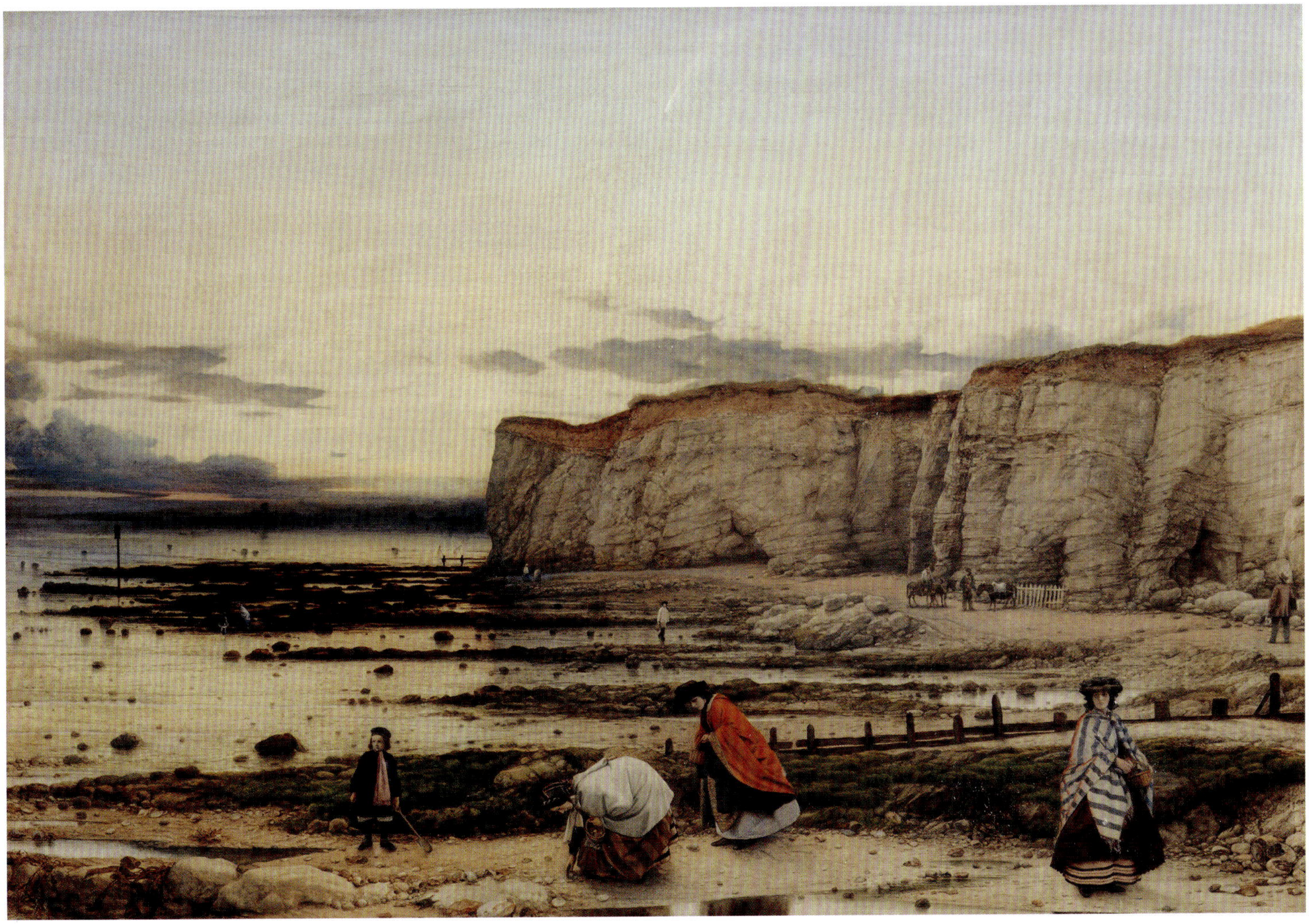

subjects in his murals in the Palace of Westminster and the interrogation of the relationship between science and art in his published lecture *Theory of the Fine Arts*. Again, however, Dyce's aesthetic in his Arthurian paintings differed from the Pre-Raphaelites' own medievalism, while his conclusions on the relationship of the arts to science were different too, as in his view the arts did not belong in 'the province of physical science', although they could advance what he called, somewhat vaguely, the 'science of morals'.[65]

It was only in the 1850s, once Pre-Raphaelitism was underway, that Dyce's style changed under their influence, incorporating a far greater intensity of detail than in his earlier paintings. *Pegwell Bay* typifies this minute detail. First exhibited at the Royal Academy in 1860, it has become a standard point of reference in surveys of Pre-Raphaelite painting.[66] Dyce certainly does pay a remarkable degree of attention to natural forms in this painting, particularly the shells and pebbles on the beach and the stratification of the cliffs. But in other regards it is unlike the paintings produced by the Pre-Raphaelites themselves. For one thing, Dyce's working methods were different. Like earlier landscape painters such as Constable, he made a watercolour study or studies of the landscape and then painted the oil painting itself back in the studio rather than on site. The painting's full title – *Pegwell Bay, Kent – a Recollection of October 5th, 1858* – affirms precisely this practice, as it is painted from memories and souvenirs rather than direct from life. Rosenfeld has pointed out that the picture is based on 'a rigid geometric armature', suggesting that compositionally Dyce is thinking in the very terms that the Pre-Raphaelites took to be an obstacle to truth and progress in art.[67] Dyce's palette is far more muted than theirs too. Pre-Raphaelite paintings are typically intensely colourful, where *Pegwell Bay* seems almost drained of colour. Its narrative element is likewise oddly muted relative to Pre-Raphaelite art, which typically foregrounds moments of significance, moral gravity and emotional intensity in the lives of its subjects. The ostensible subject of the painting and its source – a family holiday at the seaside – is more typical of precisely the kind of academic genre painting that the Pre-Raphaelites disparaged than of Pre-Raphaelitism itself, although *Pegwell Bay* departs from this tradition too by draining not only the colour but the feeling from its subjects, who, as Timothy Hilton observes, 'are a family group, and yet not a group, for they stand apart from each other, and nobody looks at anyone else'.[68]

Pegwell Bay is exonerated from the charge of a 'lack of ideas' that William Rossetti would later level at the art the Pre-Raphaelites rebelled against by its concerted engagement with science.[69] In an influential essay published in 1978, Marcia Pointon interprets Dyce's painting as a meditation on time through references to geology and – in the presence of Donati's comet, seen in the sky above the cliffs – to astronomy.[70] Since then, the painting's engagement with science has become a persistent feature of its interpretation. The women in the foreground of the picture are amateur naturalists and collectors, gathering shells and fossils. The painter himself, seen on the extreme right holding his portfolio, is looking up, perhaps at the record of time layered in the rocks of the cliff face, perhaps at the comet, emblematic of the cyclical patterns of celestial time. Both geology and astronomy seemed to dwarf the human realm. On the other hand, Alison Smith points out that Pegwell Bay was the site of St Augustine's landing in Britain on his mission to convert the English to Christianity, while the comet recalls pictorially the star in the sky from representations of the Nativity.[71] This ambivalent symbolism has led to universal agreement among critics that the painting engages with science, but much less agreement as to how to interpret it. For Tim Barringer and Rebecca Bedell, the ambivalence itself suggests religious doubt, fuelled by the publication of Darwin's *On the Origin of Species* in 1859 between the picture's conception and its completion.[72] For Rosenfeld, the comet is rather a sign of 'what was, for the deeply religious Dyce, unquestionably God's heaven'.[73] The muted tones and small figures of Dyce's painting further deepen the uncertainty over what precisely he intends his viewers to make of his allusions to science. Rosenfeld sees the comet as 'clearly visible' and interprets the artist in the painting as 'staring at the wondrous comet', yet for Hilton the comet goes 'unnoticed', while for Smith its 'near-invisibility' casts 'doubt on the traditional affinity between heavenly and human affairs', suggesting instead that 'nature is seen to operate according to its own invariable laws'.[74]

Wherever it may lead, the direct meditation on the findings and implications of science exemplified by *Pegwell Bay* is almost entirely alien to Pre-Raphaelite art beyond the few poems in *The Germ* which orient the Pre-Raphaelite project as aligned to science. The engagement with science teased out by these critics is essentially a literary one. Dyce is responding to the theories that he has read and talked about, perhaps in books such as Darwin's *Origin* or Gideon Mantell's *Thoughts on a Pebble*, perhaps in periodicals such as the *Westminster Review* or *The Athenaeum*.[75] For the Pre-Raphaelites, by contrast, the artist should, in William Rossetti's words, 'act upon his own perception and study of Nature'.[76] A painting such as *Ophelia* aspires to be scientific in its approach to the natural world and to its subject, but Millais does not take science as his subject, nor does he interpret nature through the lens of pre-existing science. The two paintings are comparably detailed, but the natural details they record reflect this distinction. Where *Pegwell Bay* records the immense reaches of time revealed by geology through the meticulous depiction of the cliffs and the pebbles on the beach, *Ophelia* concentrates our attention on a single instant in Ophelia's life and the life of the riverbank around her. The date and even in effect the time of Dyce's *Recollection* are recorded, but the landscape itself is changing only at a geological pace. Ophelia and the riverbank are changing instant by instant, as she sinks and the river flows. The robin that is perched above her head may take flight at any moment. For all that his subject matter is contemporary, Dyce is concerned with the past, and what science can tell us about the past, and by implication the future. Millais, by contrast, concentrates all his attention on the present, even in a picture derived from a fictional past. His landscape is dense with life in rich colour, Dyce's with stone, fossils and dead seashells in shades of grey which leach the life out of the cheeks of the women and child in the foreground. Ophelia may be dying, but Dyce's figures seem petrified, their orange capes and scarves notwithstanding, as if they were half-way to fossils themselves.

When Stephens takes geology as a standard of accuracy in 'Modern Giants', he does not mean that painters should take its discoveries and controversies as their theme, nor that they should concentrate in particular on painting rocks. Instead, he is setting geology up as an example of science as a byword for precise observation. *Pegwell Bay* is a response to issues raised and discussed by scientists such as Mantell and Darwin. Excepting occasional early poems and a few very late paintings (discussed in Chapter 9), Pre-Raphaelite art largely ignores these debates. Instead, it takes its own scientific look at the phenomena of the natural and human world – the same phenomena studied separately by scientists themselves. *Puck* and *The Hireling Shepherd* are the product of observations concordant with Darwin's on the one hand and Davy's and Brewster's on the other, but they owe nothing to their research nor engage directly with it. And while the Pre-Raphaelites modelled their art in part on science, the results of their attentiveness to nature are complementary to, not identical with, those of science itself. As Gillian Beer has observed, 'Discovery is a matter not only of reaching new conclusions but of re-describing what is known and taken for granted.'[77] Through painting, sculpture and poetry, the Pre-Raphaelites discover and reveal phenomena that otherwise pass undiscerned. They explore and interpret interactions between material nature and the mind, between people and animals and their environments. They even propose explanations and causes for phenomena and behaviours. On the other hand, because their concentration is on specifics – particular landscapes, particular people, particular moments – they do not typically generalise as science does from the individual instance to the law or cause. As Stephens makes clear in 'The Purpose and Tendency of Early Italian Art', their concern is with 'fact... not theory', whereas it is theory more than fact that concerns Dyce and his interpreters in *Pegwell Bay*.

Pre-Raphaelite art was a technique – or rather a group of distinct but interrelated techniques – for studying the natural world in the broadest sense. It was, in effect, a new scientific technology for examining as well as recording phenomena. Much has been made by recent critics of a comment made by the Newcastle-based painter William Bell Scott in his *Autobiographical Notes* that 'the seed of the flower of Pre-Raphaelism [*sic*] was photography.'[78] Scott is not a wholly reliable source. The art critic Sidney Colvin, who got to know many of the Pre-Raphaelites in the late 1860s, noted that he had found *Autobiographical Notes* to be 'almost unfailingly inexact in every one of its statements that I have had means or occasion to check'.[79] By his own account, Scott's identification of Pre-Raphaelitism with photography was his own conclusion on thinking over his first meeting with Hunt and Rossetti, in response to seeing Hunt's *Rienzi* and Rossetti's *Girlhood of Mary Virgin* as works in progress. He was not yet intimate with the Pre-Raphaelites and their art, and, as Diane Waggoner has noted, none of the other Pre-Raphaelite painters ever wrote about their work as deriving from photography.[80] For his part, Ruskin poured scorn on the idea that the Pre-Raphaelites worked from photographs, either literally or as a model, challenging those who thought that they did 'to produce a Pre-Raphaelite picture, or anything like one, by themselves copying a photograph'.[81]

Rather than being a derivative of photography, Pre-Raphaelite painting was a rival and in many ways superior technology. As Stephens's comments on *The Hireling Shepherd* make abundantly clear, painting was a far more subtle mechanism for conveying the complexity, variety and even the brilliance of light than early photography. Not only – and obviously – could it capture colour, it could also convey more subtle effects, such as the reflection of sunlight off a hidden cliff in Hunt's next painting, *Our English Coasts* (fig. 12), which creates what Stephens calls 'a whitish, hazy glare, which

12 William Holman Hunt, *Our English Coasts, 1852* (detail of fig. 24), 1852. Tate Gallery, London.

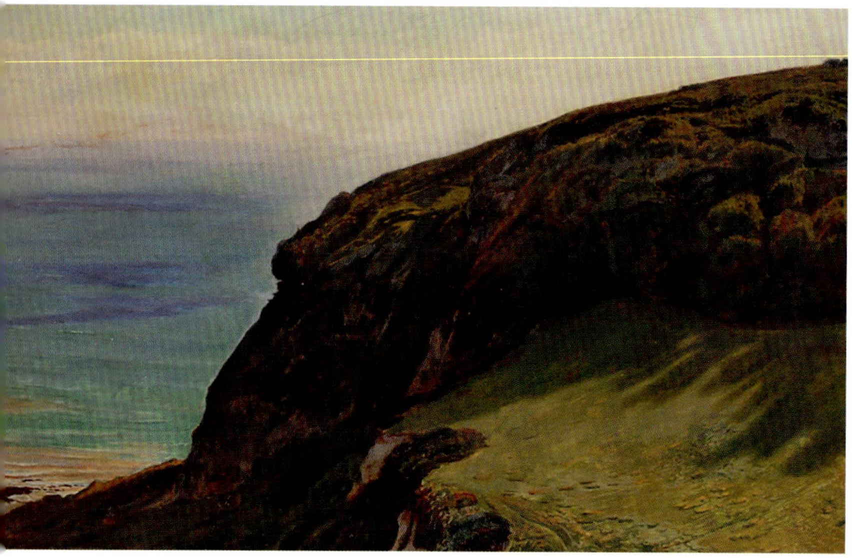

showed above and around it, though the cause was invisible'.[82] As Allen Staley notes – with reference, as it happens, to *Pegwell Bay* – Pre-Raphaelite painting 'not only equalled, but outdid the camera in clarity and thoroughness'.[83]

Pre-Raphaelite painting could study movement too with a precision that Victorian photography, with its long exposure time, could not rival. The clearest example of this is Millais's portrait of Ruskin painted at Glenfinlas in Scotland in 1853 and 1854 (fig. 13). Barringer argues that, in representing the movement of the stream, Millais 'utilizes an effect previously known only through photography' as 'the long exposure time registers the stream flowing over the rocks with a distinctive frothy blur'.[84] Yet what is distinctive about Millais's painting is precisely the lack of blur.[85] Millais records with extraordinary precision the different types of movement within the water. In some parts of the stream you can see the colour of the rock beneath, while elsewhere the turbulence leaves an almost pure white. Not all the water is flowing at the same pace, as it is channelled over more or less steep drops and into pools and inlets secluded from the main flow. At the upstream edge of the painting some of the water of the narrow stream spreads out across a shallow plateau, giving a thin, almost glassy surface. Beneath this in the picture, a waterfall drops into a pool hidden behind the rock where Ruskin himself is standing. To be more precise – as Millais is – some of the water is propelled over the drop with the energy of the flow, arcing down through the air. To the right of it as we look, the spray dampens the rock, forming trickles that run down at their own, slower pace, while to the left another, less vigorous channel overspills into the same pool, dripping down vertically. At the downstream edge of the painting in the lower left-hand corner, Millais gives us a similarly discriminating taxonomy of the bubbles on the stream below the rapids, from the fast-moving froth in the pool immediately below the main falls to the pattern of smaller bubbles given off by the miniature falls below it, revealing the topography of the rocks immediately beneath the surface. At the very bottom corner he includes two or three larger, individual bubbles, intensifying the sense that what we are seeing is a precise moment in the flow of this stream, as where the other details should be much the same at any time when the water level was at the same height, these rogue bubbles come and go in an instant and effectively at random. At the same time, like a glass vase in a seventeenth-century still life, the bubbles are a final virtuoso touch, Millais's assertion of his technical skill as a painter as well as an observer.

If painting enabled the Pre-Raphaelites to simulate and examine precise instants in time, as Millais does in both *Ophelia* and *John Ruskin*, poetry gave them a medium in which they could explore movement through time and the perception of time itself. On 27 September 1849, at half past one in the afternoon, Dante Gabriel Rossetti and Holman Hunt left London on a train to Folkestone. This was the first leg of a tour which would take them to Boulogne (where Rossetti wrote 'From the Cliffs'), through Amiens to Paris, and then on to several sites and cities in Belgium. Rossetti recorded his experiences in a series of verse letters and discrete lyrics and sonnets that he sent back to his brother. After Gabriel's death William published these as a single sequence under the title 'A Trip to Paris and Belgium'. In 'Modern Giants', Stephens calls for artists to engage with 'the poetry of the things about us; our railways, factories, mines, roaring cities, steam vessels, and the endless novelties and wonders produced every day' – the technological products, that is, of modern science.[86] Yet like scientific theories, modern technology did not emerge as a major theme in

13 John Everett Millais, *John Ruskin*, 1853–4.
Ashmolean Museum, Oxford.

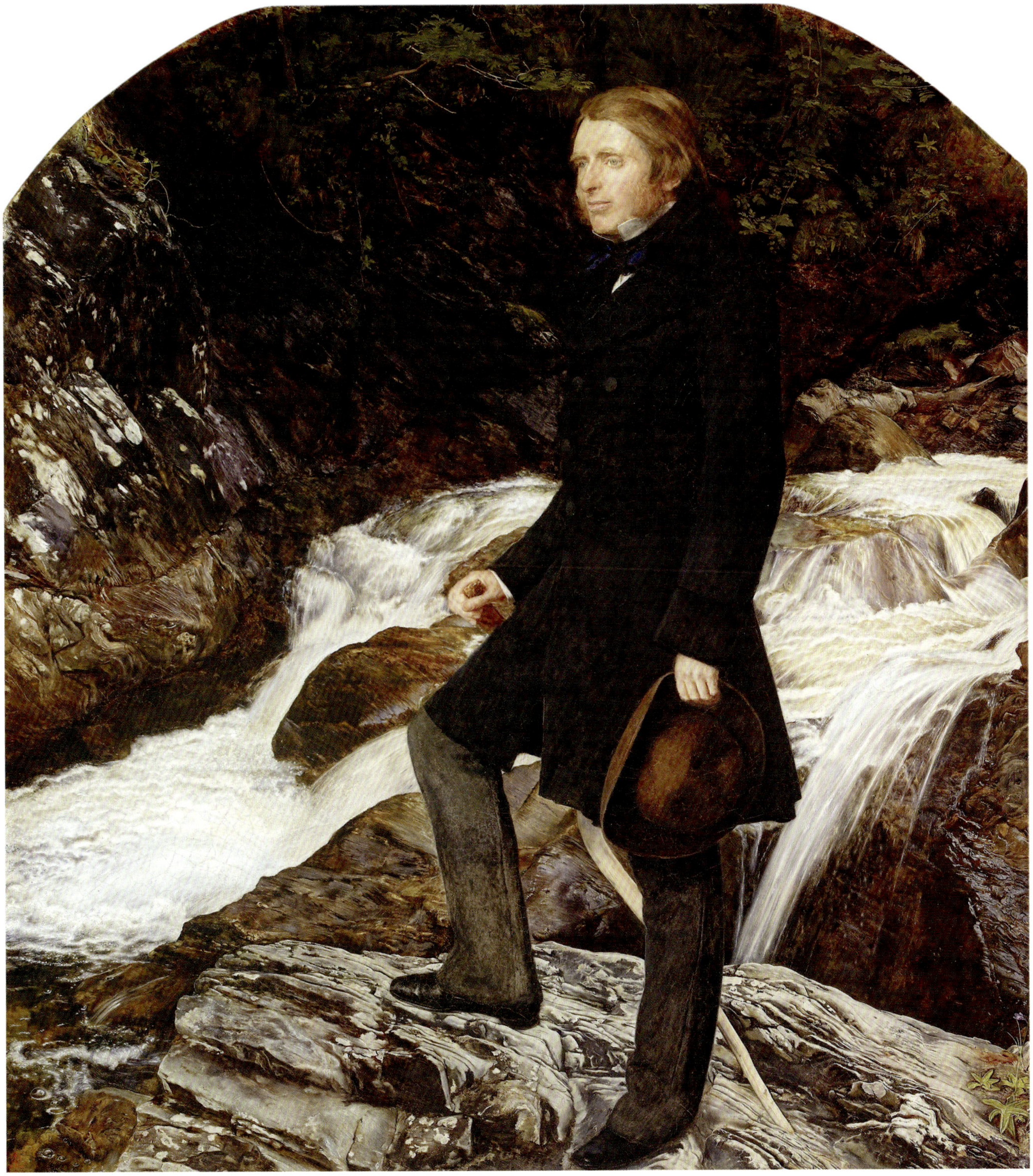

Pre-Raphaelite art. Scott's well-known painting *Iron and Coal on Tyneside*, like Dyce's *Pegwell Bay*, is ubiquitous in discussions of Pre-Raphaelitism and industrial technology not because he was a central figure in its development but because the work stands almost alone as a treatment of this theme by a painter associated with the movement.[87]

Rossetti does engage with railways and steam travel in his verse letters, however, particularly in terms of how they affect perception. The reason why we know precisely when Rossetti and Hunt set off is that the opening section of his first verse letter is headed 'Thursday, 27th / LONDON TO FOLKSTONE / ½ past 1 to ½ past 5'. These precise times and places establish documentary specificity as a defining feature of these poems. Rossetti does not claim to describe this journey as it always is, but rather as it was on those dates at those times. At the same time, they enable us to identify the poem as a record of an 'event experiment', defined by the critic Robert Mitchell, following the historian of science Peter Dear, as 'dated observations of nature as it acted in a specific place'.[88] Rossetti does not claim that this rail journey is necessarily typical of train travel in general. Instead, he offers an account of it as a particular instance from which, through comparison with other journeys, a knowledge of the characteristics of rail travel might be derived by induction.

Rossetti catalogues what he finds distinctive about this experience of rail travel in the opening lines of this first letter:

> A constant keeping past of shaken trees,
> And a bewildered glitter of loose road;
> Banks of bright growth, with single blades atop
> Against white sky; and wires – a constant chain –
> That seem to draw the clouds along with them;
> (Things which one stoops against the light to see
> Through the low window; – shaking by at rest,
> Or fierce like water as the swiftness grows);
> And, seen through fences or a bridge far off,
> Trees that in moving keep their intervals
> Still one 'twixt bar and bar; and then at times
> Long reaches of green level, where one cow,
> Feeding among her fellows that feed on,
> Lifts her slow neck, and gazes for the sound.[89]

As Marcia Werner notes, Rossetti records 'the random, fleeting, impressionistic effects of... landscape glimpsed from a moving train'.[90] The medium he chooses for this record is blank verse. This allows him to track 'a constant chain' of impressions, each itself recurrent if not continuous, of which the wires here are at once an example and a symbol or type. While blank verse gives him this syntactic and rhythmical fluidity, it limits the potential for onomatopoeic effects, such as the use of rhythm to imitate the rhythmic quality of train travel itself, as in Robert Louis Stevenson's well-known poem for children, 'From a Railway Carriage'. After the unobtrusive alliteration of the first few lines has set the train in motion, Rossetti does not court these effects, even allowing for the limitations of blank verse. Instead, he takes advantage of the neutrality of the medium to enhance his poem's claim to observational accuracy and detachment. He registers the 'Strong extreme speed, that the brain hurries with', as he puts it when setting off from Boulogne towards Amiens, but he does not attempt to replicate that effect in the reader through rapid and insistent verse rhythms.[91]

Over successive poems, Rossetti extends the range of his observations to include not merely the mundane details visible from the train window but the properties of nature that can be registered through close observation. One of the most remarkable of this series of poems describes a journey at night by river and rail between the two medieval Belgian cities Antwerp and Ghent:

> We are upon the Scheldt. We know we move
> Because there is a floating at our eyes
> Whatso they seek; and because all the things
> Which on our outset were distinct and large
> Are smaller and much weaker and quite grey,
> And at last gone from us. No motion else.
>
> We are upon the road. The thin swift moon
> Runs with the running clouds that are the sky,
> And with the running water runs – at whiles
> Weak 'neath the film and heavy growth of reeds.
> The country swims with motion. Time itself
> Is consciously beside us, and perceived.
> Our speed is such the sparks our engine leaves
> Are burning after the whole train has passed.

14 Joseph Mallord William Turner, *Rain, Steam, and Speed – The Great Western Railway*, 1844. National Gallery, London.

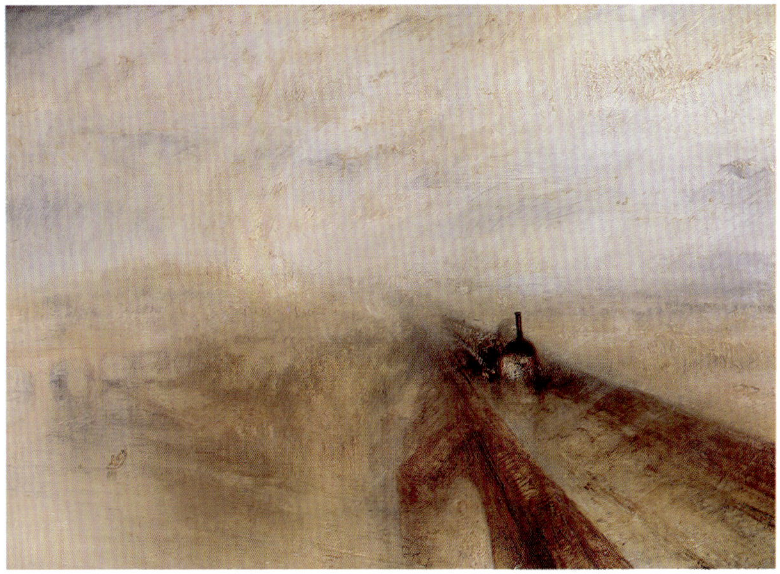

The darkness is a tumult. We tear on,
The roll behind us and the cry before,
Constantly, in a lull of intense speed
And thunder. Any other sound is known
Merely by sight. The shrubs, the trees your eye
Scans for their growth, are far along in haze.
The sky has lost its clouds, and lies away
Oppressively at calm: the moon has failed:
Our speed has set the wind against us. Now
Our engine's heat is fiercer, and flings up
Great glares alongside. Wind and steam and speed
And clamour and the night. We are in Ghent.[92]

This poem is a remarkably precise record of observations under changing conditions. Rossetti contrasts travel by river with travel on the railway. In so doing, he identifies motion as a crucial marker of time. In one of the poem's very few internal echoes, the first paragraph details the 'motion' of 'floating' on the River Scheldt. But the 'floating' Rossetti refers to directly is a visual distortion caused by this movement, not the movement itself. Indeed, what strikes him about river travel is that it is only through visual markers that 'We know we move' at all, rather than through any physical sensation. This is recorded in the poem through the gradual receding of the details of the landscape. Rossetti does not concern himself here with what those details are, but with what happens to our perception of them, as, in the evening light, they become first 'smaller', then 'weaker', then 'quite grey' until finally, in the next line, they are 'gone from us'. This is also the last line describing this stage of the journey, its ultimate if not necessarily its final point.

When travelling slowly by water at night, the changing size, clarity and colour of features of the landscape can only be registered over intervals of time. At any given moment, the sense of time passing is in abeyance. When travelling by train, on the other hand, even things that would not otherwise appear to be moving – the moon, the clouds, the sluggish river on which Rossetti and Hunt have been travelling up to this point – seem to 'run'. Rossetti captures this sense of energy in the recurrence of 'runs' or 'running', both in the present tense, four times in two lines. The impression given by river travel is of 'No motion', mirrored in the slow movement of the syntax in the first paragraph. But when travelling by rail the whole 'country swims with motion', as does Rossetti's verse, moving rapidly from one short sentence or clause to another. Over the course of the second and third paragraphs, the poem transforms into a verbal equivalent of Turner's *Rain, Steam, and Speed* (fig. 14), exhibited at the Royal Academy in 1844. Rossetti alludes to this painting in the list that marks the crescendo of the poem in the last two lines before their eventual and sudden arrival in Ghent. The steam train is so loud that you can only register other sounds if you happen to spot their source, so that sound itself is apprehended by sight, not ear. At the same time, the 'cry' of the engine and the 'roll' of the wheels on the tracks are so constant that intense speed and noise come to seem like a 'lull'. Rossetti even catches the flick of your head as you turn to look at something out of a train window just as it disappears from view. The shrubs have been replaced by trees before you can turn, while the rapid scanning motion of your eye is replicated as it moves from one line of poetry to the next.

On the river, motion is known only by induction. On the train, it is known by direct experience. Yet both of these impressions are characterised by illusion. Rossetti and Hunt are in motion on the river, though they seem not to be.

The countryside around them is not, even when they travel by train. Even 'Time itself' is recognisable, indeed definable, only through motion. Rossetti's poem is as much about the process of observation as it is about the observations themselves – hence the lack of specific detail in the first paragraph. It is here, in this analysis of observation, that the poem's Pre-Raphaelite precision resides. Like Hunt and Millais in their paintings, Rossetti directs our attention to phenomena that otherwise might escape our attention, but which, when we do attend to them, reveal themselves to be remarkable, even extraordinary. From his poem we gain a heightened awareness of our own perceptions of motion and time, as we do of the flow of water in *John Ruskin*, of living nature in *Ophelia*, and of the effects of light in *The Hireling Shepherd* and *Our English Coasts*.

SCIENTIFIC METHOD AND THE MORAL PURPOSES OF THE ARTS

The arts for the Pre-Raphaelites were a mode of 'investigation', in William Rossetti's word. They might even be a means 'to make real scientific discoveries', as Tupper mooted in his diary as early as June 1841.[93] But they were no mere adjunct or helpmeet to science. They had their own ends. As Stephens makes clear in 'The Purpose and Tendency of Early Italian Art', the Pre-Raphaelites' aim in imitating the methods of science was not primarily to contribute to science itself, but to further 'the moral purposes of the Arts'. William Rossetti probes the benefit of close attention to nature for its own sake in 'Sheer Waste', a poem published in the second issue of *The Germ* alongside Stephens's essay. In so doing, he provides a demonstration of how a scientific approach can indeed advance the moral ends of poetry:

Is it a little thing to lie down here
 Beside the water, looking into it,
 And see there grass and fallen leaves interknit,
 And small fish sometimes passing thro' some bit
Of tangled grass where there's an outlet clear?

And then a drift of wind perhaps will come,
 And blow the insects hovering all about
 Into the water. Some of them get out;
 Others swim with sharp twitches; and you doubt
Whether of life or death for other some.

Meanwhile the blueflies sway themselves along
 Over the water's surface, or close by;
 Not one in ten beyond the grass will fly
 That closely skirts the stream; nor will your eye
Meet any where the sunshine is not strong.

After a time you find, you know not how,
 That it is quite a stretch of energy
 To do what you have done unconsciously, –
 That is, pull up the grass; and then you see
You may as well rise and be going now.

So, having walked for a few steps, you fall
 Bodily on the grass under the sun,
 And listen to the rustle, one by one,
 Of the trees' leaves; and soon the wind has done
For a short space, and it is quiet all;

Except because the rooks will make a caw
 Just now and then together: and the breeze
 Soon rises up again among the trees,
 Making the grass, moreover, bend and tease
Your face, but pleasantly. Mayhap the paw

Of a dog touches you and makes you rise
 Upon one arm to pat him; and he licks
 Your hand for that. A child is throwing sticks,
 Hard by, at some half-dozen cows, which fix
Upon him their unmoved contented eyes.

The sun's heat now is painful. Scarce can you
 Move, and even less lie still. You shuffle then,
 Poised on your arms, again to shade. Again
 There comes a pleasant laxness on you. When
You have done enough of nothing, you will go.

Some hours perhaps have passed. Say not you fling
 These hours or such-like recklessly away.
 Seeing the grass and sun and children, say,
 Is not this something more than idle play,
Than careless waste? Is it a little thing?[94]

Rossetti, who was *The Germ*'s editor, published 'Sheer Waste' as the fourth of a series of his own poems grouped together as 'Fancies at Leisure'.[95] Like Deverell's 'The Garden' and other early Pre-Raphaelite poems such as Dante Gabriel Rossetti's sonnet 'Idle Blessedness', written at Woolner's in August 1849, and Tupper's 'Ah, to lie down', a proto-Pre-Raphaelite poem written in May 1841, the poems in this series take the opportunity for leisure provided by a Victorian middle-class lifestyle as a starting point for examining the world around us and the processes and value of perception and observation.[96]

The very idea of leisure in these poems is a marker of their modernity. As William Rossetti recalled fifty years later, 'Sheer Waste' was 'chiefly the outcome of an early afternoon spent lazily in Regent's Park'.[97] This is a poem for an age in which a young government clerk can take the time to enjoy such self-indulgence, and when public parks have been opened to provide for it. Regent's Park itself had opened its gates to the public as recently as 1835. Leisure carries with it an implied contrast to everyday work. Rossetti comically embodies this air of indolence in the middle of the poem as a half-hearted choice to move – 'You may as well rise and be going now' – is succeeded by an equally lackadaisical submission to the urge to 'fall / Bodily' back on to the grass, the ground reconnecting with the body with a satisfying blow as one line moves into the next.

Rossetti's poem is as modern in its syntax and form as it is in its situation. The entire poem is written in the second person. This is a strikingly original move. Through it, the poem's readers become its subjects. It is not Rossetti himself who is lazing in the park, but us. The poem is written in the present tense too, so it takes place in our own moment as readers, whenever that may be. This device keeps it fresh and its observations current. Another stylistic marker of its modernity is Rossetti's disregard for the convention of matching a poem's syntax to its form. The sentences in 'Sheer Waste' run over line-endings, even over the break between stanzas. This freedom aligns Rossetti's poetics with Robert Browning and Arthur Hugh Clough, the avant-garde poets he praises in his regular poetry reviews for *The Germ*.[98]

Rossetti's poem may seem to revel in laziness, yet it asks us to consider whether it is in fact a 'sheer waste' of time, a mere 'little thing', to spend an afternoon lounging in a park. 'Sheer Waste' may seem indolent, but it is dense with observations. Here two elements of the poem's modernity – its scenario and its poetics – combine with a third – an alignment with science. In his reviews, Rossetti evaluates poetry using the same scientific criteria that Stephens applies to painting. He praises Clough's *The Bothie of Toper-na-fuosich* for its general 'aspect of *fact*' and criticises it when he departs too far from 'a strict observance of nature'.[99] In an elegant pun, observation becomes a form of observance, an article of the Pre-Raphaelite creed that must be followed. Rossetti praises Matthew Arnold's poems on the same grounds, noting with satisfaction that the 'reflective' tendency of his thought 'does not absorb the due observation or presentment of the outward facts of nature', 'absorb' here meaning subsume or overwhelm.[100] The 'aspect of fact' in 'Sheer Waste' itself is almost palpable. In his *Preliminary Discourse*, Herschel remarks, 'To the natural philosopher there is no natural object unimportant or trifling.'[101] This is the spirit in which Rossetti pays attention to nature, recording the effect of the breeze on insects, leaves and grass, the calls of rooks, the behaviour of damselflies ('blueflies'), dogs, children, cows. In themselves these details may appear trivial, but like Herschel's natural philosophy the poem asks us to consider their significance.

Following Bacon, Herschel insists, 'As our senses are the only inlets by which we receive impressions of facts, we must take care, in observing, to have them all in activity, and to let nothing escape notice which affects any one of them.'[102] Like Deverell in 'The Garden' and his own brother in 'Antwerp to Ghent', Rossetti takes account of several senses at once, not only sight and sound, but also touch and a more general awareness of 'energy' or effort. In 'The Garden', Deverell states his observations as simple facts, qualified and expanded by more precise observation as the poem goes on. In 'Antwerp to Ghent', Dante Gabriel Rossetti records his findings in the first-person plural as 'we'. Literally this refers to himself and Hunt, but it implies that we too may be able to generalise from them. In 'Sheer Waste', William Rossetti's second-person address takes this a stage further, inviting us to imagine that we ourselves are making the observations attributed to us by the poem. Like scientific data, these observations are at once precise and provisional, even conditional. This is 'perhaps' what we experience – it might be otherwise. It is for us to verify Rossetti's account ('These hours') by testing it against our own experiences ('or such-like'). According to Herschel, part of the appeal of science and especially natural history is that it can 'be enjoyed . . . in the intervals of the most active business'.[103] Through the agency of the poem, not only the poet himself but his readers too become naturalists, attuned to noticing revealing details of natural history in their own leisure time.

The bonds the poem establishes between the poet writing the poem, the second-person subject within the poem, and our own observations beyond the poem recreate the community of careful and scrupulous observers which Bacon's mid-Victorian editor James Spedding identified as central to the Baconian programme.[104] At the same time, through his use of the second person and the present tense, Rossetti trains us to observe ourselves closely as observers. In this reflexive attention to the process of observation, 'Sheer Waste' realises both Rossetti's aim to raise poetry to the level of science as a mode of investigation and Stephens's ambition to further the moral purposes of the arts through this same strategy. In reading the poem, we become aware of the effects of bright sunshine on our powers of observation, as the strength of the contrast blinds us to the presence of the damselflies 'where the sunshine is not strong'. We realise the physical sensation of lassitude, as even pulling up grass becomes an effort. We are conscious too that our observations are made in time, as they succeed one another through the poem ('And then . . . After a time . . . having walked for a few steps . . . and soon . . .').

In the penultimate stanza, Rossetti completes our day of leisure for us. Each sentence creeps haltingly closer to the end of a line and the decision, finally, to move on. This slow progress is recorded in the typography as well as the sounds of the poem, as the full-stops inch to the right. Then, in the final stanza, Rossetti returns to the question he raises at the outset. He invites us to consider whether and how we might benefit from the heightened awareness, self-knowledge and sympathy that come with precise, collective observation of the natural world. Rossetti's questions are leading, but he resists the urge to preach, which would run counter to the Pre-Raphaelites' rejection of teaching by authority. Instead, his closing questions set us the task of determining for ourselves what precisely it is that we have gained by 'Seeing the grass and sun and children' – what the value of the experience attributed to us might be, whether experienced directly or vicariously through reading the poem. In contesting the premise of his own title – that to idle away time lounging in a park is 'sheer waste' – he exposes the irony of his and our sense that we have 'done enough of nothing'. Far from doing nothing, we – you – have been observing acutely. In the process, you have increased your knowledge and understanding of nature, and deepened too your knowledge of yourself.

Rossetti's moral authority in writing his poem comes less from the form itself and the status it gives him as a poet – precisely the kind of authority Pre-Raphaelitism calls into question – and more from the rigour of his observation. To reverse Stephens's question in 'Modern Giants', he has kept his observation awake and alert, and so he knows nature, including human nature. The poem's very detail establishes its author's credentials as an acute observer and investigator. The same is true of Pre-Raphaelite paintings. The moral authority of a painting such as Hunt's *Awakening Conscience* (fig. 15) resides precisely in its painstaking attention to detail. In a notice of this painting, which was exhibited at the Royal Academy in 1854, William Rossetti writes: 'We rejoice to find the earnestness and unflinching thoroughness of the Præraphaelite school applied to subjects of our own day, which, properly treated, are among the highest of all; supremely difficult, but free from uncertainties, and possessing for succeeding times a value such as no other subjects can pretend to, – a truth to us, and to our children a record.'[105] In this painting Hunt examines a social and moral problem – the prevalence of the sexual exploitation of poor women by wealthy men in Victorian society – with the same 'earnestness and unflinching thoroughness' with which he paints every detail of the scene. In other words, the thoroughness which bears witness to his careful attention and acute perception is transferred by analogy from the material to the moral sphere. Hunt's credentials are further strengthened by the fact that he has been working with comparable rigour on a range of subjects for several years. This rigour gives him the right 'to meet full in the front the moral evil of

15 William Holman Hunt, *The Awakening Conscience*, 1853. Tate Gallery, London.

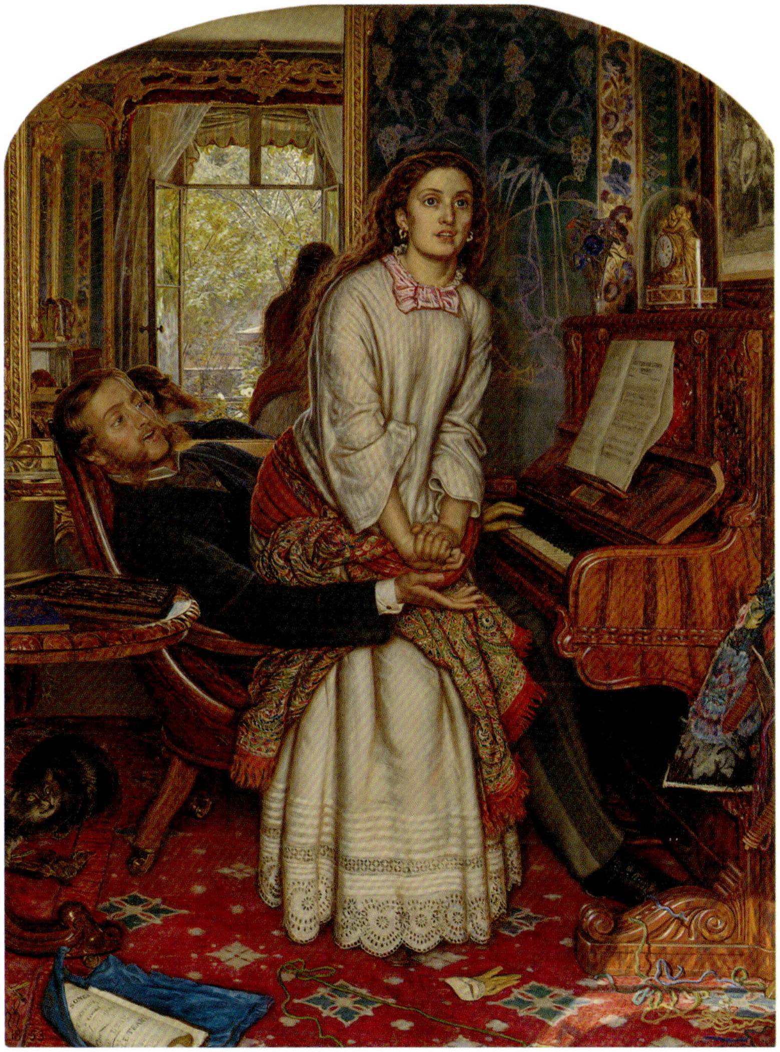

the age', as Ruskin put it in a letter to *The Times* in support of his painting.[106]

Hunt's moral study in *The Awakening Conscience* is robust. As Rossetti remarks, it is, of all that year's pictures, 'the one which most challenges, and most is painted for the express purpose of challenging, grave practical thought in the beholder'.[107] This is a situation that needs to be faced, even at the cost of declaring our own shame. But Hunt's record of sexual exploitation in this painting is also an investigation into the nature of the kept woman's predicament. Like Rossetti's poem, it challenges us as viewers to join in the investigation by examining the problem for ourselves. Both Ruskin and Stephens participate in this act of collective study as they interpret the painting in their reviews. Both of them see the expression on the woman's face as one of abject horror as she realises the evil of her situation. In Ruskin's eyes, she is 'the lost girl', her eyes 'filled with the fearful light of futurity, and with tears of ancient days'.[108] For Stephens, too, her eyes 'seem as if they saw Hell open before her', while her 'fingers are intertwined with spasmodic force'. But this hellish vision is a sign of 'the evil which has befallen her', not the evil she has done. In Stephens's eyes, she is merely a 'light woman', not a lost one.[109] An alternative interpretation, however, would be that her expression is one of rapture, not horror. In awakening her conscience, Christ – famously painted by Hunt in the pendant painting to this one, shown at the same exhibition, as *The Light of the World* – has effaced her guilt and provided her with a glimpse not of hell but of heaven. Hunt's paintings are surprisingly generous, given his reputation for didacticism. Through their close and attentive study of their subjects, both materially and morally, they often reveal a greater degree of subtlety and openness than Ruskin's more puritanical severity would allow. Stephens's judgement is already a good deal more forgiving – her sin is carnal but not mortal. But when looked at afresh, the painting reaches beyond forgiveness to redemption. On this interpretation, Hunt's investigation succeeds because it frees itself from conventional Victorian attitudes to sex, and conventional assumptions about Victorian attitudes to sex – conventions which, like those of art, the Pre-Raphaelites and their viewers are bound to interrogate by their own ethic of art as independent investigation.

As when reading 'Sheer Waste', it remains ultimately up to us to retrace Hunt's investigations in his paintings through our own moral engagement with them, and to see for ourselves where they lead us. Some of the concerns of *The Awakening Conscience* are anticipated in *The Hireling Shepherd* (fig. 16), one of Hunt's most subtle and suggestive moral explorations. On the face of it, the moral of this painting too seems clear. The sexual dalliance of the shepherd and the young woman in the painting leads him to neglect his duty, with the result that some of the sheep are wandering off into the neighbouring cornfield, while others appear

16 William Holman Hunt, *The Hireling Shepherd*, 1851–2. Manchester Art Gallery.

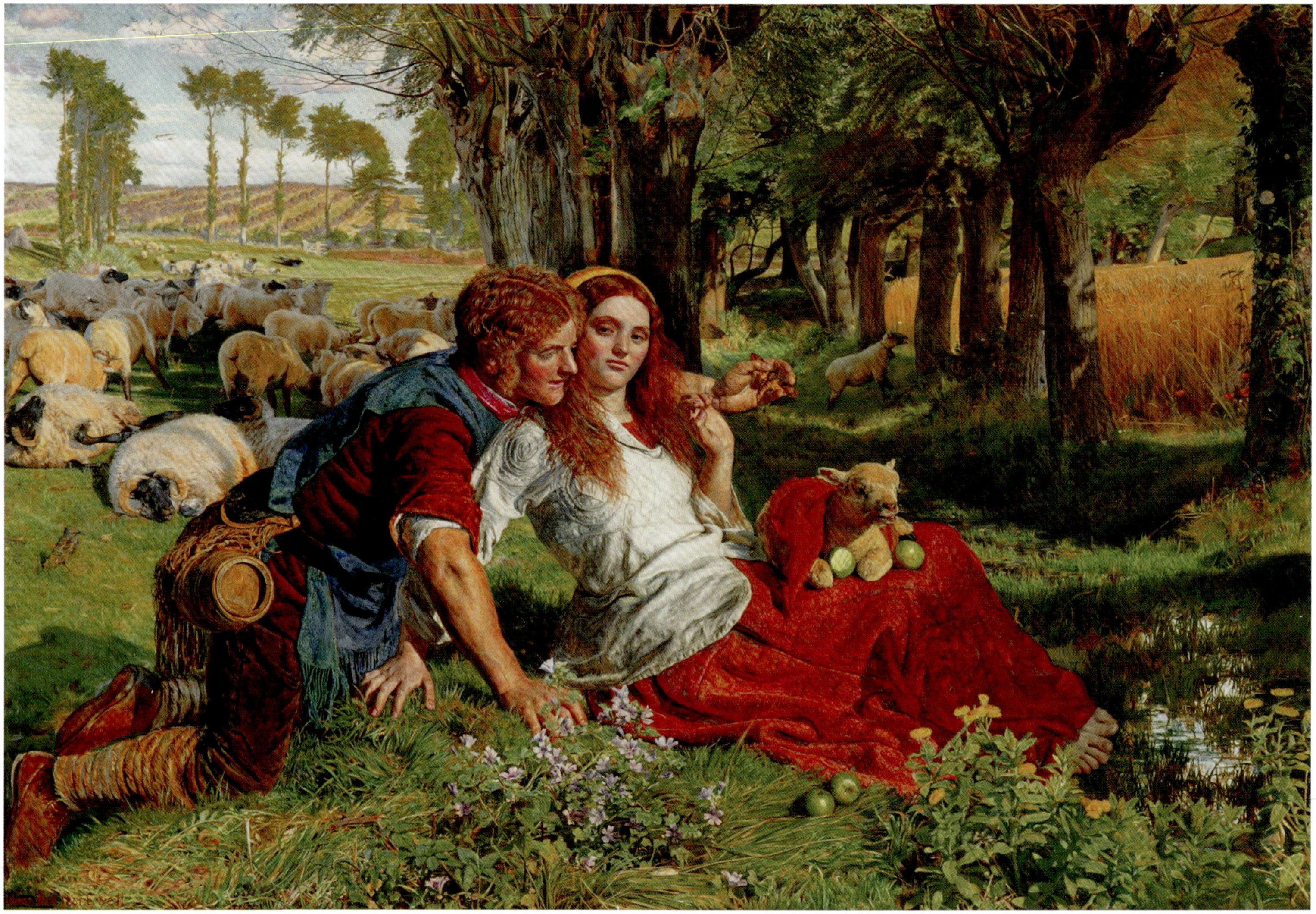

bloated with overeating. This neglect of duty is unquestionable. It is reinforced by the epigraph from *King Lear* which accompanied the painting ('Sleepest or wakest thou, jolly shepherd? / Thy sheep be in the corn: / And, for one blast of thy minikin mouth, / Thy sheep shall take no harm.') and by the title itself, which echoes Christ's distinction between himself as 'the good shepherd' and the negligence of a 'hireling'.[110] But there are other elements in the painting that complicate the moral picture. The title identifies the shepherd as a paid labourer. One crude moral reading of this is to see it as a warning against trusting the working classes to be responsible for property that is not their own. Treating the painting not as a statement of a prejudice but rather as an investigation of a phenomenon, however, it could equally be read as a critique of labour relations after the fashion of Thomas Carlyle, who was an admirer of Hunt's painting.[111] Perhaps the problem is not the unreliability of the rural working classes per se, so much as the breakdown of social relations resulting from the move from a feudal to a wage-based system of labour, or from shepherds owning their own sheep to sheep farmers paying other men to shepherd for them.

The Hireling Shepherd interrogates sexual morality too alongside labour relations. The shepherd's sexual desires are a distraction from his duty. That much is clear. But are they a moral fault in themselves? And if they are, who is to blame? The symbolism within the painting is ambiguous. The shepherd shows the woman a death's head moth. This may seem 'macabre', as Rosenfeld remarks, even ominous.[112] Yet the Pre-Raphaelites themselves were fascinated by insects, as we shall see in the next chapter. In sharing his observation of the moth with the young woman, the shepherd is enacting the procedures of Pre-Raphaelitism itself. Carol Jacobi has pointed out that the upturned gesture of her left hand 'registers as . . . the coy refusal conventional in pastoral courtship scenes', but on closer inspection it turns out that she is 'running a skein of hair between forefinger and thumb, a vanitas motif that matches the moth' and is 'suggestive of entrapment'.[113] The bitten apple on her lap is a clear allusion to Eve and the fall. On the other hand, the lamb, which she has been cherishing, could be taken to be an allusion to Christ in his inviolable innocence. The pairing of the two could compound her wickedness – she has tempted even the lamb – or mitigate it. Finally, it is at the very least striking that her headscarf resembles a nimbus in its colour, position and arc, all of which must have been chosen and arranged by Hunt himself.

Hunt's symbolism insists on the moral questions that his scene provokes – the issue of wilful temptation, for instance, as against innocent play – but it does not provide us with a decisive answer. To probe the question further, we could consider the bodies of the lovers, or potential lovers, themselves. Both William Rossetti and Stephens were drawn to what Rossetti called the 'healthy-blooded rusticity of the figures'.[114] As Sophia Andres has remarked, their 'unconventional sexual equality' is striking, with both of them shown 'intimately enjoying sensuous pleasure'.[115] This is suggested particularly by the way they lean in towards one another, their hands taking the strain, assured in their contact with one another's bodies. Rather than enacting temptation, the woman's face registers her own desire, a desire over which she has control, revealed in the poise of her left hand and her evaluative gaze. The effect is frankly, one might say healthily, erotic.

Prettejohn notes that Hunt deliberately chose a local woman from Ewell, where he had painted the background, as his model.[116] The painting can be seen as a study of her sexuality. The journalist Marian Evans – soon to become the novelist George Eliot – reproached Hunt for spoiling a 'landscape of marvellous truthfulness' by placing a sentimentalised 'pair of peasants' in the foreground, writing: 'We want to be taught to feel, not for the heroic artisan or the sentimental peasant, but for the peasant in all his coarse apathy, and the artisan in all his suspicious selfishness.'[117] Evans advocates undeceived realism here, but her terms of reference rehearse her own class prejudices. Hunt's peasants are in fact more 'coarse' than she recognises, but at the same time he celebrates and empathises with their sexuality – in particular the woman's sexuality – even as he is wary of its possible consequences. Setting *The Hireling Shepherd* alongside *The Awakening Conscience*, we can see Hunt drawing out the differences between middle- and working-class attitudes to sex in Victorian England, and particularly to women's sexual agency. He finds in favour of working-class frankness and freedom over middle-class convention and double standards.

Paintings such as *The Awakening Conscience* and *The Hireling Shepherd* reveal how scrupulously detailed study modelled on science enabled the Pre-Raphaelites to probe with rigour, independence and authority complex moral issues at the heart of Victorian society. In pursuing their investigations, they paid equally close attention to the environments shaping their subjects and to the expression of their character and emotions through their bodies. Like Rossetti's second-person address in 'Sheer Waste', the subtlety of these paintings requires us to repeat that process of investigation for ourselves to appreciate their full richness. In the process, we too become moral scientists in the Pre-Raphaelite mould, interrogating the facts of the case rather than unconsciously replicating conventional attitudes and values. As we shall see over the next two chapters, these investigative strategies in both painting and poetry led the Pre-Raphaelites to sophisticated new insights into the natural world and the human body and mind, insights that, in their time, were beyond the range of science itself.

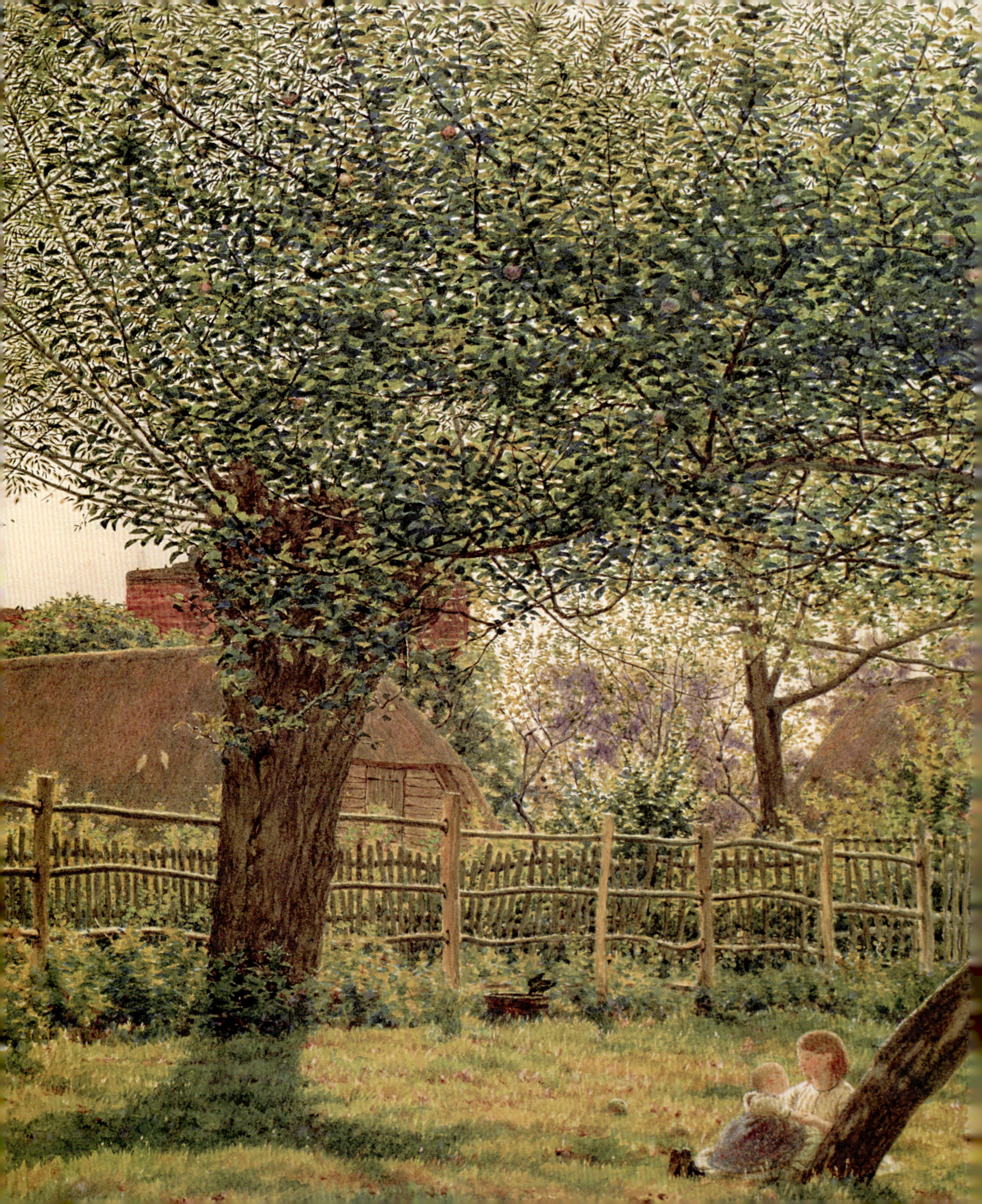

2 · To Know What Has Grown

THE ECOLOGICAL VISION OF PRE-RAPHAELITE ART

In May 1849 William Rossetti began keeping a log of the artistic progress of the Pre-Raphaelite Brotherhood and their comrades-in-arms. The 'object' of the *P.R.B. Journal*, he later recalled, was 'to be a record, from day to day, of the proceedings of *all* the Members, so far as these were of a professional or semi-professional character'.[1] Rossetti recorded the rise and fall of *The Germ*, the composition of poems, the creation and exhibition of paintings and sculptures, the responses to Pre-Raphaelitism in the press, and the beginnings of his own and Stephens's careers as art critics. He also recorded detailed observations of natural history. On 24 May 1849, he watched a small moth climbing a window pane, noting 'that it has not the power of walking along glass, as a fly does, but always slips down in the attempt sooner or later'. The day before he had been discussing Millais's ambition to paint 'a river-sparrow's nest, built, as he says they are, between three reeds'. Four months later, on holiday on the Isle of Wight on 28 September, he recorded: 'I watched for some time a grasshopper with its body broken quite in two and scarcely hanging on at all, which continued walking about, tho' hopping less briskly than is usual.'[2] Rossetti's record of these precise observations of animal behaviour is one of many testimonies to the Pre-Raphaelites' lively interest in natural history. According to the biographies written by their children, Woolner and Millais were both enthusiastic amateur naturalists.[3] Natural history was a collective enterprise for the Pre-Raphaelites, as we can see from Rossetti's conversation with Millais about the reed warbler, which he calls a river-sparrow. On the same day that William was studying a half-dead grasshopper, his brother and Hunt were 'Wading for starfish' at Boulogne in France, as Gabriel put it in one of his verse letters home.[4]

Facing page Detail of fig. 29.

Rossetti's detailed observations of natural history are not included in the *P.R.B. Journal* arbitrarily. The study of natural history was integral to the Pre-Raphaelite project itself. In *The Romance of Victorian Natural History*, Lynn Merrill argues that Pre-Raphaelitism reflected a pervasive mid-Victorian enthusiasm for natural history. Looking in more detail at how the Pre-Raphaelites conducted natural history through their art, and the conclusions that can be drawn from these investigations, we can see that their approach to nature reached beyond the usual limits of mid-Victorian natural history to become fundamentally ecological, both scientifically, in their understanding of how individuals and communities of different species interact, and ethically, in the value they placed on the living world.

NATURAL HISTORY AND VALUE IN PRE-RAPHAELITE POETRY AND PAINTING

In 'The Garden' and 'Sheer Waste', Walter Deverell and William Rossetti encourage us to observe in detail the living world around us. In the poems published alongside 'Sheer Waste' as 'Fancies at Leisure' in the second issue of *The Germ*, Rossetti extends this same invitation in a more intimate setting. Each of these poems stages a companionable walk between two people in a broadly natural landscape. In the first, 'Noon Rest', the speaker and his companion pause to take a nap in the middle of a walk along a riverbank at a warm midday. In the second, 'A Quiet Place', they find their way to a shady stream or 'runnell', overgrown with grasses and toadstools. In the third, 'A Fall of Rain', the speaker recalls how, in the morning, he had imagined going for a moonlit walk that night with Mary, presumably his sweetheart, only to be forestalled by the rain, which has left them intimately alone indoors instead. Aside from this recollection, all three poems are written in the present tense. The sense of immediacy this gives is enhanced by the specificity of their locations. In 'A Fall of Rain', for example, Rossetti imagines walking where 'The moon makes chequered chestnut-shade / *There* by the south-side where the vine / Grapples the wall'.[5] Details such as these help to establish the authenticity of the speaking voice and his relationship with his companions. He knows both his friends and these places intimately.

'Noon Rest', 'A Quiet Place' and 'A Fall of Rain' are among several poems Rossetti published in *The Germ* that were written to the game *bouts-rimés*, where the rhyme words are supplied by a second party and the poet then has to write a poem to fit them. Rossetti recalled that the rhymes for 'Noon Rest', for example, were given to him by Millais 'on one of the days in 1849 when I was sitting to him for the head of Lorenzo in his first Præraphaelite picture from Keats's "Isabella"'.[6] As artistic collaborations, these poems participate in the Pre-Raphaelites' collective pursuit of natural history. Their overall title 'Fancies at Leisure' marks them out as imaginary, as they virtually have to have been, given the circumstances in which they were written. Yet they are characterised by a very high level of verisimilitude, bearing witness to the close attention Rossetti paid to nature, as recorded in the *P.R.B. Journal*.

Inherent in all three of them is Rossetti's desire to share his observation of nature with those he loves, and by extension with his readers. 'A Quiet Place' captures this most effectively:

> My friend, are not the grasses here as tall
> As you would wish to see? The runnell's fall
> Over the rise of pebbles, and its blink
> Of shining points which, upon this side, sink
> In dark, yet still are there; this ragged crane
> Spreading his wings at seeing us with vain
> Terror, forsooth; the trees, a pulpy stock
> Of toadstools huddled round them; and the flock –
> Black wings after black wings – of ancient rook
> By rook; has not the whole scene got a look
> As though we were the first whose breath should fan
> In two this spider's web, to give a span
> Of life more to three flies? See, there's a stone
> Seems made for us to sit on. Have men gone
> By here, and passed? or rested on that bank
> Or on this stone, yet seen no cause to thank
> For the grass growing here so green and rank?[7]

Rossetti's poem has the air of a shared discovery. The speaker points out the features and inhabitants of one very particular spot by a stream to his friend, casually yet with a precision that anticipates the specificity of Millais's observations of comparable scenes in *Ophelia* and in the background to his portrait of Ruskin. Each detail is unique, in itself and in the men's relation to it. They are 'here', 'upon this side' of the stream, not 'on that bank'. The stone they are tempted to sit on is 'this stone', the spider whose web they blow on is 'this spider', the bird they startle 'this ragged crane'. Rossetti's speaker does not know whether he and his friend are the first men to visit this spot, although he imagines that they might be. If anyone else has been here before, it remains unrecorded, outside the parameters of the poem. At the same time, the poem raises the possibility that others might have 'gone / By here, and passed', or stopped and sat to contemplate the natural scene around them. In the process, it encourages us to imagine ourselves as potential visitors to the same spot.

Eavesdropping on the conversation between the speaker and his friend, we share in their appreciation of an otherwise undistinguished, even unappealing, landscape. The poem's flora and fauna – pulpy toadstools and rank grass, spiders and flies, 'ancient' rooks and the ragged crane – might seem ugly, even disgusting, more fit for a Gothic landscape than an idyllic one. Yet Rossetti celebrates and cherishes them. The crane's fear is 'vain' not because it cannot escape, but because it has no need to be afraid. Far from harming animals, the speaker even frees flies from a spider's web. The poem's final rhyme is especially revealing. 'A Quiet Place' is a *bouts-rimés* poem, so the rhyme-word, 'rank', will have been supplied for Rossetti, not taken of his own choosing. As an adjective 'rank' is usually negative, associated with excessive profusion and corruption, yet Rossetti returns it to its earlier meaning as a term of richness. In the intense and appreciative pairing 'so green and rank', rankness becomes a quality to be relished rather than reviled. Even as growth turns a cusp to become decay, as the toadstools rot the trees, the old crane's wing gets ragged and the green grass grows rank, still the sheer life of nature is there to be treasured and enjoyed.

Merrill argues that the 'particularity' in Pre-Raphaelite poetry is 'an end in itself', valued for 'the same reason that a specimen in a cabinet is there – for its intrinsic beauty'. As she sees it, Pre-Raphaelitism is characterised both in poetry and in painting by this 'cabinet-specimen approach'.[8] A specimen is not merely a treasured object, however, but a type. It signifies kinds and existences beyond itself, and its value lies not only in its beauty but in these meanings. The most powerful articulation of a case for 'scientific representation' of rocks, plants and animals in art on these grounds was made by Ruskin in the preface to his second edition of *Modern Painters* in 1844. Ruskin insisted: 'Every herb and flower of the field has its specific, distinct, and perfect beauty; it has its peculiar habitation, expression, and function. The highest art is that which seizes this specific character, which develops and illustrates it, which assigns to it its proper position in the landscape, and which, by means of it, enhances and enforces the great impression which the picture is intended to convey.'[9] Ruskin contests the common assumption that it is legitimate to generalise natural forms for aesthetic reasons in painting. 'As if it were possible to generalize things generically different,' he exclaims: 'It is just as impossible to generalize granite and slate, as it is to generalize a man and a cow'.[10] Art for Ruskin must convey truth. Without the careful and accurate study of natural forms, it is inevitably false.

The Pre-Raphaelites shared Ruskin's firm commitment to truth in art. As Stephens put it in 'The Purpose and Tendency of Early Italian Art', 'it is by truth alone that the Arts can ever hold the position for which they were intended, as the most powerful instruments, the most gentle guides.'[11] As for Ruskin, so for Stephens, this truth can best be learnt through natural history. He exhorts readers of *The Germ*: 'Believe that there is that in the fact of truth, though it be only in the character of a single leaf earnestly studied, which may do its share in the great labour of the world.'[12] The animals, plants and fungi that Rossetti attends to in his poems and in the *P.R.B. Journal* are humble, not beautiful in any conventional sense, but to know them is to gain a deeper understanding of the kinds of life that make up the natural world and a fuller and more tangible sense of our own relationship to it.

It is the very life of these other creatures that is the source and object of this understanding and of the pleasure that follows from it. By definition, this life distinguishes the natural

17 Charles Allston Collins, *Convent Thoughts*, 1850–51. Ashmolean Museum, Oxford.

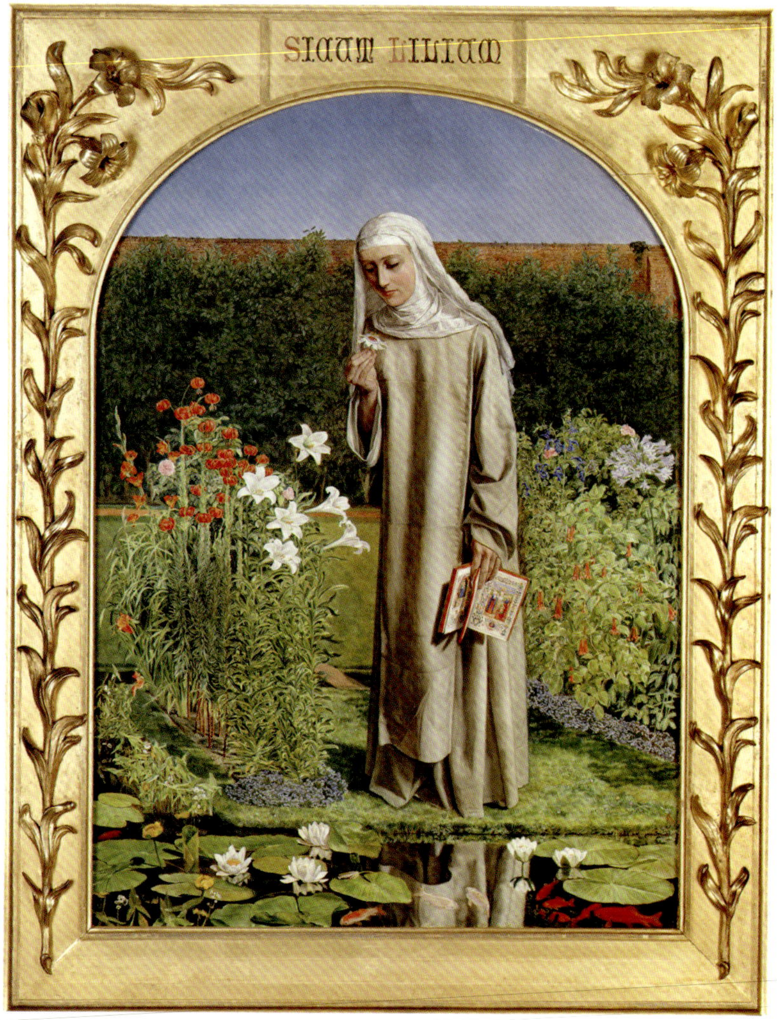

history in Rossetti's poems, like that in Millais's *Ophelia*, from specimens in a case. Here too Pre-Raphaelite practice chimes with Ruskin's values. In the first of the letters he wrote to *The Times* championing the Pre-Raphaelites in May 1851, Ruskin praised Millais's close friend Charles Collins's painting *Convent Thoughts* (fig. 17), then on show at the Royal Academy, as 'invaluable' even 'as a mere botanical study'.[13] Yet as Ruskin made clear in his 1844 preface to *Modern Painters*, the artist could and should take botany to a higher level than the botanist himself. According to Ruskin, what distinguishes the painter or poet from the botanist is the fuller knowledge he gains of his subjects as living beings. The botanist 'counts the stamens, and affixes a name, and is content'. By contrast, the artist, whether a poet or a painter:

> observes every character of the plant's colour and form; considering each of its attributes as an element of expression, he seizes on its lines of grace or energy, rigidity or repose; notes the feebleness or the vigour, the serenity or tremulousness of its hues; observes its local habits, its love or fear of peculiar places, its nourishment or destruction by particular influences; he associates it in his mind with all the features of the situation it inhabits, and the ministering agencies necessary to its support. Thenceforward the flower is to him a living creature, with histories written on its leaves, and passions breathing in its motion.[14]

In studying the natural world, the Pre-Raphaelites do not murder to dissect or even to collect. Instead, as Ruskin proposes, they study nature directly for themselves in the field.

If observation yields a true knowledge of nature, then making art that records and synthesises these observations is at once a disciplined method for conducting research and the best means to convey that knowledge. As Tupper puts it in a poem written as early as 1841:

> Ah, to lie down
>   On the glad green grass,
> To feel your hair blown,
>   To feel the wind pass;
> And to know what has grown
>   This daisy was
> Never known![15]

Prettejohn has pointed out that where the 'aim of botanical illustration is to represent such features as will permit identifying the type or species of plant', the 'plants in a Pre-Raphaelite painting . . . are particular plants, and much more detail is required to give them individual identity.'[16] To become knowledge, a 'living creature' – 'This daisy' – must be comprehended by the artist. In the work itself, it must be precisely fixed so that knowledge of it – or, as in Tupper's poem, the knowledge of knowing it – can be shared with others.

In *Ophelia* and 'A Quiet Place', Millais and Rossetti fix their observations of the natural history of riverbanks through painting and poetry, so that we as viewers and readers can share the knowledge they have garnered through these observations. In line with Stephens's arguments in *The Germ*, such scrupulous observation within Pre-Raphaelite art not only yields knowledge but also brings moral gains. Deverell spells this out in the conclusion to his poem 'A Modern Idyll', published in the magazine's last issue:

So in your ramblings bend your earnest eyes.
High thoughts and feelings will come unto you, –
Gladness will fall upon your heart like dew, –
Because you love the earth and love the skies.[17]

Deverell's overt moralising here reads like a clunky imitation of one of Wordsworth's more didactic poems. Rossetti by contrast allows us to reach this conclusion for ourselves. Writing in the *Westminster Review* in 1856 – in the same essay in which she took issue with *The Hireling Shepherd* – Marian Evans argued that 'the greatest benefit we owe to the artist, whether painter, poet, or novelist, is the extension of our sympathies.'[18] In 'A Quiet Place', Rossetti intensifies our sympathies for our friends through his celebration of shared experience, while extending them too to birds and even insects. Ruskin's prose makes this process explicit as he reaches unapologetically towards anthropomorphism, speaking of the flower as experiencing love, fear and passions. Rossetti resists this direct projection of the author's and reader's minds into the being of other living creatures. Instead, he conveys his own appreciation and care for them. The pleasure that we gain from observing nature through his eyes is not a selfish but rather a generous one, as we share in his warm-hearted joy in the natural world.

The catalogues of observations of different living creatures in Rossetti's 'Fancies at Leisure' mirror the list of the aspects of a single plant observed by the artist in Ruskin's *Modern Painters* and the proliferation of vegetation in Millais's *Ophelia* and Collins's *Convent Thoughts*, where each plant is precisely delineated in arresting detail and vibrant colour. As well as being a scientific description, this enumeration of detail is, as Deverell suggests, an act of love. In Rossetti's poems and Ruskin's prose, this is conveyed directly through their sympathetic engagements with other living things. In Pre-Raphaelite painting, it is born out by the very detail itself. Alongside painstaking *plein-air* painting and jettisoning geometrical arrangement, one of the most radical innovations of Pre-Raphaelite art was the refusal to prioritise the figures over the landscape. As Prettejohn remarks, this was 'bizarre or even perverse' according to the conventions of figure painting, overturning 'the clear conceptual hierarchy, found in [Sir Joshua] Reynolds's *Discourses* and other works of academic art theory'.[19] Where this traditional hierarchy asserted the primary importance of human subjects, the Pre-Raphaelites paid equal attention to plants and animals. As Ruskin recognised, this procedure required an 'enormous cost of care and labour'.[20] In spending months painting their foregrounds and backgrounds inch by inch, recording every natural detail they could see, before filling in the figures later in the studio, Millais, Hunt and Collins set down in oil their testimonies to the value they placed on living things.

The Pre-Raphaelite Brotherhood saw themselves as principally history painters, but Pre-Raphaelitism as a movement had a major impact too on landscape painting.[21] Two paintings by artists who applied Pre-Raphaelite principles to landscape painting illustrate how detail can serve as a testimony to value even in the absence of a contrast – or rather lack of contrast – between the setting and human subjects. John Brett discovered Pre-Raphaelitism in 1852 through reading Ruskin's pamphlet.[22] His conversion was swift. As he wrote in his diary in May 1853, he was already 'going on fast towards preraphaelitism'.[23] Brett's older sister and close collaborator, Rosa, converted to Pre-Raphaelitism even more quickly. He records her making 'first rate sketches on Preraffaelite principles' as early as September 1852.[24] Where John became known for expansive landscapes captured in astonishing detail, Rosa applied the same principles of scrupulous attention to nature in small-scale studies of hedgerows, gardens and small animals such as cats and rabbits.[25]

John Brett exhibited *The Val d'Aosta* at the Royal Academy in 1859. Rosa Brett exhibited *Thistles* two years later, under the pseudonym Rosarius.[26] These two paintings demonstrate the Pre-Raphaelite principle of detail as value on widely divergent scales. *Thistles* is a small oil painting depicting three orangey-pink thistle flowers (fig. 18). These loom with a hyper-real intensity in the foreground of an English landscape, towering above some pale blue scabious flowers, grasses and other field-edge plants. The view recedes in

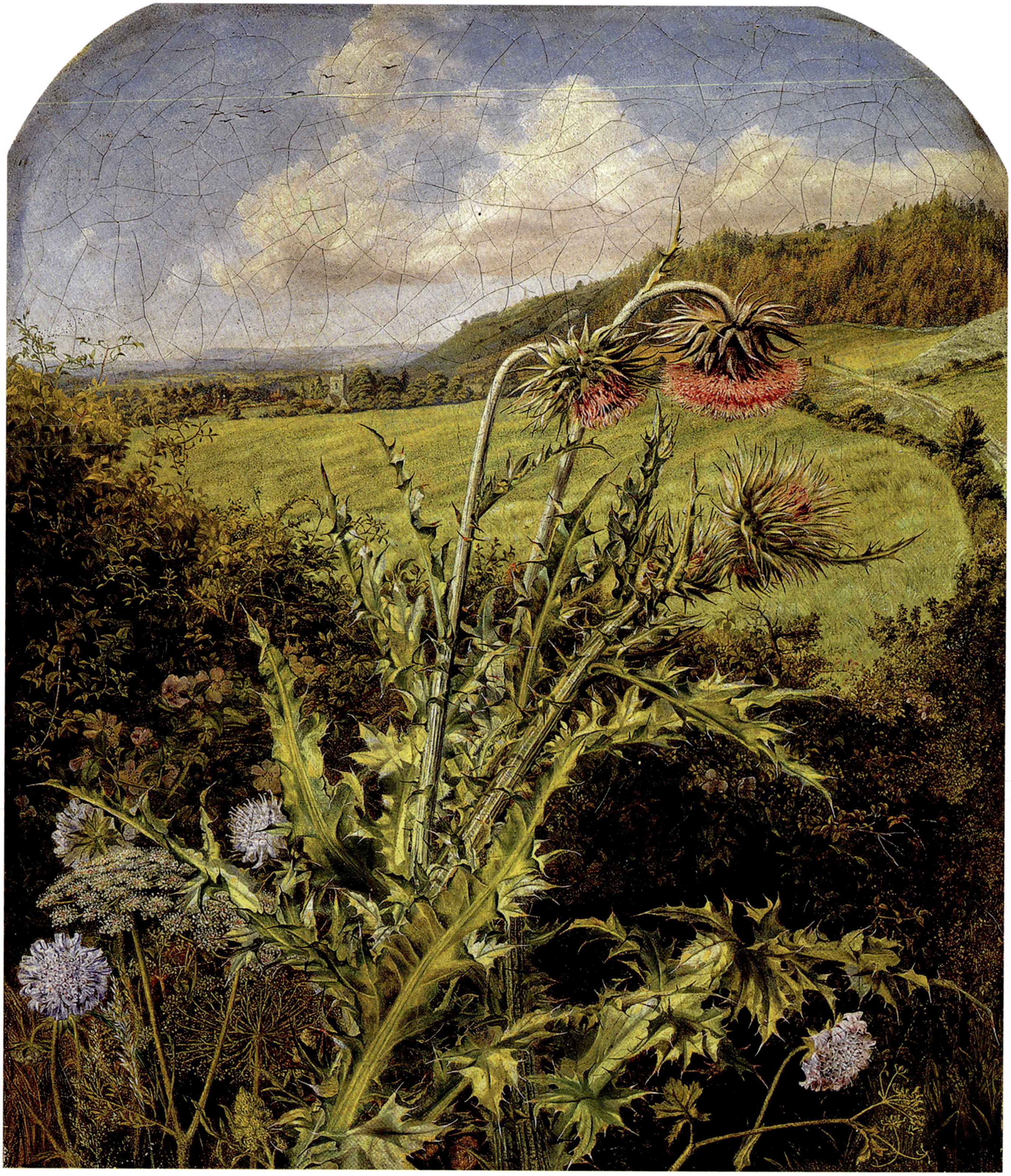

(*facing page*) 18 Rosa Brett, *Thistles*, 1860. Private collection.

successive layers. Behind the flowers and grasses is a hedge which tapers off and downhill to the right of the canvas before swinging round to form the boundary to the field lying fallow on its far side. This field forms the third layer of the landscape. In the gentle undulations of the grass Brett suggests the movement of the breeze and the contours of the landscape. Behind it stand a stone church with a square tower and some woods, dominated by oaks and elms on the flat and by conifers – perhaps larch, as the leaves seem to be turning – on the hillside to the right. This fourth level of the landscape dwindles into the distance before it too gives way to a fifth, the blue plateau on the horizon.

The precision of the second, third and fourth levels is impressive and striking – the individual leaves and blossoms at different stages of growth and decay in the hedge, the waving grasses of the field, suggested by individual brush-strokes subtly shading into stippling, the identifiable form of each distant tree. But most striking of all are the thistles themselves. They stand out so boldly that they seem palpable. The weight of two of the three flowers is bending their stalks. The leaves are prodigiously thorny and tangled, while the third flower seems barely able to outgrow their hold upon it. Like Tupper with his daisy, Brett has determined to know these particular thistles, and to impress that knowledge upon us. At the same time, like Rossetti with his rank grass, she invites us to appreciate the beauty in their ugliness, the vibrancy of their lives even as they enter into the final stages of growth and then collapse.

If *Thistles* directs our attention to the unremarkable and neglected, *The Val d'Aosta* shows us a remote landscape on a grand scale (fig. 19). In *Pre-Raphaelitism*, Ruskin encouraged artists to travel so as to be able to bring back to England records of landscapes unseen at home.[27] Brett does precisely this, as Ruskin recognised when he wrote about his painting in his *Academy Notes* for 1859: 'Here is, at last, a scene worth painting – painted with all our might... And here, accordingly, for the first time in history, we have, by help of art, the power of visiting a place, reasoning about it, and knowing it, just as if we were there, except only that we cannot stir from our place, nor look behind us.'[28] Indeed, Ruskin himself had publicly exhorted Brett to paint just this scene in his review of his painting *The Stonebreaker* the previous year, writing: 'If he can paint so lovely a distance from the Surrey downs and railway-traversed vales, what would he not make of the chestnut groves of the Val d'Aosta!'[29] Yet although he found Brett's painting to be 'unimpeachable' as 'a wonder of toil and delicate handling', and although it fulfilled to the letter his own precepts, Ruskin remained dissatisfied, unable to detect any 'real love' of the scene in the painting. 'I never saw the mirror so held up to Nature,' he concluded, 'but it is Mirror's work, not Man's.'[30]

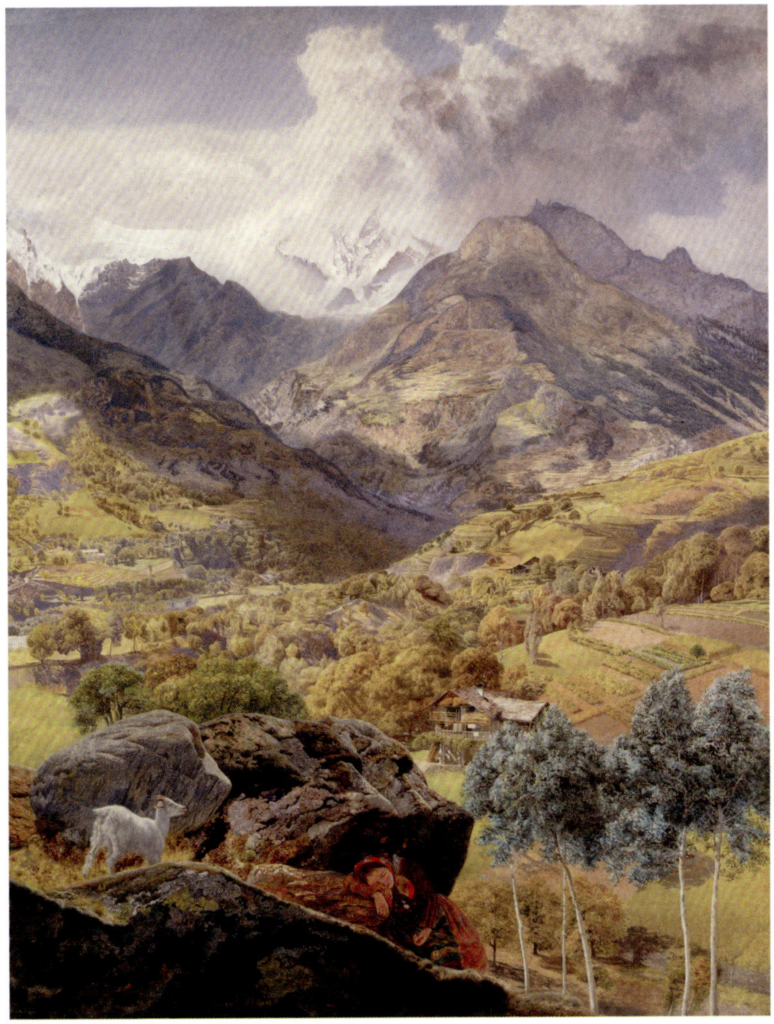

19 John Brett, *The Val d'Aosta*, 1858. Private collection.

Ruskin cannot discern the characteristic Pre-Raphaelite love of nature in *The Val d'Aosta* because Brett's care is so absolutely egalitarian. Although the foreground contains a human figure in the form of a shepherd girl, she is little more than a picturesque detail, subsumed into the outcrop of rocks beneath which she sleeps. Brett's concern is rather with the view, which reaches deep into the landscape and high up into the mountains. By contrast with *Thistles*, with its distinct layering, in *The Val d'Aosta* there is a gradual transition as the eye moves further into the landscape. Brett captures perfectly the changes in colour tones as the landscape recedes, but he refuses at any point to compromise on detail. At no point is the extraordinary visual clarity that became his stock-in-trade abandoned. Brett's work is badly served by reproduction, as in his paintings themselves this visual clarity is complemented by a remarkable precision of texture that captures not only the appearance but even the feel of different components of a landscape. In *The Val d'Aosta*, his paintwork enables us to simulate within our own imaginations the colours, forms and textures of the lichen and moss on the rock in the nearest foreground, the scattered buildings, pastures and cultivated terraces of the settlements in the middle-ground, the increasingly bare mountain meadows and scree slopes, and the crags and snowfields towards the summits.

Brett's work on this painting was prodigious. Christiana Payne calculates that he spent the best part of five months on it.[31] But its claim to value does not rest merely on his labour. Rather the painting itself asserts the value of everything seen within it, indiscriminately and whatever its appeal. *The Val d'Aosta* is an invitation to join Brett in his minute investigation of this landscape in all its variety. His painting enables the viewer to take in the landscape at a sweep first, but because of its remarkable detail we can linger on, examining closer and closer this rock face, crevice or snowline, that tree or cottage or vegetable garden. Ruskin's infamous judgement on his painting does not reflect Brett's failure to value what he sees, but rather Ruskin's own failure to enter into the richness of studying what he shows him.

FROM NATURAL HISTORY TO ECOLOGY

The Pre-Raphaelites celebrate other living creatures in their art. But the Val d'Aosta, the riverbanks in *Ophelia* and 'A Quiet Place', and the Surrey fields in *The Hireling Shepherd* – like the fishpond in *Convent Thoughts*, Regent's Park in 'Sheer Waste' and Deverell's garden – are more than simply different environments populated by diverse animals and plants. To use the vocabulary of Rossetti's poems, each is a 'whole scene' where all the different elements are 'interknit'. Even Rosa Brett's thistles and Tupper's daisy are growing among communities of other species including ourselves. In grounding their study of individual animals and plants in an investigation of the wider communities of which they are members, the Pre-Raphaelite painters and poets move still further beyond the 'cabinet-specimen approach' to natural history described by Merrill and towards the discovery of a new science – ecology – in which living beings themselves constitute and transform one another's environments.

Ecology was identified as a scientific discipline in the mid-1860s by the German biologist Ernst Haeckel, who coined the term 'Oecologie' to refer to 'the investigation of the total relations of the animal both to its inorganic and to its organic environment'. For Haeckel, this amounted to 'the study of all those complex interrelations referred to by Darwin as the conditions of the struggle for existence'.[32] In *On the Origin of Species*, Darwin gave detailed accounts of how the interactions of different species within a particular environment could lead to the evolution of new forms through selective pressures and adaptation. But while historians of ecology agree that Darwin's work provided the theoretical framework it needed to crystallise into a science in its own right, they see him as only one among many whose practical research laid its foundations.[33] In his recent study of the roots of ecology, Frank Egerton traces the history of natural history from antiquity to the nineteenth century. Egerton shows that what we understand as ecology today required the breakdown of the previous conception of the economy of nature defined by Carolus Linnaeus in the mid-eighteenth century as 'the all-wise disposition of the Creator in relation to natural things, by which they are fitted to produce general ends, and reciprocal uses'.[34] The essential fixedness of Linnaeus's model collapsed under the burden of evidence that life on earth had changed radically over time, whether that change was interpreted as evolution or a succession of distinct creations. At the same time, the study of relations between plants and animals in different environments across the globe by Alexander von Humboldt at the very beginning of the nineteenth century, by Darwin during the *Beagle* voyage in the 1830s, and then by the naturalists who would cluster around him in the 1850s and 1860s, including Joseph Hooker, T. H. Huxley, Alfred Russel Wallace and Henry Walter Bates, provided growing evidence for more complex and fluid ecological interrelations.[35]

Karl Kroeber, James McKusick and Onno Oerlemans, among others, have argued that the Romantic poets anticipated aspects of this new ecological thinking in their poetry.[36] While the poetry of Wordsworth and Coleridge is often profoundly ecological in the looser ethical and political sense of the term, John Clare's poems anticipate scientific ecology more directly in their close observation of interactions between animals and plants.[37] Within the visual arts, Diana Donald has likewise traced a halting transition from the fixed order of natural theology and Linnaean taxonomy to a Darwinian emphasis on the struggle for survival in British natural history painting from the mid-eighteenth to the late nineteenth centuries.[38] For Donald, while earlier artists such as George Stubbs and John James Audubon anticipate the violence of Darwin's vision of nature in works such as Stubbs's series of paintings of lions attacking horses or Audubon's *Birds of America*, it was Joseph Wolf, a German natural history painter based at the British Museum in London, who produced 'the first truly ecological vision of life in the wild . . . in the wake of *Origin of Species*'.[39]

There are clear affinities between the Pre-Raphaelites and the poets and artists who have been identified as ecological by previous scholars. They share Clare's close attention to natural interactions, for example. Their ties to Wolf were more direct. When Wolf moved to London from Germany in 1848, Woolner recognised his 'high originality and vivid truthfulness' and immediately began championing his work.[40] The Rossetti brothers too were enthusiastic admirers of what William called his 'delicate and scientifically exact work' and his mastery of a 'zoologically natural treatment' of animals in art.[41] Indeed, the Pre-Raphaelites as a group saw Wolf as a fellow worker in their own aesthetic project (fig 20), inviting him to join them in their first collective exhibition in Russell Place in 1857, where a painting of a woodcock's nest by Wolf was exhibited alongside work by Brown, Hunt, Millais, Rossetti, Collins, Siddal, John Brett and others.[42]

At the same time, Pre-Raphaelite painting and poetry make their own distinct contribution to the emergence of an ecological understanding within the arts. Unlike Clare's poems, early Pre-Raphaelite poetry is part of a sustained campaign to model art on science, and its discoveries are the result of deliberate practices of investigation. The Pre-Raphaelites admired Wolf because his animal paintings showed the same rejection of conventional styles and commitment to careful study that they aspired to in their own art, yet their work reached far beyond his in its attention to the interactions between animals, plants, people and the landscapes they share. The premise of ecology is that these interactions are played out on different scales of space and over time, gradually transforming relationships between organisms, populations, species and even landscapes. This is what distinguishes ecology from earlier static and circumscribed models of natural history, more than a specific commitment to Darwin's idea of the struggle for existence, though that would become its crucial explanatory framework. It implies too that animals, in particular, are agents in their

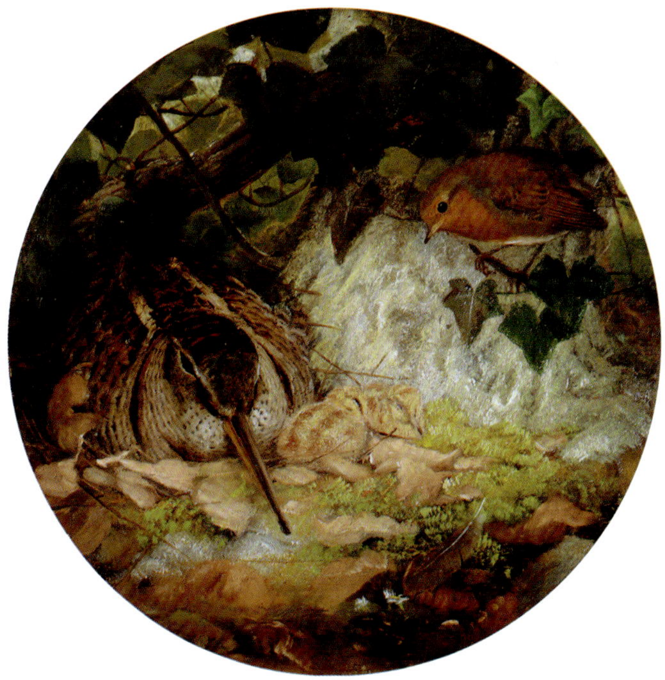

20 Joseph Wolf, *A Woodcock with its Young and a Robin*, ?1857. Private collection.

21 John Everett Millais, *Ferdinand Lured by Ariel*, 1849–50. The Makins Collection.

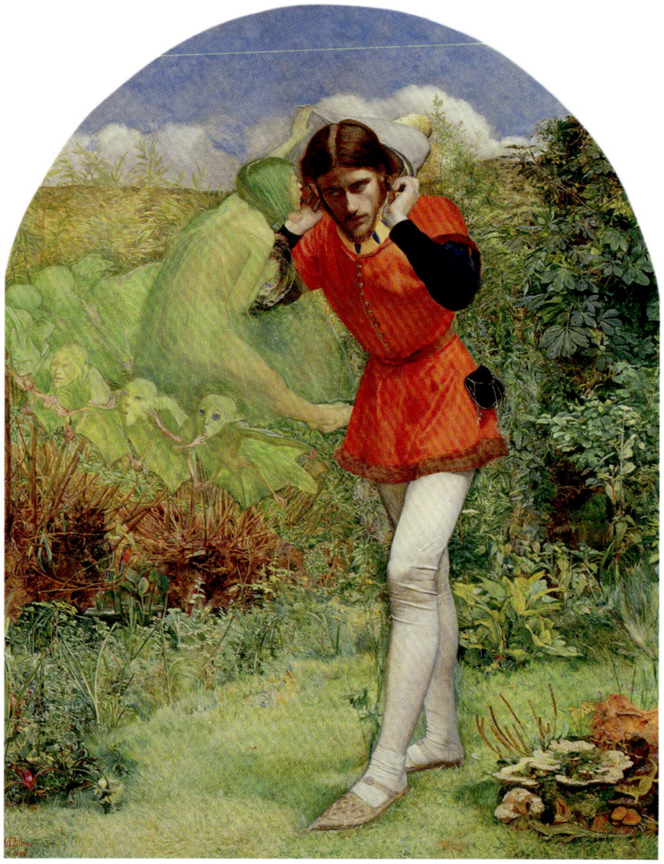

own destinies whose choices and actions have widening ramifications.

Reaching beyond the more narrowly focused work of Clare or Wolf, Pre-Raphaelite art registers and explores the 'complex interrelations' identified by Haeckel as the subject of ecology, prior to and independent of his specifically Darwinian conception of nature. Recent critics have begun to trace the ecologies revealed in Pre-Raphaelite painting. Paul Barlow has identified 'the natural dynamic of eco-systems' as a specific preoccupation within what he calls 'Millais's distinctively ecological vision of nature'.[43] Writing on *Ophelia*, Barlow notes the 'claustrophobic density of the composition, its entangled, overgrown network of plants, all forcing themselves forward to the surface' as 'plants overlie other plants' and 'dead and dying flowers give way to younger ones'.[44] Like William Rossetti in 'A Quiet Place' and Rosa Brett in *Thistles*, Millais's investigation of the natural life of a riverbank leads him to a vision of nature characterised, as Barlow points out, by 'cycles of life, death, decay and regeneration'.[45] Barlow finds the same vision anticipated in Millais's earlier painting *Ferdinand Lured by Ariel* (fig. 21). The vocabulary of Rossetti's poem is apt to both these paintings. Barlow draws attention to the 'proliferating fungi, which feed on a decaying stump' in the foreground in front of Ferdinand, taking them as a sign of 'the strange richness of nature' which complements the supernatural strangeness of Ariel and his sprites.[46] Millais's fungi set Rossetti's 'pulpy stock / Of toadstools' before our eyes (with added lizards) in what is indeed a rich variety of forms and colours. For their part, the reeds growing in the river where Ophelia is drowning are 'green and rank' in precisely the sense that Rossetti articulates in his poem.

The meticulous observation of early Pre-Raphaelite painting encourages the viewer to see each scene as representing a moment in real time. Where poetry records observations sequentially, painting represents them simultaneously, but by showing different stages of growth and decay in the same picture and even in the same plant, Millais can nevertheless introduce time as a dimension into his paintings. Brett does the same, not only in how she depicts the thistles themselves, but in the hedge that forms the immediate backdrop to them, where wild roses chart a path from unfolding buds to exhausted flowers with only one or two petals left attached. In this way the Pre-Raphaelites suggest not only the life of an individual plant, with, as Ruskin says, its 'histories written on its leaves', but also the ongoing history of the ecological community to which it belongs.

Looking back on the beginning of Millais's career thirty-five years later, Stephens saw another early painting as capturing with particular clarity and poignancy this view that nature has its own history. *The Woodman's Daughter* (fig. 22) was Millais's second Pre-Raphaelite painting in the open air, after *Ferdinand* and before *Ophelia*. It illustrates a poem of the same name by Coventry Patmore, an established poet who came to know the Pre-Raphaelites through Woolner. Patmore contributed two short lyrics and an essay on *Macbeth* to *The Germ*, encouraged Ruskin to write his decisive letters to *The Times*, and championed Pre-Raphaelite art himself

22 John Everett Millais, *The Woodman's Daughter*, 1850–51. Guildhall Art Gallery, London.

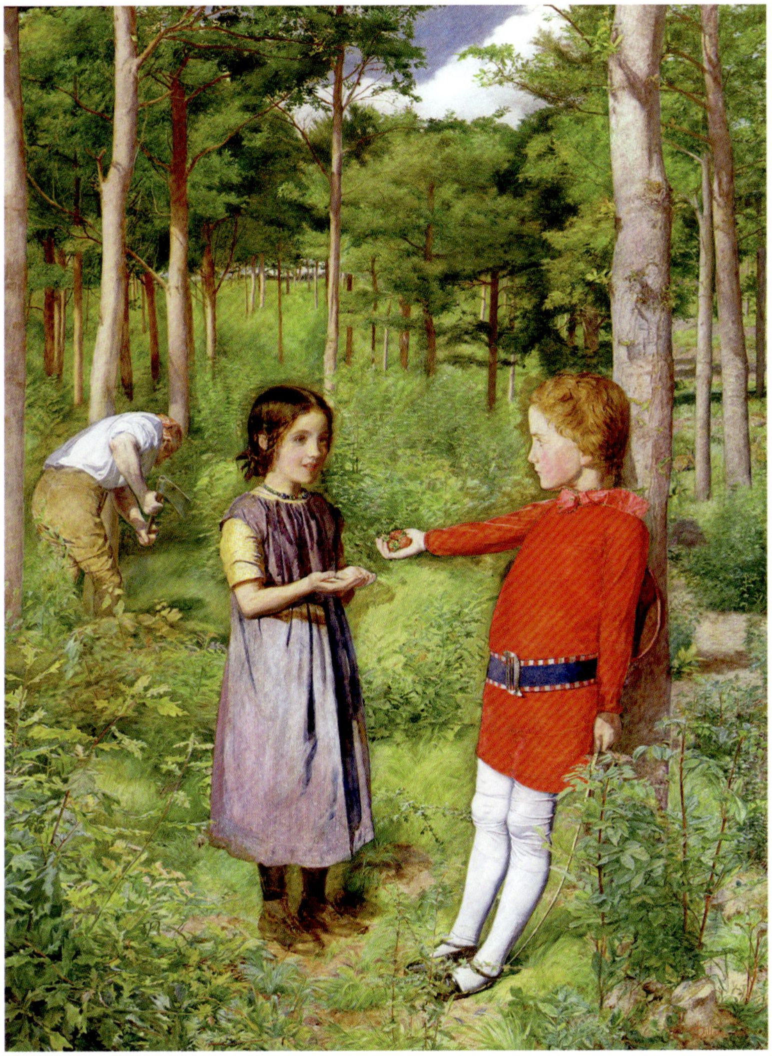

in the *Saturday Review*.[47] His poem, first published in 1844, tells the story of Maud, a woodman's daughter, who is befriended as a child by the son of the local squire but seduced or perhaps raped by him as an adolescent. Her father repudiates her, and Maud, who is only sixteen years old, drowns her baby out of shame and despair in a pool near a stream. The story itself is melodramatic and even formulaic, but Patmore's telling of it is at once vivid and elliptical, giving the poem an intense atmosphere that satisfied the Pre-Raphaelites' yearnings for moral seriousness and imaginative power in the arts.

Millais's painting was at least the second *plein-air* illustration to this poem planned by the Pre-Raphaelites. The first entry in the *P.R.B. Journal* records Millais recommending a pool on Wimbledon Common to William Rossetti as a suitable location for his own projected design on the same theme, entitled *Maud's Madness*.[48] Rossetti meant to illustrate the poem's ending, foreshadowing the theme of Millais's *Ophelia*. For *The Woodman's Daughter*, Millais takes a moment earlier in the poem when the young boy is watching Maud and her father at work:

> And sometimes, in a sullen tone,
> He offered fruits, and she
> Received them always with an air
> So unreserved and free,
> That shame-faced distance soon became
> Familiarity.[49]

The boy in Millais's painting looks physically awkward, his right arm too straight and out of proportion. This awkwardness is a visual corollary for his sullenness in the poem. Maud's ingenuousness too seems exaggerated in the painting, as if to intensify the contrast between them that Patmore sets out here. Both poem and painting are stark in their morality and grim in their proleptic irony. Patmore's rhyme of 'free' and 'Familiarity' is knowingly out of key. It jars in the same unsettling way as the boy's gift of strawberries in Millais's painting does with the switch he grasps in his other hand, as his hostile expression and embarrassed blush do with his apparently generous gesture, and as his bright red tunic and white stockings do with the greens of the wood around him. His very presence is an intrusion, a disturbance. He does not belong here.

The boy's incompatibility with the scene around him is, as Stephens observes, a problem of history. He is the product of 'generations of refinement'.[50] Maud's history, by contrast, is the history of the wood itself, encapsulated for Stephens in the detail of Millais's observation: 'The scene is a wood, of which every branch and leaf, and tuft of herbage, the grey, green, and ashy-coloured trunks, and every stone and lump of earth within range of the artist's vision, have been painted with unflinching faith, and marvellous delicacy. By this means we have the history of the shaw, and of every

23 John Brett, *The Glacier of Rosenlaui*, 1856. Tate Gallery, London.

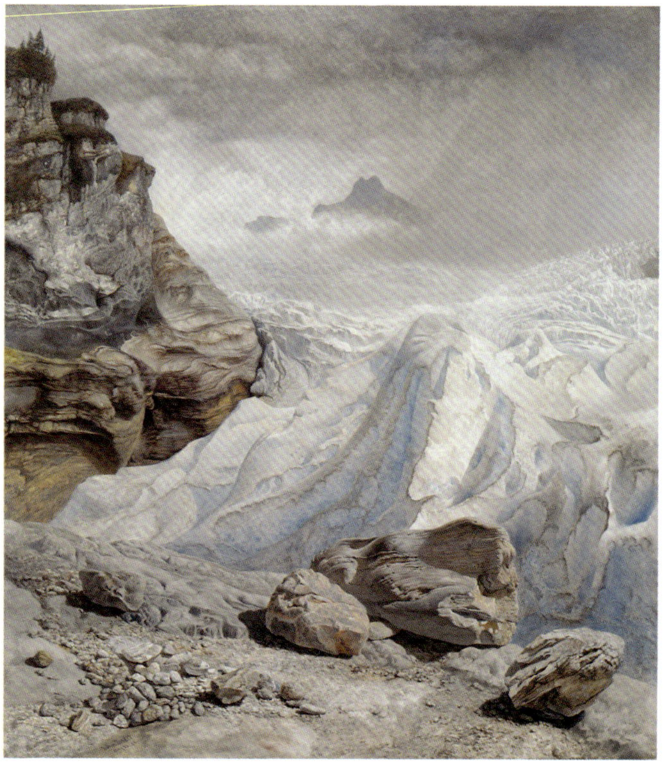

tree within it.'[51] By contrast, as Stephens notes, the boy's strawberries are 'the outcomings of generations of fortune, skill, and taste'.[52] Where they have been selectively bred from plants drawn from across Europe and the Americas, the woodman's oaks, beeches and other trees are natives with a deep history in the English landscape.[53] The contrast between the fruit and the trees mirrors that between the boy and the girl. He is an aristocrat of Norman descent, as both his hair colour and his haircut suggest. She is Anglo-Saxon, a growth of English soil. Her scuffed boots take on the colour of the earth around them; her dark stockings rise straight up from the ground like the trees themselves. Where he lives through pleasure, she and her father live through work. His father may own the land, but he only enters the wood as a spectator and an interloper, where the woodman and his daughter are an integral part of its ecology.

One of the most profound results of the Pre-Raphaelites' ecological investigations is their realisation that environments are the collective creation of all the organisms that inhabit them. The wood in *The Woodman's Daughter* is not a cultivated environment, but it is a managed one. Its history is the history of the interactions between the plants and people that live there. What Stephens sees in Millais's painting is a history that is at once local – the history of each individual tree, as per *Modern Painters*, and of this specific wood – and part of a wider national history that has shaped and been shaped by the ecology of the English landscape. Both these dimensions are suggested in his use of the word 'shaw', an Old English word for a copse or small wood that by the nineteenth century survived mainly in dialect and the archaisms of literary tradition.[54]

Pre-Raphaelite landscapes bear witness to human habitation but exceed human control. This is apparent on a grand scale in *The Val d'Aosta*, where human impact lessens the further up the mountain and the canvas we look, and where our significance dwindles beside the record of geological forces and time written into the rocks and forms of the Alpine landscape. In John Brett's earlier Alpine painting, *The Glacier of Rosenlaui*, painted in 1856 and exhibited in 1857, this diminishment of humanity when set alongside the forces of geology is even more stark, as the glacier has scoured and crumbled the rocks, leaving no scope for human habitation and no sense of the human scale (fig. 23).[55] In *Thistles*, an equivalent but in this case organic process can be seen in miniature, as weeds assert their independence in front of a cultivated landscape. Pre-Raphaelite paintings and poems recognise too that non-human animals transform the landscapes they inhabit. The spider's web that catches the speaker's attention in 'A Quiet Place' is an example. It may seem not worth noticing, but Rossetti notices it all the same, both singling it out for attention and attributing it to a particular spider, depending how you stress the phrase 'this spider's web' in reading the poem.

Another unobtrusive but revealing example is the paths made by the sheep on the lower slope of the Fairlight Downs near Hastings in Holman Hunt's *Our English Coasts, 1852* (fig. 24). The principles that Wolf applied to the study of wild animals, the Pre-Raphaelites applied to the study of the domesticated animals living alongside human beings. Like the 'unmoved contented' cows in Regent's Park in 'Sheer Waste', sheep in Pre-Raphaelite paintings interact

24 William Holman Hunt, *Our English Coasts, 1852*, 1852. Tate Gallery, London.

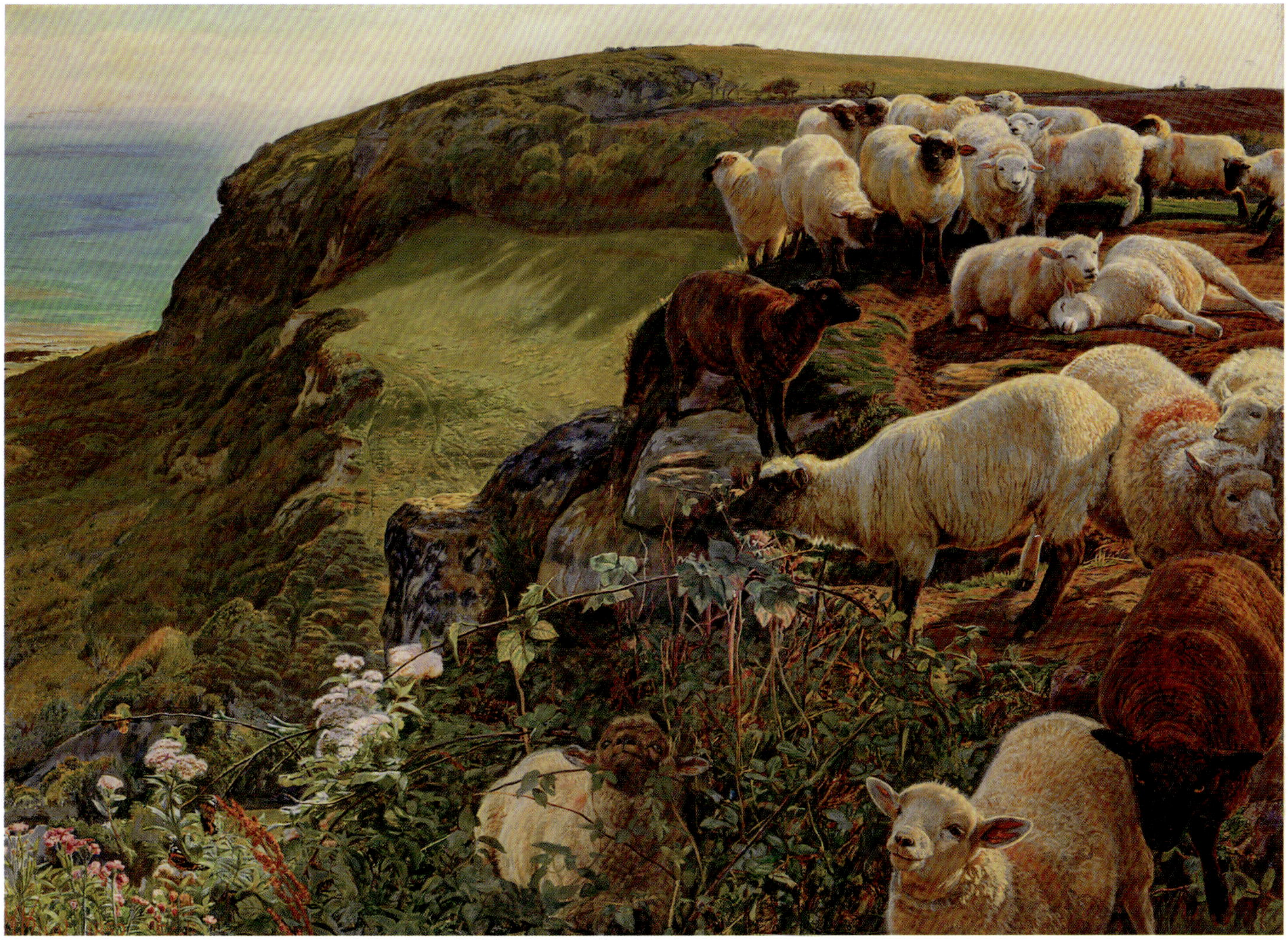

with humans on their own terms. The Pre-Raphaelites' investigations of nature led them too to recognise animals' individuality. Some of the sheep in *The Hireling Shepherd* are curious and head off into the corn. Others are quarrelsome, others contentedly lazy. These are qualities of the moment – we do not know that the same sheep will always behave in the same way, any more than people do – but they are also signs of what current studies in sheep behaviour recognise as their individual temperaments.[56] The animals in Brown's *Pretty Baa-Lambs* show the same individuality (see fig. 8). Of the two lambs in the foreground, one is keen to receive the flower Emma Brown is offering it, while the other shows no interest. Neither pays attention to her baby, who looks on with her own apprehensive curiosity, unable to comprehend that the bunch of flowers she is offering to the lamb is way out of its reach.

Our English Coasts is the richest study of animal individuality in Pre-Raphaelite painting. It has typically been interpreted through analogies, either as a satire on the vulnerability of England to invasion, or as an allegory for the dangers of the Church neglecting its flock. Both these interpretations were advanced by Stephens in his retrospective essay 'The Two Pre-Raphaelitisms' in 1856.[57] But Hunt's painting was commissioned and executed expressly as a natural history study. When the naturalist William Broderip bought *The Hireling Shepherd*, his cousin Charles Maude, who had not been able to afford the asking price of 300 guineas himself, offered Hunt 70 guineas to do a copy of the sheep from the original painting. It was Maude who had put Hunt in touch with Broderip, so once the deal on *The Hireling Shepherd* had been clinched and Hunt was no longer anxious about his finances, he offered instead to paint Maude a new painting of the same subject.[58] In the essay in which he recalls the origin of the painting, Hunt comments that 'while I make mystic subjects bear reserved meanings, I have no unusual intention when treating natural facts.'[59] *Our English Coasts, 1852* is thus first and foremost a picture of sheep in a landscape, with the date in the title functioning to anchor the painter's observations in time as well as space, as the equivalent dates do in Dyce's *Pegwell Bay* and Dante Gabriel Rossetti's verse letters from his trip to Paris and Belgium. The painting may be 'suggestive of its finer analogies', as William Rossetti remarked in his review of it, but, as he also noted, it is 'the individual expressions of the strayed flock' that give this picture 'as deep a human interest as any in the gallery'.[60]

Rossetti rightly observes that Hunt's attention to sheep in *Our English Coasts* created 'a new experience in art'. At the same time, it is a 'triumphant vindication' of 'the Præraphaelite principle' of modelling art on science.[61] Hunt's practice in painting sheep was notoriously thorough. Millais records in his diary that he and his servant had to hold them down so Hunt could paint them for *The Hireling Shepherd*.[62] It is hard to dispute Hunt's biographer Anne Clark Amor's remark that these animals were 'ill-treated in the cause of artistic truth'.[63] The reward for this rough treatment, however, is that each sheep is recognised in Hunt's paintings as a unique individual. Unlike Edward Landseer, the most popular and famous Victorian animal painter, Hunt does not anthropomorphise his subjects. Instead, he confines himself, in Rossetti's comically pedantic phrase, 'to strictly ovine expression'.[64] But Hunt recognises, as Darwin did, that animals' expressions, in particular how they carry their bodies, bespeak emotions. Indeed, in the case of sheep he saw this more clearly than Darwin himself, who commented twenty years later in *The Expression of the Emotions in Man and Animals* that 'cattle and sheep are remarkable from displaying in so slight a degree their emotions or sensations, excepting that of extreme pain.'[65]

Hunt studied the sheep he painted with extraordinary attentiveness. Looking at *Our English Coasts* with the same attentive eye, it is possible to detect a wide range of emotions on show, including affection, contentment and ebullience. The sheep at the top left looking out to sea may be curious. Perhaps it has noticed the plume of smoke in the distance from a steam ship – the only unequivocal sign of modern life in the painting besides the title itself. Possibly it is even appreciative, enjoying the beauty of the landscape as the viewer does. Certainly, this particular sheep is independent-minded, in that it has chosen to turn its head and look over its shoulder and into the distance rather than at its immediate companions or surroundings. Stephens attributes bewilderment to the sheep who appears to be entangled in the vegetation over the brow of the cliff. As he sees it, this sheep is making no 'effort to free himself', while those above are either 'following him into the same difficulty, blindly, as men do' or indolently frisking 'in profound unconcern for the troubled ones before their eyes'.[66] This lapse into conventional allegorising is an uncharacteristic failure of Pre-Raphaelite principles that blinds Stephens to the true nature of the interaction between the sheep themselves. Sheep are remarkably sure-footed animals and thick-skinned. The entangled sheep may well be oblivious to its entanglement because it is not in any serious difficulties.

Stephens is clearly thinking of the sheep as principally symbolic of people, so he may also be expecting more anthropomorphism in their expressions than Hunt provides. Unlike primates, sheep and other ungulates do not have especially expressive faces. Their posture is often more revealing. The sheep among the brambles is craning its neck. It may be trying to free itself after all, but it may also be listening, as the sheep immediately above it on the clifftop is bleating. If so, Hunt's interpretation of sheep behaviour is again confirmed by subsequent science. In *The Expression of Emotions* Darwin qualified his own view of sheep as inexpressive animals when he noted that 'when a flock of sheep is scattered, the ewes bleat incessantly for their lambs, and their mutual pleasure at coming together is manifest.'[67] In the two pairs of sheep seen lying down and nuzzling, Hunt

25 Thomas Sidney Cooper, *Midday Repose*, 1866. Laing Art Gallery, Newcastle upon Tyne.

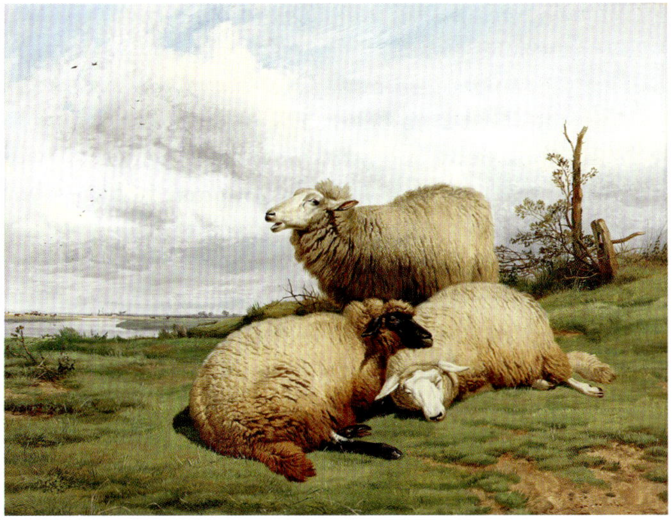

26 Joseph Farquharson, *The Shortening Winter's Day is Near a Close*, 1903. Lady Lever Art Gallery, Liverpool.

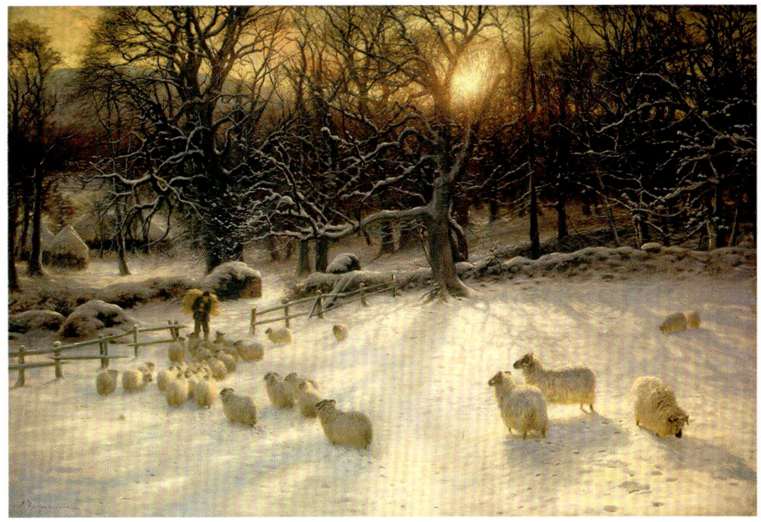

anticipates Darwin's insight that ewes and lambs experience pleasure in contact with one another. In the sheep on the cliff edge, the one above bleating, the one below listening to its voice, he prefigures the more specific finding of a series of papers published in the journal *Applied Animal Ethology* in the 1970s that sight is the primary sense by which ewes and lambs recognise one another, and that sheep use 'vocal cues if visual cues are not available'.[68]

Hunt's achievements in *Our English Coasts* becomes still more evident when his painting is set alongside other Victorian paintings of sheep. Hunt is unique in taking the question of how sheep experience their own lives seriously. Conventionally, sheep contribute to landscape paintings either as picturesque details or as the signifiers of pastoral, whether real or idealised. The sheep in these pictures are typically painted in broad strokes and never individualised. More detailed representations can be found in the demotic tradition of livestock painting typified by the work of Thomas Sidney Cooper, who lived from 1803 to 1902 and made a lifetime's speciality of painting sheep and cattle. But Cooper's paintings, such as *Midday Repose* (fig. 25), painted in 1866, capture none of 'the texture of the sheep's wool . . . the close pile and clefts . . . its various colours in the sunlight' that Stephens rightly praises in Hunt's work.[69] The body language of Cooper's sheep is dumb too, suggesting nothing of the companionship that Hunt's sheep enjoy in their proximity to one another.

A closer comparison with Hunt's painting is with the work of Joseph Farquharson, a late nineteenth- and early twentieth-century Scottish landscape painter who specialised in scenes of flocks of sheep, particularly in the snow. Like Hunt's, Farquharson's sheep are an active part of the landscape, wandering through it at will, taking shelter where they can, and leaving tracks and wearing down paths. Farquharson characteristically gestures towards the interaction between sheep and sheep-farmers in the presence within his paintings of anonymous shepherds or buildings, walls and tracks. Like Millais in *The Woodman's Daughter*, he shows the landscape to be the product of the interaction of people and other creatures. But for all that he gives the sheep their due in this regard, he still does not recognise their individuality as Hunt does. In a typical painting such as *The Shortening Winter's Day is Near a Close* (fig. 26), painted in 1903, he shows one or two sheep that are attending to their own business or holding back, but the majority are simply following the rest. This flocking behaviour is typical of the sheep in his paintings. This contrast between Hunt's and Farquharson's paintings does not show that either of them is more correct in his view of sheep behaviour. Sheep do flock, after all. Nor are they necessarily incompatible. While Farquharson's sheep are in motion, responding to a specific stimulus – the promise of food in this case – Hunt's are at rest, so their behaviour is different. Nevertheless, where Farquharson is repeating

required. For Collinson, the fallen world is not, in the words of the natural theologian William Paley, 'a happy world after all', but rather one in which the struggle to survive is brutal and its consequences at times barely bearable.[79]

THE ECOLOGY OF MODERN LIFE

In his essay on 'The Subject in Art' in *The Germ*, Tupper argues that in order 'to become a more powerful engine of civilization', painting and sculpture 'should be made more directly conversant with the things, incidents, and influences which surround and constitute the living world of those whom Art proposes to improve'.[80] Ford Madox Brown's famous painting *Work*, begun in the early 1850s and finally completed and exhibited in 1865, is the fullest realisation of Tupper's ideal in social and economic terms. But as his phrase 'the living world' suggests, Tupper's conception of modern life is also fundamentally ecological. Together with early Pre-Raphaelite poems such as Rossetti's 'Fancies at Leisure' series and 'The Child Jesus', *The Woodman's Daughter*, *Our English Coasts*, *The Val d'Aosta* and *Thistles* reveal the interaction between humans, animals, plants and geology in shaping environments. Other Pre-Raphaelite paintings are more explicit in tracing the ecology of modern life specifically. Two series of Pre-Raphaelite landscape paintings, in particular, embody this ecological vision of contemporary life. The first is a group of landscapes painted by Brown in the 1850s in various North London suburbs where he lived. The second is a longer series of watercolours of houses, farms and mills in villages on the Thames in Berkshire and Oxfordshire painted by George Boyce – a close friend of Dante Gabriel Rossetti and another contributor to the 1857 Pre-Raphaelite exhibition – from the late 1850s through to the end of the 1860s.

By the canons of conventional landscape painting there is an element of perversity to both of these sets of paintings. When Ruskin challenged Brown over the ugliness of his oil painting *An English Autumn Afternoon, Hampstead – Scenery in 1853*, asking why he had chosen such an unpromising subject, he replied that it was because it lay out of his back window.[81] A watercolour Brown sent to an exhibition in New York in 1857 was explicitly titled *Looking towards Hampstead: From the Artist's Window*.[82] In both cases the selection of the landscape depended on where Brown himself happened to be living rather than any intrinsic quality of the scene itself, such as its aesthetic qualities or historical significance. Collins pioneered this approach to landscape painting in *May, in the Regent's Park*, painted in 1851 and exhibited in 1852 (fig. 27).[83] Both painters made the same apparently arbitrary choice to paint what happened to lie out of the window, but while Collins's window opened straight on to a landscaped royal park – the scene of William Rossetti's 'Sheer Waste' – complemented by a well-tended garden, the more open parkland of Hampstead Heath that Brown could see from his windows lay in the distance beyond the haphazard sprawl of suburban buildings.

Boyce's subjects – cottages and other buildings in an English rural landscape – were more conventional in themselves, but he too chose angles of vision that cut across the tradition of picturesque landscape painting.[84] Rather than being painted from the front and carefully positioned within the wider landscape to suggest an approach to the building, Boyce's houses and working buildings are typically painted from the rear or the side. One example is *Mapledurham House: Early Morning*, painted in 1860 and shown at the Hogarth Club in 1861. In this work we as viewers are positioned so close to the house and its outbuildings that they exceed the frame of the picture. Even so, they are not quite as close to us as the perimeter wall, which obscures the lower portion of each building from view. In *At Binsey, near Oxford*, painted in 1862 and exhibited the year after, the cottages, again seen from the rear, are half-hidden behind wooden fences, two looming willow trees, and the rich growth of an apple tree coming into the frame from the right-hand side.

In both *Looking towards Hampstead* (fig. 28) and *At Binsey* (fig. 29), the observational attention paid to the trees and their place in the foreground of each picture argue for their importance within each artist's vision of day-to-day life. Brown's angle of vision draws attention to the density of vegetation even in a London street. Indeed, the trees in his painting come close to concealing the suburban townscape entirely. We glimpse the streets and houses of Kentish Town through branches and over the tops of the lower trees. The only people in the picture – a woman and child walking (*top*) 27 Charles Allston Collins, *May, in the Regent's Park*, 1851. Tate Gallery, London.

(*bottom*) 28 Ford Madox Brown, *Looking towards Hampstead: From the Artist's Window*, 1857. Delaware Art Museum, Wilmington.

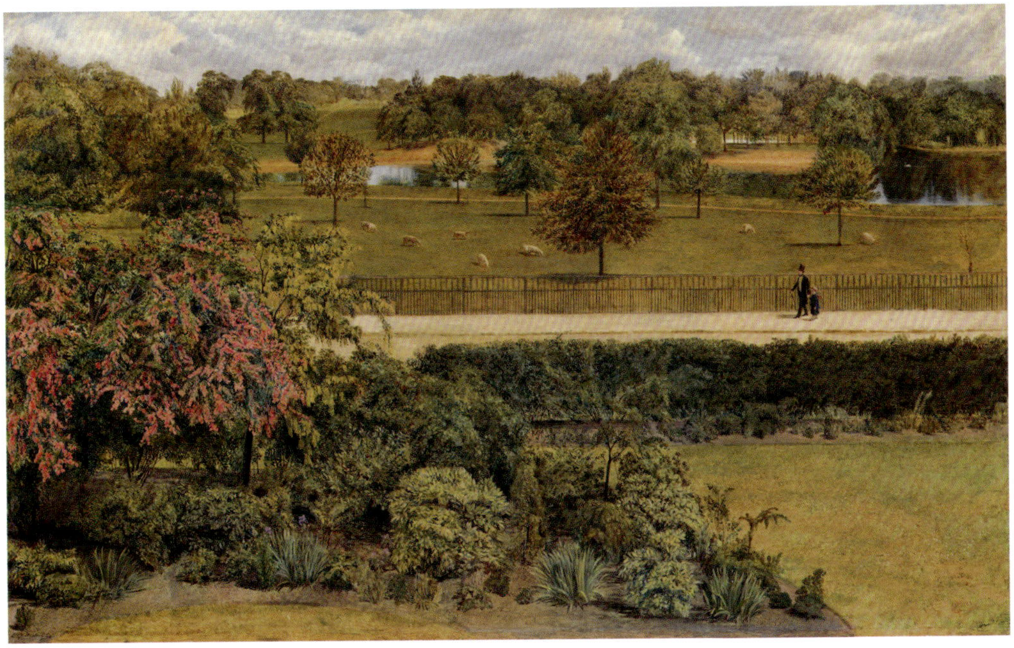

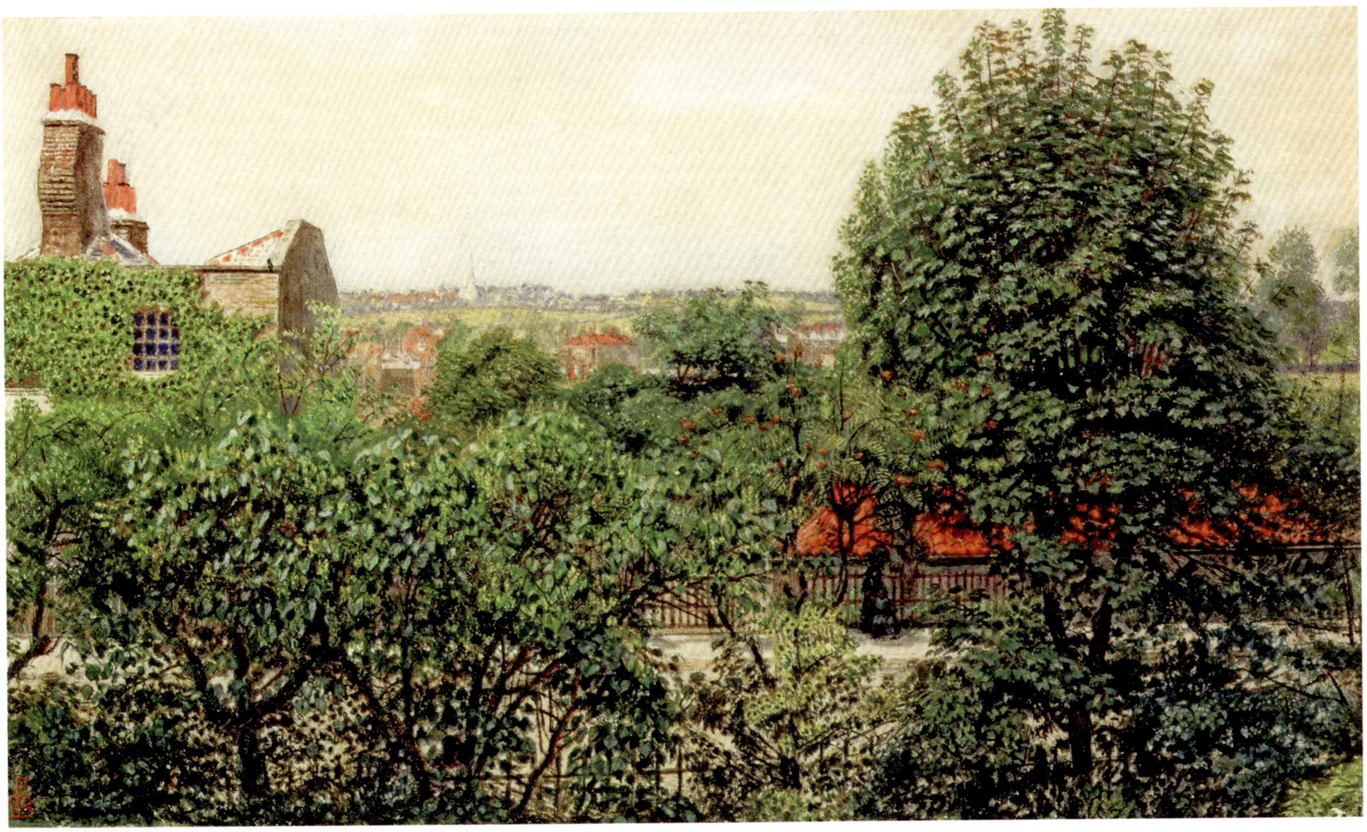

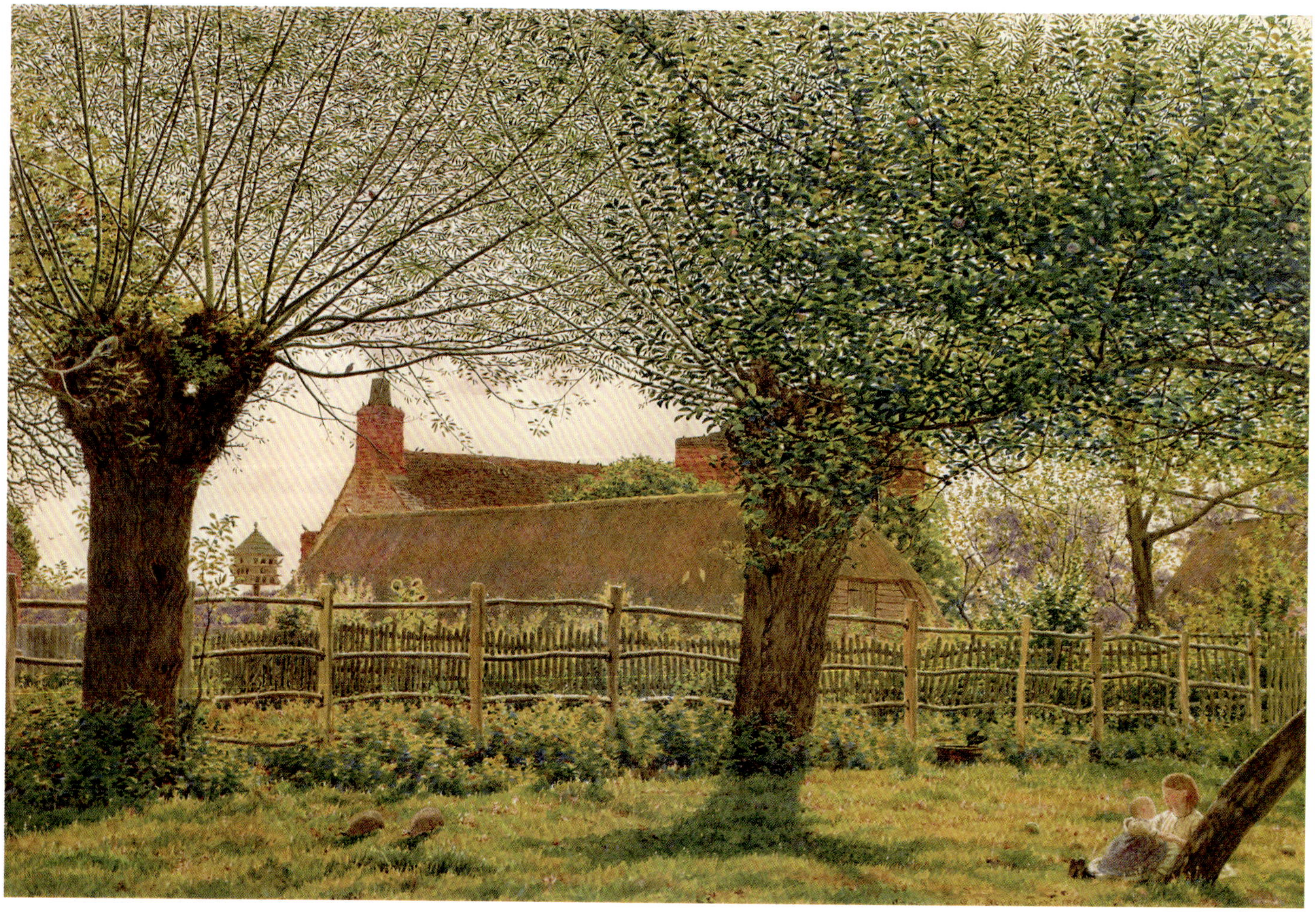

hand-in-hand, visible beneath the rowan tree in Brown's front garden – look at first glance just like more of the foliage. The nearest house – again, the only one that is distinctly visible – is covered in a thick growth of ivy. More patches of deep green are interspersed among the red roofs of the houses in the distance until the pale green of the sunlit Heath opens up between Kentish Town and Hampstead itself. In *The Woodman's Daughter* the boy's red tunic stands out as not belonging in the wood tended by Maud's father. Here the mellower reds of the roof tiles and chimney pots maintain an easy interplay with the greens of the trees and fields. They come together in the rowan, whose leaves and berries unite these two colours in one plant. Each patch of red in this area of the canvas sets a visual puzzle – is it a clump of berries or a glimpse of the low roof of the building across the road? The rowan itself is an emblem of the same interplay of humanity and the natural world. A native of northerly and mountainous regions – its other name is the mountain ash – it has clearly been planted and cultivated here for the pleasure it brings. So have all the trees in the picture. The modern suburban environment, Brown shows us, is built for people but sustains itself through plants.

Brown's brushwork in *Looking towards Hampstead* is relatively free. He shows us every leaf on the trees in the foreground, but not with the precise clarity of many Pre-Raphaelite paintings. Boyce by contrast traces every last leaf and twig in *At Binsey* with minute precision. On the left, birds perch in the branches that radiate out from the bole of the first willow tree. These branches interweave with those of the second willow, forming patterns of remarkable intricacy and delicacy that are at once organic and geometric. On the right, the thicker leaves and fruit of the apple overlie those of the willows with a denser growth that sometimes catches and sometimes blocks the light. The sun is directly behind *(facing page)* 29 George Price Boyce, *At Binsey, near Oxford*, 1862. Higgins Art Gallery & Museum, Bedford.

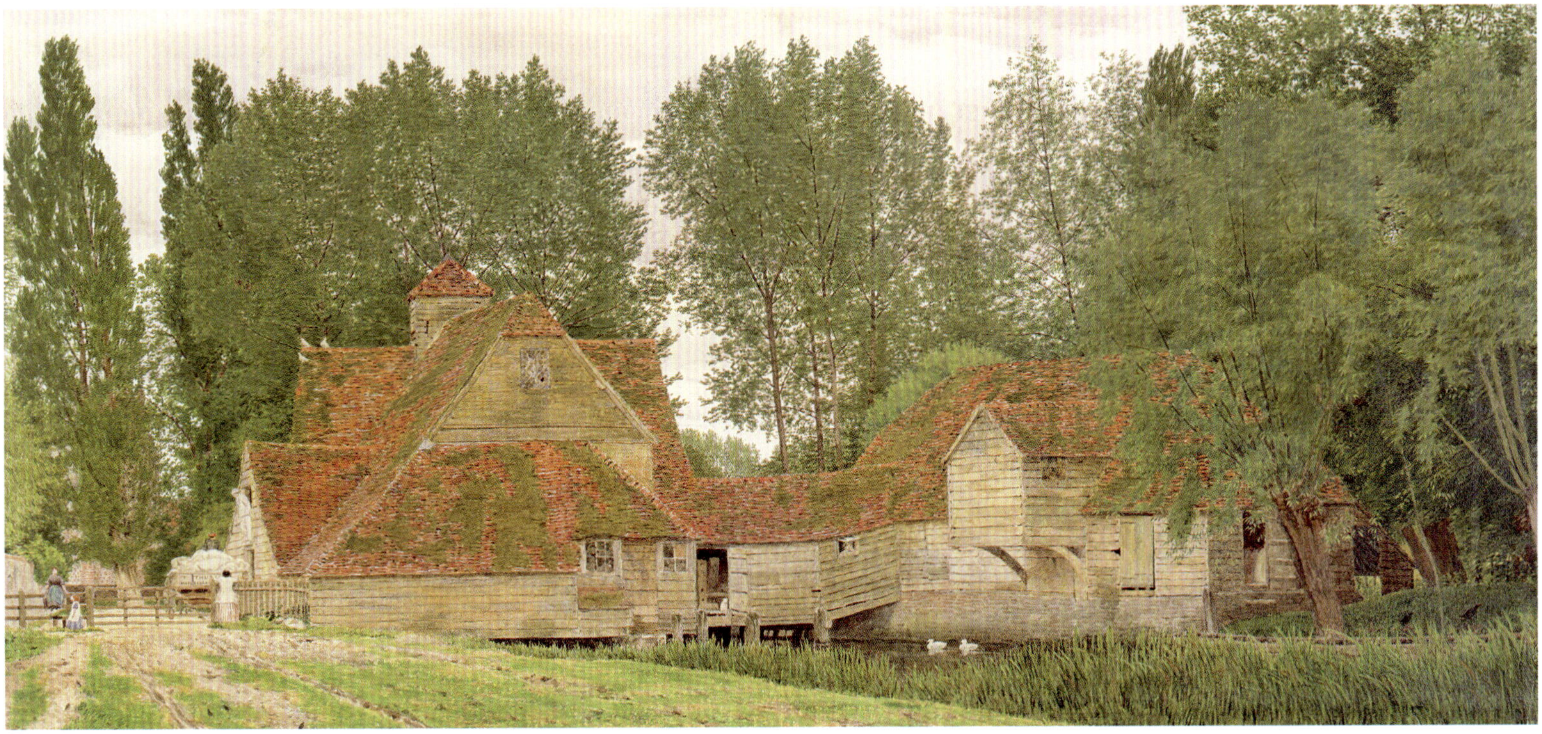

30 George Price Boyce, *The Mill on the Thames at Mapledurham*, 1860. Fitzwilliam Museum, Cambridge.

the central willow, so its shadow simulates its roots, further intensifying the sense that Boyce is celebrating the rich growth of these trees. As in Millais's managed wood and Brown's suburban landscape, the trees in Boyce's painting bear witness to the symbiotic relationship between people and plants. The willows are pollarded, the apple cultivated. Like the trees, the birds in the painting, including a robin, doves and guinea-fowl, are sources of both pleasure and food. The human environment, providing both for their needs and their exploitation, includes a small dovecot and a bucket which, whether by accident or design, holds water for the birds to drink.

The buildings in Boyce's picture are old. There is nothing about the landscape that has obviously changed in two hundred years. But as Hunt incorporates a steamship into his landscape in *Our English Coasts*, so Boyce includes a marker of modern life in the clothes of the young woman and baby who sit at the foot of the apple tree. Their presence at the bottom right-hand corner of the picture reminds us that life need not necessarily be urban or even suburban to be modern. Boyce himself was a Londoner, who painted back views of London in watercolours as he did of rural Oxfordshire. One of these showed the backs of a row of houses in Soho as viewed from a back window in 1866, with trees in the foreground as in Brown's view towards Hampstead. This sketch was a particular favourite of Dante Gabriel Rossetti's, who exchanged one of his own sketches of the model Alexa Wilding for it.[85] But in his studies of the Thames countryside Boyce relishes an alternative vision of the contemporary as a persistence of rural life that foreshadows William Morris's vision of a rural socialism in his futuristic romance, *News from Nowhere*. Unlike Morris, however, and like Rossetti, Boyce prefers relaxation to vigorous work. Boyce's paintings often depict working buildings – *The Mill on the Thames at Mapledurham*, for instance, or *Old Barn at Whitchurch, Oxon* – but rarely rural labour in action. His interests are rather in the relationships that have grown up between people, animals, plants and the vernacular architecture of the English countryside.[86]

In *The Mill on the Thames at Mapledurham* (fig. 30), painted in 1860, these are revealed as a pattern of continuities and

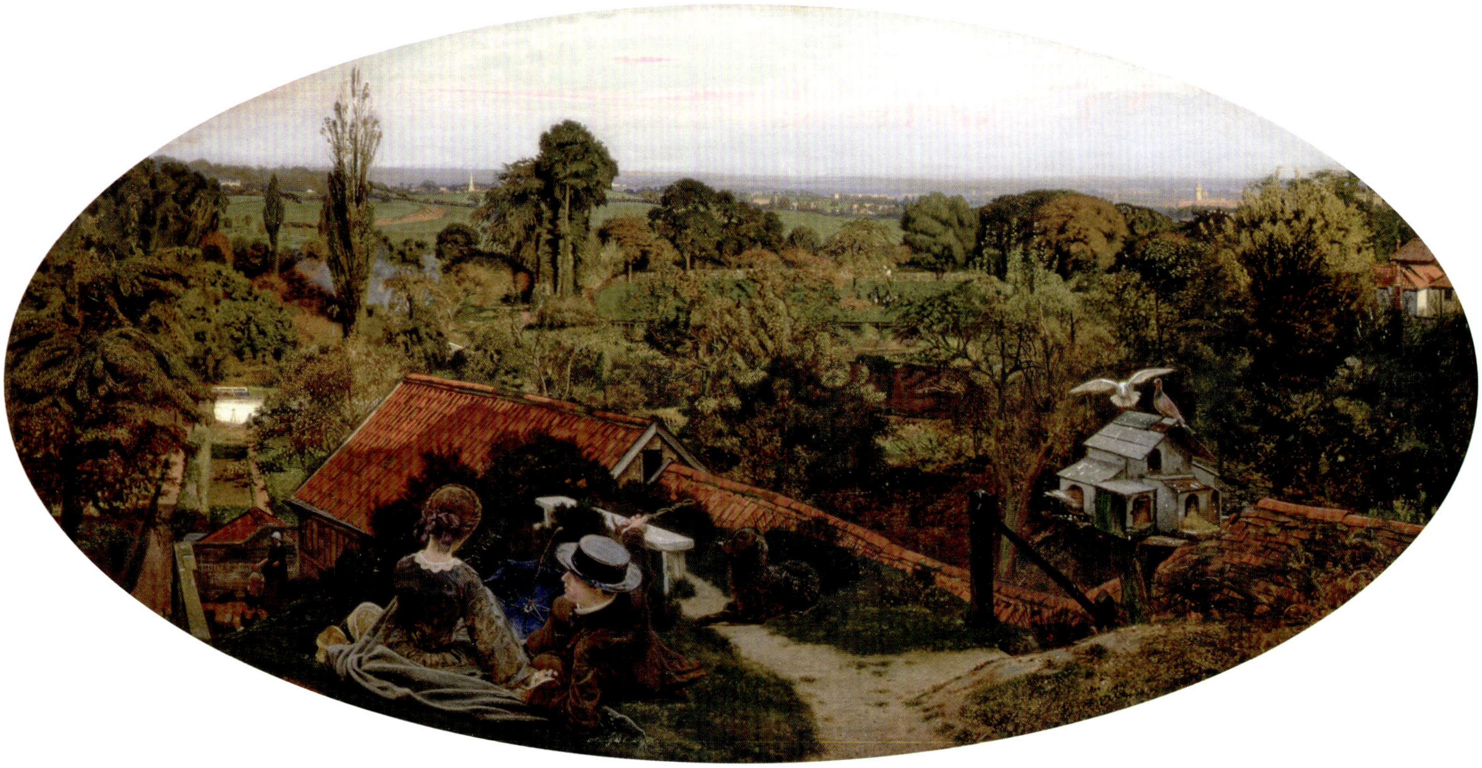

echoes between the man-made mill and the landscape that contains it. The granary on the far bank of the river is partly obscured from view by the trees – pollarded willows again – that grow beside it. By the time the eye reaches the far end of the building it requires some care to distinguish its wooden corner support, which frames a patch of lush vegetation glimpsed through the building, from the trees themselves. Not only does its shape resemble a tree trunk, but the colour of the mill's wooden walls darkens in the shade to become virtually indistinguishable from the bark of the willow trees. Looking back across to the other end of the building, the line of the mill's gable end is picked up by the branches of the tall trees behind it. The moss that grows on the roof merely adds another tone to the picture's rich range of greens, while the doves that have settled on it complement the ducks swimming beneath it. This is a building that has been fashioned from the surrounding ecology, depends upon it, and is itself a portion of it.

There is a nod to the mill's continued function as a working building in Boyce's picture, as we see the miller watching as sacks of flour are laid flat on a cart. This activity is a sign that the mill is not a picturesque relic but persists into the modern world. At the same time, this is a world in which work can pause, as the two girls chatting in the bottom left-hand corner, one standing by the fence, one sitting on it, reveal. In *At Binsey*, the young mother's relationship to the environment is characterised not by labour at all but by leisure, much as it is for the young couple looking over the outskirts of Hampstead in *An English Autumn Afternoon* (fig. 31) and for the father and daughter walking along the road beside the park in *May, in the Regent's Park*. Like Rossetti's 'Fancies at Leisure', these paintings celebrate the companionable pleasure that people can take in the presence of other living things, whether in the apparently unchanged countryside or on the edge of a great modern city. In Boyce's and Brown's paintings, in particular, the environment is integral to this companionship itself. The mother and baby in *At Binsey* are clearly happy, engaging with one another beneath the trees where she has chosen to sit. The young woman and man in *An English Autumn Afternoon* too have chosen this spot to sit down and enjoy each other's company. Ruskin may have thought it ugly, but they like it here. It is conducive to their relaxation and their intimacy. Rather than sit stiffly on the bench in front of them, they have half-sat, half-lain down together on the rough grass by the path, and are looking out over the trees to the Heath and further suburbs

31 Ford Madox Brown, *An English Autumn Afternoon, Hampstead – Scenery in 1853*, 1852–5. Birmingham Museums and Art Gallery.

beyond. These are modern, urban young people, uninhibited and informal, yet they thrive in the sight of open spaces and tall trees, on the ramshackle edge of town. For all that their object is pleasure, not labour, they do not intrude on the landscape as the boy does in *The Woodman's Daughter*. Instead, the colour of their clothes blends in, while their postures lead them to shadow visually the line of the bushes immediately in front of them. Where his pleasure is predatory, theirs is in harmony with the environment around them, growing in it without taking from it.

As Julian Treuherz has pointed out, the full title of Brown's painting affirms both the Englishness of this 'unpretentious mixture of back gardens, sheds and orchards' and its validity as 'scenery'.[87] It also assigns it a precise date. Where Hunt's dating of *Our English Coasts, 1852* is best explained as straightforward documentation, Brown's dating of his *English Autumn Afternoon* to 1853 is more freighted. Barringer interprets the oval framing and off-centre perspective of the latter as signalling 'an experiment in seeing' in which the artist becomes at once 'a scientific apparatus conveying empirical truth' and 'a poetic visionary finding value and pathos in the quotidian world'.[88] Alastair Wright and Allen Staley have both discussed the fact that, at the time when Brown was living in Hampstead, the Heath was under threat of development.[89] Although there is no record of Brown's direct response to this threat, it looms over the landscape in the painting, cutting a bare scar through the green countryside in the distance, as Staley remarks. In his pamphlet *Pre-Raphaelitism*, Ruskin urged artists to record 'all objects of historical interest, or of natural beauty existent at the period' before they were 'swept away in the approaching eras of revolutionary change'.[90] In painting and dating his landscape, Brown sets down its claim to be both historically interesting and beautiful, for all that Ruskin himself failed to see these qualities in it. By declaring it to be quintessentially 'English' and 'scenery', he lodges an implicit plea too for it to be worth preserving. Like the wood in *The Woodman's Daughter*, this is a landscape with a history that marks both deep continuities and breaks with the traditions of English life. The fruit trees, poultry and vernacular architecture which Boyce was to find in the Oxfordshire countryside still persist here. Moreover, it is a landscape that sustains people, both literally and psychologically. By dating his painting, Brown enables future viewers to compare what they see in it with what they can see of the same landscape in their own day. *An English Autumn Afternoon* is thus not only a record but a scale by which we can measure both the physical and the psychological impact of environmental change in London since the 1850s.

Neither Brown nor Boyce denigrate the modern urban landscape, but they affirm the benefit of living alongside plants and animals, and with access to green spaces and the time to enjoy them. Jonathan Bate argues that Wordsworth and other Romantic writers anticipated modern ecological campaigners who 'move from theoretical description to an applied science of healthy and happy living'.[91] In their own investigations of nature, the Pre-Raphaelites too came to see an alienation from the rest of the living world as profoundly unhealthy. In a recurrent pattern across several of their paintings and poems, the natural life of plants, animals and the people who are at ease among them is juxtaposed against wilful perversity and social convention, particularly in the context of love and sexual relations. In the poem which opens the first issue of *The Germ*, Woolner's 'My Beautiful Lady', the two lovers bond over the shared pursuit of natural history, looking at animals and plants with the intensity and precision characteristic of Pre-Raphaelite poetry and painting. So do the shepherd and his companion in Hunt's *Hireling Shepherd*. The young man and woman in *An English Autumn Afternoon* again form a pair, whether or not they are strictly speaking a couple. Brown himself said that they were 'hardly lovers – more boy and girl, neighbours and friends'.[92] 'Hardly' here is ambiguous. It could imply that they are not lovers at all, or that they have as yet barely become lovers, having started out as neighbours and friends. Either way, they make for a positive image of relaxed intimacy between a young man and a young woman, again associated with the open air and with plants and animals who live alongside them in a shared environment.

Set against these images of healthy and happy alfresco sexuality are a number of paintings from the 1850s which depict scenes of sexual oppression or distress in claustrophobic interiors. In Millais's *Mariana* (fig. 32), exhibited at the Royal Academy in 1852, Mariana from Shakespeare's *Measure for Measure* is shown as an effective prisoner in an old, dark

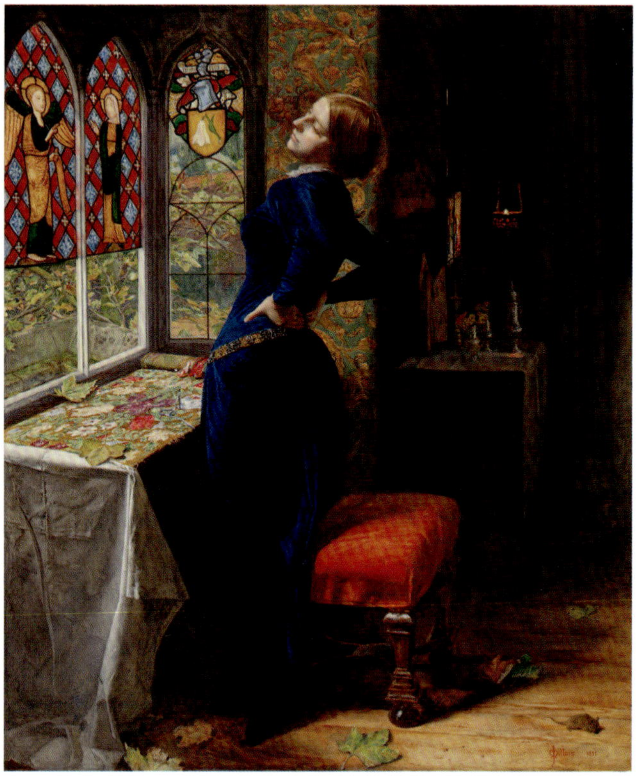

32 John Everett Millais, *Mariana*, 1851. Tate Gallery, London.

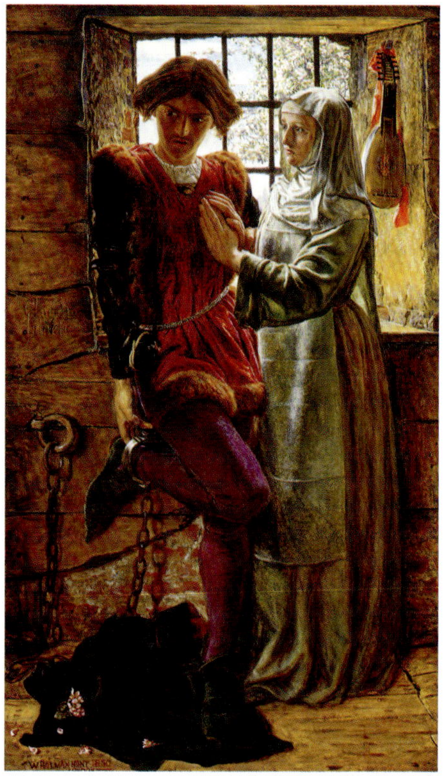

33 William Holman Hunt, *Claudio and Isabella*, 1850–53. Tate Gallery, London.

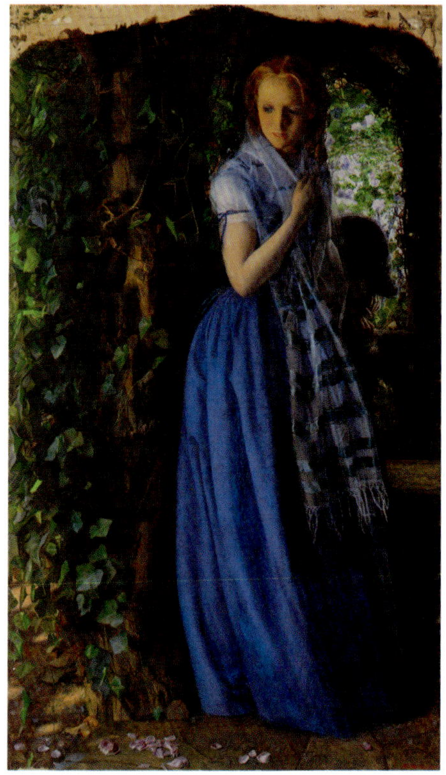

34 Arthur Hughes, *April Love*, 1855–6. Tate Gallery, London.

grange as she waits unhopefully for her lover Angelo, who has callously abandoned her since her dowry was lost at sea. In Hunt's *Claudio and Isabella* (fig. 33), exhibited the following year and based on the same play, Claudio, chained in a dungeon, sullenly refuses to accept the comfort offered him by his sister Isabella, a novitiate nun, after she has refused to sleep with the same Angelo in exchange for her brother's freedom. In *April Love* (fig. 34) by Arthur Hughes, who fell in with the Pre-Raphaelites in the early 1850s and became one of their most dedicated followers, two lovers are seen in a dark bower at the moment when 'Love is made a vague regret' and only 'Idle habit links us yet', to quote from the epigraph from Tennyson which the painting was given when it was exhibited at the Royal Academy in 1856.[93] In all three paintings the dark, oppressive interior is contrasted with the profuse life of trees and flowers glimpsed through a window.

The implied symbolism in each of these paintings is essentially the same. Exploitative and unhappy sexual relations are equated with darkness and confinement, while freedom from such constraint is equated with light and natural growth. Hunt makes this symbolism explicit in *The Awakening Conscience* (see fig. 15). To return to the moral judgement Hunt passes in this painting, the sight that prompts the woman to realise she has been entrapped is daylight shining on and through the leaves of a tree. Like those in *Looking towards Hampstead*, this tree stands in the front garden of a London house. The visual effect of Hunt's painting looks forward to the intricate patterns Boyce discerns in the trees in *At Binsey*. Within the woman's room, by contrast, the only plants are cut flowers and the heavily stylised grapes and stalks of wheat on the wallpaper. Like the female sparrow almost caught by the cat under the table, if she can make for the outside world there is at least a hope that she can be free of the sexual exploitation. If not, she runs the same risk as Millais's Ophelia, another victim of sexual exploitation and subsequent neglect, who drowns with her own cut flowers even as the life of the natural world is rampant around her.

As the shoe poking out beneath her dress shows, the kept woman in *The Awakening Conscience* is already taking a step towards her escape. Unlike the cat, who has seen where the bird is heading, and who is either preparing to pounce or possibly cowed by whatever vision the woman herself has seen, the man in Hunt's painting has yet to see what she has seen or where she is going. We ourselves only see through a mirror on a wall, like the shadows of the truth on the walls of Plato's cave or the glimpse of heaven seen through a glass, darkly, in Paul's second letter to the Corinthians. It seems unlikely that the man will give up his claim on her readily, but at least, Hunt suggests, there is a countervailing impetus towards natural growth which may afford her other, healthier possibilities for her life, especially if we as Victorian viewers intervene to realise the vision of a world we too have glimpsed in the mirror and in other Pre-Raphaelite paintings where women and men can be free to interact naturally and on a footing of sexual equality.

In Pre-Raphaelite art, an alert and meticulous engagement with the natural world deepens human interaction, whether friendship or sexual love or the bond between mother and child. It increases our knowledge and understanding of ourselves and establishes sympathies which reach across the barrier between one species and another. It is enabled by modern life, with its opportunities for leisure, and at the same time reminds us that modern life needs to accommodate nature if it is to thrive. It even helps us to face death. In 'Of my Lady. In Death', the sequel to 'My Beautiful Lady', the only respite from the otherwise pervasive grief of Woolner's poem comes in the careful description of the ecology of the churchyard where the lady is buried:

Small birds twitter and peck the weeds
  That wave above her head,
  Shading her lowly bed:
Their brisk wings burst light globes of seeds,
  Scattering the downy pride
  Of dandelions, wide:
Speargrass stoops with watery beads:
  The weight from its fine tips
  Occasionally drips:
The bee drops in the mallow-bloom, and feeds.[94]

Between the two poems the lover has watched his lady die of a wasting disease, perhaps consumption. His imagination is leading him to mental disease, as he dwells on her corpse and how 'Now o'er its perfect forms / Go softly real worms', but acts of observation like those the two lovers shared in life distract him from this morbid obsession with death.[95] All around him the birds and bees feed, helping plants to spread seed and pollen, beginning a new cycle of life, as Woolner hints in the cycle of the rhymes themselves. The 'weeds' produce the 'seeds' on which the birds feed even as the bee 'feeds' on the mallow. As the 'brisk wings burst' the dandelion seed-heads, the alliteration and internal echoes capture a momentary energy which disturbs and revivifies the gentler waving of the weeds. In the same way, the 'drips' of the beads of water are echoed with more impetus as the bee 'drops' into the mallow flower to feed.

Wordsworth famously condemned the physician or natural philosopher who 'would peep and botanize / Upon his mother's grave'.[96] For Woolner, even by the graveside close attention to nature has a therapeutic benefit. It affirms the continuity of life alongside death. It revives too the bond between the lover and his dead lady which first came to fruition in the previous poem in their shared exploration of the natural world, as shown in the upper panel of Hunt's etching. Dante Gabriel Rossetti praised Woolner's paired poems to Scott, commenting on their 'delicate study of nature'.[97] This study is both their technique and their subject. Even after the lady's death, her lover can find the emotional reserves to love nature because of the joy they took in it together. As in other Pre-Raphaelite poems and paintings, it is not especially rare or attractive animals that revive these feelings, but small, everyday, unappealing creatures – 'Flies, buzzing, strengthened with the morn' and knocking on window panes, 'swarms of insects' which 'revel / Along the sultry level'. 'My lady's death', the mourner declares, 'makes dear these trivial things.'[98] Through their remarkable close attention and their unprejudiced, scientific gaze, the Pre-Raphaelites discovered the value of such 'trivial things' as trees, sheep and insects. In their art, they communicated an ecological vision of nature which prized these plants and animals both for their own sake and for ours.

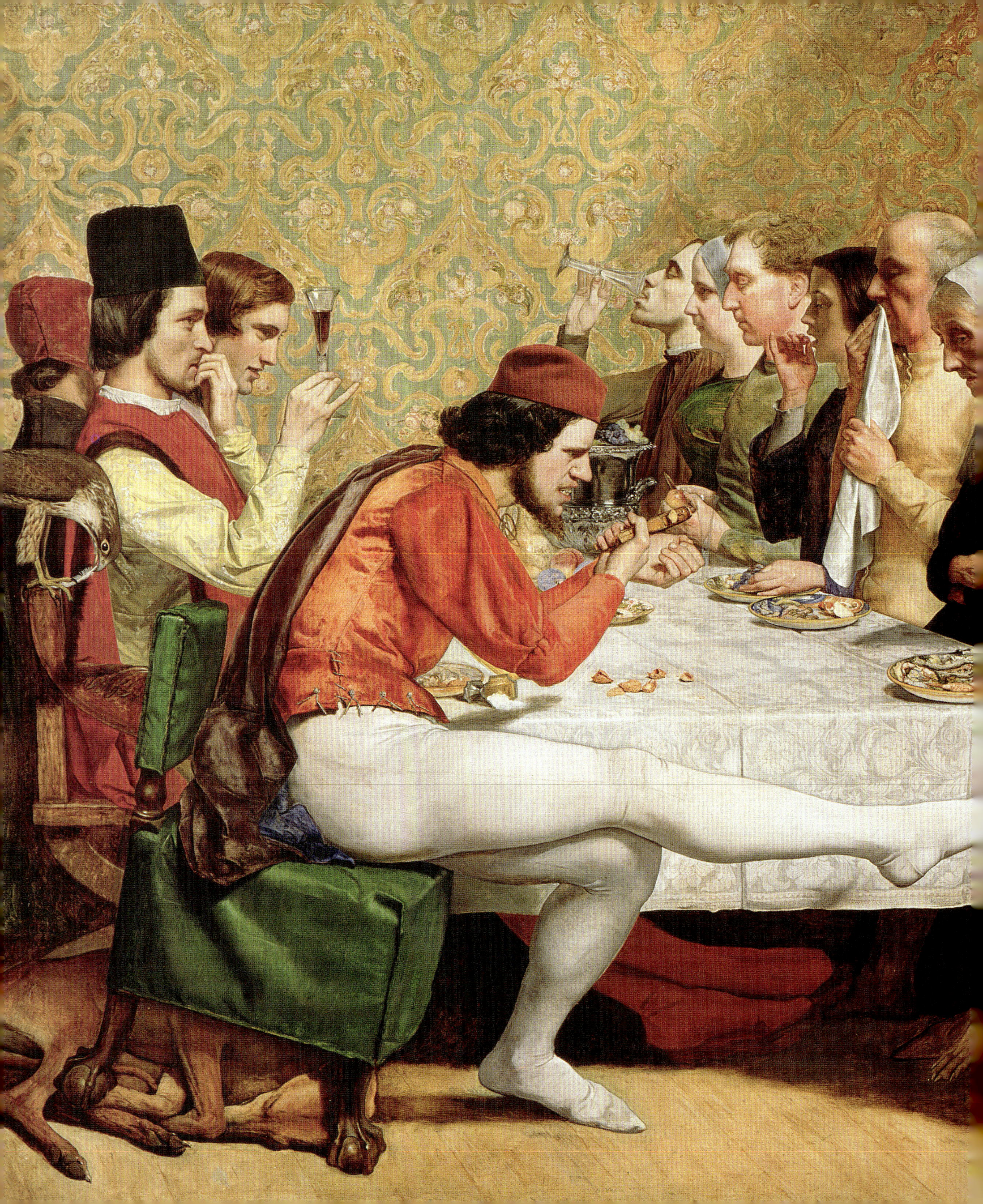

3 · The Knowing Hand of the Anatomist

EMBODIED PSYCHOLOGY IN PRE-RAPHAELITE PAINTING AND POETRY

Pre-Raphaelitism had a profound impact on English landscape painting in the 1850s and 1860s. It helped to cement the pre-existing preference among English landscape painters for the 'pure transcripts and faithful studies from nature' identified by Stephens in *The Germ* and celebrated by Ruskin in *Modern Painters*, and it set the standard for the precise and vivid pictorial style which artists including Rosa and John Brett and George Boyce would apply to these ends.[1] For the men who launched the Pre-Raphaelite project themselves, however, its 'first impulse' was to transform historical painting.[2] They achieved this in part through their radical new approach to the relationship between the figures within a painting and the landscape they inhabit. But they also applied a new rigour to depicting human subjects themselves. Here too they modelled their practice on the principles of science. Just as Pre-Raphaelite landscapes constitute investigations into the natural world, so Pre-Raphaelite history painting conducts investigations into psychology and human interaction. But where the study of ecology in Pre-Raphaelite paintings and poems chimes with wider trends within mid-Victorian natural history, their history paintings and psychological poetry mount a challenge to the burgeoning science of psychology itself.

In 1850, when *The Germ* was launched, psychology was on the cusp of becoming a serious scientific discipline in Britain. The *Journal of Psychological Medicine and Mental Pathology* had been launched only two years before in 1848 – the same year as the Pre-Raphaelite Brotherhood was formed. It was, Sally Shuttleworth notes, 'the first journal devoted to the study of mental disorders'.[3] The field was to grow substantially both in scope and standing, within medicine and in the wider intellectual arena, over the next few years.

Facing page Detail of fig. 36.

The early 1850s saw the publication of a string of textbooks on this new science. In 1852 William Carpenter added a new section on physiological psychology to the fourth edition of his influential *Principles of Human Physiology*. Henry Holland's *Chapters on Mental Physiology* appeared the same year, followed by J. D. Morell's *Elements of Psychology* and Daniel Noble's *Elements of Psychological Medicine* in 1853, Benjamin Brodie's *Psychological Enquiries* in 1854, and two landmark works, Alexander Bain's *The Senses and the Intellect* and Herbert Spencer's *Principles of Psychology*, in 1855.[4] The *Journal of Psychological Medicine* had by this point been joined by the *Journal of Mental Science*, which began publishing in 1853.[5] This rapid emergence of psychology as a science coincided with the Pre-Raphaelite project, but it was not yet highly visible or fully established when the Pre-Raphaelites set out to reform art along scientific lines.

For Stephens, writing in 1850, psychology is less a science than a pseudoscience. Stephens makes the case for the arts as an alternative and superior route to a scientific understanding of mental and emotional life in his second essay for *The Germ*, 'Modern Giants'. Without naming him, he singles Robert Browning out as the 'greatest, perhaps, of modern poets' who exemplifies the scientific application of the arts to the problem of psychology.[6] Browning, Stephens writes:

> has looked into the heart of man, and shown you its pulsations, fears, self-doubts, hates, goodness, devotedness, and noble world-love; this is not done under pretty flowers of metaphor in the lispings of a pet parson, or in the strong but uncertain fashion of the American school; still less in the dry operose quackery of professed doctors of psychology, mere chaff not studied from nature, and therefore worthless, never felt, and therefore useless; but with the firm knowing hand of the anatomist, demonstrating and making clear to others, that the knowledge may be applied to purpose.[7]

Like Tupper in 'The Subject in Art', Stephens contrasts two sciences. The one – anatomy – is rigorous and instructive. The other – psychology – is as yet no more than quackery. The difference between them lies in the close attention to nature that the Pre-Raphaelites hold up as the standard by which to judge both art and science.

Stephens's argument here parallels that in his earlier essay 'The Purpose and Tendency of Early Italian Art'. Like an anatomist, Browning studies the human organism with care directly from nature. Recording his results and demonstrating them to others in his poems, particularly in the dramatic monologues collected as *Dramatic Lyrics* in 1842 and *Dramatic Romances and Lyrics* in 1845, he achieves the moral purposes of the arts by the same means as the anatomist achieves his own scientific ends. Judged by these properly scientific standards, where the poet passes, the 'professed doctors of psychology' fail, as they do not ground their own supposed science in sufficiently robust observation of the actual world: in this case of people's individual behaviour, character and experience of living. For his part, the 'pet parson' is like a Royal Academician, repeating received wisdom rather than investigating the question for himself, while the American school – Poe, Emerson and Longfellow, according to the list of Immortals – are working along the right lines but without the precision of Browning himself.

In a recent study of psychology in Victorian poetry, Gregory Tate has shown how the reflexive analysis of the processes of the mind became a dominant theme within Victorian poetry, forming a parallel and complementary tradition to that of Victorian psychology itself.[8] Tate's thesis would seem to vindicate Stephens's, but there are striking differences in the ways in which they each characterise the mental science pursued in poetry. For Tate, the key means by which Victorian poets such as Browning sought to understand psychology was through introspection, while the key problem they had to face was 'whether the essence of being was located in a transcendent and immaterial soul or a contingent and material mind'.[9] Tate's key words, which he examines with great care and sophistication as they recur across Victorian poetry, are 'mind', 'soul' and 'brain'. But for Stephens, Browning is not an expert on the mind so much as the 'heart'.[10] His poetry reveals its 'pulsations'. Psychology as the Pre-Raphaelites understand it is more than just the science of the mind. It is the study of the emotional and moral make-up of the whole person. This is expressed through the entire body, for which the heart stands as a synecdoche, and not merely in the mind, whether embodied in the brain or disembodied as a soul. For all their admiration for Browning and indeed

Tennyson, Pre-Raphaelite poetry stands largely apart from the tradition of introspective poetry that Tate traces in his book – 'the dialogue of the mind with itself', as Matthew Arnold famously dubbed it.[11] Instead, it is a poetry of the body.[12] But the body expresses and even determines mental and emotional states.

For the Pre-Raphaelites, if we are to understand the mind, we need to comprehend the body objectively, as the anatomist does, and through fully apprehending too what we ourselves feel, as the quack psychologists so abjectly fail to do. In using anatomy as an analogy for poetry, Stephens reveals too his grounding in painting. Throughout Pre-Raphaelite poetry and painting, visual details of people's countenances and posture reveal depths of their personalities and emotions which would remain otherwise hidden and inarticulate. As their close attention to living creatures and environments leads the Pre-Raphaelites to a new ecological understanding of the natural world, so their equally close scrutiny of human feelings leads them to a more profound understanding of the ways in which changing bodily states and impulses bear on questions of motivation and moral responsibility, of self-knowledge and the knowledge claimed by science itself.

SEARCHING FOR DRAMATIC TRUTH: PSYCHOLOGY AND ANATOMY IN PRE-RAPHAELITE ART THEORY

Stephens's likening of Browning to an anatomist of the human heart is not merely a flourish. Rather, for the Pre-Raphaelites, the study of anatomy was a prerequisite for the analysis of psychology. In this they followed part way the lead of Sir Charles Bell, a Scottish surgeon whose work on the *Anatomy and Philosophy of Expression* went through several editions under various titles from 1806 onwards. Bell argued for the importance of teaching anatomy in training for art, on the grounds that it was through the form of the face and the body that emotion and character were expressed. The artist who did not understand anatomy could not properly apprehend these fundamental subjects of art nor truthfully represent them. As Lucy Hartley explains in *Physiognomy and the Meaning of Expression in Nineteenth-Century Culture*, 'Only through a knowledge of the internal structure of the body could the responses of the body to various emotions not only be accurately observed but also translated into a visual language.'[13] Hartley's suggestion that Bell was an important influence on the Pre-Raphaelites is borne out by the biographical sources. Tupper read Bell as early as 1840.[14] Hunt followed suit, recommending him to Millais in 1848.[15] Dante Gabriel Rossetti too read up on anatomy, although his brother does not name his textbooks in his biography of him.[16] Bell's theory postulated a chain of reactions by which, as Hartley summarises them, 'an emotion disturbs the heart, activates breathing, and produces expression', so Stephens's stress on the pulsations of the heart as the source of expressed emotion in his discussion of Browning suggests that he was familiar with Bell's work as well.[17]

The Pre-Raphaelites did not simply adopt Bell's physiognomic system, however. Within Bell's account, emotions expressed repeatedly over time gave permanent form to the face; hence physiognomy became the key to character. In Hartley's words, he 'derived uniformity from variety and, as a result, reduced the individuality of an expression to its typological form'.[18] Here the Pre-Raphaelites departed from Bell, as they did from the yet more rigid fixity of phrenology, which mapped the shape of the brain and thereby the form of an individual's character through the pattern of bumps on his or her skull. Codell and Grilli have identified what they see as phrenological types in several early paintings by Millais and Hunt, including *Isabella*, *Christ in the House of his Parents*, *The Hireling Shepherd* and *The Awakening Conscience*.[19] It is certainly true that the Pre-Raphaelites took an interest in phrenology, but it was a decidedly sceptical one. In November 1849, not long before Stephens denounced the 'quackery of professed doctors of psychology' in *The Germ*, William Rossetti and Woolner had their heads read by the craniologist Cornelius Donovan, founder of the London School of Phrenology. Rossetti recorded in the *P.R.B. Journal* that, for all that his own reading was 'very acutely judged and generally correct', Donovan's analysis of Woolner was in some regards 'decidedly mistaken'.[20] Writing some fifty years after the fact, Hunt was tickled to recall that, after reading Millais's head, Donovan had advised him against pursuing a career in painting, and that he had sagaciously informed Dante Gabriel Rossetti that Keats and Dante owed their poetic faculty to scrofula.[21] Tupper too doubted the judgements of phrenology, arguing instead, as Thomas De Quincey had done in his recent essay 'Suspiria de Profundis', serialised in *Blackwood's Edinburgh Magazine* in 1845, that 'the key to a true psychology' lay rather in the analysis of 'dreams, delirium, &c'.[22]

The Pre-Raphaelites' scepticism towards phrenology chimes with that of a growing number of scientists, led by Carpenter, who started to critique its claims publicly in the mid-1840s during his stint as the Fullerian Professor of Physiology at the Royal Institution.[23] It chimes too with their scientific attitude to knowledge in general. Rather than taking phrenology's findings or those of Bell's physiognomy as authoritative, they subject them to their own sceptical scrutiny. As often as not, they find them wanting. Given that both phrenology and physiognomy stood and fell as received scientific systems, not as a collection of independently corroborated observations of nature, it would be surprising if the Pre-Raphaelites had followed them in their paintings with any consistency. Phrenology and physiognomy each offered a key to what Codell calls 'a taxonomy of identifiable human types'.[24] But as Prettejohn has pointed out, this codification of human kinds according to external characteristics is 'altogether alien to the Pre-Raphaelite emphasis on observing every individual entity for itself'.[25] For this reason Pre-Raphaelite painting is an exception to Barbara Larson's observation that 'references to cranial and physiognomic difference were everywhere' in Victorian art, even if there are occasional correspondences between individual figures and physiognomic or phrenological types.[26]

Tupper theorised the emphasis on the individual rather than the type that would become characteristic of Pre-Raphaelite art as early as 1841. In his diary entry for Saturday, 15 May 1841, he challenges Joshua Reynolds's insistence that the nature of humanity is to be found and represented in the ideal form of the species, not in the ways in which individuals depart from that form. Tupper ventriloquises Reynolds, asserting that 'a very little reflection will serve to show us that these particularities cannot be Nature; for how can that be the nature of man, in which no two individuals are the same?' Tupper replies, in his own voice:

> How? Easily: it is the Nature of the genus homo that no two individuals of that genus should be the same. This is really a characteristic of man since it is a law in physiology that the higher the genus is, the greater will be its individual differences. You, Sir Joshua, have here fabricated a concrete of human perfections, which is purely factitious, and mistake it for something which exists, possibly for that aggregate humanity, which is anything but perfect in your sense.[27]

The specific characteristic of the human species for Tupper is our variability. Human anatomy has, as William Rossetti explains in his *Spectator* essay, an 'ordinary plan' which the artist is obliged to follow unless he wishes to depict 'a monstrosity'.[28] But it does not have a uniform type, or even, as the physiognomists and phrenologists claimed, a taxonomy of different types according to character. Instead, humanity is an 'aggregate' made up of all individual human beings, living and dead. To understand human character and behaviour, we must set aside 'factitious' ideals and pseudo-scientific preconceptions – 'the dry operose quackery of professed doctors of psychology', in Stephens's words – and look at each individual as an individual.

The Pre-Raphaelites rejected Bell's codification of character types. Even so, they agreed with him that the study of anatomy was integral to the study of psychology through art. In his reminiscences of his training as an artist, Tupper implicitly links his expertise in anatomy to his ability to investigate the emotional lives as well as the physical form of individual human beings. Of all the Pre-Raphaelites, Tupper pursued the study of anatomy most rigorously. As well as Bell, he read the influential eighteenth-century anatomists Lavater and Albinus, the work of the Scottish surgeon John Lizars, and *The Dublin Dissector* by the Irish surgeon Robert Harrison – not to mention Quain, Arnott, Smee and others, as mentioned in Chapter 1.[29] But he stressed that it was the practical discipline of dissection that contributed most to his artistic development, teaching him to distinguish true from false representations of the human body in art.

Tupper's account of his growth as an artist centres on his encounter with the Elgin Marbles as a student. In this, he echoes, perhaps imitates, the elder artist Benjamin Robert Haydon's account of his own awakening to the power of naturalism in art through studying these same ancient sculptures. As Prettejohn explains in her recent study *The Modernity of Ancient Sculpture*, the recovery of the Elgin Marbles and their arrival in London led to a new perspective on ancient sculpture. The classical statues that had been held up as ideals in the eighteenth century were rapidly superseded as the Parthenon sculptures were discovered to be more precisely attentive to the true anatomy of the human form.[30] By his own account, according to a lecture he gave in 1840, Haydon was 'the first artist here to investigate the science of their imitation of life'. He records beginning his study of the marbles in 1808, after spending around two years dissecting humans and animals.[31] By the time Tupper did so himself, thirty or so years later, Haydon was sending his own students to look at the marbles in order to see and imitate their remarkable anatomical accuracy.

Haydon, who killed himself in 1846, was one of Hunt and Rossetti's Immortals. The language in which he writes about the Parthenon sculptures resembles closely the discourse of Pre-Raphaelite art theory in its allusions to science and its insistence on accuracy. Tupper pushes Haydon's account still further. Tupper casts himself in his reminiscences as 'anatomist in ordinary to Theseus' – a sculpture from the east pediment of the Parthenon now more commonly identified as Dionysos (fig. 35).[32] He 'made drawings and models of the Theseus, and dissected to verify its anatomy'.[33] Like Haydon, he found that his scientific understanding of the human body exposed the inadequacy of the ideals of art held up and taught by the Royal Academy, such as the 'turgid and inflated curve' of flesh found in Roman sculpture and its derivatives.[34] Where Haydon declared – relating his younger self's encounter with the marbles in the third person – that 'so completely alive was he, from conviction and reason, of their perfection – he would joyfully have died in their defence,' Tupper gleefully joked, 'I dissected, and would have vivisected, my countrymen, to refute the anatomy of Rome.'[35]

As Tupper relates it, the first stage of his training in anatomy was to teach him to reject false knowledge. But it also enabled him to grasp new knowledge in turn through art. Working on Greek sculpture, he and his fellow students could make their own 'discoveries in anatomy':[36] 'We were philosophers investigating *nature*. These were not factitious products, whereof the *amount is known*. We believed in the *infinite marble*, doubting not that muscles might reveal themselves on Thursday which defied our scrutiny on Tuesday. We made *discoveries* in the Theseus; laid our hands on the awful shoulders, between the great collar-bones, and felt for the beating of his heart.'[37] As Prettejohn shows, the sense that the Elgin Marbles were, in Haydon's words, 'instinct with life' rapidly became almost a critical commonplace.[38] Tupper pushes the logic of this familiar motif further, however, by literalising it. If Browning shows us the 'pulsations' of the 'heart of man' in his poems, so for Tupper the Theseus reveals its own 'beating . . . heart' to those acute and knowledgeable enough to apprehend it.

Tupper's self-portrait as an art student recalls directly William Rossetti's ideal of the artist as an investigator, and not just, as in Haydon's account, an investigator of the science of art itself. In the boldness of his investigations, he discovers for himself not only physical facts of science but metaphysical truths about the vitality of art and its ability to embody felt emotions – the beatings of the heart – that lie well beyond the realm of ordinary perception. Ford Madox Brown articulates the precise method by which Pre-Raphaelite painting investigates psychology through attention to anatomical detail in his essay 'On the Mechanism of a Historical Picture: The Design'. Where Tupper stresses the variability of human beings, the logic of Brown's argument complements this recognition of individuality with a reaffirmation of a

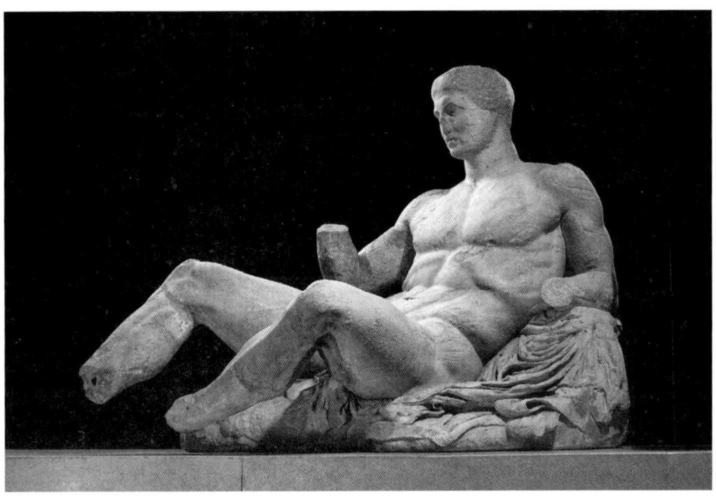

35 Marble statue from the east pediment of the Parthenon (identified as Dionysos), 438 BC–432 BC. British Museum, London.

shared humanity. In painting a historical picture, Brown argues, the artist should seek 'to enter into the character of each actor, studying them one after the other, limb for limb, hand for hand, finger for finger, noting each inflection of joint, or tension of sinew, searching for dramatic truth internally in himself, and in all external nature, shunning affectation and exaggeration, and striving after pathos, and purity of feeling, with patient endeavour and utter simplicity of heart.'[39] Brown's language here, with its emphasis on patience, commitment and objectivity – 'purity of feeling', 'utter simplicity of heart' – recalls the moral and scientific earnestness of Stephens's essays published alongside Brown's in *The Germ*. In order to grasp and so express the psychology of an individual, the artist has to undertake a meticulous study of human anatomy in action, right down to the inflections and tensions of joints and sinews.

To do this requires not only a thorough knowledge of anatomy, but also the reflexive study of the observing subject himself. As we have seen, a reflexive attention to observation is a recurrent theme in early Pre-Raphaelite poetry – in 'Antwerp to Ghent', for example, or 'Sheer Waste'. Like these poems, Brown's essay operates on the principle that we can empathise with and comprehend one another's experiences across the boundaries of our own individual personalities. Through imaginative introspection, the artist can test the expression of feeling in the models and painted subjects in front of him against how he himself would literally *feel* – against his sense of his own body – were he in their place. Forty years after Brown published his essay, Oscar Wilde, one of the Pre-Raphaelites' most enthusiastic admirers, wrote in the 'Preface' to his novel *The Picture of Dorian Gray*, 'From the point of view of form, the type of all the arts is the art of the musician. From the point of view of feeling, the actor's craft is the type.'[40] Brown's phrase 'dramatic truth' shows that for him too the artist must also be an actor, able to simulate within himself the full range of emotions and character types, so as to portray them accurately in the bodies of his subjects.

Where some early Pre-Raphaelite poems draw attention reflexively to acts of observation, others employ the same method to study how temperaments and states of mind are expressed through gesture and posture. Several passages from Dante Gabriel Rossetti's early poem 'Bride-Chamber Talk', later retitled 'The Bride's Prelude', read as verbal cartoons for the kind of historical picture envisaged by Brown in his essay. Rossetti worked on this long narrative poem over several months in late 1849 and early 1850 with an eye to publishing it in *The Germ*, but he did not finish it. After two or three later attempts at completing it, he finally published it unfinished in the second edition of his *Poems* in 1881.[41] The pretext of the poem is that an aristocratic medieval bride, Aloÿse, is confessing her sexual experience and shame to her younger sister Amelotte in the hour before her wedding. It transpires that the man with whom she shares this past is Urscelyn, her cousin and the very man she is to marry. Yet where this might seem a positive proposition – one that might clear both parties after the fact of any sexual wrong-doing, like the marriage at the end of 'Stratton Water', another of Rossetti's early medieval narratives – it is actually an appalling one. Aloÿse recalls that Urscelyn, known for his medical skill, cared for her after an injury. He declared 'a passionate fierce love' for her, and she realised that she loved him too.[42] She became pregnant, but before the baby could be born Urscelyn deserted her and her family in the face of impending war. Her brothers threatened to kill her for her shame, and though their father prevented them, the family stripped her baby from her nonetheless. When Urscelyn returns from the war a successful soldier of fortune, they find it expedient for her to marry him after all, even though she now loathes him and them with equal fervour.

The poem is a tissue of horrific ironies, successively unfolded by Rossetti to build up our sympathy for Aloÿse and disgust at her mistreatment. Rossetti fosters his readers' sympathy for Amelotte too, confronted with this grim revelation. Through detailed descriptions of her bearing, gestures and expression each time her sister pauses in her narrative, we see the extent to which Aloÿse's experiences have a traumatic effect on Amelotte as she learns of them. Early in the poem, as Aloÿse has revealed that she has kept a shameful secret from her, Amelotte offers up a prayer on her behalf before assuming a particular listening posture:

> She bowed her neck, and having said,
> Kept on her knees to hear:

> And then, because strained thought demands
> Quiet before it understands,
> Darkened her eyesight with her hands.
>
> So when at last her sister spoke,
>   She did not see the pain
> O' the mouth nor the ashamèd eyes,
> But marked the breath that came in sighs
> And the half-pausing for replies.[43]

When Aloÿse pauses again after confessing that she fell in love with Urscelyn, she looks across to her kneeling sister, who is at last so agitated that she cannot stay still in this position:

> Her fingers felt her temples beat;
>   Then came that brain-sickness
> Which thinks to scream, and murmureth;
> And pent between her hands, the breath
> Was damp against her face like death.
>
> Her arms both fell at once; but when
>   She gasped upon the light,
> Her sense returned. She would have pray'd
> To change whatever words still stay'd
> Behind, but felt there was no aid.
>
> So she rose up, and having gone
>   Within the window's arch
> Once more, she sat there, all intent
> On torturing doubts, and once more bent
> To hear, in mute bewilderment.[44]

The precision of these postures suggests pictorial composition, a staging of models as actors, to use Brown's term. Yet Rossetti also takes great care to explain why it is that Amelotte covers her eyes to listen, for example, and why in turn her hands drop abruptly, and how these actions at once express and affect her emotional state. Rossetti enters into sympathy with her through an act of indirect impersonation, thinking and feeling his way through her trauma and her needs as she experiences them. Strikingly, he engages the full range of senses in the process, not just studying her visually, but hearing what she hears and recording her physical sense of her own body. He even crosses the line into synaesthesia as she seeks mental 'quiet' through blocking out not sound but light, so that she is left seeing her sister's pain not in how she looks but in the fluctuations in her voice. Amelotte cannot sustain this self-imposed calm. It ruptures as she becomes aware that her own temples are beating through feeling them with her fingers, that she is suffering a 'brain-sickness' through her inability to scream, that her breath is trapped through its dampness on her face. Her arms drop involuntarily to allow her to breathe air but also light, restoring her mental balance – her 'sense' – but also her sense of the world beyond her own mental anguish. But because she cannot change the cause of that anguish, all she can do is physically change her own position, moving across the room only to have to sit once again and subject herself further to her sister's demands on her for sympathy.

The link between Amelotte's sensory experience of her mental pain and its expression in her movements is traced with such exactitude that actions that might seem on a first reading theatrical, even histrionic, are shown and crucially felt to be true to the intensity of her suffering. The process by which Rossetti achieves this enacts Brown's method for historical painting, combining the artist's external observations of bodily form and posture with reflection on his own dramatic impersonation of the character. For the Pre-Raphaelites, the physical body is a vector for the expression of emotion, not because physiology determines character, as it did for the phrenologists, nor because character forms physiology, as it did for the physiognomists, but because the mind is necessarily embodied. All emotion, all psychology, is experienced as bodily sensation – tensions in the sinews, the pulsations of the heart, an ache in the brain, even lassitude and the urge to fall back on the grass – and expressed in the minute details of our changing gestures and postures.

PROBING PSYCHOLOGY: PAINTING THE EMBODIED MIND

Between them, Brown's approach to painting, Tupper's study in sculpture and Stephens's account of Browning's poetry set out a programme for the arts to undertake the close empirical study of the embodied mind in the absence of a

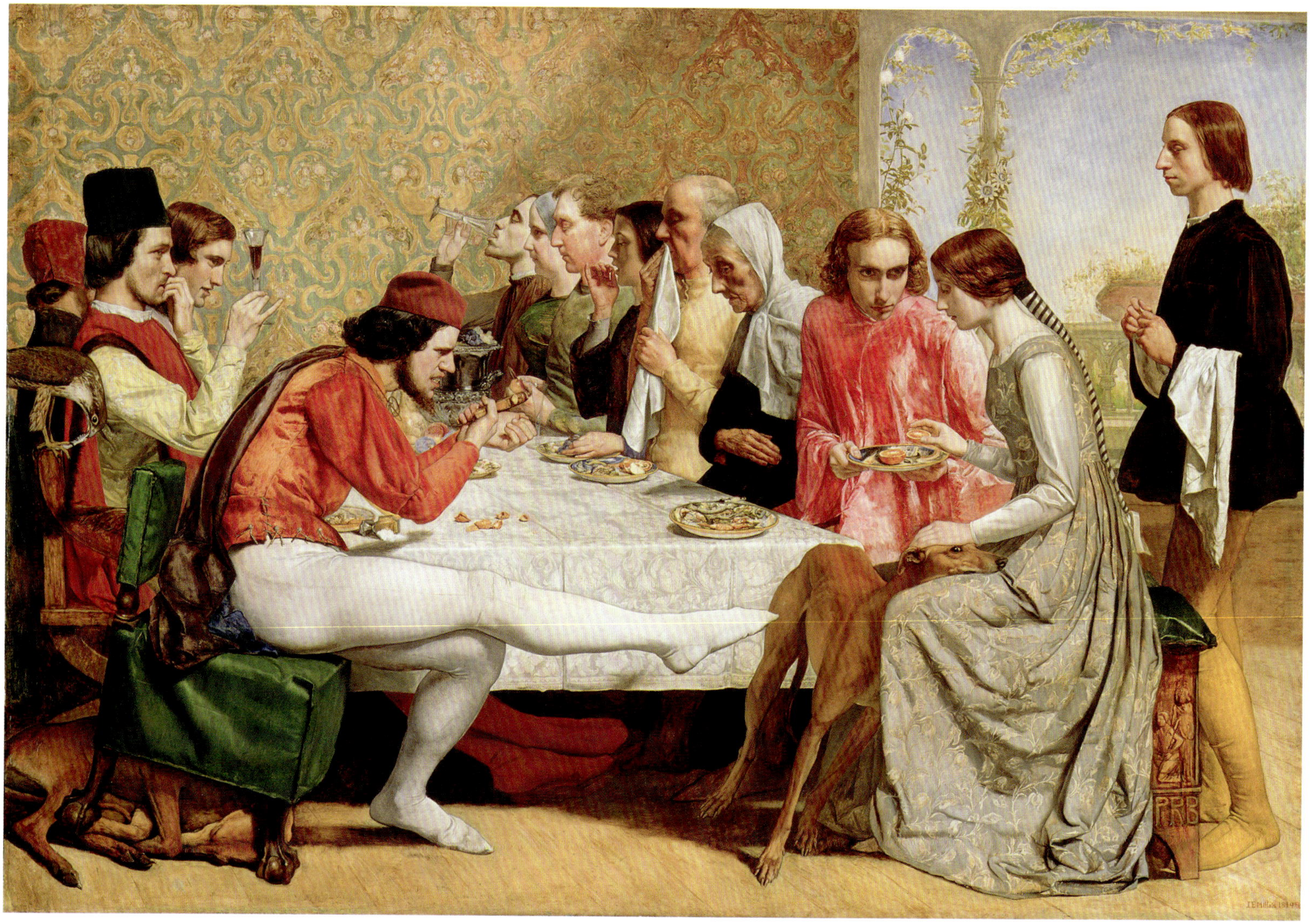

persuasive science of psychology. Fulfilling this programme, Pre-Raphaelite paintings, like Rossetti's poem, show a remarkable subtlety, care and complexity in their treatment of psychology as it is revealed through their subjects' bodies. These processes of psychological investigation can be traced most clearly by comparing successive imaginings of the same scene or character. In moving from a design to a finished work, or from one painting to its sequel, an artist can reinterpret its characters, probing their moods and motivations, and testing out with greater precision how these might be expressed in the barely conscious gestures of a given instant. In this section I want to examine how four Pre-Raphaelite painters worked through their studies of their characters' minds in four specific cases: Millais's *Isabella*, exhibited at the Royal Academy in 1849; Hunt's *Valentine Rescuing Sylvia from Proteus*, exhibited, again at the Royal Academy, in 1851; Rossetti's studies of the Virgin Mary across two paintings and several sonnets between 1848 and 1850; and Collinson's depiction of the same family before and after emigrating to Australia in *Answering the Emigrant's Letter* and *The Emigration Scheme*, exhibited respectively at the Royal Academy in 1850 and the National Institution and the Liverpool Academy in 1852. In each case the process of composition and recomposition allows the painter to probe the characters' psychology through subtle changes of posture and gesture.

Millais's *Isabella* (fig. 36) set the standard for luminescent colour, precise lines and meticulous detail among the early Pre-Raphaelite paintings. In working it up from a design (*facing page*) 36 John Everett Millais, *Isabella*, 1848–9. Walker Art Gallery, Liverpool.

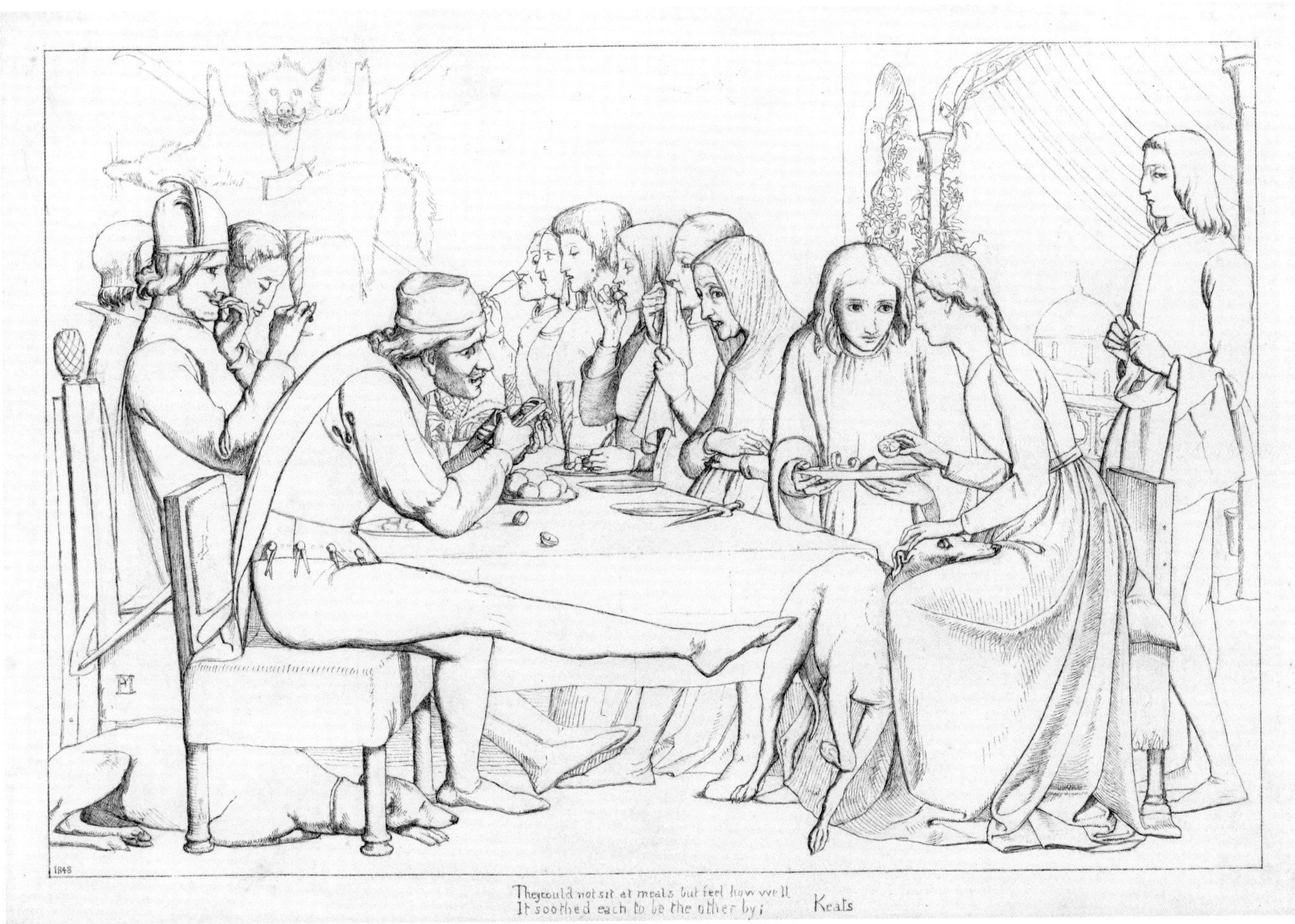

37 John Everett Millais, *Design for 'Isabella'*, *c.*1848. Fitzwilliam Museum, Cambridge.

to the finished painting, Millais demonstrates his commitment to exploring psychology through subtle details of gesture and expression. These details are so precise that viewing the painting becomes itself an ongoing process of investigation. A narrative that seems straightforward reveals itself to be progressively more complicated as we look more and more closely. On the face of it what we are witnessing is the anger of Isabella's brothers on the left sparked by Lorenzo's attentiveness to their sister on the right. Looking at the design for the painting in the Fitzwilliam Museum in Cambridge (fig. 37), that is indeed what we see. The brother in the foreground is looking round the lip of the table at Isabella's greyhound as he reaches out to jab at it with his foot. The second brother is looking across the table at Lorenzo and Isabella and biting his fingernails in suppressed rage. Isabella herself is inclining slightly towards Lorenzo while the greyhound itself is raising one forepaw, suggesting that it is flinching away from her brother's coming kick.

The finished picture bears a close enough resemblance to the drawing as to suggest at first the same interpretation of events, especially as it appears to correspond to Millais's source, Keats's poem 'Isabella, or The Pot of Basil', in which the brothers go on to murder Lorenzo. But Millais has made several refinements to his design. In the painting, neither brother is looking at the couple across from them. While the one behind stares abstractedly at the wine in his glass, the one in the foreground directs his gaze – and with it his anger – on the almond in the nutcracker he holds in his

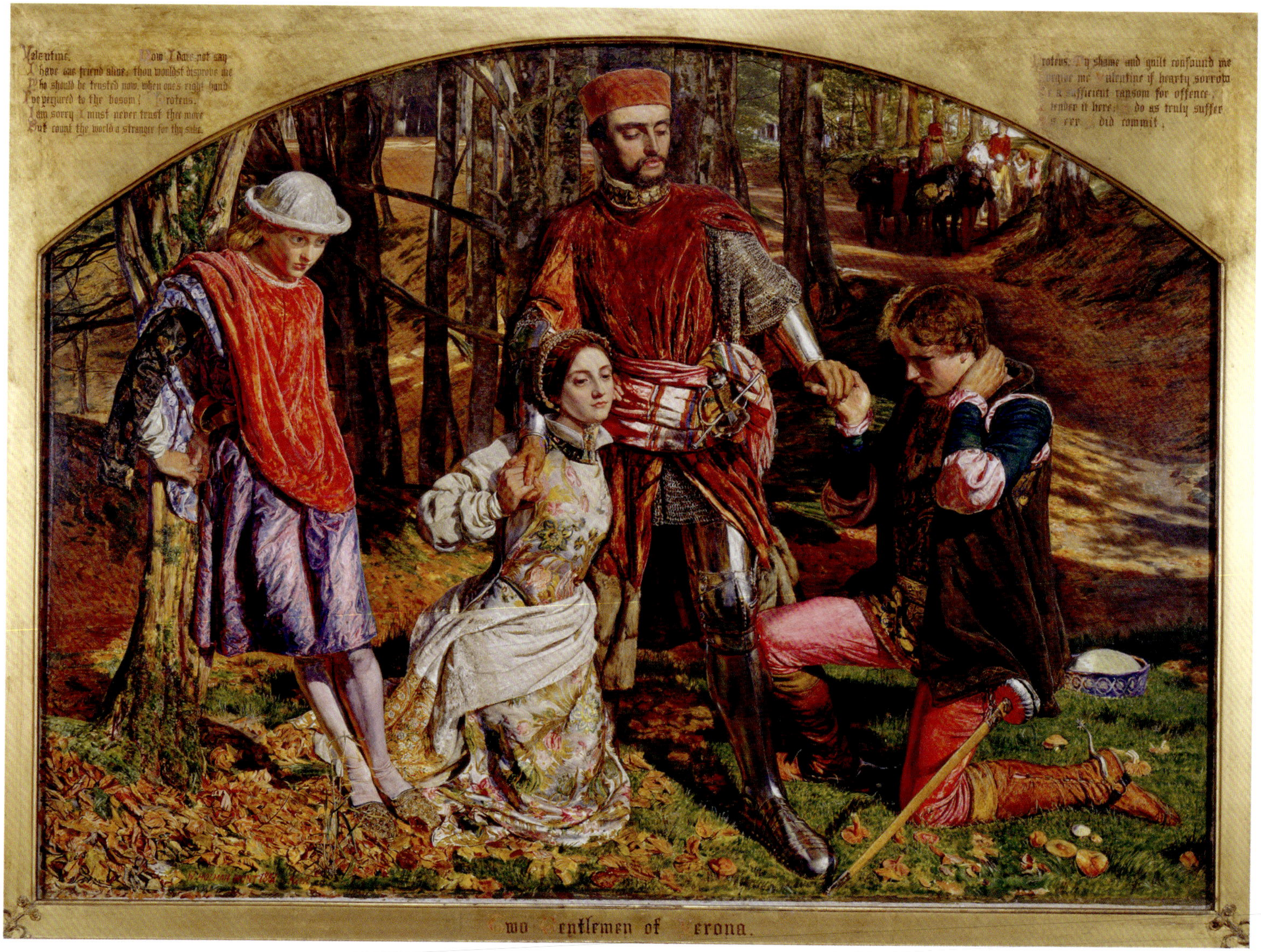

hand. Unlike the nuts on the table, this one is refusing to crack. He extends his leg not deliberately, as in the design, but kicking out in frustration in a sudden movement which rocks his chair. But he is not kicking the dog here at all, and the dog does not flinch. The only person who is looking at Lorenzo and Isabella is a third young man on the left, who seems more amused than shocked by their courtship. Perhaps he is smiling at the unselfconscious intensity of Lorenzo's gaze, or at Isabella's studied refusal to acknowledge it, as in the painting, unlike the design, she does not incline her body or her head towards him but remains chastely disengaged. Alternatively, perhaps he is just pleased with himself for having spotted what is going on between them when those around him appear not to have noticed. Either way, his smile seems benevolent, even indulgent.

In moving from the drawing to the painting, then, Millais probes each character's psychology more deeply. The first brother's hostility towards Lorenzo, which is displaced on to the dog in the design, is redirected to the nut in the painting, making that displacement unconscious, not deliberate, and showing how it betrays itself through sudden bursts of disproportionate anger. The second brother's stagey gesture of biting his fingernails while staring in abject hatred at Lorenzo is made more revealing as it becomes more muted, a displaced reminder that this hatred is still playing nervously on his mind even as he ostensibly focuses his (*facing page*) 38 William Holman Hunt, *Valentine Rescuing Sylvia from Proteus*, 1850–51. Birmingham Museums and Art Gallery.

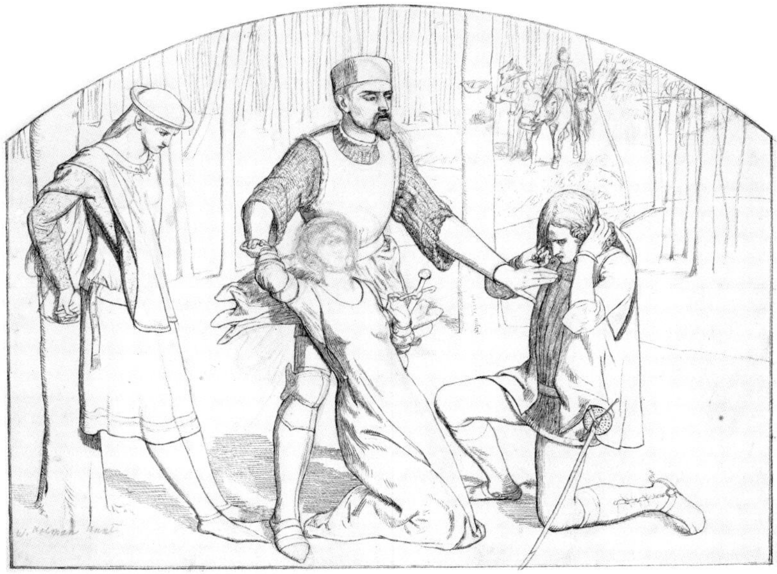

39 William Holman Hunt, *Study for 'Valentine Rescuing Sylvia from Proteus'*, 1850. British Museum, London.

attention on his wine. Isabella's lack of engagement with Lorenzo subtly suggests her consciousness of propriety and her calm suppression of her own feelings in public, particularly in her brothers' company. In each case a trivial, apparently inconsequential and largely unconscious action, expressed in posture and the direction of the gaze, reveals the subject's mood and underlying character. These gestures reveal how these men and women feel towards one another without ever taking the form of a direct interaction. In these small but precise changes from the original design, Millais registers the results of his sustained investigation into how personality and emotions express themselves through the body.

The anonymous young man's smile is revealing too, as a reminder of the possibility of different perspectives on the same event. Where the brothers see a violation of their status and ownership, and we, knowing Keats's poem, see a tragedy in the process of unfolding, he sees the beginning of a light and winsome romantic comedy. We stand outside both the time and the frame of the picture. Looking at it from this angle, we have scope to interpret all the main characters in the full knowledge of how events will unfold. By contrast, he sits within the painting itself, only noticing part of the interaction around him – Lorenzo and Isabella's love, not the brothers' hatred – and lacking any privileged insight into the future. He is a reminder of our partial vantage point on our own lives, and of the need to turn a comprehensive, steady, scientific gaze on others and ourselves if we are to gain the deeper understanding of emotions and reactions that Pre-Raphaelitism offers. At the same time, as a sympathetic and appealing figure in the painting itself, he reminds us of the need to respect and take account of other perspectives in building up our own, rather than discounting them as ill-informed or naive.

The acute psychological probing of inadvertent gestures and expressions that Millais sustains so effectively and so early in *Isabella* is characteristic of the paintings of the Pre-Raphaelite Brotherhood, setting them apart from the more crudely legible emotional register of much conventional mid-Victorian history and genre painting. Hunt's re-examination of characters from Shakespeare's comic play *The Two Gentlemen of Verona* in *Valentine Rescuing Sylvia from Proteus* (fig. 38) is another case in point. If the young man in Millais's painting gives us a fleeting glimpse of Lorenzo and Isabella's tragedy as a comedy, Hunt reimagines the Shakespeare comedy as, if not exactly a tragedy, then as a sorry episode in the characters' real lives. The precise scene he depicts comes near the very end of the play, when Valentine's friend Proteus has tried to force himself on Sylvia. Carole Silver notes an apparent disjuncture between the title of Hunt's painting and the scene itself, as 'it appears that, rather than rescuing Sylvia, Valentine is uniting the kneeling couple' were it not for his leg extended between them. In fact, as Silver remarks, he is doing both, at once protecting Sylvia from Proteus's attempted rape and offering her back to him in spite of his own supposed love for her.[45] The psychology of this scene in the painting is at once closer to Shakespeare's play and more complex than Hunt originally had in mind, as a study for the painting shows Sylvia recoiling from Proteus as Valentine interposes not a leg but an arm, clearly warning Proteus off (fig. 39). By the painting she is calm, while Valentine is holding not just her hand, reassuringly, but Proteus's as well.

In the play Valentine's offer to Proteus, 'All that was mine in Sylvia I give thee' – heralded by the rhyming line, 'And that my love may appear plain and free' – is preposterous, while the rapidity with which Proteus switches his desires back to Julia after his attempt to thrust himself upon Sylvia is farcical.[46] Hunt strips away this comedy to explore the character and emotions of people who might actually behave in this way. Valentine in Hunt's painting takes himself seriously, but unlike in the play we are not obviously being invited to laugh at him. His face is hard to read, partly because of his downcast eyes. In the sketch Valentine is glowering at Proteus, but in the painting he does not seem angry or even severe, in spite of Proteus's offence. Indeed, his calmness seems rather sanctimonious, tallying with his self-righteousness in the play. Proteus's own expression is easier to read. In the play he professes 'shame', 'guilt' and 'hearty sorrow', all of which are plainly legible, along with apprehension, on his face in the drawing.[47] But his body language in the painting suggests rather that his dominant emotion is embarrassment. He looks down and away, nervously rubbing his neck in another unconscious physical displacement of emotional unease. Julia has watched the man she loves attempt to rape another woman. She looks bewildered, even dazed. But the most intriguing expression is Sylvia's, whose face is not yet fleshed out in the outline drawing. She has regained her composure, but through abstraction. She does not look at Proteus, and seems barely conscious that she is leaning on Valentine. It is as if she is no longer mentally engaged with what is going on around her at all. In this extreme detachment – perhaps shock or a defence against it – Hunt provides a rationale for the fact that Sylvia does not speak at all in the play after her rescue, even though she is transacted back and forth between Valentine, Proteus, Thurio and her father, the duke.

Isabella and *Valentine Rescuing Sylvia from Proteus*, like *Mariana*, *Claudio and Isabella* and *The Awakening Conscience*, expose the violence that underlies male control of female sexuality. The most profound and disturbing Pre-Raphaelite treatment of this theme is Rossetti's *Ecce Ancilla Domini!*, the second of his oil paintings of the Virgin Mary, exhibited at the National Institution in London in 1850. Stephens singled out Rossetti's two early oil paintings for the subtlety and delicacy of their handling of 'psychological problems'.[48] Rossetti's investigation into Mary's inner life began in 1848 with his first oil painting, *The Girlhood of Mary Virgin* (fig. 40), and two sonnets written to accompany it. In one of these sonnets Rossetti imagines her response to the Annunciation, a narrative moment which crucially does not occur in the painting itself:

> Till one dawn, at home,
> She woke in her white bed, and had no fear
> At all, – yet wept till sunshine, and felt awed;
> Because the fulness of the time was come.[49]

In pairing the poem with the painting, Rossetti anticipates a sudden and radical shift in Mary's sense of herself that is not even glimpsed in the circumscribed experience of her childhood as it appears before us. In moving from painting to poem, Mary herself changes from placid to awestruck.

By referring to the 'fulness of time', Rossetti conflates the Annunciation with Jesus's birth, as Mary realises both the full pregnancy of the moment and her own literal pregnancy. In this one phrase, Rossetti hints at an indissoluble bond between Mary's motherhood and her apprehension of the significance of her son's life. Rossetti explores this bond in detail in another sonnet published in the fourth and last issue of *The Germ*:

> Mystery: God, Man's Life, born into man
> Of woman. There abideth on her brow
> The ended pang of knowledge, the which now
> Is calm assured. Since first her task began,
> She hath known all. What more of anguish than
> Endurance oft hath lived through, the whole space
> Through night till night, passed weak upon her face
> While like a heavy flood the darkness ran?
> All hath been told her touching her dear Son,
> And all shall be accomplished. Where he sits
> Even now, a babe, he holds the symbol fruit
> Perfect and chosen. Until God permits,
> His soul's elect still have the absolute
> Harsh nether darkness, and make painful moan.[50]

In *The Germ* this sonnet is entitled 'A Virgin and Child, by Hans Memmeling; in the Academy of Bruges'. In the letter Rossetti wrote to his brother to tell him about the paintings he had seen in Bruges, he mentions 'some most stupendous works of Memling – among them one of a Virgin & child, quite astounding', now generally regarded to be an outer wing (fig. 41) of an altarpiece by Memling's younger

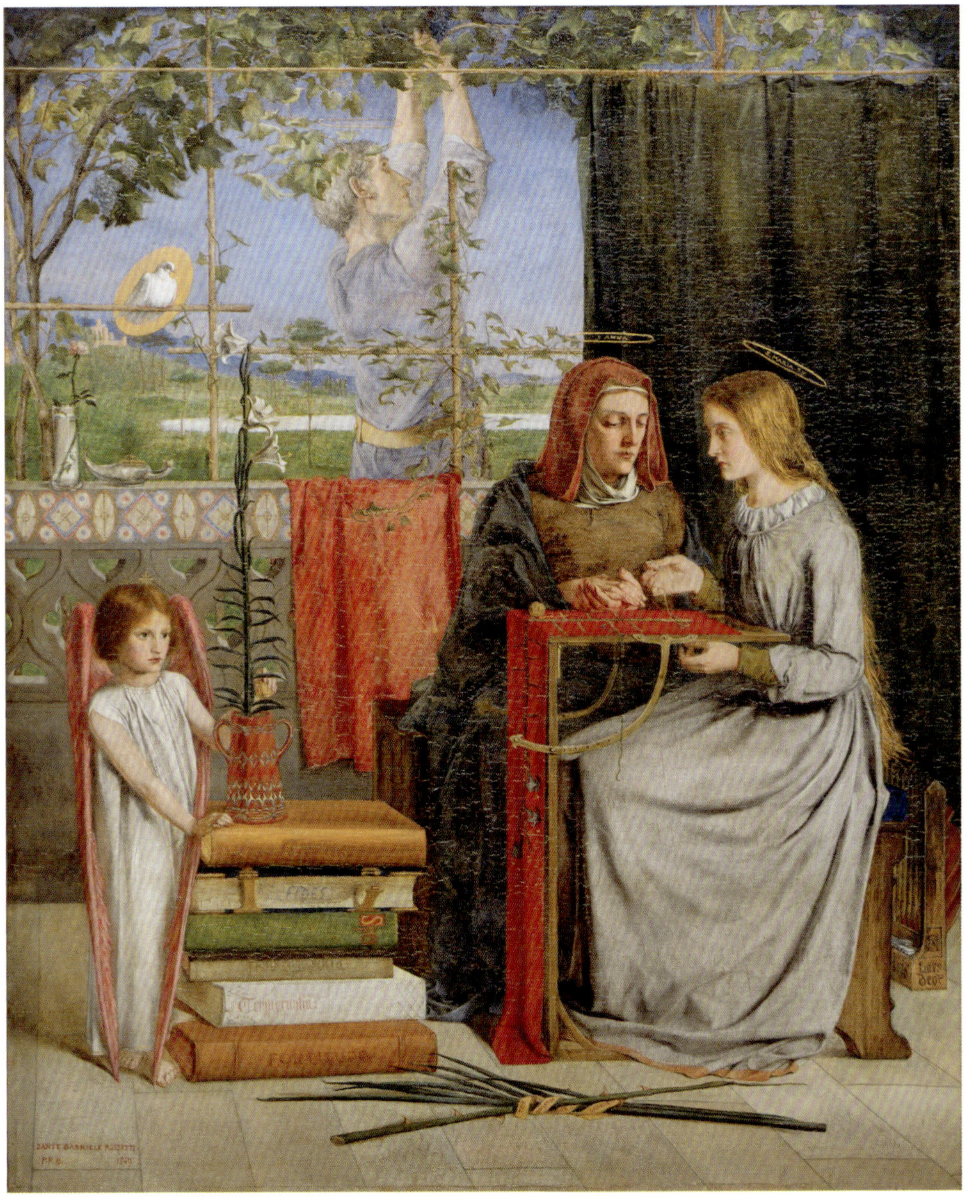

40 Dante Gabriel Rossetti, *The Girlhood of Mary Virgin*, 1848–9. Tate Gallery, London.

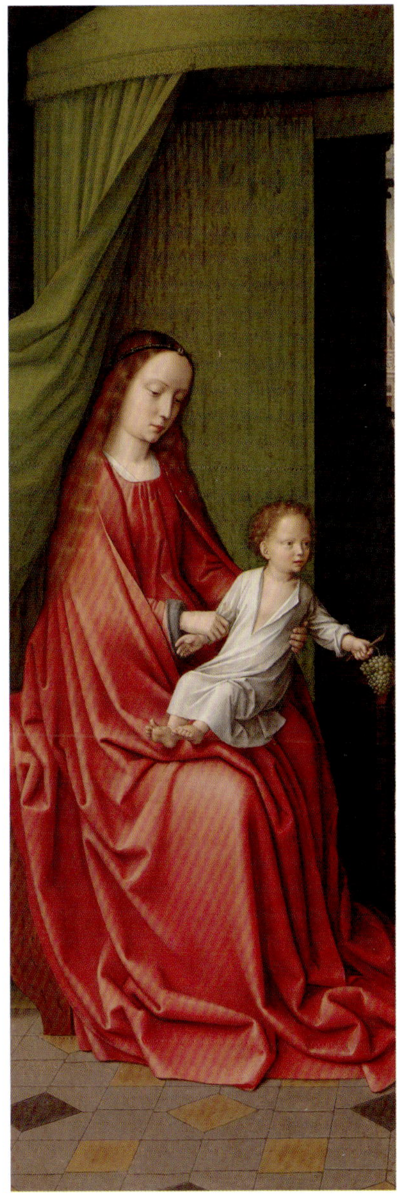

41 Gerard David, reverse of the left wing of *Triptych of Jan Des Trompes*, 1506. Groeninge Museum, Bruges.

contemporary Gerard David which was attributed to Memling at the time.[51]

Rossetti did not tell his brother in his letter why he found this painting to be so 'astounding', but the sonnet suggests that it gave him an insight into Mary's own perception of her role in the incarnation. As in his sonnet on his own painting, Rossetti draws an intense narrative out from a picture which on a first impression is calm and static. As he interprets it, the scene represents a moment very soon after Jesus's birth. After hinting at the pain of childbirth in the phrase the 'ended pang of knowledge', Rossetti makes a striking attempt to empathise with this himself in the light of his own recent experiences. In 'Antwerp to Ghent', written only days before he saw the painting, he observes how per-

42 Dante Gabriel Rossetti, *Ecce Ancilla Domini!*, 1849–50. Tate Gallery, London.

ceptions of time and motion can be distorted by different means of travelling.[52] Here he imagines that childbirth leads to a similar distortion of perception. Mary is weakened after a long labour lasting 'the whole space / Through night till night'. This may be no more than 'Endurance oft hath lived through', as Mary's experience in bearing a child is in one sense the same as that of other women throughout time. Yet that double stress on the 'whole space' draws out every mother's need for endurance in childbirth, as time becomes reimagined as a physical space that needs to be crossed. There is no break either at the end of the line, while the break in the middle of the next line only marks Mary's growing weakness, before she has to endure another enjambment pushing on till the end of the octet. Towards the moment of the birth – the heavy flood hints at her waters breaking – time seems to speed up, and the 'darkness' runs on, as it does tumultuously in 'Antwerp to Ghent'. Yet even at the moment of the birth itself, Mary remains in darkness, as the souls of God's elect are until Christ redeems them at the end of the sextet.

Mary's darkness is a condition of physical and mental suffering, but it is not a state of ignorance. Since the Annunciation, Rossetti asserts, 'She hath known all.' This 'all' is reiterated at the beginning of the sextet, projecting both backward into the past – it is the 'all' she has already been told – and forwards into the future, as 'all' that she knows 'shall be accomplished'. Interpreting her expression in the painting, Rossetti imagines that, now that Jesus has been born, Mary bears her knowledge of God's incarnation within him calmly and without pain. This is in marked contrast with her realisations of the truth and the grief that accompanies them in Collinson's poem 'The Child Jesus'. There is a similarly marked contrast with Rossetti's earlier sonnet, written to accompany his own painting. The two sonnets share the same emphasis on fulfilment, but in the first that fulfilment comes with the Annunciation, in the second with the birth. Both bring relief, but the responses to that relief are different, as in the earlier sonnet Mary breaks down in tears where in the latter she is placid and content. This calm is a newfound state, replacing the 'anguish' that her knowledge brought her up to that point, perhaps intensified, as the word 'pang' hints, in the days and hours leading up to Jesus's birth.

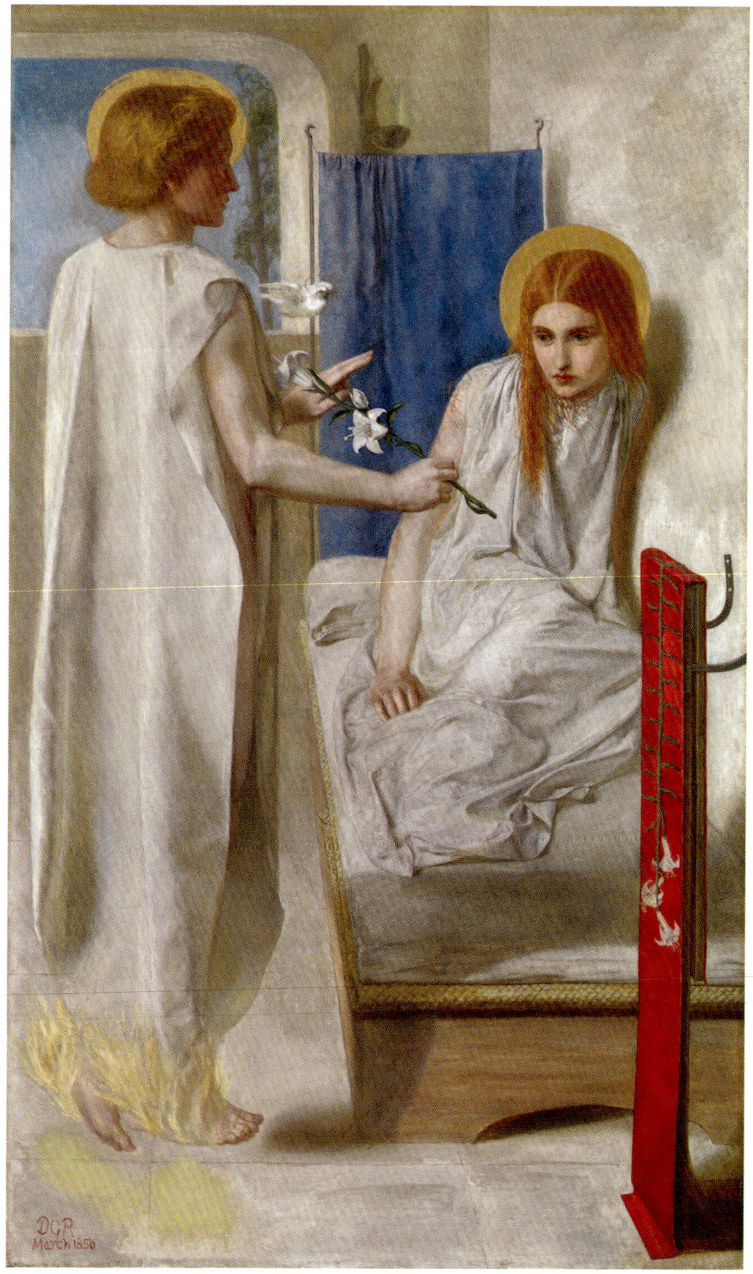

Looking at the painting, Rossetti finds traces of this pain still inscribed on her 'brow', in the slight tension in her temples or eyebrows, perhaps, or her dark, downcast eyes.

Rossetti saw the Bruges *Virgin and Child* on 25 October 1849.[53] Exactly a month later he began work on his sequel

43 Dante Gabriel Rossetti, *Study for 'Ecce Ancilla Domini!'*, 1849. Birmingham Museums and Art Gallery.

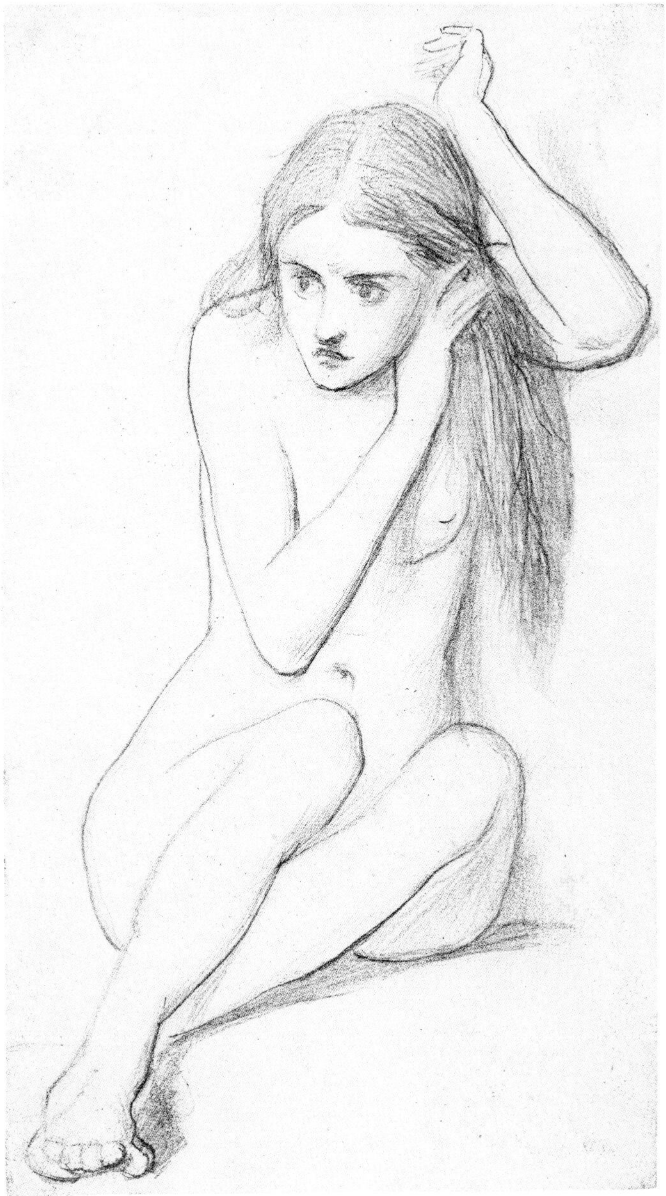

to *The Girlhood of Mary Virgin*, his famous Annunciation *Ecce Ancilla Domini!* ('Behold the Handmaiden of the Lord!'), which was exhibited at the National Institution in London in 1850 (fig. 42).[54] This is, as Elizabeth Helsinger has remarked, 'a thoroughly psychologized and sensuously realized presentation of the annunciation'.[55] Once again, Rossetti asks the question how Mary may have responded to learning the part God had determined for her to play in the incarnation. The staging for the scene is as projected in Rossetti's first sonnet for his earlier picture. It is morning, and Mary is 'at home . . . in her white bed'. But we do not see her weeping, and her reaction does not suggest either relief or awe. Instead, she is in shock at acquiring in one instant 'all' the knowledge that Rossetti attributes to her in his sonnet on the Bruges painting. The suddenness of this realisation leaves her dazed, cast into an interior world her mind can barely cope with. Correlating the painting with the sonnet, it will only be with the birth that she will be able to attain the calm Rossetti found in the Bruges painting.

While the language in the Bruges sonnet is of pain, in the painting Mary's experience is of fear. In a nude study, the model is dramatically posed to express this fear (fig. 43). Her knees and right arm are drawn up in front of her in a gesture of self-protection, while her left arm is raised back above her head as she recoils in horror. Her bodily movements are largely inadvertent. Her right arm crosses her body to move her hair away from her face, for example, not to protect herself. But her facial expression, with her eyebrows low over her eyes in a frown and her lips set tight, suggests close attention to the threat in front of her, and she may yet hit out defensively against it. As in *Isabella* and *Valentine Rescuing Sylvia from Proteus*, Rossetti's psychological study becomes subtler and more profound in reworking the sketch for the painting. Here Mary's fear is captured in a less posed, more instinctively apprehensive movement as she edges back up the bed. In the sketch Mary retains a certain strength. She is clearly capable of independent action, of a struggle against the aggressor. But in the painting her fear is more disturbing because it is accompanied by incapacity, even abjection. She wants to escape. Like Christ asking God to relieve him of his duty of sacrifice in the Garden of Gethsemane, she wants the knowledge she has gained to become unknown again, and untrue. Yet she has nowhere to go. Where Mary's body in the painting is subtly but acutely expressive, her face is tellingly inexpressive of conscious thought. Her blank expression confirms her state of mind. In the study, the tension in her face shows that she is alert, watching closely. In the painting, her wide yet sunken eyes seem only half-focussed, as if what she sees, and then only darkly, is an interior image rather than anything directly in front of her. It is something in her own imagination she is unconsciously recoiling from, something resulting from the angel's visitation. She may even be partly asleep still, or catatonic, unable to distinguish the reality of her vision from the paralysing fear of a dream.

Over the course of sustained meditations across his paintings and sonnets, Rossetti seeks to grasp the psychological reality of the experience undergone by Mary according to Christian tradition and told only very baldly in the first chapter of Luke's Gospel.[56] The conclusion he reaches in *Ecce Ancilla Domini!* suggests a shocking moral insight. The male control of female sexuality exposed in Millais's and Hunt's paintings is shown by Rossetti to lie at the heart of the story of the incarnation itself. As Alison Smith notes, Mary is 'drawing up her legs as if recoiling from a sexual advance'.[57] The angel does not make any direct advance on her himself, but he is nonetheless clearly naked beneath his robes. Even as we can see and appreciate the smooth whiteness of his skin, so the muscles and veins in his arm affirm his own bodily presence alongside Mary and dominating over her. He is pointing at her womb with the lily stem and holding up his left hand in a gesture which, from his erect posture and unemotional expression, suggests an assertion of authority rather than reassurance. Smith is surely right to point out that the colours of the painting too add to this 'psychological charge', as its pervasive whiteness creates a 'vertiginous feel' which draws us to empathise with Mary, even as the ascetic lack of furnishings or decoration keeps us focussed on her response.[58] In *Ecce Ancilla Domini!* Rossetti combines vivid anatomical detail with psychological acuity to create something that is entirely new in the history of paintings of the Annunciation, and that he himself only discovers through the act of painting itself. Viewed not as an affirmation but as a critique, *Ecce Ancilla Domini!* demonstrates how the sexual violence towards and domination over women that are overt in classical mythology underlie no less the central myth of Christianity.

To move from paintings as famous and commanding as *Isabella* or *Ecce Ancilla Domini!* to Collinson's muted domestic scenes may seem like a falling off, but what Collinson achieves at his best is to bring the new psychological acuteness of Pre-Raphaelitism to bear on what may seem to be mundane subjects of contemporary life. In *Answering the Emigrant's Letter* (fig. 44), Collinson addresses what was then a highly topical issue – the promises and risks of emigration, increasingly seen at once as an opportunity for poorer families such as the one depicted in the painting to enrich themselves and as a way of extending around the globe the values and bonds of a Greater Britain, as J. R. Seeley would call it towards the end of the century.[59] Emigration was on the Pre-Raphaelites' minds at least as early as May 1849, when William Rossetti began keeping the *P.R.B. Journal*.[60] A few years later Woolner took part in the movement himself, leaving England for Australia in 1852 and returning in 1854, his hope of growing rich on gold mining disappointed. Bernard Smith, another sculptor associated with the P.R.B. in its early days, made the same journey and stayed on as a colonist.[61] Inspired by their departure, Brown painted one of his most famous works, *The Last of England*.

Answering the Emigrant's Letter is the first of two paintings in which Collinson examined for himself the mental and emotional strains of emigration. One curious detail of the scene it depicts is that it is the eldest child, not the father, who answers the letter the family has received from Australia. It is not impossible that the son writes because the father cannot. New funding for elementary schools had been introduced by Parliament in 1833, so perhaps the son has benefited from an education his father was denied. On the other hand, marriage registers reveal that already two-thirds of married men were able to sign their own names by 1840.[62] It is not clear what this family does to earn a living, but the man's waistcoat is well cut from a fetching red cloth, and his shirt clean and crisp, for all that his breeches may be somewhat old-fashioned. He holds a letter in his hand and the map on his knee, suggesting that he can read, and while many more working-class men and women could read than write, the ease and poise with which he holds a quill pen suggests that he could write the letter himself if he chose to.[63]

Perhaps the boy, reluctant as he is, has been asked to write part or all of the reply merely for practice. Compositionally, though, it frees the father's gaze from the act of writing to allow him to share a thought with his wife. In both *Isabella* and *Valentine Rescuing Sylvia from Proteus*, the interaction

44 James Collinson, *Answering the Emigrant's Letter*, 1850. Manchester Art Gallery.

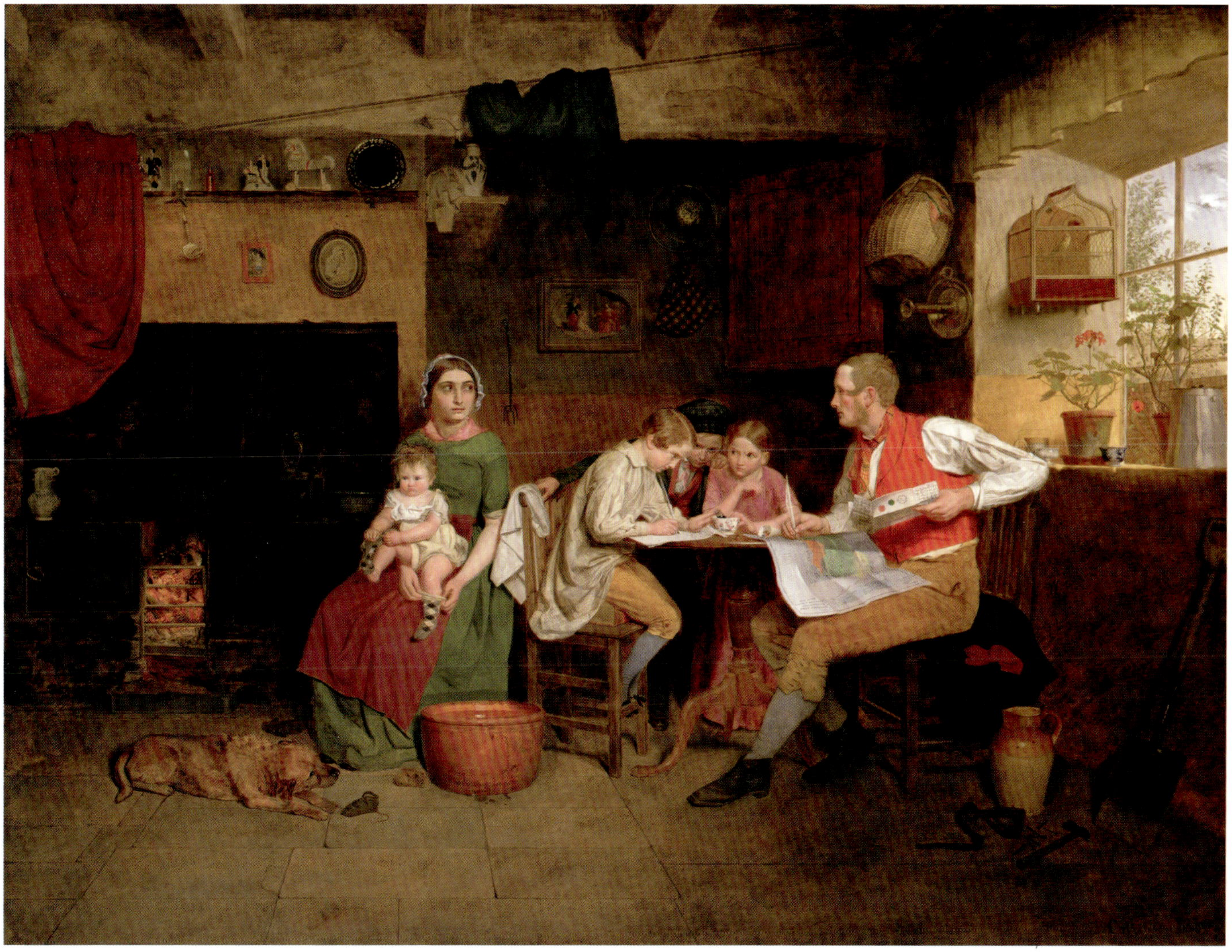

between the characters is revealed in how they avoid engaging with one another. In *Answering the Emigrant's Letter*, lines of sight reveal a more direct but equally subtle engagement. No one else in the room, except possibly the dog, has noticed that the man and the woman are exchanging glances. From the map on his knee, the father seems to be entertaining the possibility that they too should emigrate, though whether this is a decision that has been taken, to be written down by the boy scribing the letter, or whether it is an idea that remains unspoken is not clear. The boy's own discomfort supports each conclusion equally well. He may be unhappy at having to write something he wishes were not true: that they are going to leave home. Alternatively, he could be embarrassed at having to write at all, under the gaze of his parents and siblings, while the very fact of his writing could be read to suggest a partial censorship of the message: his parents may have asked him to write precisely because they are not expressing themselves in full in the letter.

Either way, the wife's sidelong look across at her husband registers the same possibility of emigration, but she seems more apprehensive. Beyond this, their faces are relaxed, making it hard for them or us to be certain where the other's thoughts

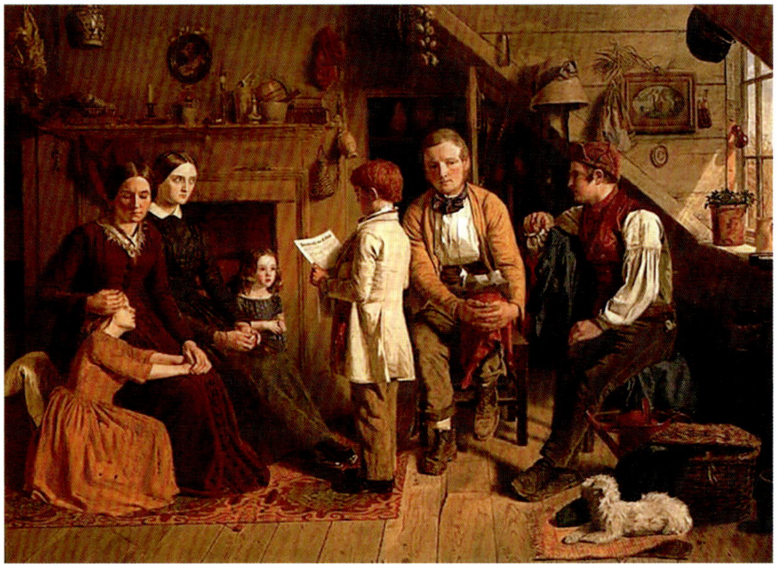

45 James Collinson, *The Emigration Scheme*, 1852. Private collection.

are tending. Perhaps the man is set on going, his impassive countenance reflecting his determination. Perhaps they are both waiting for a chance to discuss the matter away from the children. Perhaps it is only in this moment that they both recognise that they too might prosper through emigration. Collinson does not resolve these different possibilities into a single interpretation. Instead, he keeps them suspended as equally plausible explanations for what we see in his picture. But whatever the precise explanation for their shared glance, he conveys the seriousness with which they themselves are considering this prospect and with which we too need to examine it, if not for ourselves, then as a social issue. He does this not through melodramatic gestures or intense emotions, but through the same minute attention to unconscious details of expression – to careful self-control and the psychological indicators which nevertheless elude it – as Millais and Hunt do in their paintings.

Collinson also pioneers what will become a familiar Pre-Raphaelite promise of freedom in the glimpse we see of trees outside the cottage window. As William Rossetti noted in the *P.R.B. Journal*, he painted 'the trees outside to a pitch of the extremest minuteness'.[64] It is unclear from this brief reference whether Collinson painted in the open air, or whether Rossetti is merely specifying where the trees are in the painting. Either way, these trees play the same symbolic role within this painting as living nature glimpsed through a window does in Millais's *Mariana*, Hunt's *Claudio and Isabella* and *The Awakening Conscience*, and Hughes's *April Love*. In each of these paintings, a claustrophobic, constricting interior is juxtaposed with the promise of life and growth symbolised in trees and other plants seen in clear, full light. Of all these paintings, Collinson's is the earliest, and the symbolism here lacks the specifically sexual resonance that it would go on to acquire. At the same time, Collinson's use of this same motif provides a template for these later paintings and a further affirmation of the ecological leanings of early Pre-Raphaelitism, as well as an emblem of the possibilities for a better, less stunted, life overseas.

Answering the Emigrant's Letter is an open-ended consideration of the aspirations and anxieties of emigration. In his second painting, Collinson resolved many of these narrative uncertainties. As a result, *The Emigration Scheme* (fig. 45) is in some ways more conventional. Much of this conventionality follows from the story Collinson chose to tell across the two paintings. The family have joined their relatives in Australia, but they have not prospered. The elder daughter is ill, her dress tattered and torn. The elder son, who reads the letter in the original painting, is absent. As the mother is dressed in black and the father looks as though he is now habitually downcast, we may assume he has died. The look of apprehension which the wife gives her husband in the earlier painting is recalled in a look of resentment bordering on stark hatred in the sequel. Both looks reach across the canvas from the left to the right, but the husband no longer meets her gaze. Instead, he looks down at the floor – his gaze is literally downcast – with a hint in his face of tears repressed so long they have dried up. The original painting is recalled in several other ways too. The stone cottage in England has been exchanged for a wood cabin in Australia but, aside from a couple of rugs on the floor, no better furnished. Where the green leaves of the trees outside the window at home promised life and freedom, the foliage that is visible outside the equivalent window in their new house is yellow and dry.

Between painting *Answering the Emigrant's Letter* and *The Emigration Scheme* Collinson left the Pre-Raphaelite Brotherhood. It would be possible to see the more melodramatic and sentimental elements of the later painting as a reversion to the more obvious characterisation of the tradition of genre painting from which his work derives.[65] In fact, Collinson sustains his Pre-Raphaelite attention to the unconscious expression of emotion through physical gestures and postures into this second painting. The wife's resentment of her husband manifests itself not only in her look but also in the way in which she is disconsolately fiddling with her wedding ring. For his part he clasps his knee in both hands, pulling it towards him in self-comforting embrace. Though the situation and emotions of the family themselves are plain to see, in part through these unconscious gestures, Collinson has preserved an element of ambiguity too in the faces of the couple they have gone out to join. The woman comforts the sick, perhaps dying girl, while the man looks diametrically across the picture as the husband did in the original. As if to highlight their equivalence, he is even dressed as the husband was in the earlier painting in a white shirt and red waistcoat. Neither the man nor the woman betrays intense emotion, yet each seems profoundly and sympathetically aware of the misery that what we presume was their example, perhaps even their invitation, has brought upon their friends or relations. Where in the original painting Collinson asks us to imagine the thoughts and feelings of the family at the prospect of emigration, here he invites us to consider how the couple who went out before them feel in turn about its consequences. In both paintings, just as in Millais's, Hunt's and Rossetti's better-known works, unconscious and conscious gestures and expressed and concealed emotions combine to create a rich and complex portrayal of individual and group psychology as it is inscribed on the faces and bodies of its subjects.

THE ANATOMY OF DESIRE: PRE-RAPHAELITE NARRATIVE POETRY

Rossetti's Mary, Hunt's Sylvia and the kept woman in *The Awakening Conscience* all reveal in different ways the trauma of sexual violence. Through narrative poetry, the Pre-Raphaelites could explore more thoroughly the bodily experience of desire and the violence, emotional and physical, that it brings with it. We are drawn in by these paintings, invited, even compelled, to imagine these women's physical experiences and mental conditions. But we know still more precisely how it felt to Aloÿse to awake to her own sexual desire because she tells us or, more precisely, we overhear her as she tells her sister. Before sleeping with Urscelyn, Aloÿse becomes aware of how desire changes how her own body feels. 'Strange quick heat', she tells Amelotte, 'Made all the blood of my life meet.' Her newfound sexuality makes her conscious not only of previously unfelt sensations but of her body itself. While she feels a rush of heat and life within her, the surface of her skin too develops a new and sharper sensitivity to sensation. As he kissed her, she recalls, 'His hot lips stung my hand.'[66]

In 'Modern Giants' Stephens suggests that, in examining the emotional life of the body, poetry can comprehend the human mind both more fully and more scientifically than pseudo-scientific psychology. This principle is put into practice in two major Pre-Raphaelite poems composed in 1849, though neither was published until twenty years later. Each is a case study in insanity brought about by sexual desire. In Dante Gabriel Rossetti's 'A Last Confession', a young man is driven by suppressed desire to murder a girl he has fostered since childhood. In William Rossetti's 'Mrs. Holmes Grey', a doctor's wife, dies of the physical and emotional strain of her unrequited yearning for another man.[67] Both poems are set in 1848. The uprising against Austrian rule in Lombardy is the context for 'A Last Confession', the barricades in Paris a point of reference in 'Mrs. Holmes Grey'. Both poems are written in blank verse too. Their genres are noticeably different, however. Where 'A Last Confession' is a dramatic monologue spoken by the killer as a confession to a priest, 'Mrs. Holmes Grey' is a detached narrative which examines the minds of its subjects not from within but from without. In line with Brown's essay, Gabriel searches for the dramatic truth of his scenario within himself and thereby within his persona. William by contrast explores what we can and cannot know of other people's motivation from their external nature.

For all the difference between their approaches, both Rossetti brothers take a lead from Browning, Stephens's model poet. Gabriel uses the dramatic monologue to examine the mind of a sexually possessive murderer as Browning had

of control at his fear that she too may have become what to his distorted vision is a sexual obscenity. Yet the gift he means to give her is the very knife he stabs her with, introduced in the opening lines of the poem:

> Our Lombard country-girls along the coast
> Wear daggers in their garters: for they know
> That they might hate another girl to death
> Or meet a German lover. Such a knife
> I bought her, with a hilt of horn and pearl.[75]

In looking to give her this knife, he is already making assumptions about her sexual activity and linking sex to murder. The motivations he attributes to her – sexual jealousy and resistance to Austrian oppression – are his own. It is he, not she, who hates a girl to death, and not ultimately on account of any change in her – as he admits at the end of the poem, he does not know that she has yet changed in the way he fears at all – but because of changes in him.

'A Last Confession' is an extraordinarily subtle and minute study of one man's susceptibility to perverse desire and his failure to suppress it. It exposes too the possibility for the wilful misinterpretation of one's own bodily impulses and of the emotions and desires inscribed on other people's bodies. These same subjects recur in 'Mrs. Holmes Grey'. This poem contributes to the Pre-Raphaelite project of exploring how far art can make use of the model of science in two seemingly conflicting ways. On the one hand, it is a studied experiment in scrupulous realism which asserts its own scientific objectivity as the poet seeks to eliminate himself from his narration as entirely as a scientist would in writing up his own experiments. On the other hand, the story it tells calls the objectivity of science itself – in this case, medical psychology – into question. In 'Sheer Waste' Rossetti takes science as the closest approximation to objectivity that can be reached. In 'Mrs. Holmes Grey' he suggests that, by combining the investigative gaze of science with the humane and open-minded sympathy of poetry, the poet can help his readers to reach a more comprehensive understanding that cannot be achieved through the reductive procedures of science on their own.

'Mrs. Holmes Grey' follows Stephens's proposal that the moral purposes of the arts can be advanced by harnessing the methods of science. In the process, it exposes the failure of science to comprehend questions of morality. Rossetti's central concern in this poem is with how knowledge is established and what authority it has. More precisely, he addresses the key problem of psychology as a science: how sure can we be of our knowledge of the mental conditions of others? The poem opens at sunset at the end of a rainy day in October 1848 in an English seaside resort. John Harling, a Tory Englishman, has returned home after several years abroad to escape the revolutionary struggles in Paris. Catching a glimpse of an old friend, the eminent surgeon Holmes Grey, at a window, he calls on him, only to find Grey nervous and morose. Reluctantly, Grey invites Harling in. He shows him into a bare room where the corpse of his wife, Mary, is lying in state ahead of her funeral the next day. Grey explains that they had Edward Luton, a medical colleague, to visit that summer. During his visit, Mary began to show signs of distress, which she claimed were the result of dwelling on the death four months previously of their infant daughter. Luton curtailed his visit, and soon after she too left, supposedly to spend some time recuperating with her sisters. Three weeks later, 'A letter on black-bordered paper came' from Luton telling Grey 'That I was "to prepare myself" – some trash,' as Grey himself remarks, dismissing Luton's euphemisms.[76] Grey arrived to find his wife dead and himself called to the coroner's inquest as a witness.

At this point, rather than tell any more of the story himself, Grey hands Harling the newspaper report of the inquest. At over 300 blank-verse lines of studiedly commonplace courtroom journalism, this is the tour de force of Rossetti's realist experiment. It transpires from this report that Mary had known Luton before her marriage, unbeknownst to her husband, and that the romantic attachment she had felt for him then had been reawakened by his visit that summer. Unable to suppress her feelings, she left her husband in an attempt to persuade Luton to take her up as his mistress. According to Luton's testimony, he resisted her increasingly desperate pleas until one evening she forced her way into his house. He commanded her to go, threatening to write to her husband, but before he could leave the room she had a fit and fell dead. The jury at the inquest accept Luton's account, as does Harling, but Grey cannot stifle the hatred he now feels towards him, blaming him for his wife's desertion and death, and planning to use his professional influence to cripple his career.

The poem ends on a rare and highly effective symbolic note:

> He shut the sash. Harling had seen the night,
> Equal, unknown, and desolate of stars.[77]

The desolation here is to be expected. That the night seems 'equal' implies that it looks blank. To Harling, at least, it has no features which can be distinguished from one another, any more than the misery and moral predicaments of the principal characters in Grey's narrative can be teased apart or ranked. Most significantly, it is 'unknown'. Bald and realistic as it seems, Rossetti's poem is an acute exploration of the difficulty of gaining the true, objective knowledge which Baconian science set as a standard and which realism itself appears to embody. 'Mrs. Holmes Grey' focusses attention on three of the main means by which society at large seeks to establish the truth: news reporting, the courts and science. Yet in spite of their collective authority, and of the broad agreement among them, Harling ends the poem not convinced of having any knowledge but with a profound, even existential, sense of ignorance.

Rossetti's interrogation of science is one of the subtlest aspects of his poem. According to the newspaper report, Grey's 'scientific eminence / Is fully and most widely recognized.'[78] Yet when a juror asks him, after he has testified to the 'symptoms' of Mary's distress, 'Could your scientific skill / Assign some cause for this debility?' he replies, 'No. I believed it was occasioned (so / She intimated) by a domestic grief / Quite unconnected with the present case.'[79] Luton too is a surgeon. With better information than Grey, he recognises the danger Mary's mental condition poses to her. When she began to loiter around his house at dusk, he declares, 'I knew / By this how dangerous it must be to wait / For a day longer.'[80] But although he writes to her at once telling her to return to her husband, she continues to come round to his house for six more days. It is only the fear of gossip that eventually leads him to bar the door to her, which in turn precipitates her death. Where Grey fails in his diagnosis, Luton fails in his duty of care.

We may doubt, as Grey clearly does, the veracity of Luton's evidence. Pressed by a juror – supported by the coroner in spite of Grey's objection, and reported blandly but no doubt with an underlying prurience by the newspaper – 'Did no acts / Of familiarity occur between / Deceased and you?' Luton replies, 'She would clasp / Her arms around me in speaking tenderly, / And kiss me. She has often kissed my hands. / Not beyond that.'[81] The stark reporting of this dialogue gives little clue as to whether these last two short sentences are a rapid attempt on Luton's part to correct any misapprehension that she kissed his lips, or whether he is talking his way out of an accidental confession. We do not learn more. When the juror asks, 'And did you / Respond –' the coroner intervenes, insisting, 'The witness should, / I think, be pressed no further. He has given / His painful evidence most creditably.'[82] By insisting on Luton's credit, the coroner declines to let it be put to the test, in spite of his own earlier insistence that the juror's line of questioning was legitimate. In the process, the credit of the inquest itself is undermined.

The medical men in the poem fail to prevent Mary's death, but they agree as to its cause. Luton's deposition ends:

> I had her carried home without delay,
> And a physician called, whose view concurred
> With mine – that instant death must have ensued
> Upon the rupture of a blood-vessel.[83]

This view is indeed corroborated by Dr Wallinger, the last witness in the case:

> I was called in to see
> The body of deceased upon the sixth:
> Life then was quite extinct; the cause of death,
> Congestion and effusion of the ventricle.
> Death would be instantaneous. Any strong
> Emotion might have led to that result.[84]

In his summing up, the coroner advises the jury that they should 'decide / Upon their verdict from the testimony / Of the professional witness – which was clear, / And seemed to him conclusive.'[85]

The death of a would-be adulteress from a busted blood vessel may seem a classic case of moralising Victorian melodrama, but the sparse language of courtroom testimony gives it a certain credibility, which would have been reinforced by contemporary medical opinion. In her study of accounts of female insanity in the period, Helen Small notes that of the 562 cases treated at Bethlehem Hospital (better known as Bedlam) recorded by Alexander Morison in the 1848 edition of his *Lectures on the Nature, Causes, and Treatment of Insanity*, twenty-five were put down to 'disappointed love', of whom twenty were women.[86] Rossetti took pains to ensure that the autopsy too was in accord with medical opinion. His brother, writing from Paris during his trip with Hunt, recommended that he consult an expert on this section of the poem: 'I would advise you, if practicable, to show it to any medical man at hand, – Dr. Hare for instance. He might discover some absurdity which escapes us, or suggest something of value to the story.'[87] Charles Hare was the physician who attended Christina Rossetti from 1845 to 1850.[88] In the end, William did not show the poem to Hare. Instead, he read it to Tupper, who in turn asked 'a Medical Man about the scientific requirements for the death of the woman' in the poem, as Rossetti recorded in the *P.R.B. Journal*. The medical man confirmed that 'congestion and effusion of the ventricle' was indeed 'the right term'.[89]

The care taken by the Pre-Raphaelite circle to ensure the accuracy of the autopsy within the poem is another indication of their commitment to scientific precision. But it stands too as a sign of what modern science can and cannot grasp. Mary's physical condition falls within its knowledge and understanding, but her mental state is not so easy to determine. The poem lays out her psychological condition to us very plainly, taking on what D. M. R. Bentley rightly calls a 'forensic quality' in documenting her symptoms.[90] We learn what distressed her, how her 'strong / Emotion' grew, how in the end it killed her. Yet the characters in the poem fail to apprehend her condition until it is too late. The knowledge we gain comes through poetry and narrative, not through medical science.

This may seem an unfair contest. The scientists exist within the poem, while the poet has the licence to create it. Besides, we too do not understand Mary's mental condition until after she has died. Even so, the poem suggests that poetry can grasp psychological problems with a greater subtlety than science, the law or the newspapers. The crucial difference between poetry and these other means to knowledge is that the poet is not obliged to reduce the complexities of the case, which the doctors, the courts and the newspapers all do as they seek to determine a specific cause for Mary's death. Medical science, the law and the papers, Rossetti suggests, are incapable of apprehending changes and contradictions within an individual's psychology which cannot be classified. Similarly, they cannot understand in anything but reductive terms how psychology and morality are inseparable – how mental states include individuals' attempts to control their own minds, and are affected by the courses of action which they choose, and their own and others' moral judgements upon them.

Within the poem, the moral judgements that are made are absolute. Grey is sure that Luton has transgressed against him; Harling equally convinced that he has not. The coroner makes it clear to the jury that 'the loss of life had followed such / A glaring sacrifice of duty's claims' on Mary's part, steering them to bring in the chilling verdict 'Died by the visitation of God'.[91] A trial that proceeds on strictly secular principles, with medical science putting forward the 'conclusive' evidence, ends with a verdict of divine intervention. Rossetti scrupulously avoids direct authorial intervention himself, but the change in register is sudden. The verdict itself is stark, economical and uncompromising, abruptly framed by the initial stress on 'Died' at the beginning of the line and the firm caesura after 'God', before the reporter moves on to wind up his account of the trial in a continuation of the same verse line but a new verse paragraph.

It seems that we are being invited to reach our own verdict on the verdict itself, to decide whether the kind of inference that is made, from material effects to divine causes, is legitimate. As readers, we are aware of the woeful inadequacy of all these judgements, in part because they are not sustained by the evidence, but more because they fail to recognise the moral struggles of the men and women involved. Mary's collapse follows on weeks wracked by a threefold conflict between her desires, her will and her social responsibilities. Luton too is damaged by his own moral weakness, while Grey is reduced to a vengeful obsessive. It is a grim irony that all his scientific eminence amounts to at the poem's close is a means of ruining his one-time friend. For Rossetti, as for Stephens, it is poetry, not such morally and intellectually bankrupt 'quackery', that can lead us to a proper understanding of psychology.

'Mrs. Holmes Grey' suggests that poetry which adheres to the standards of science can be more rigorously objective than science itself. Like the Pre-Raphaelite painters as he describes them, Rossetti conducts his own investigations into hitherto settled issues – in this instance, the familiar verdicts of medicine and the law. As promised, he gives an 'unflinching avowal' of their results. Because poetry is not reductive in its methods, these results, Rossetti implies, can be more comprehensive and ultimately truer than those of science itself.

By adopting undemonstrative blank verse as his medium and detached realist narrative as his genre, he collapses the distance between his poem and the real world it professes to represent. Presenting his record as starkly as possible, he stakes his claim to the veracity of his observations and his own moral authority. Yet at the same time, as we are continually reminded by the artifice of verse itself, these investigations are fictional, not factual. Where prose might be a documentary account, blank verse can only ever be art, even if what we are reading is a newspaper report. For Rossetti, the fact that the poem is fiction does not compromise the value of its insights when the artist is 'working from absolute data of fact', as he puts it in his *Spectator* essay.[92] Instead, by transforming data – accounts of legal and medical judgements – into verse, Rossetti elevates them from single instances into an exemplary narrative.

In setting the data before us, William Rossetti places us in a position analogous to Harling's within the poem, so we can investigate the case for ourselves. But he is careful to ensure that we have all the data, and not merely the medical report or the court testimonies, to go on. In reading Dante Gabriel Rossetti's 'A Last Confession', by contrast, we stand in for the priest, privy to only one prejudicial account of the case. It is beholden on us to determine for ourselves the truth behind the murderer's self-deceptions, but at the same time to recognise that he is a fellow specimen of a fallen humanity. In holding back from drawing direct conclusions in an authorial voice, both poets give us the freedom to judge for ourselves. As with Pre-Raphaelite paintings, it is up to us to infer the psychological insights and the moral conclusions which they leave implicit. Yet these conclusions are themselves answerable to the strictures of science. Transmuted into a poem, they stand or fall on their truth relative to human behaviour at large and not merely to these specific cases. In 'Mrs. Holmes Grey', William Rossetti suggests, as Stephens does in 'Modern Giants', that medical science on its own is inadequate without the guidance of the poet's knowing hand. Only through the arts, practised as anatomy, can we hope to comprehend fully the 'pulsations, fears, self-doubts, hates, goodness, devotedness, and noble world-love' of the embodied mind.

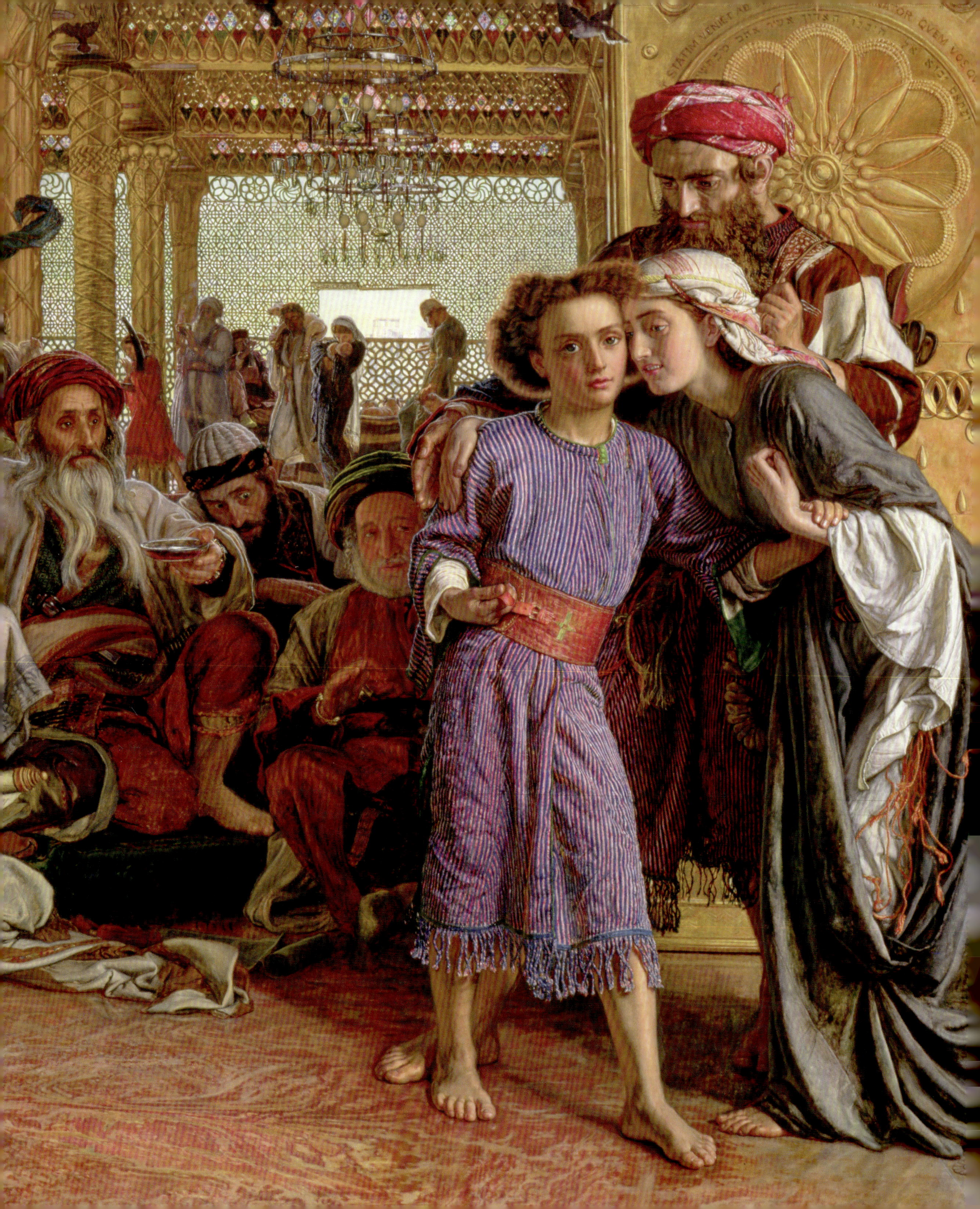

4 · Looking the Matter Through and Through

SACRED ART ON SCIENTIFIC PRINCIPLES

In 'The Purpose and Tendency of Early Italian Art', Frederic George Stephens declares: 'Truth in every particular ought to be the aim of the artist. Admit no untruth,' he goes on, 'let the priest's garment be clean.'[1] This powerful exhortation comes at the end of the same paragraph in which he encourages artists to follow the methods of science in their 'search after truth'. For Stephens, the Pre-Raphaelites' mission as artists is at once scientific and religious. Stephens's equation of the Pre-Raphaelite artist-as-scientist with a priest has a High Church ring to it. They were drawn to the ritualism and sacramentalism of the Oxford Movement, and Collins, Collinson and Dante Gabriel Rossetti all created works that were self-conscious examples of what Rossetti called the 'Art-Catholic'.[2] Even Hunt's *Light of the World* was scorned by Carlyle, who had admired his earlier paintings, as a 'papistical fantasy'.[3] For all his priestly rhetoric, however, Stephens closes his essay by casting the artist rather as a questing pilgrim who can only fully bring to bear 'the real power of the Arts' by holding always in mind the motto '*No Cross, no Crown*'.[4] Both the image and the motto are resonant of the seventeenth-century prose of Samuel Purchas, John Bunyan and William Penn. In invoking this tradition, Stephens aligns Pre-Raphaelitism not with Roman Catholicism but rather with a radical Protestantism. William Rossetti too implicitly identified Pre-Raphaelite art as a Protestant project. In an important essay on 'The Externals of Sacred Art', published in the *Edinburgh Weekly Review* in 1857 and reprinted the following year in *The Crayon*, Rossetti describes Protestantism as depending on 'the assertion of the right of private judgment (or, more strictly, the irresponsibility for private judgment save to God alone)'.[5] As for Stephens, so for Rossetti, there is an affinity between Protestantism's

Facing page Detail of fig. 51.

insistence on the individual's right to judge for themselves, independent of authority, and the Pre-Raphaelites' Baconian conception of science which stresses the individual's duty to investigate each question by testing it against their own close study of the world around them.

The link between science and religion in Pre-Raphaelite art is clearest in those paintings where they turn their attention to sacred subjects, in particular in those which turn away from the received aesthetic of the Oxford Movement and pursue instead their own independent study of biblical subjects. As Ruskin commented in his Oxford lectures in the 1880s, the '*material* veracity' which led him to identify Pre-Raphaelite art with science was especially striking in their sacred art. According to Ruskin, it enabled them 'to feel and illustrate the relation of spiritual creatures to the substance and conditions of the visible world' as few artists had done before.[6] As we have already begun to see in Collinson's 'Child Jesus' and Rossetti's *Ecce Ancilla Domini!*, Pre-Raphaelite art brought its emergent sense of ecology and its determined attention to psychology, along with other sciences including archaeology and physical geography, to bear on interpreting the sacred truths of the Old and New Testaments.

PRE-RAPHAELITE PRINCIPLES AND PRACTICE IN SACRED ART

In 'The Externals of Sacred Art', William Rossetti sets out to examine 'whether the artistic feelings and sympathies of a Protestant people of the nineteenth century are best met by a typical or by a narrative expression of religious subjects'.[7] His answer to his own question is that the modern painter would do better 'to understand a narrative, and translate it into form, rather than to symbolize a conception for translation by others'. This is because, in a century which tends to place its trust in, 'we will not say material, but eminently positive' or empirical knowledge, Rossetti argues, 'Men like to see a proposition put into a concrete form, and to have a truth enforced rather by example than by precept.'[8]

Rossetti anticipates the terms which Ruskin would use to define Pre-Raphaelite sacred art twenty-five years later. At the same time, he recalls the verdict passed on the didactic art of an imaginary medieval painter in his brother's famous short story 'Hand and Soul', published a few years earlier in the first issue of *The Germ*. This highly suggestive story charts the career of a fictional painter called Chiaro dell'Erma whose finest work was supposedly painted in 1239, a year before Cimabue – the first recognised master of Florentine art and the supposed teacher of Giotto – was customarily believed to have been born.[9] Chiaro makes two false choices in art before he meets and paints a vision of his own soul in the form of a woman, enabling him to create the remarkable work which the modern narrator of the story professes to have seen in Florence in 1847. At first Chiaro wins fame painting beautiful Madonnas, before realising that 'he had misinterpreted the craving of his own spirit' and 'that much of that reverence which he had mistaken for faith had been no more than the worship of beauty'. This realisation can be read as a warning against mistaking the aesthetics of Catholicism for the substance of religious truth – a warning, in effect, against embracing Tractarianism too readily. But this is matched with a second false choice: to attempt a stark moral didacticism in art without any regard for the need for empathy on the part of the viewer. Chiaro's own historical context is necessarily Roman Catholic, yet this second impulse suggests the iconoclasm and rigidity of Puritanism. Again it fails, this time with the public as well as failing to satisfy Chiaro himself. Part at least of Chiaro's mistake, we are told, was that 'he did not choose for his medium the action and passion of human life, but cold symbolism and abstract impersonation.' As a result, 'the people ceased to throng about his pictures' and 'The works he produced at this time have perished – in all likelihood, not unjustly.'[10]

Both Rossetti brothers see narrative as an essential component of successful sacred art. For Gabriel, this narrative may be implied in the aspect of human figures even in a symbolic scene with no obvious narrative component, such as David's altarpiece (see fig. 41) or *The Mystic Marriage of St Catherine* by Hans Memling, for which he wrote another sonnet published in *The Germ*.[11] For William, in modern art and in a Protestant country, the most effective sacred paintings will be those expressly illustrating stories from the

46 Dante Gabriel Rossetti, *The Passover in the Holy Family*, 1856. Tate Gallery, London.

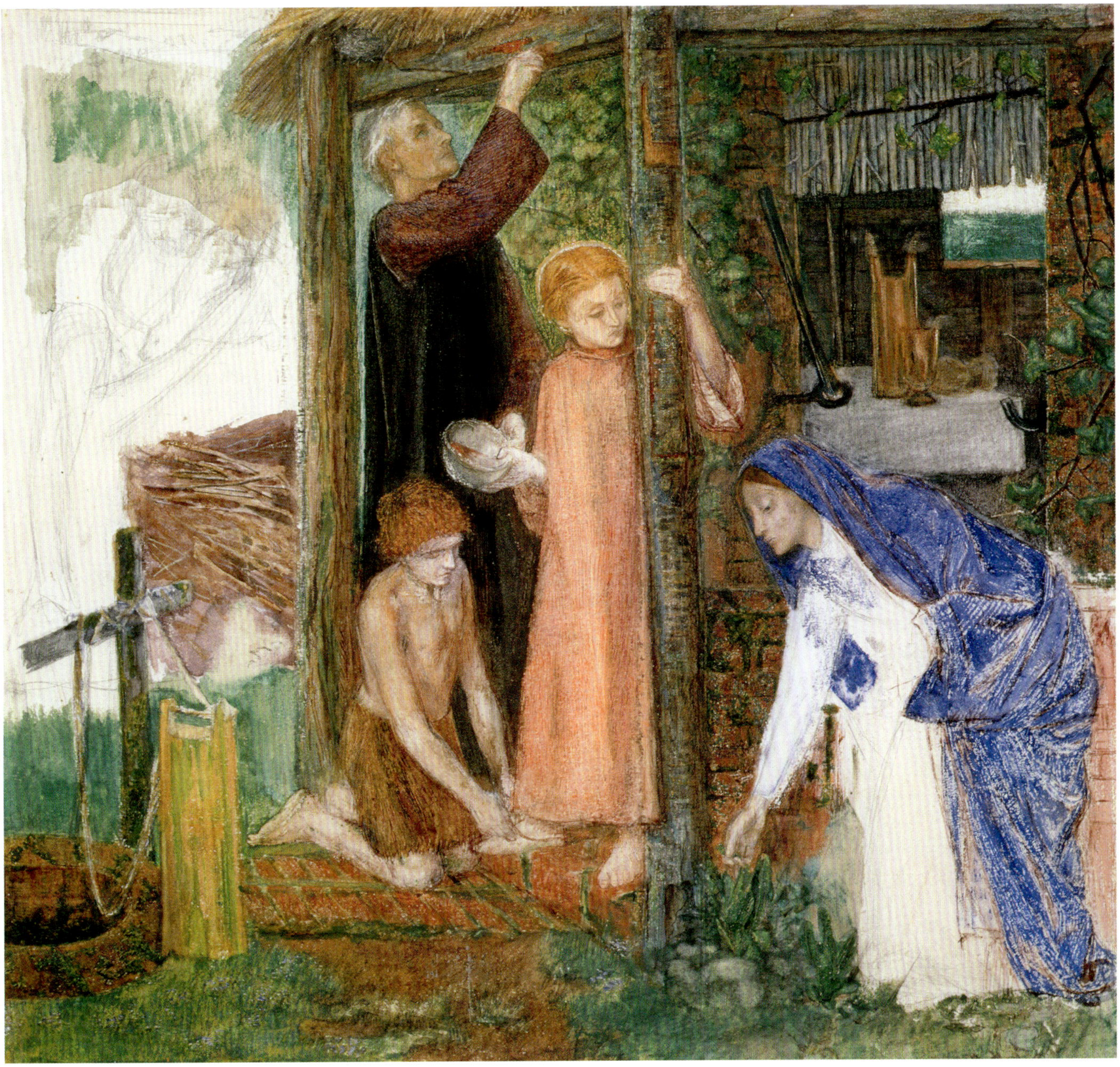

Bible. With a few exceptions, most notably *The Light of the World*, the sacred art of early Pre-Raphaelitism adheres to this precept as set out in William Rossetti's essay. Collinson's 'Child Jesus' and its accompanying etching, Rossetti's *The Passover in the Holy Family* (fig. 46), Millais's *Christ in the House of his Parents* and Hunt's *Finding of the Saviour in the Temple* all imagine or reimagine episodes from Jesus's childhood, while, as we have seen, Rossetti returns time and again to the Annunciation and the birth of Christ in his paintings and sonnets. Each of these works at once captures

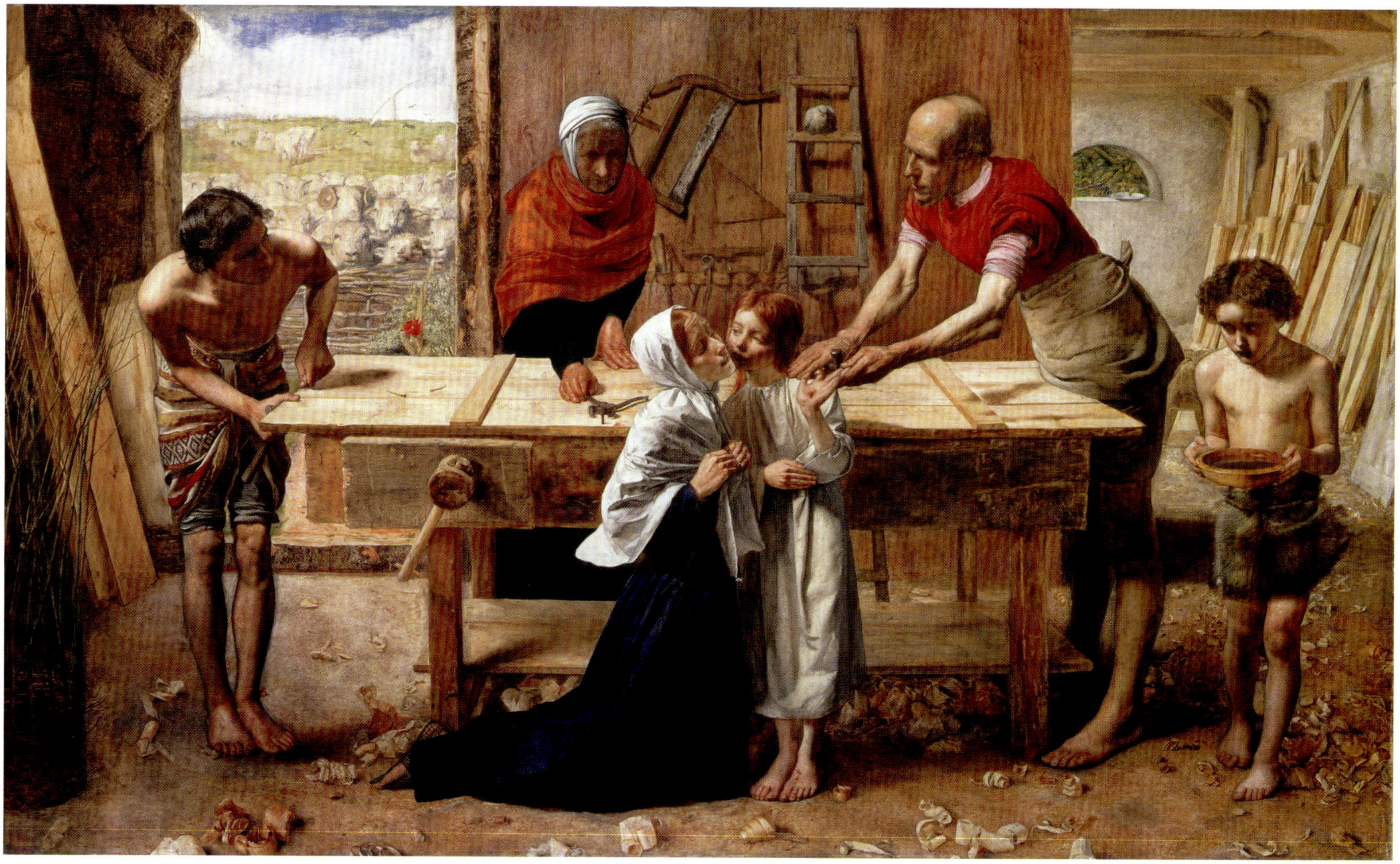

an individual narrative moment and anticipates emblematically the trajectory of its subject's life.

The Pre-Raphaelites delve into the interior life of Jesus and Mary, not to deny their claims to divinity or sanctity, but better and more fully to understand the incarnation, whether as truth or myth. In this sense, their sacred art shows the same scientific commitment and methods as Pre-Raphaelite secular art, with the caveat that, as William Rossetti puts it in his essay, 'sacred art has utterly failed, unless it produces a sacred impression.'[12] This tension between scientific realism and sacred power is evident in Millais's *Christ in the House of his Parents* (fig. 47), exhibited to much hostility at the Royal Academy in 1850. This painting remains astonishing for its attention to mundane detail. Millais places Jesus and his family in the deliberately everyday scene of Joseph's carpenter's shop. Every wood shaving on the floor of the workshop is picked out. So too are the veins in Joseph's arms and the grime on his legs. The scene was painted from a contemporary carpenter's shop, Joseph's body from a working carpenter.[13]

Millais's picture is dense with religious symbols, many of them anticipations of Christ's later life – the bowl of water carried by the young John the Baptist, the beams of the wellhead in the distance hinting at the cross, the blood dripping from Jesus's palm onto his foot to suggest stigmata. But like the setting, the appearance and reactions of the family themselves are decidedly ordinary. Joseph, St Anne and the carpenter's assistant are each moderately but not unduly concerned and solicitous. This is, after all, only a small wound. Mary and John both look more anxious, but their facial expressions do not give more away than this. They are not obviously prescient. Jesus himself is even harder to read. He is unusually calm for a child who has just hurt himself. He allows his parents to examine his injury, but he is not obviously *(facing page)* 47 John Everett Millais, *Christ in the House of his Parents*, 1849–50. Tate Gallery, London.

48 John Everett Millais, *Study for 'Christ in the House of his Parents'*, *c.*1849. Tate Gallery, London.

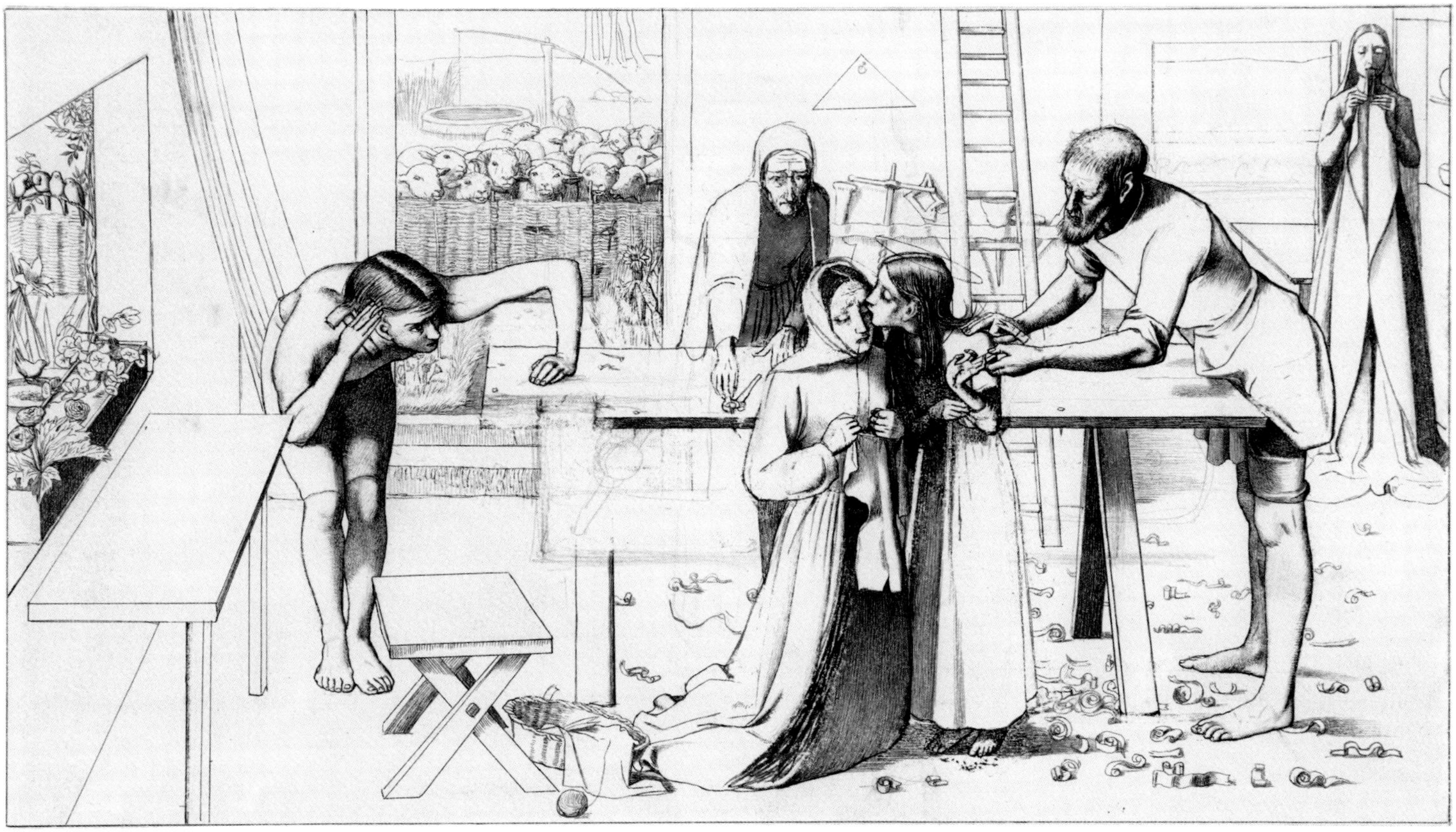

in pain. Aside from this, it is hard to gauge his character or to tell what he is thinking. He appears to be looking closely, even short-sightedly, at his mother, rather than evidently contemplating the significance of the event itself.

Taken together, the mental states of the figures in Millais's painting reinforce their normality. These are kindly but not obviously remarkable people. As Grilli has pointed out, Joseph's head was painted from Millais's own father.[14] This was in line with Pre-Raphaelite naturalism as outlined by William Rossetti in another essay, with particular reference to sacred art:

> To say that St Peter was a 'common fisherman,' and that, therefore, a common fisherman is the right model for St Peter on the naturalistic principle, is so small a fragment of truth as to become plain falsehood. St Peter was an extremely *un*common fisherman. The *essence* of Peter is that he became a fisher of men; the accident of Peter is that he was a fisher of fish. The essence of a naturalistic representation of him therefore, is to express the qualities of his soul – the fiery zeal, the worshipping love, the stumbling of human weakness. For supplying the artist with these, the right model is a man possessing, in development or potentially, the same characteristics as nearly as may be. Unless these are given, by one means or another, the St Peter of the artist is not a naturalistic St Peter.[15]

Joseph was a carpenter by trade, so Millais needed to study a carpenter's body to be sure in his own words that he would 'get the development of the muscles right'.[16] At the same time, he needed a face that would convey the essence of undemonstrative paternal care which he saw to be Joseph's key quality and which he recognised in his own father. The result, like many Pre-Raphaelite figures, is a composite, but one studied carefully from each model on the principle that both the history and the personality of the individual are inscribed on the physical body.

The very ordinariness of the Holy Family, together with the everyday form into which the symbols are translated and the workaday space in which the scene takes place, make Millais's painting extraordinary. It marks a radical new departure in sacred art.[17] Yet in making such a radical innovation, Millais sets himself a challenge which he does not fully meet: how to produce the 'sacred impression' which, as William Rossetti insists, is a *sine qua non* in sacred art. For all its bold originality, *Christ in the House of his Parents* is liable to fail as a work of sacred painting because it does not match the supposed significance of the event that it depicts with sufficient emotional intensity. That is to say, its very success as a mundane narrative limits its typological power, as while the viewer is bound to recognise the foreshadowing of future events in Jesus's life, the painting's everyday scene and muted emotional tone tend to diminish rather than to exalt their significance. In contrast, Collinson achieves precisely the opposite effect in 'The Child Jesus' by intensifying Jesus's and Mary's emotional response to seemingly trivial incidents.

In the painting itself, Millais choses realism at the expense of intensity, but an early design for it shows a significantly different and potentially more moving realisation of the same scene (fig. 48). The difference inheres partly in the style of drawing. Colin Cruise has suggested that the study is 'more outrageously anti-academic than the exhibited work'.[18] With its elongated figures and distorted perspectives, it belongs to the Gothic phase of Pre-Raphaelite medievalism typified by drawings such as Millais's *My Beautiful Lady* and *The Disentombment of Queen Matilda* and Rossetti's *Dante Drawing an Angel on the First Anniversary of the Death of Beatrice*, and by paintings such as *Isabella*, *The Girlhood of Mary Virgin* and Collinson's *Elizabeth of Hungary*, all conceived in the late 1840s. The style itself marks out the scene as significant through its eye-catching refusal of realism. It provides a lens too through which everything is heightened and intensified. In this version of the picture, the Virgin is not merely concerned, but grief-stricken, while Jesus himself is actively looking after his mother. The concern of both Anne and Joseph is more intense too, while in place of John the Baptist this early design has an unexplained and ghostly figure of a woman playing a pipe in the background, reinforcing the sense of the otherworldly.

The emotional intensity and otherworldliness of Millais's drawing are visual corollaries for the significance of the event itself, as they are in 'The Child Jesus' and *Ecce Ancilla Domini!* (see fig. 42). Millais represents a trivial injury as having a profound psychological impact on Mary. It is as if she has had a sudden premonition of her son's fate. In his own concern for her rather than himself, we can see that Jesus knows what she knows. We can see too the selfless sacrifice that is his very nature. As in the paintings discussed in the last chapter, Millais uses details of expression and gesture to reveal with remarkable economy the inner lives of his subjects. These inner lives preoccupy Collinson and Rossetti too. Like Millais, they examine the psychological impact of premonition through its expression in and on the body.

Millais's uneasy choice between physical realism and symbolic intensity in *Christ in the House of his Parents* suggests an unresolved problem within early Pre-Raphaelite sacred art. Once William Rossetti has determined that modern sacred art should ground itself firmly in biblical narrative, he moves on to consider how best to represent those narratives, deciding in favour of a scrupulous realism. Specifically, Rossetti asks 'whether in the narrative form of sacred art, strict accuracy of national type and accessory should be adhered to. Is our Madonna to be a Jewess, our biblical costume oriental, our scenery that of the Holy Land?'[19] The question Rossetti poses bears on the application to sacred subjects of the principles that the Pre-Raphaelites identified within their own art as scientific. He maps out the range of possible answers. The artist, he suggests, has to make two successive choices. The first is whether to follow tradition and imitate Raphael, or whether to approach the subject naturalistically. If the artist opts for naturalism, then a second choice presents itself: whether to imitate the practice of artists before Raphael and depict biblical characters in the clothes and spaces of contemporary life, or whether to attempt to realise as far as possible the original landscape, people and customs of the time and place in which the stories really occurred.

Of these three main options, the first, Rossetti argues, is obviously false. 'The traditionalist', he writes, 'dresses his figures in blankets which were never worn, puts a bit of Judaism here for the "characteristic" heads, and bit of Anglicism there (supposing him to be a Briton), a bit of classicism, and an entirety of nothingism.' The result is a painting which is 'not true to any vital conception of the fact', with the result that 'his means command no sympathy, and awaken no associations, save of the artistic (not to say artificial) kind.'[20] The second is in principle more legitimate, but is in practice little better, again because it cannot satisfy the 'positive' or scientific tendencies of the age. As Rossetti explains:

> The medieval artist worked wholly without *arrière pensée*. Very possibly, to many, it never so much as occurred that the men whom they painted had dressed differently, and lived under different conditions; and, even if it did, the artist neither knew nor cared what the difference verily amounted to. *Now* the case is quite the reverse. It would not be an act of good faith, but the opposite, for the artist to paint St Peter's costume from a Newhaven fisherman's, or St Joseph's from that of his easel maker. The learning of the age may be a benefit to its art, or a misfortune; but it is a fact, and cannot be ignored.

An artist must be bound not only by his own good faith, but also by the expectations of his viewers. 'If the medieval artist murdered chronology with a complacent conscience,' Rossetti remarks, 'the medieval spectator had still less feeling of foul play; the modern spectator, on the contrary, would hold his nose at the corpse.'[21]

These first two approaches fail for Rossetti because they depart too glaringly from the positive facts of the stories being represented. The third option, by contrast, seeks not only to reproduce these facts insofar as they are known but also to discover for itself the truth behind the biblical accounts. The implication is that this is the only way to reignite the sacred spark within contemporary painting. To achieve this, the artist needs to get to know 'the life and the scenery and detail of the biblical regions at the present day' at first hand, not merely as objects to be copied, but so that he becomes 'thoroughly in sympathy with the alien nature'.[22] On Rossetti's term, if Millais fails to produce a 'sacred impression' in *Christ in the House of his Parents*, it is not because he moved away from the Gothic unreality of his early drawing, but because he did not move far enough towards immersion in the true reality of the scene he was painting.

Millais pioneers the naturalistic treatment of sacred subjects within Pre-Raphaelite painting in *Christ in the House of his Parents*, as Collinson does within Pre-Raphaelite poetry through the vividly realised details of 'The Child Jesus'. The painting in particular 'exceeded in realism of the most uncompromising kind', as Stephens later put it, and staunchly Protestant realism at that, he noted, 'where the Virgin, Joseph, and Christ utterly reversed Italian types of the beautiful, happy, and young'.[23] But neither Millais nor Collinson had been overly concerned with the external details that Rossetti would later home in on as of crucial importance. In his strict naturalism and his rejection of Raphaelesque tradition, Millais's practice laid the foundations for Rossetti's later arguments, but beyond fitting a piece of oriental-patterned cloth over the carpenter's assistant's shorts, he paid relatively little attention to historical or geographical authenticity. Both his landscape and his figures are unselfconsciously English. Nor did Collinson in 'The Child Jesus'. He realises the landscape very vividly in this poem, but – a 'sunny grape vine' notwithstanding – the landscape itself is for the most part an English countryside of 'hedge-rows' and 'hawthorn trees'.[24] He also imagines Nazareth as a coastal settlement. As Rossetti noted with amusement fifty years later, 'The Præ-raphaelites, with all their love of exact truth to nature, were a little arbitrary in applying the principle; and Collinson seems to have regarded it as quite superfluous to look into a map, and see whether Nazareth was near the sea or not.'[25]

Though he does not mention him by name, Rossetti's model for how Pre-Raphaelite sacred art was to accomplish its mission was Holman Hunt. Hunt had set off for Jerusalem in January 1854 on the first of four trips to paint biblical scenes in situ. He returned two years later with the unfinished canvas of *The Finding of the Saviour in the Temple*, which he would eventually exhibit in 1860. Rossetti's essay is a justification of his remarkable commitment to painting sacred art on scientific lines. When Hunt and Dante Gabriel Rossetti had mooted going out to Jerusalem together to paint biblical scenes in October 1850, William had been decidedly sceptical.[26] But Hunt's triumphant return convinced him that he had been right to aspire to geographical and historical accuracy, and that this marked an important new stage in the history of sacred art. As far as the logic of Pre-Raphaelitism was concerned, however, Hunt's practice was consistent with its pre-existing principles. Like John Brett's immersive visits to the Alps, Hunt's trips to Jerusalem and beyond put into practice Ruskin's injunction to bring home accurate records of the geography, geology and natural history of landscapes which exceeded the narrow limits of his English viewers' experience.[27] At the same time, Hunt followed Brown's advice that the painter should 'make himself thoroughly acquainted with the character of the times' as well as the 'habits of the people, which he is about to represent'.[28] As well as living and travelling in what was then Syria, he steeped himself in archaeology both at first hand and through the scholarly literature in order to understand the culture of the ancient Near East and to calibrate in fine grain the distance between the modern and ancient worlds. His object, throughout, was to create a new form of sacred art that could communicate to its viewers the full truth of revelation by locating it in the most precisely realised vision yet of the Holy Land at the time of Christ.

GEOGRAPHY, ECOLOGY, ARCHAEOLOGY, ANTHROPOLOGY: HOLMAN HUNT IN THE HOLY LAND

In the paintings he made on successive trips to Jerusalem from the mid-1850s onwards, Hunt followed the Pre-Raphaelite principles of scientific research and the objective study of external nature in order to recreate Christ himself, his contemporaries and the world in which he lived as exactly as possible. His project was motivated by his own resurgent faith. Looking back from the 1880s, Hunt recalled that he had flirted with an irreligious materialism in the late 1840s. He attributed this brief wayward turn to materialism in part to science, specifically to his reading of the controversial evolutionary cosmology *Vestiges of the Natural History of Creation*. In the longer run, however, the injection of materialism that came with *Vestiges* inoculated Hunt, as it did Tennyson, against the more virulent strain that would appear in the 1860s in the wake of *On the Origin of Species*. Hunt claims that it was scientific materialism, underpinned by the scepticism of Voltaire and complemented by the radical poetry of Byron and Shelley, which gave him the courage to scrutinise the claims of religion directly. Yet ultimately this scrutiny reaffirmed his religious convictions rather than undermining them.[29]

Hunt's own experience is an individual instance of a major trend within liberal Christianity in the mid-Victorian period. His approach to testing and verifying the truths of the New Testament in the light of modern knowledge was analogous to that of the contemporary movement in biblical scholarship known as the 'higher criticism'.[30] Originating in Germany, the higher criticism sought to establish the truth behind the biblical narratives through comparative historical research and through cross-referencing revelation and modern science, with deference to science where the two did not square with one another. This movement began to take a firm hold in England in the 1840s with the publication of Marian Evans's translation of David Strauss's *Life of Jesus*, which rigorously excluded miracles from its account of Christ's life, interpreting Jesus rather as a human subject. It reached a watershed in 1860 with *Essays and Reviews*, a collection of seven essays by influential Anglican clergymen and academics arguing for the need to reconfigure Christian belief in the light of historical and scientific research. The Pre-Raphaelite project, which coincides precisely with the rise of the higher criticism in England, undertook its own parallel investigation into the truth of Christianity and the Bible. Indeed, Hunt, who declared himself 'not afraid of looking the matter through and through', goes beyond the higher criticism itself by critically comparing its book learning with his own detailed observations in situ.[31] This rigour allows him to declare with some authority that Strauss's theories were in his estimate mere 'romance' which 'present far greater obstacles to faith than the original Gospels offer'.[32] But if his faith remained intact throughout his investigations, his conception of Christianity became more exacting, personally, in the demands he made on himself and, intellectually, in the demands he made on the viewers of his paintings.[33]

Hunt's impulse to record what he saw – or more precisely what struck him as remarkable among what he saw, remarkable principally for its difference from English landscape, history

49 William Holman Hunt, *The Scapegoat*, 1854–5.
Lady Lever Art Gallery, Liverpool.

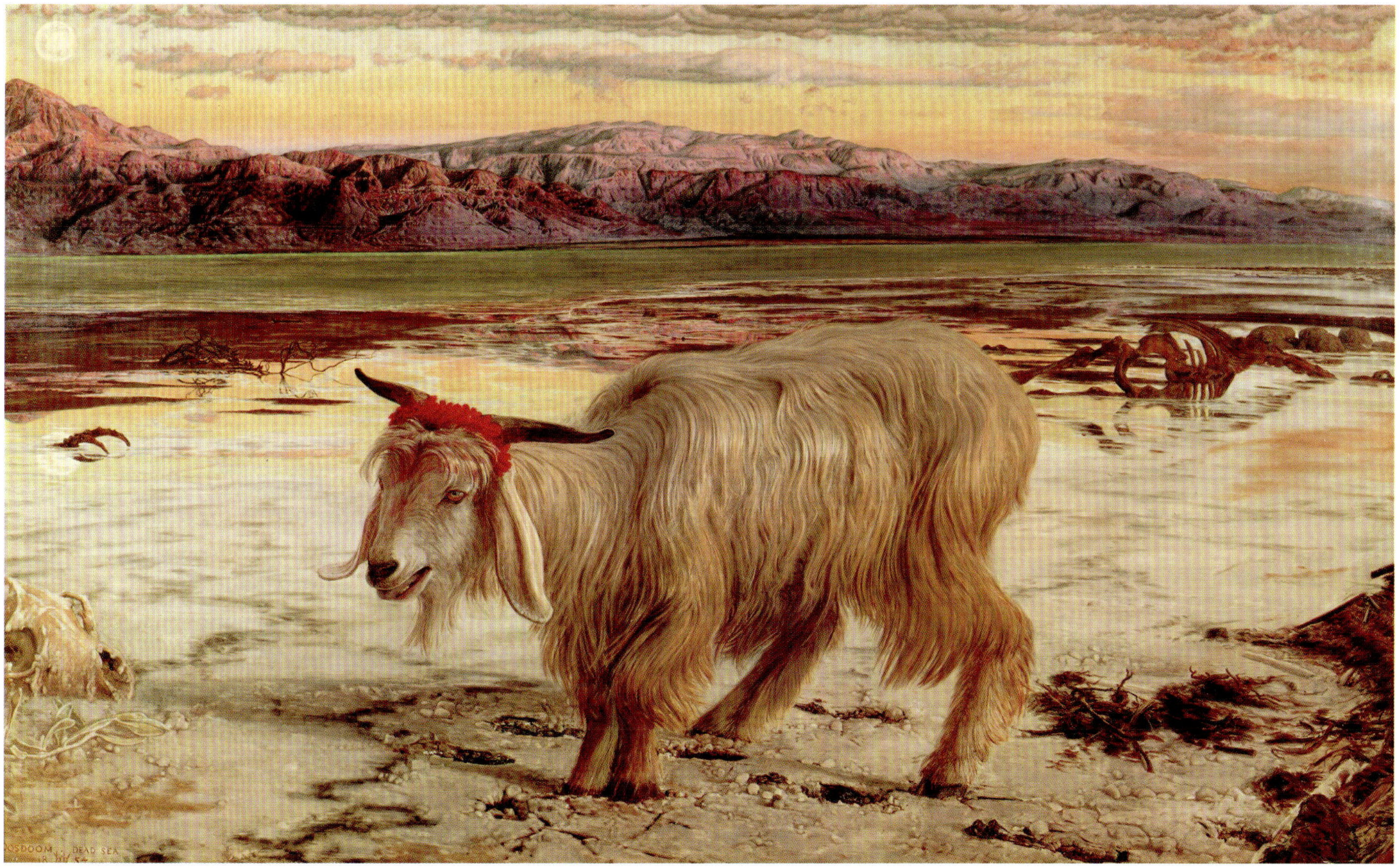

and culture – can be seen across the breadth of the works that he painted during his first expedition. He started in Egypt, where he painted among other things street life in Cairo (*The Lantern-Maker's Courtship*), ancient monuments (*The Sphinx*) and the local fauna (*The Haunt of the Gazelle*). Crossing to Syria, he painted several watercolours of Nazareth, Jerusalem and the landscape around them. But the two major paintings that he undertook on this trip were most radically original as works of sacred art painted on scientific principles: *The Scapegoat*, sent back to England and exhibited at the Royal Academy in 1856; and *The Finding of the Saviour in the Temple*, begun in Jerusalem in 1854 but completed in London and finally exhibited in 1860.

Hunt's methods diverged in painting these two works, but in each case they were founded on the scientific principles that characterised early Pre-Raphaelite art. The story of how he painted *The Scapegoat* (fig. 49) was much mythologised by Hunt himself.[34] His scientific method in this painting was a combination of physical geography and experimental natural history. The site was chosen because it was generally believed to have been the location of the biblical city of Sodom. In his memoir Hunt provided a map of the Dead Sea to detail this geographical specificity.[35] He explained this geography in detail in a letter to his friend and patron Thomas Combe, in reply to a letter of Combe's reporting that the painting had arrived safely in England:

> The Mountains on which you are looking are part of the Moab range of mountains which bound the Jordan and Dead Sea plain on the East, and which connect the low plain still lower with the Red Sea by bordering all the intermediate level in the same uniform chain. the extreme

51 William Holman Hunt, *The Finding of the Saviour in the Temple*, 1854–60. Birmingham Museums and Art Gallery.

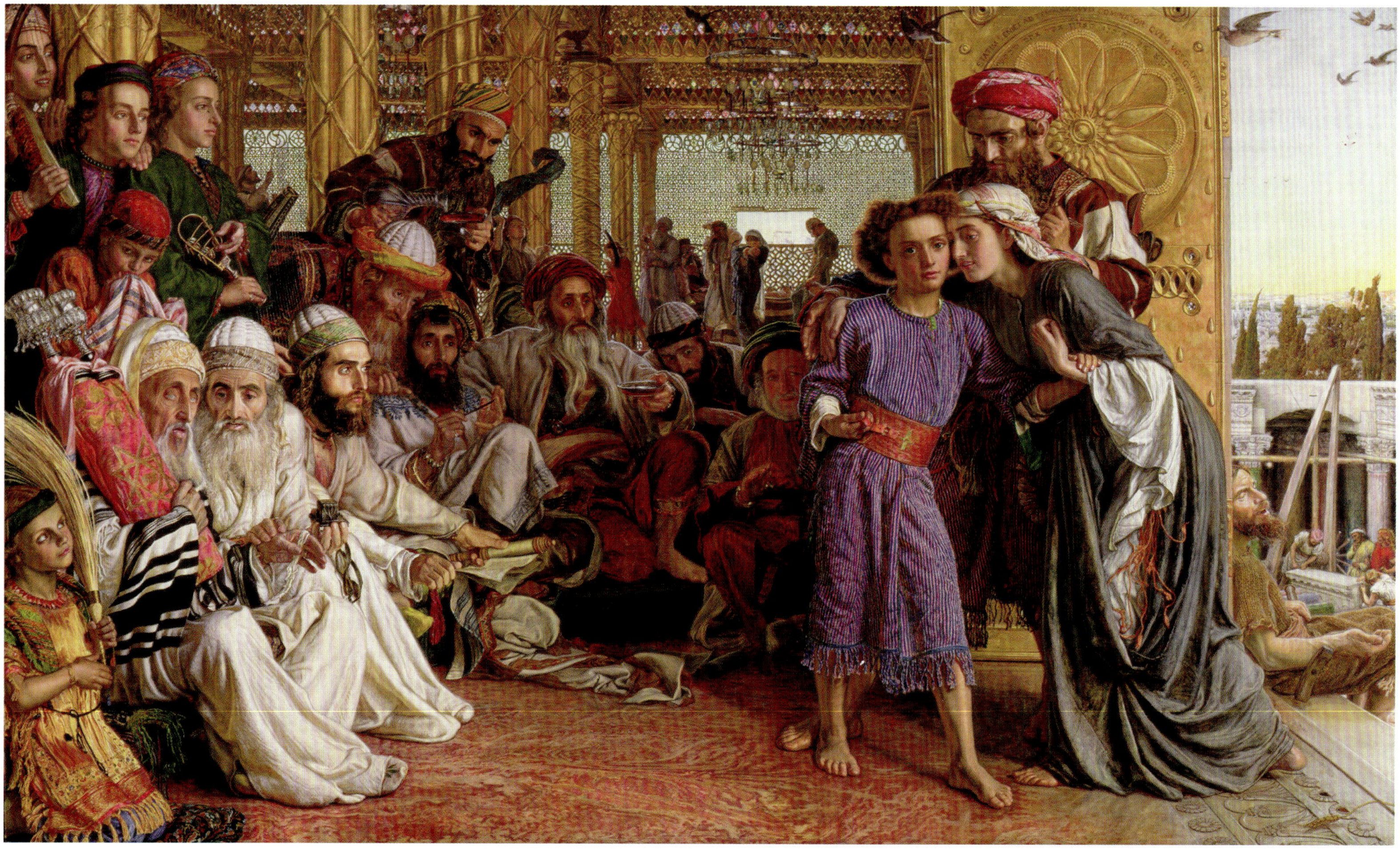

In keeping with ancient Egyptian and Assyrian practice, they meet the beams of the ceiling at right angles. The whole, Stephens insists, is not only meticulously researched, but also 'perfectly practicable, and could actually be constructed from the design'.[53]

Hunt's retellings of how he painted *The Scapegoat* and Stephens's account of his friend's methods in *The Finding of the Saviour in the Temple* are both constructed narratives.[54] Their descriptions of Hunt's methodology, his practice and their relation to the finished results are selective and idealised. Figures largely modelled by 'Jews in the East' were finished off from English Jews and Christians as required, while some of the final details of the Temple were imitated from the recreation of the Alhambra at the Crystal Palace in Sydenham, although as J. P. Birch has remarked this would have seemed less of a flagrant anachronism then that it does now 'as Islamic architecture was thought to have derived from ancient sources'.[55] But the image they project of the Pre-Raphaelites as rigorously scientific in their art is one Hunt not only aspired to and but held himself to as a standard as far as he practicably could.

Many Victorian viewers of Hunt's paintings accepted his terms of reference, judging them by these same scientific standards. *The Finding of the Saviour*, in particular, was repeatedly critiqued and defended on expressly scientific grounds. Stephens notes in his pamphlet that doubt has been cast on whether the columns of the Temple would really have been gilded. In Hunt's defence, he cites a 'singular

52 William Holman Hunt, 'Study of Jew', from *Pre-Raphaelitism and the Pre-Raphaelite Brotherhood*, 1905.

53 William Holman Hunt, 'Examples of Jewish Type', from *Pre-Raphaelitism and the Pre-Raphaelite Brotherhood*, 1905.

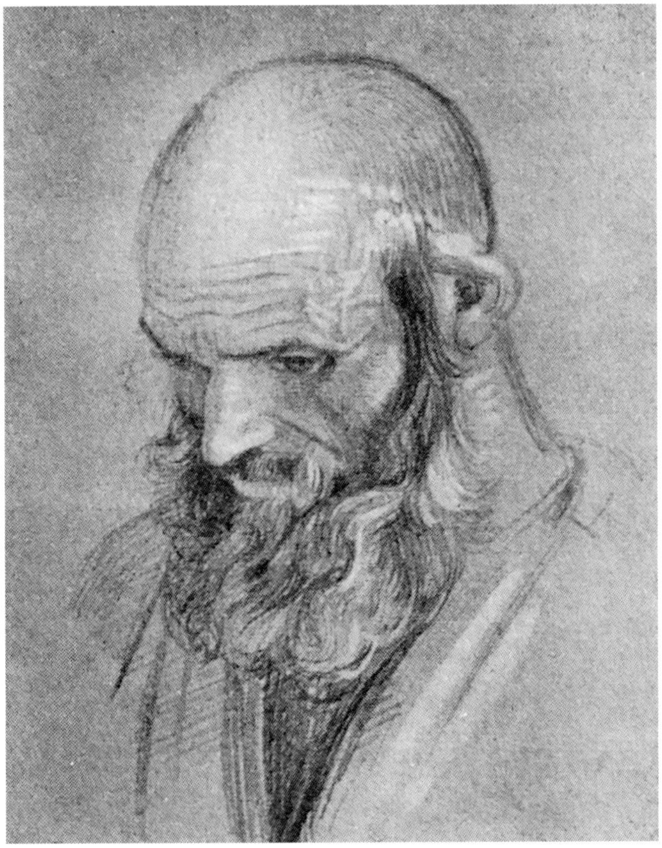

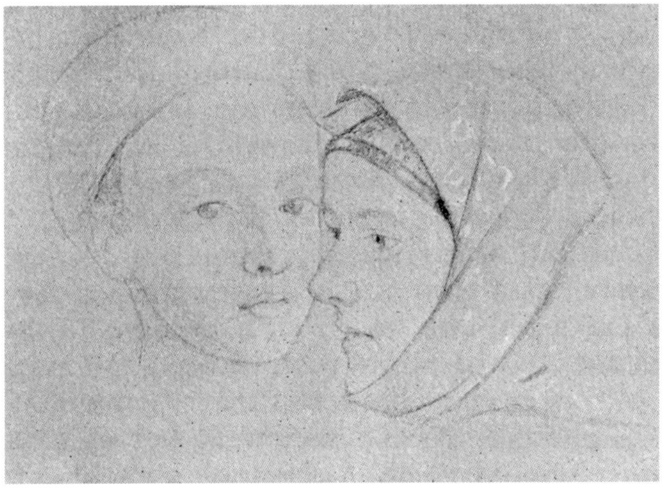

discovery' by the French archaeologist Victor Place of 'some thin plates of copper, moulded in the form of the trunk of a palm-tree', at Khorsabad in Iraq. 'Close beside these', Stephens explains, 'were found some strips of thin gold, which fitted exactly over the raised ornaments of the copper.' On these grounds, he concludes that it is 'probable' that the Temple columns too 'were coated with copper to receive a thin covering of solid gold'.[56] An anonymous essay on 'Binocular Vision and the Painter's Art', which appeared in the *Saturday Review* in 1862, took Hunt to task for misunderstanding optics. Specifically, it charged him with failing to recognise that were 'a living witness' actually viewing the scene he depicts, he could not possibly 'have *seen* the elaborate architectural background, the distant landscape, and the subordinate personages presented with that clearness and minuteness of detail which they possess in the picture'.[57] This time it was Tupper who defended Hunt, exposing the scientific fallacies in the *Saturday Review* article's account of binocular vision and pointing out that the details in Hunt's painting itself are only crisp when we look directly at them specifically and not at some other area of the canvas, just as in nature.[58] J. W. Jackson, writing in the *Anthropological Review* in 1867, both praised and critiqued Hunt's painting on scientific grounds, lauding the artist for his thorough 'antiquarian research' and his success in finding the right 'ethnic models', but proclaiming the picture to be a failure nonetheless because of his neglect of the principles of physiognomy and phrenology in his depiction of Christ and the Virgin.[59] That both Hunt's detractors and his advocates looked to science to evaluate his painting testifies to the Pre-Raphaelites' success in making scientific validity a key standard by which to judge the arts.

In *The Finding of the Saviour in the Temple*, Hunt uses ethnology and archaeology to depict as authentically and realistically as possible a scene from Christ's life. It is, as Prettejohn remarks, 'an extraordinary attempt to prove the "truth" of Scripture by empirical means'.[60] In *The Scapegoat*, Hunt adopts the same scientific rigour in the full knowledge that the effect will appear at first glance wholly unreal. Both in his diary and in a letter home to William Rossetti he casts the painting as a means to spread knowledge of a strange and unfamiliar landscape.[61] Part of the painting's power lies in the testimony Hunt's extraordinary attention to detail provides to the actual existence of the unearthly

54 William Holman Hunt, *The Shadow of Death*, 1870–73. Manchester Art Gallery.

world we see in the painting. The very colours defy our expectations. The Dead Sea is a dead green, like a thick mould. The mountains are pink and purple, their reflection on the salt water several different tones of bloody red. The sea here has a crust, while the bones of everyday animals take on eerie resemblances. As the alien scene takes possession of our imagination, the ibex's horns may become monstrous claws, the dead camel the exposed fossil of a plesiosaur.[62] These visions challenge us with a return of the repressed, in this case the self-same materialism which Hunt encountered in *Vestiges of Creation* and which led Tennyson to his horror of the 'Dragons of the prime' extinguished at the hands of 'Nature, red in tooth and claw' in *In Memoriam*.[63] The very landscape itself seemed to Hunt 'accursed by God', at once beautiful and horrible, a physical symbol of sin.[64]

The scapegoat itself is another such return, a reminder of human brutality, as well as a type of the sacrifice made of Christ, rather than by him. Like the sheep in *Our English Coasts* and the doves in 'The Child Jesus', the scapegoat is a sentient being. It is the victim of humanity's propensity to pass the guilt for our own crimes on to others. As one of the captions Hunt inscribed on the frame declares, 'And the Goat shall bear upon him all their Iniquities unto a Land not inhabited.'[65] Yet where the sacrifice of the goat was designed to force it away from people's company, and their sins with it, we are forced by Hunt to stare the goat in the face as it dies. In a subtle geographical pun, Hunt has placed Oosdoom – Sodom – 'behind me and the spectator', as he pointed out in his letter to Combe.[66] In his memoir, Hunt recalls how, when his party arrived at Oosdoom, '"Remember Lot's wife," we shouted, and round every hill and cranny we heard the words murmured in confusing mockery, until distinct in every syllable "Remember Lot's wife" was returned to us.'[67] The painting does not shout these words, but it whispers them, in the signature 'Oosdoom. Dead Sea', traced by Hunt in the bottom left-hand corner, and in the encrustations which appear to be slowly turning the camel's skull into salt. If we turn our back on the goat, like Lot's wife we are implicitly turning back to sin and away from salvation, and we will suffer the consequences.

How does the painting seek to effect our salvation as we stand and look at it? Prettejohn remarks that, in the absence of a direct narrative, 'the possibility is left open for an encounter between the Saviour and the spectator' in *The Light of the World*.[68] *The Scapegoat* – Hunt's next major sacred painting – does have an implied actual narrative as well as a symbolic one. As the Royal Academy catalogue for 1856 made clear, the goat, tormented and driven away from its village, is now on the shore of the Dead Sea and will soon die.[69] But it depends for its effect on the same imaginative engagement between the viewer and the subject. The pupil of the goat's left eye is just visible beneath its eyelid. It is looking up at the viewer, and Hunt invites us to return its gaze, to reach out to it in its suffering and to take a portion of that suffering and of the responsibility for it upon ourselves. As in the paintings and poems discussed in Chapter 2, the ability to extend the range of our sympathy beyond our own kind to other animals is a marker of a profoundly humane morality for the Pre-Raphaelites. In 'The Child Jesus' it becomes a sign of Christ's own moral perfection. Through entering into a compact of suffering with the scapegoat, we release it from the imputation of iniquity and imitate Christ in taking the sins of others upon ourselves. This is the demand that Hunt's painting makes on its viewer, which is all the more intensely challenging by its alienating strangeness. But in surmounting that challenge, the viewer, like Hunt himself – and like the young Jesus faced with the death of the fledgling bird in Collinson's poem – can revitalise faith and solidarity in the face of the evil that exists both in nature and in man.

In its detailed study of an unfamiliar landscape and its empathic engagement with a non-human subject, *The Scapegoat* is the most clearly ecological of Hunt's sacred paintings. But the principles of ecology underwrite his treatment of human subjects too. From *The Scapegoat* onwards, Hunt's religious paintings insist that to grasp the truth of Christ's life and message we need to understand him in his own time and place, to know his origins, the culture which formed him and the environment in which he lived – to know him, that is, ecologically. This is clearest in *The Shadow of Death* (fig. 54), painted during Hunt's second trip to the Holy Land in the late 1860s and early 1870s. Here we see Jesus in his own home, or more precisely in the carpentry workshop in the house of his parents, as Millais had painted him twenty

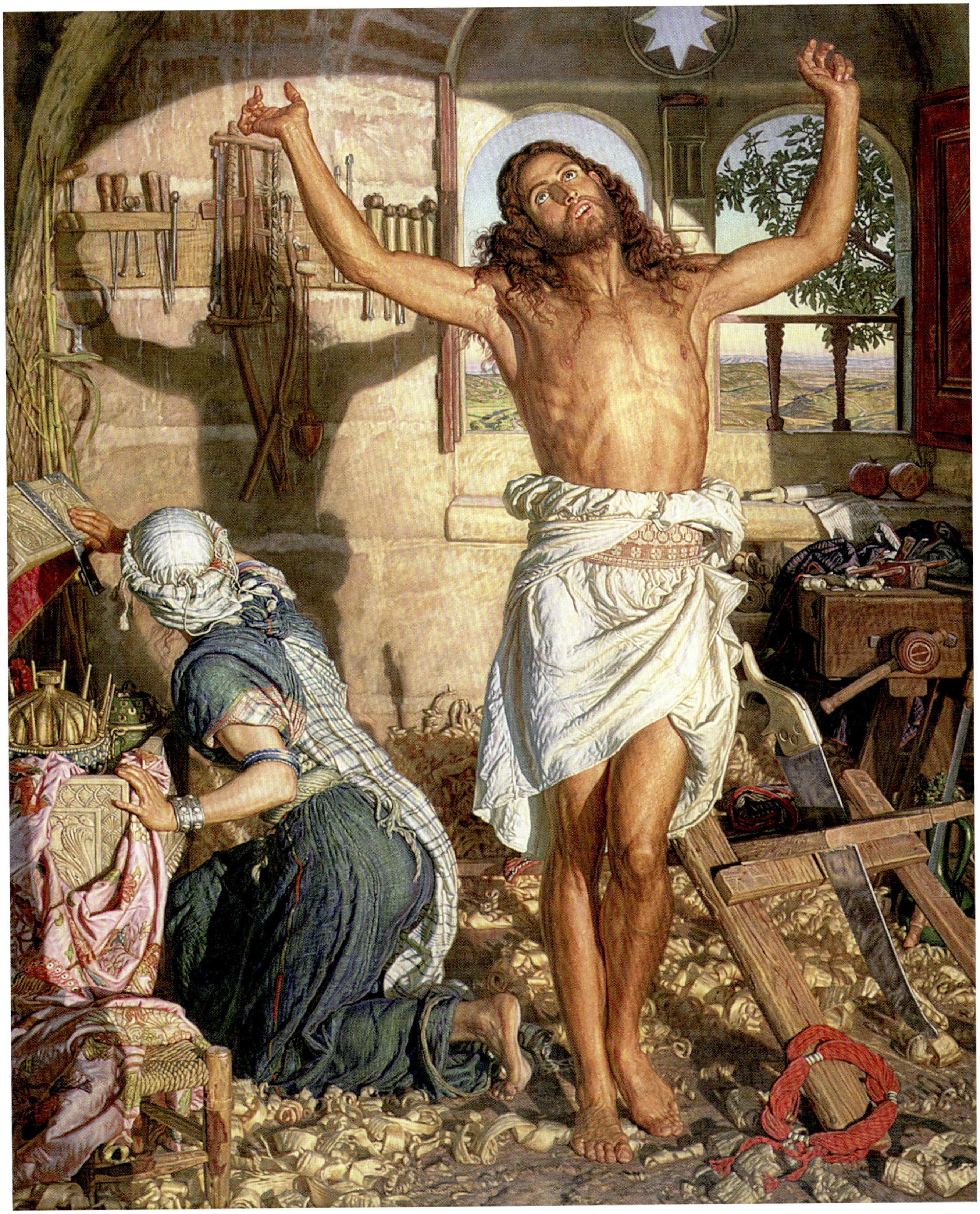

55 William Holman Hunt, *The Plain of Esdraelon from the Heights above Nazareth*, 1870, 1877. Ashmolean Museum, Oxford.

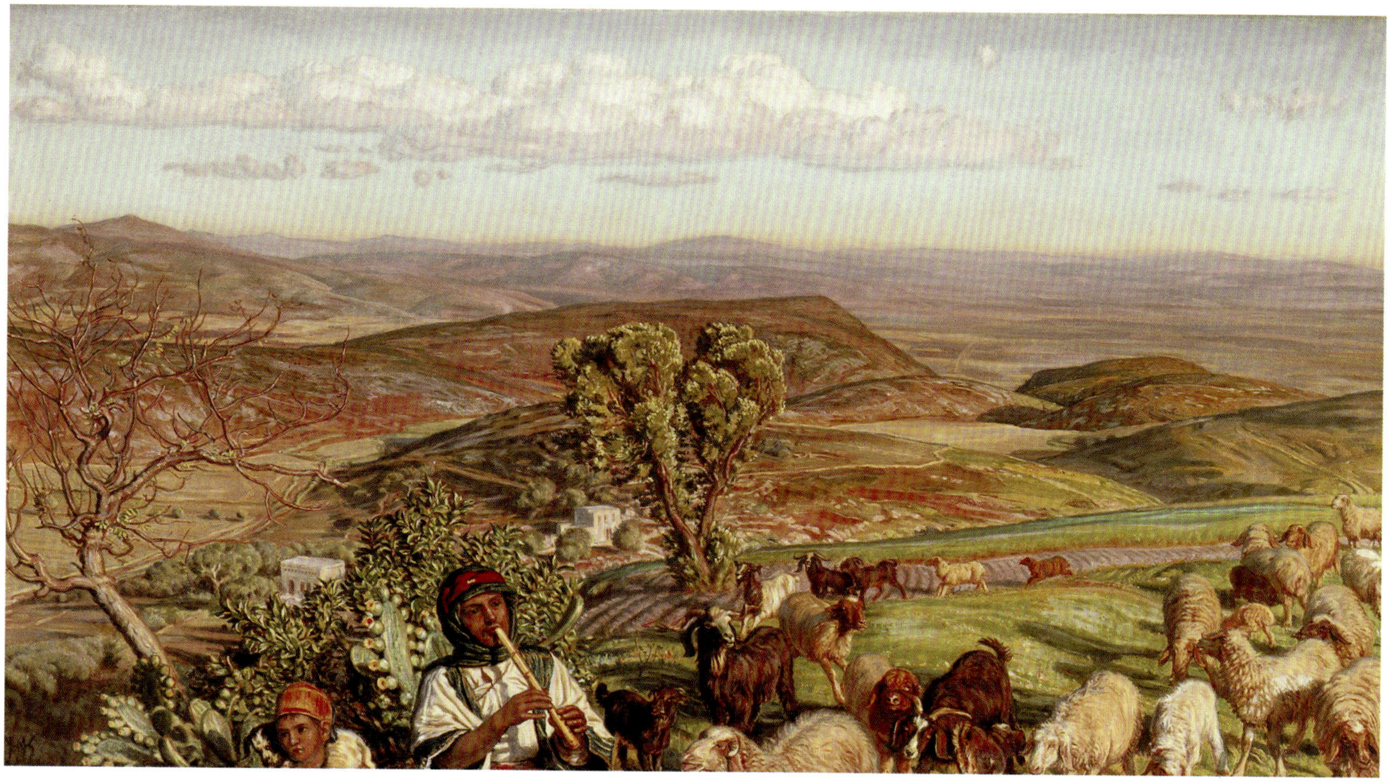

years earlier. A comparison of their two paintings shows both Hunt's homage to Millais and how much further Hunt took their shared commitment to realism. Where Millais limited himself to the meticulous recreation of a carpenter's shop and his family, every detail of Hunt's painting is carefully researched for authenticity.

In the pamphlet that accompanied the painting's exhibition in 1873 Hunt details his sources for the saw ('early Egyptian representations of this tool and the form of the modern Oriental implement'), the other tools ('a collection of ancient carpenters' implements bought at Bethlehem… most of which appear in Egyptian paintings of a date long anterior to the time of the Saviour'), the red headdress or 'aghal' ('portrayed in both the Assyrian sculptures, and in Egyptian pictures of the Jews led away as captives' and still 'worn by the Bedouin of Syria'), the crown, supposed to be one of the gifts from the magi ('a combination of the forms of coronets of the dynasties of Antiochus and Herod, and of the ancient and modern Persian monarchs'), the design on the ivory box ('almost an exact copy from the ornamentation of capitals of columns still existing at Persepolis') and the rounded arch of the windows ('justified by more than one example of buildings of the Christian era, discovered at Jerusalem'). The landscape visible through the window, he explains, 'represents the hills of Galilee, with Gebel-al-Covvies, the Hill of Precipitation, and, further off, the plain of Jazreel, and, beyond this, the mountains of Gilboa'.[70]

Hunt is very clear about his working methods here. As in *The Finding of the Saviour*, the central principle is that every feature of the painting should be deduced from the comparison and correlation of ancient archaeology with modern life in the Middle East. The landscape is presumed not to have changed significantly, at least in historical if not geological time, and the contours are recognisably the same as those of Hunt's contemporary pastoral landscape *The Plain of Esdraelon from the Heights above Nazareth* (fig. 55). On the face of it, Hunt is guilty of the orientalist presumption that Middle Eastern civilisation and character too had become fixed centuries before, still holding their ancient form in the modern world.[71] Yet he is scrupulously careful to check all the cultural details of the scene as far as possible against archaeological as well as contemporary evidence.

Carol Jacobi is surely right that the 'historical and visual accuracy of this painting is an attempt to answer fears about the subjective, non-empirical, nature of faith'.[72] Hunt's Pre-Raphaelite sacred art lives by the standards of science, not in the teeth of it. But what he offers us is also, as Jacobi again notes, emphatically 'a Christ for a post-Darwinian age'. In presenting 'a naked Jesus of extraordinary physical presence', he sets out to counter the 'doubts raised by archaeology, geology and science on their own materialist terms' – to give us the determinedly 'material veracity' that Ruskin so valued in Pre-Raphaelite art.[73] Although we see Mary's moment of foreboding in the painting – as we do in the much earlier paintings, designs and poems by Millais, Rossetti and Collinson – Hunt does not offer an insight into Jesus's specific psychology through its expression on his body. What he does do is show us a palpably real 'Man Christ'.[74] Again, Stephens insisted in his review of the painting for *The Athenaeum*, 'the extraordinary solidity of the flesh, which looks sound and round as life,' is down to Hunt's 'scientific' study of every detail of his anatomy.[75] Looking at Hunt's painting, we can almost feel Christ's beating heart, as Tupper felt the heart of the Theseus.

As well as being a thoroughly material repost to materialism, Hunt's Christ in *The Shadow of Death* is an intervention in the ongoing debate over the ramifications of the higher criticism. Alison Smith notes that Hunt painted it after reading the French scholar Ernest Renan's *Vie de Jésus*, first published in 1863.[76] Like Renan, and like Strauss before him – and indeed like Millais in the painted version of *Christ in the House of his Parents* – Hunt excludes any directly supernatural happenings from his painting. But where Renan sought to co-opt Jesus for an anti-Semitic racialism by claiming that he had managed to shed his Jewishness and become an Aryan, Hunt dismissed Renan's account as a 'sentimental travesty' and insisted that Jesus must be recognised as ethnically Jewish.[77] To this end, as he explained in his exhibition notes – and following his own practice in *The Finding of the Saviour in the Temple* – he composed his figure of Christ from twenty Jewish models, including several 'natives of the tribe of Judah, the tribe to which our Lord belonged'. He made a point too, he says, of living for a time among them.[78] The logic of this practice was that, by living alongside the people alive today who are genealogically and culturally closest to Jesus, it would be possible to get to know them and through them to know Christ himself. As well as pursuing the physical anthropology typical of the 1860s, then, Hunt also pioneered a form of what we would now call social anthropology as yet another scientific component of his artistic method.

Hunt is sometimes denigrated as an orientalist, a bluff Victorian convinced of his own superiority to the perpetually frustrating and at times disgusting Arabs.[79] Undoubtedly, he often comes across in this way in his own journals. Yet his scientific mode of painting often enabled him – and us – to see things which he was casually blind to in the day-to-day prejudices he shared with many of his contemporaries. The healthy sexuality of the young woman in *The Hireling Shepherd* is a case in point. So too are the strength of character caught in the answering gaze of the fellah woman in *The Afterglow in Egypt* and the social politics of *The Lantern-Maker's Courtship* (fig. 56). Here the flirtatious interaction of a young man and a young woman in a Cairo street scene is presented with good humour and without passing judgement. Indeed, this scene, like *The Hireling Shepherd*, suggests once again that Hunt was accepting towards consensual intimacy on equal terms. At the same time, the figure of a white man in a top hat mounted on a donkey and beating at some Arab boys in the background, far from being 'Hunt's own idealised self-image', is a sign that he was also aware of the violence intrinsic to the claims laid by white men like himself to superiority over other races and cultures.[80]

Hunt did not treat Jews with the same degree of prejudice as he did Arabs in his day-to-day life. His depictions of Christ as a Jew in *The Finding of the Saviour* and *The Shadow of Death* register the distance between Jesus and modern English Christians. At the same time, they promise that, through careful research and imaginative empathy, it is possible to reach across that divide. Hunt's paintings themselves exemplify this implicitly ecumenical move, extending a hand some way at least across the rift between Judaism and Christianity. Drawing on Stephens, Barringer suggests that

56 William Holman Hunt, *The Lantern-Maker's Courtship*, 1854–6, 1860–61. Manchester Art Gallery.

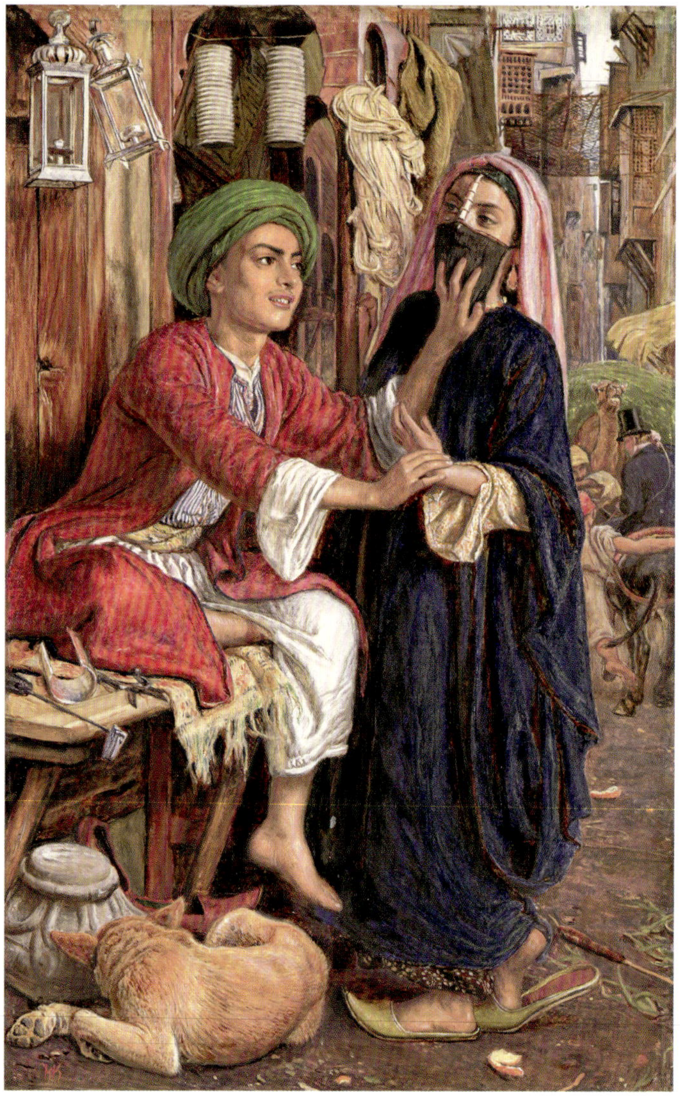

Hunt's depiction of the rabbis in *The Finding of the Saviour* is a deeply 'hostile portrayal of Jewish culture' as sensuous, superstitious and self-satisfied.[81] But as in his interpretations of *Our English Coasts* and *The Awakening Conscience*, Stephens is here more hidebound than Hunt himself. In the key Hunt wrote to accompany a later engraving of his painting, he details each of the ritual objects and practices documented in it without at any point denigrating them as superstitious or lax.[82] As for the rabbis themselves, far from each one being 'a caricature in the Hogarthian tradition' as Barringer sees them, they are presented as serious and thoughtful.[83] Of course, they are conservative in their attachment to their own religion – the biblical story requires this of them, and besides they are rabbis and therefore learned in and inevitably attached to their own faith. But with one literal exception – the 'blind, imbecile, and decrepit' old priest, to use Stephens's unkind words, near the left-hand edge of the scene – they are not blind in this attachment, but rather keen-eyed and curious.[84]

Hunt himself is kinder than Stephens to this old man. In the other old rabbi's gesture of gently touching his hand, he registers that he is pitiable but not contemptible. These old men may be unable to accept or even recognise Jesus's challenge to their beliefs, but their care for one another is nevertheless a sign of their humane sympathy. The two younger doctors seated next to them are giving Christ's arguments serious consideration. Hunt shows them both gazing into the middle distance in thought, the one neglecting the scroll in his hand, the other twiddling a pen. The older, orange-bearded man leaning over them too is remarking on something Jesus has said or done. None of them makes direct eye contact with the viewer as Jesus himself does, and Hunt's Christian theology inevitably confines them to their own moment, a moment Christ alone transcends. But within their own time they are presented with the same sympathetic attention that characterises the earlier Pre-Raphaelite group scenes discussed in Chapter 3. In the impish little girl with the whisk in the bottom left-hand corner who exchanges a sardonic glance with the boy behind the old men, Hunt locates a wry satirical impulse within the society itself, rather than subjecting it to satire from without. In the young mother kissing her baby in the distance even as her husband takes a lamb to be sacrificed to redeem the baby, he shows his affectionate sympathy for the people in this alien culture even as he repudiates their religious practices.

The biblical scenes Hunt painted in situ embody to the full the Pre-Raphaelite project of grounding art in the practices of science. In *Ecce Ancilla Domini!* and *Christ in the House of his Parents*, Rossetti and Millais had already brought about extraordinary new departures in sacred art, the one through unwavering psychology, the other through a defiant realism. Building on Millais's achievement in particular, Hunt was able to achieve a third innovation in religious painting. In keeping with the Pre-Raphaelite project as a whole, Hunt's paintings of the ancient Near East insist that morality must be grounded in a humane sympathy that can reach across cultures and religions, places and times, and even across the divide between humans and other animals. At the same time, they propose that full knowledge of a subject can only be gained by seeing it in its own time and place. Even Christ himself must be understood ecologically to be understood correctly. To achieve this requires not only the meticulous scrutiny of natural history and anatomy typical of Pre-Raphaelite art, but also geographical exploration, an up-to-date and practical knowledge of archaeology, and original research in social and physical anthropology. The three major paintings Hunt created in his first two visits to the Holy Land, as he would have thought of it, are at once a triumphant vindication of the claims the Pre-Raphaelites made in *The Germ* to model their art on science and a culmination of the programme that they formulated in that early magazine.

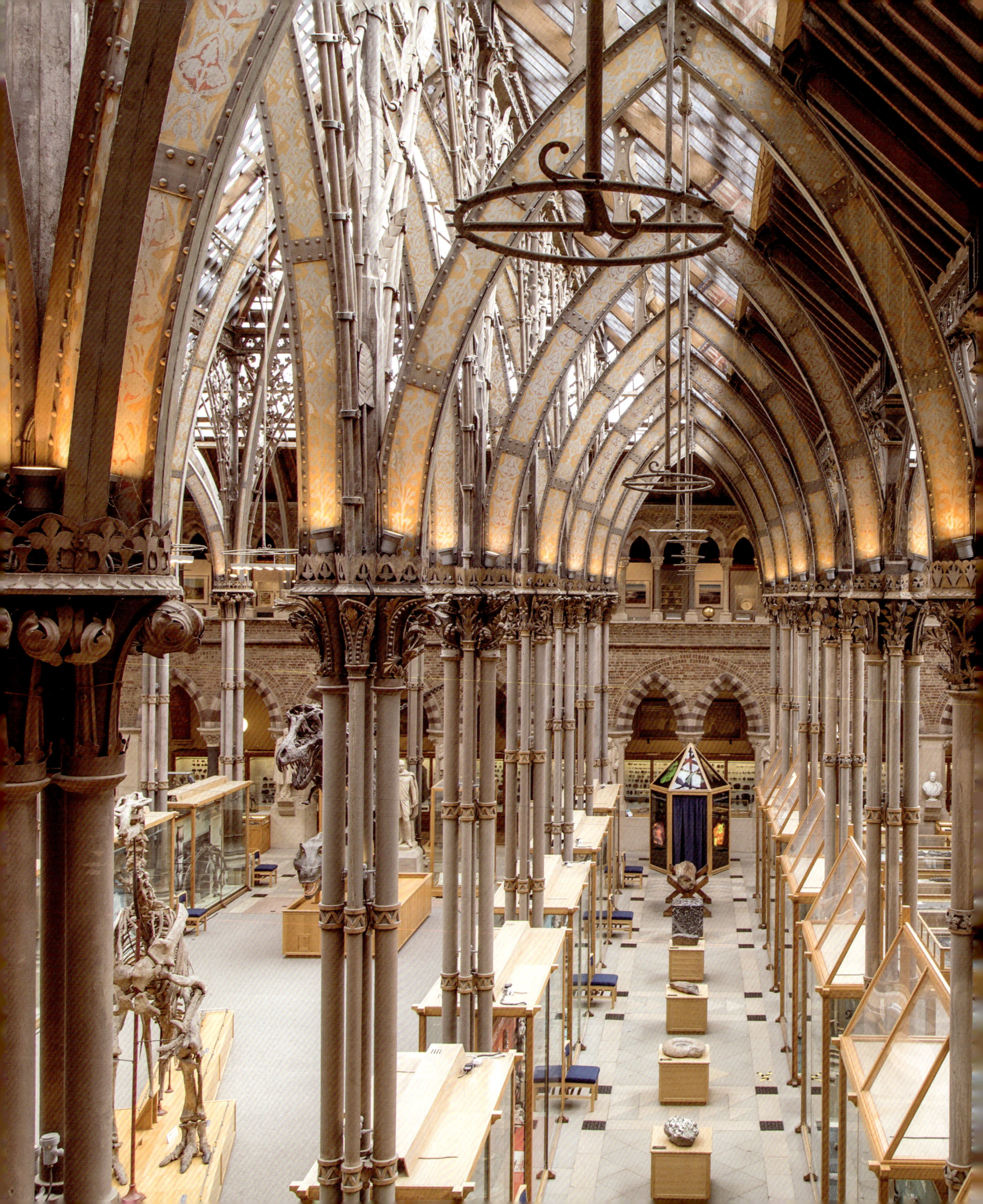

5 · An Experiment in Architecture

PRE-RAPHAELITISM AND SCIENCE AT THE OXFORD MUSEUM

So far I have been examining how the Pre-Raphaelites modelled their artistic practice in painting and poetry on science. This was a remarkably thoroughgoing project across their art. It began with the formation of the Pre-Raphaelite Brotherhood in the late 1840s and reached its fullest expression in Hunt's working visits to the Holy Land. As well as effecting a major shift within art itself, through their poems and paintings the Pre-Raphaelites were able to achieve a deeper understanding of ecology, on the one hand, and of psychology, on the other, than many contemporary scientists. These insights remain striking and valid, contributing to the enduring power of their work. But the influence of the Pre-Raphaelite experiment in creating art on scientific principles was not limited to art alone. It had a profound influence too on the material and intellectual culture of Victorian science itself.

The naturalist William Broderip, the palaeontologist Richard Owen and the physician Henry Acland were among the Pre-Raphaelites' early patrons and supporters, at a time when the press and the artistic establishment seemed set on crushing them. The relationship between Victorian scientists and Pre-Raphaelite artists went well beyond mere patronage, however. Writing in 1856, Tupper remarked that there had never been 'so broad a road of intercourse and fraternization between artist and [natural] philosopher, as now': 'There are archaeologists, palaeontologists, microscopists, histologists, anthropologists, psychologists, and transcendental anatomists, all ready to run into the arms of the artist, if he is only strong enough to bear the embrace.'[1] Just such a collaboration between scientists and artists had begun only a little over a year earlier when the scientists of Oxford University, led by Acland, had invited Ruskin and the Pre-Raphaelites to work

Facing page Detail of fig. 61.

57 Deane and Woodward, Oxford University Museum, 1855–60.

with them on designing and decorating a new university museum of natural history. Conceived as an entire science faculty, the museum building was to be an experiment in transforming architecture into a representational art form which could communicate the true nature of nature itself. This experiment would entail geologists, botanists and anatomists working closely together with architects, designers, sculptors and even poets. Both through this collaboration and as an avowed policy, Pre-Raphaelite aesthetics came to define the environment in which science was studied and taught at Oxford University in the second half of the nineteenth century, and how and where it was communicated to the wider public who visited and still visit the museum. As for the museum itself, it has a strong claim to be the greatest single work of Pre-Raphaelite art. This chapter aims to explain how this comes to be: how the collaboration between the scientists and the artists came about; how architecture became one of the Pre-Raphaelite arts; how the façade, forms and materials of the museum set out a vision of the natural world as God's creation; and how two groups of artists put forward their own visions of nature and science through two distinct programmes of Pre-Raphaelite sculpture at the museum.

OXFORD SCIENTISTS AND PRE-RAPHAELITE ARTISTS

Pre-Raphaelitism was always fundamentally collaborative. Initially, the Pre-Raphaelites had worked with one another, looking across to science from the vantage of the arts. But as their art drew the attention of the scientific community, so new projects took shape which involved scientists themselves working closely alongside artists. The earliest and richest of these collaborations began in December 1854 when the dons at Oxford University chose a Gothic design by the Irish firm of architects Deane and Woodward for their new museum of natural history (fig. 57).[2] The Regius Professor of Medicine John Kidd and the Keeper of the Ashmolean Museum Philip Bury Duncan had been calling for a new museum to house natural history specimens since the late 1820s, but institutional resistance in a university wedded to a classical and theological curriculum and wary of the latent materialism of science meant that these initial proposals came to nothing.[3] In the late 1840s, however, the campaign began again in earnest. In July 1847 the Professor of Chemistry and Botany Charles Daubeny, the Reader in Experimental Philosophy Robert Walker and Acland, then Reader in Anatomy, joined Duncan in signing a Memorandum to the University calling for the building of what was in effect a new science faculty, combining lecture rooms, laboratories, a library and a museum in one.[4] Their objective was to stake out a permanent place for the sciences within the Oxford University curriculum. The polymath and educational reformer Richard Greswell at Worcester College threw his weight behind the proposals, as did the Deputy Reader in Geology H. E. Strickland. Even so, it was a drawn out and frustrating process. In February 1853 the Convocation of the University at last consented to appoint a delegacy or subcommittee to look into the matter. When the delegacy reported in favour of building a new museum, Convocation voted against. In spite of scathing retorts from Daubeny and Greswell, it was not until December that a grant was finally approved to buy the land to build the museum, while the money itself was not voted until the following March.[5] An open competition for architects was launched in April 1854, the designs were exhibited in November, and the winning design chosen by a vote in Convocation on 12 December.[6] Even this was not the end of the matter, as an anonymous Member of Convocation called one last vote to stop the 'Babylon of a new Museum' in May 1855, but the University at last knew its mind.[7] The foundation stone was finally laid on 20 June 1855.[8]

The proposal for the Oxford museum was not only a matter of educational controversy. It rapidly became the focus of intense architectural debate. Greswell raised the issue as early as 1850, returning to it in a pamphlet in 1853 where he set out a proposed plan for the building. He himself was not particular about the style of architecture to be employed, whether 'Grecian or Gothic'.[9] But the Diocesan Architect for Oxford George Edmund Street put in a strong plea for Gothic, specifically the fourteenth-century English style.[10] So did another pamphleteer, possibly J. H. Parker, an Oxford printer and bookseller who went on to be Keeper of the Ashmolean. Parker had published Street's pamphlet

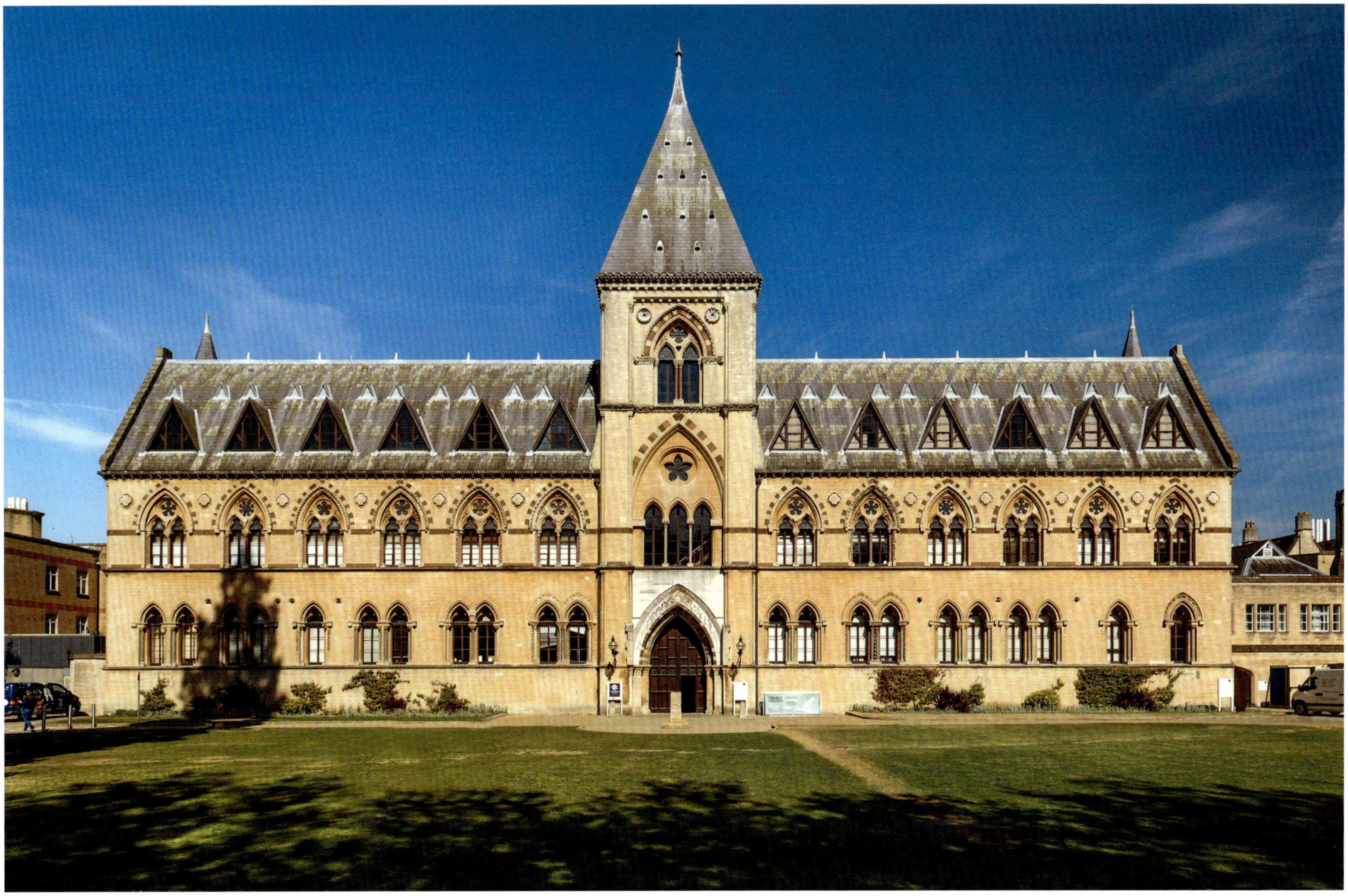

for him, but here he knocked him back from fourteenth- to thirteenth-century Gothic.[11] Like Greswell, he included his own proposed plan for the museum in his pamphlet. By his own testimony, Acland too was a lover of Gothic architecture.[12] He intervened to promote Deane and Woodward's design the very day before it went to the vote.[13]

It was primarily through Acland that the Oxford museum became not only a major Gothic Revival building but also a collective work of Pre-Raphaelite art. He and Ruskin had been undergraduates together at Christ Church in the 1830s. In 1851 he met and befriended Holman Hunt.[14] Two years later, Acland (fig. 58) joined Ruskin and Millais at Glenfinlas, where Millais was painting the background to his portrait of Ruskin (see fig. 13). Acland was converted to Pre-Raphaelite aesthetics by Ruskin's arguments and Millais's example, and the painting itself ended up hanging in his library back in Oxford.[15] When Ruskin heard that Deane and Woodward had won the contract for the museum, he wrote to Acland, saying, 'I hope to be able to get Millais and Rossetti to design flower and beast borders.'[16] Soon he became still more ambitious, writing to his friend Pauline, Lady Trevelyan, that he wanted to get all the Pre-Raphaelites involved in designing capitals and arches for the museum.[17] Ruskin did indeed involve Rossetti in the project, prompting Acland to contact him about the museum while he himself introduced Rossetti to Benjamin Woodward (fig. 59), the architect primarily responsible for the building.[18] Rossetti and Woodward became fast friends, while on Ruskin's advice Acland was engaged as Elizabeth Siddal's doctor.[19] Rossetti passed on an invitation to Millais to join the project as well, but

58 John Everett Millais, *Henry Acland*, 1853. Ruskin Library and Research Centre, Lancaster.

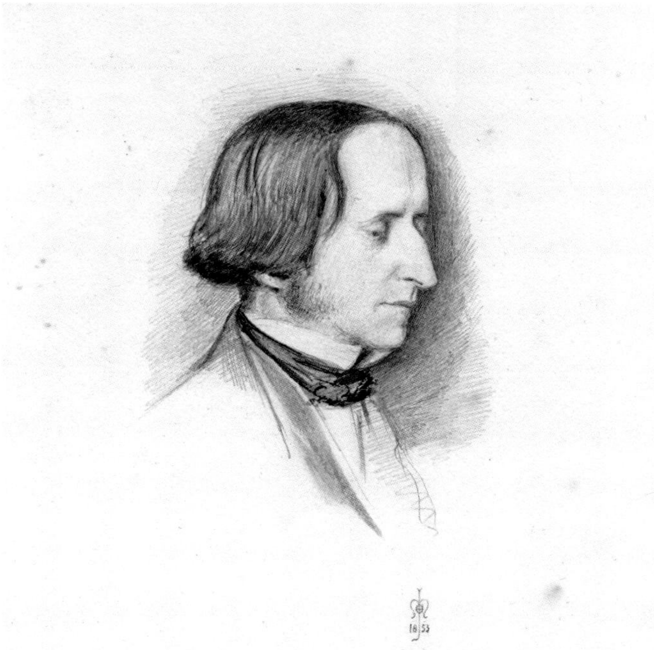

59 Alexander Munro, *Benjamin Woodward*, 1861. Oxford University Museum.

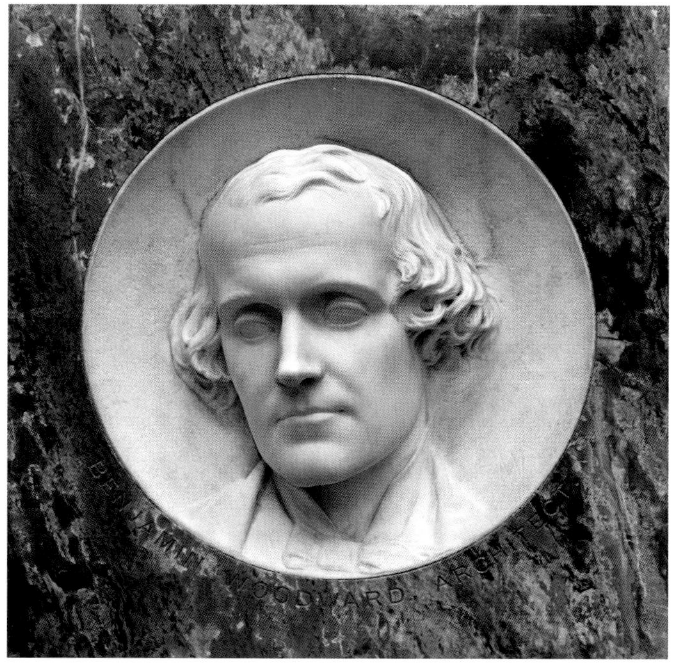

since becoming involved with Ruskin's wife, Effie, Millais had cut off all ties with him and would not agree to work with him on the museum.[20]

Woodward repeatedly pressed Rossetti to contribute decorative designs and even a statue for the museum.[21] Rossetti equally persistently flannelled. In the end he did not create any work directly for the museum himself, but he brought several other Pre-Raphaelite artists into the project. The University determined that the museum should house a series of statues of great scientists. Through Rossetti's efforts, the commissions for the statues of Bacon and Linnaeus went to Woolner and Tupper respectively, neither of whom had won commissions for public sculpture before.[22] Another Pre-Raphaelite sculptor who became very actively involved was Alexander Munro. Munro had been a close friend of Rossetti ever since he had enrolled at the Royal Academy Schools in 1847. Like Woolner and Tupper, he sought to apply Pre-Raphaelite principles in his sculpture. One of his commissions was a portrait medallion of Acland's wife, made in 1854.[23] Munro was contracted to sculpt the statues of no fewer than seven scientists, though in the end he only completed six. One further commission, for the statue of Aristotle, went to a fourth sculptor with close Pre-Raphaelite connections. Henry Hugh Armstead, who later became known for the friezes round the Albert Memorial, was a silversmith who only turned to sculpture as a career in the early 1860s. He was also a good friend of Hunt.[24] Again, *Aristotle* was Armstead's first important work on a large scale. There is no direct evidence that Hunt recommended Armstead for this commission, but given his connection with Acland, and as he too became friendly with Woodward through Rossetti, it seems likely.[25]

When the museum opened, the nine statues on display were all by sculptors grounded in Pre-Raphaelitism. They were joined over the next few years by four statues by the older, more established sculptors Henry Weekes, Edward Stephens and Joseph Durham. Four more were added between 1886 and 1914, all by Henry Hope-Pinker. In the meantime, several Pre-Raphaelite artists had been involved in designing the museum's decorative scheme. Before the building was even begun, Ruskin consulted Rossetti and Munro on the decoration for the windows. He sent Acland a portfolio of twelve of his own designs, elements from several of which were incorporated into the carvings on the museum's façade.[26]

The job of designing the arch over the main entrance was passed from Rossetti, who agreed to do it but did not, to Woolner, who drew a design which was approved by Woodward but never carved, and on to John Hungerford Pollen, whose design was partly carved, with modifications, before the funds ran out and work on the front of the building stopped in 1862.[27] Pollen, who has been described as 'Deane and Woodward's most important artistic collaborator', had been Professor of Fine Arts at Cardinal Newman's Catholic University in Dublin in the mid-1850s.[28] After returning to England in 1857, he worked on several of Woodward's buildings, including producing natural historical friezes for the long-since demolished Crown Life Assurance Office in Blackfriars, described by Rossetti as 'the most perfect piece of civil architecture of the new school that I have seen in London'.[29] By this point his association with the Pre-Raphaelites was well established. In 1850 he had worked alongside Millais in Merton College Chapel, where Pollen decorated the ceiling while Millais drew the windows for *Mariana*. The following year he met Hunt, going on to supply him with some of the embroidery he needed for *The Light of the World*. The admiration between Pollen and the Pre-Raphaelites was mutual. Like Collinson, he was sympathetic to their artistic aims where they did not conflict with his own Catholic faith, and he especially admired their attention to natural detail.[30]

In spite of Woodward's invitations, none of the close Pre-Raphaelite circle contributed directly to the decorative work on the museum's interior, although Ruskin's friend Pauline Trevelyan is known to have drawn designs for several capitals in consultation with William Bell Scott, while his protégé Revd Richard St John Tyrwhitt painted two murals in the geology lecture room.[31] Even so, the principles of Pre-Raphaelitism were central to its aesthetic. Acland stated this explicitly in a lecture he gave to delegates of the General Architectural Congress in June 1858 when he took them to see the work in progress on the museum. After reading out a letter of support by Ruskin, Acland closed his lecture by affirming that, in the spirit of truth, 'we have sought to hinder all ornament, unless that ornament be free from vicious carelessness; and to stop all professing transcript of Nature, unless it be painstaking, sagacious, and honest. Herein,' he continued, 'we owe a just debt of gratitude to the young school of Artists, called, half in jest, Pre-Raffaelites'.[32] Pre-Raphaelite attention to and interpretation of natural forms is the rule across all the decorations in the building, including the wrought ironwork of the spandrels and pillars of the roof designed and manufactured by Francis Skidmore of Coventry, the stylised decorative scheme for the entomology museum prepared by Ruskin's protégé Henry Swan, and above all the richly expressive carvings on the capitals and corbels around the central court executed by the brothers James and John O'Shea and their nephew Edward Whelan.[33]

Woodward brought the O'Shea family over from Ireland to work on the museum and the Crown Life building after the success of their earlier work for him at Trinity College Dublin. They have been at once celebrated and patronised as 'Nature's own Pre-Raphaelites'.[34] This characterisation stems from nineteenth-century attitudes to the Irish and to the rural working class, reflected in both Acland's good-humoured and Ruskin's more ambivalent and at times unkind references to James O'Shea in their accounts of the museum.[35] Because of their social standing and their medium – decorative carving – the O'Sheas are still more often thought of as artisans or craftsmen than as artists. Yet they identified themselves as sculptors in the 1861 census, and were recognised as fellow artists by the Pre-Raphaelites.[36] Rossetti commented, in a letter quoted by Alexander Gilchrist in an obituary of Woodward, that under Woodward's direction they had produced 'innumerable works of real beauty in all branches of architectural carving'.[37] Stephens devoted the best part of a review article on the museum to a detailed and appreciative account of their decorative sculpture.[38] After working on the museum, Woolner worked alongside them – and Skidmore too – in decorating the Manchester Assize Courts for Alfred Waterhouse.[39] He went on to employ James O'Shea to decorate his own house and studio.[40] The O'Sheas never joined the Pre-Raphaelites' social circles, but, like Joseph Wolf, they were recognised by them as fellow artists pursuing comparable aims in their own field and medium. In their work at Oxford they developed the exuberant natural history carving which they had been pioneering in Dublin in line with Acland's commitment to Pre-Raphaelite principles.

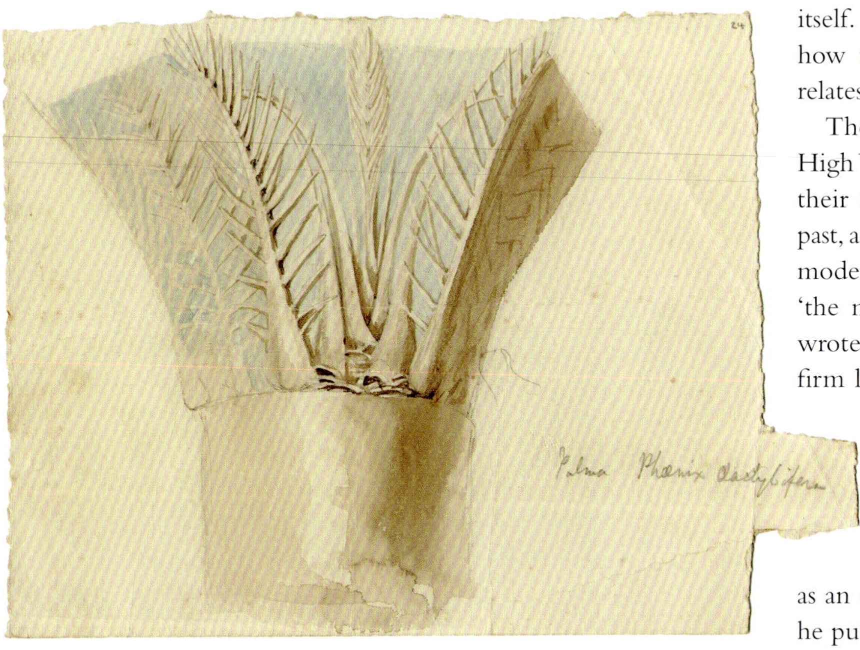

60 John Phillips, design for palm capital, Oxford University Museum, *c.*1857.

While Acland set out the museum's aesthetic principles, its decorative scheme was mapped out in detail by another scientist, John Phillips.[41] Phillips had come to Oxford in 1853 to succeed Strickland as Deputy Reader in Geology. He succeeded Duncan as Keeper of the Ashmolean too in 1854, and William Buckland as Reader in Geology in 1856, and in 1860 he became the first Keeper of the new museum. As the University's senior geologist, he selected the marbles for the columns round the central court. Each column was to be made of a single piece of British or Irish stone, labelled on the base for teaching purposes, and so extending systematically a motif which had been used before at the Museum of Practical Geology in London, where Phillips had worked for the Geological Survey, and at Trinity College Dublin, where Deane and Woodward had used a selection of Irish marbles for the columns and balustrades.[42] Phillips also decided on the botanical scheme for the capitals around the court, as he explained in a letter published alongside Acland's 1858 lecture and two letters by Ruskin, first as a promotional pamphlet and later as a guidebook to the museum.[43] Like Acland, he embraced Pre-Raphaelite principles in the execution of this scheme, even sketching out some designs for capitals himself (fig. 60).[44] Working in collaboration with Daubeny, who as Professor of Botany was in charge of the University's Botanic Gardens, he supplied the O'Sheas with specimens of the plants he had in mind for them to work from in their carvings.

The Pre-Raphaelite collaboration at the Oxford University Museum, though it included the Pre-Raphaelites themselves, thus went well beyond them, taking in the architect, the other artists employed to decorate the museum and the scientists who commissioned it. Its significance too went well beyond its place within the Pre-Raphaelites' own project. Indeed, the Oxford museum became one of the most influential buildings of the Victorian period, establishing a durable if contentious model for the communication of science through the fabric of museums, setting a new standard for modern Gothic architecture, and fusing decorative art with didactic science.

PRE-RAPHAELITE PRINCIPLES IN ARCHITECTURE

In its structure and decorative details, the Oxford museum gives physical form to distinctive and complex visions of nature as revealed by science. In the series of statues around its central court, it implicitly puts forward an idea of science itself. Before examining these in detail, it is worth asking how the architecture of a building such as the museum relates to the Pre-Raphaelite project in painting and poetry.

There is a clear affinity between Pre-Raphaelitism and High Victorian Gothic architecture. Both movements took their name and their aesthetic models from the medieval past, and yet both conceived of themselves as wholeheartedly modern. Together they formed what Eve Blau has called 'the medievalist artistic avant-garde'.[45] In the memoir he wrote towards the end of his life, Hunt sought to draw a firm line between the medievalist and the modern aspects of Pre-Raphaelitism. Rossetti and his camp had been mere Gothicists, Hunt claimed, following a revivalism that was every bit as imitative as the classicism preached by the Royal Academy. He and Millais, by contrast, had taken the early Renaissance painters as an example to be built on, not a model to be copied. As he put it in a somewhat tortuous metaphor, 'We drew from this fountain source, and strove to add strength to its further meanderings by the inflow of new streams from nature and scientific knowledge.'[46] In fact, as we have seen, the Pre-Raphaelite Brotherhood as a whole took their lead jointly from science and medieval art, albeit in different ways and proportions. Much as the Pre-Raphaelites worked creatively with the principles of early Renaissance painting, so the new generation of Gothic architects who came to maturity at the same moment took the elements of medieval architecture and transformed them into ambitious and arresting new buildings. As Michael Lewis puts it in his compelling history of the Gothic Revival, 'an energetic body of architects and patrons was released in a sudden spasm from a position of the most docile antiquarianism to one of dazzling and limitless originality.'[47]

The leading English architects of High Victorian Gothic were closely associated with one another, with the Pre-Raphaelites, and with the pioneers of the Arts and Crafts movement. Street had worked under George Gilbert Scott in the 1840s, as had George Bodley. Philip Webb and William Morris both worked for Street in the 1850s. Earlier Webb had studied with William Butterfield.[48] As an Irishman based in Dublin, Woodward was not part of this lineage, but he was a member, along with Street, Bodley, Webb and Morris, of the Hogarth Club, which was founded by Ford Madox Brown in 1858 and is generally acknowledged to have been, in Deborah Cherry's words, 'a Pre-Raphaelite venture'.[49] Aside from Millais and Collinson, the original P.R.B.s were all members, as were Boyce, Brett, Burne-Jones, Hughes, Pollen, Ruskin and Tupper.[50] Speaking to the Ecclesiological Society in June 1858, just over a week before Acland gave the lecture on the museum at which he paid tribute to the Pre-Raphaelites, Street went so far as to declare to his fellow Gothic architects, 'the Pre-Raphaelite movement is identical with our own.' Each had to contend against false artistic conventions and the ridicule of the establishment, each looked to the example of medieval art, and each insisted that the first necessary principle in all art was 'absolute unwavering truth'. In both movements, looking back to medieval art was not an end in itself, but a means, in Street's words, to 'enable men possessed of its principles to grapple with the difficulties of nineteenth century inventions and thoughts'.[51] Speaking for the Pre-Raphaelites proper, William Rossetti repaid the compliment when he remarked in an essay on the International Exhibition art in London in 1862 that 'we regard the Gothic movement as equally an endeavour after realism, or constructive use and decorative propriety, rather than as a mere dilettante vamping-up of an old form of art.'[52]

Street's call for modern Gothic architecture parallels closely Stephens's arguments regarding painting in *The Germ*. This is clearer still in the pamphlet he contributed to the debate on the architecture of the Oxford museum in 1853, *An Urgent Plea for the Revival of True Principles of Architecture in the Public Buildings of the University of Oxford*. Here Street recalls both Stephens's and Tupper's essays on art in his own sphere. He insists: 'that architecture cannot be the best which is content to forego the use of the greatest mechanical advantages and inventions.' This is at once a demand for the utmost modernity and, surprisingly, a case for Gothic. The Pre-Raphaelites' invocation of science as a model for art hints that art, like science, might progress through time. Street's argument for Gothic is explicitly founded on a narrative of progress in engineering. The greatest invention in architecture, he contends, was the arch, and the Gothic pointed arch was structurally an advance on the Roman and Romanesque rounded arch. To argue for classical principles in architecture over Gothic ones – to return to rounded arches, or worse still lintels – is to substitute redundant for advanced engineering.[53]

Street's case for the Gothic arch is paradoxically of a piece with his insistence on the use of modern materials and building techniques in contemporary architecture. This is itself analogous to Stephens's and Tupper's insistence that modern art should address modern life. In *The Germ* Stephens calls for art to engage in 'a wider range of experiment' as science does, to increase its reach and deepen its understanding in new directions.[54] Acland described the Oxford museum in these same terms at a meeting of the Oxford Architectural Society a week before the foundation stone was laid in June 1855. In their annual report, the society expressed their confidence that the museum would show 'that it is possible to use what are called the materials of the nineteenth century, at once with perfect freedom, and with perfect agreement with the English style of the thirteenth century'. On request,

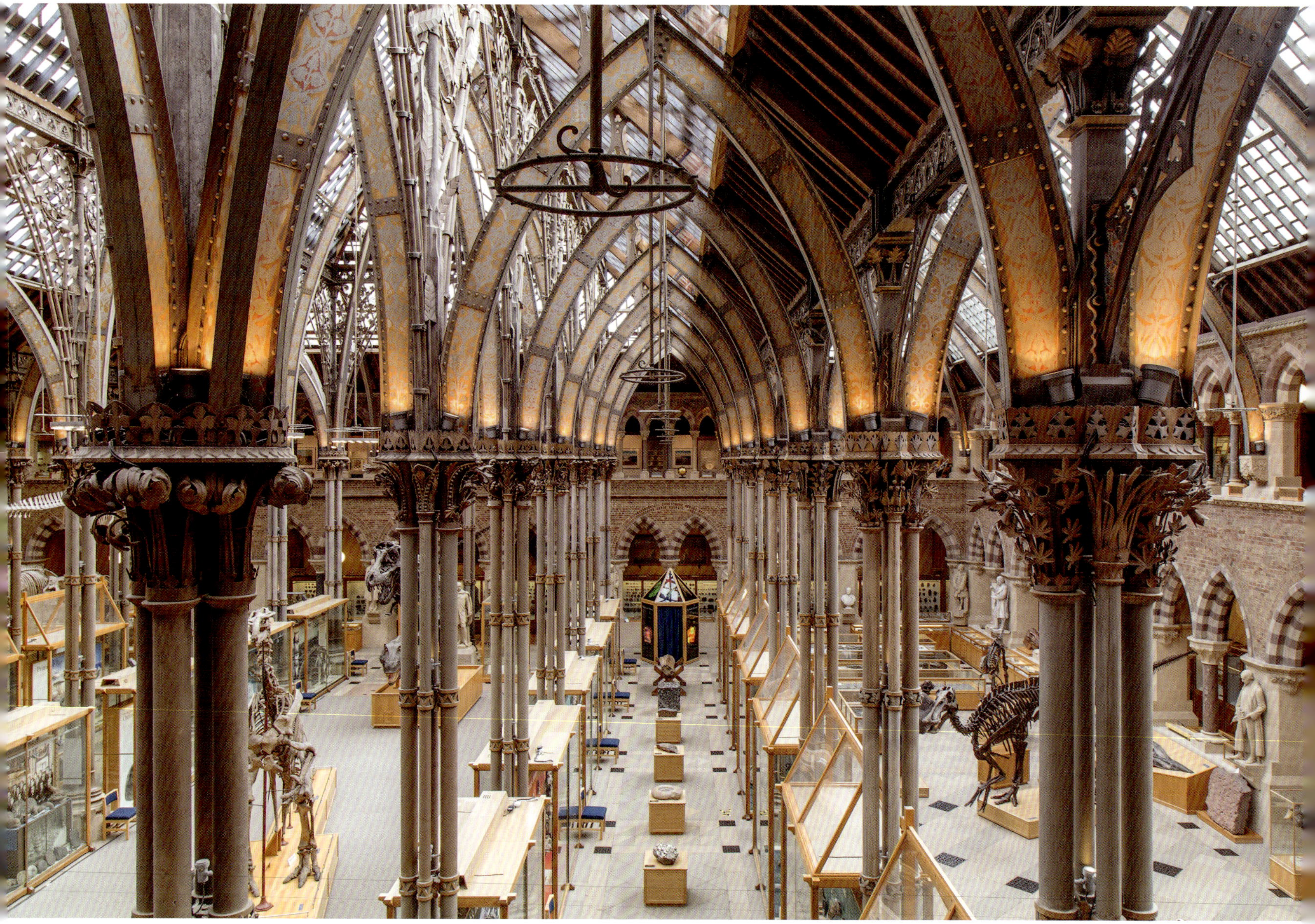

Acland set out the plans for the building in more detail, as the society recorded in its report:

> Oxford was about to perform an experiment; it was about to try how Gothic art could deal with those railway materials, iron and glass; and he was convinced, when the interior court of this museum was seen, – with its roof of glass, supported by shafts of iron, while the pillars and columns around were composed of variously coloured marbles, illustrating different geological strata and ages of the world, and the capitals represented the several descriptions of floras, – that it would be felt that problems had been solved of the greatest importance to architecture.[55]

The fusion of Gothic style with modern materials that Street had called for was to be fulfilled in Woodward's design. The pointed arch was to be refashioned in iron to support a roof of glass (fig. 61), providing natural lighting across the whole court. While the shafts which helped raise the roof were to be iron, the marble columns around the court would be practical and modern in another sense, contributing to the scientific education of students and visitors. As a whole, the

61 Deane and Woodward, Oxford University Museum, 1855–60, interior.

building would be a vision of nature as revealed by science. All this, as Acland was keen to point out, was as yet untried.

Acland's description of the museum as an experiment stresses the pioneering modernity of Woodward's building. In engineering terms, the experiment produced at first a negative result. On Skidmore's suggestion, the roof was initially supported with wrought iron spandrels. These did not prove strong enough to support the thick glass plates that Woodward had commissioned. Ironically, it was Skidmore's involvement in the project that gave the Oxford Architectural Society confidence that the building would succeed.[56] But as Frederick O'Dwyer has commented, he was 'not a structural engineer at all', but a master of ornamental metalwork, and his 'ambition seems to have outweighed his skill'.[57] In February 1858 Ruskin explained to an audience in Tunbridge Wells that machine-made, utilitarian cast iron was vicious. Wrought iron, on the other hand, was the only material properly suited to the sculptural expression of intricate natural forms.[58] It was a theme he had begun harping on nearly ten years earlier in his first major contribution to the Gothic Revival, *The Seven Lamps of Architecture*. Sadly, Ruskin's and Skidmore's ideological commitment to wrought iron was not on its own strong enough to hold up the museum's roof, which collapsed the same month that Ruskin gave his lecture.

As Acland admitted in his lecture that June, the initial 'experiment failed'.[59] On the other hand, the effect of this failure was to compel Woodward and Skidmore to push the Gothic still further into the arms of modernity, forcing them to use not just iron but cast iron, the ultimate 'railway material'. A second, cast iron skeleton for the glass roof was constructed, with the original wrought iron spandrels incorporated as decoration. For Acland, this marked 'a step, but not a final step . . . towards an harmonious union of the ironwork of the nineteenth century with the refined architecture of the Middle Ages', as he explained in the second edition of his guide to the museum in 1860.[60] As the enthusiasm of Acland's language as reported in the minutes of the Oxford Architectural Society reveals, however, the museum was as much an experiment in aesthetics as it was in engineering. The building itself was to be at once beautiful and instructive. Its final structure may have been pragmatic and contingent, but that very contingency was part of its accomplishment, while the flight of the iron arches up to the glass roof, resonating with, yet transcending, the stone arches around the cloister-like court, remain a tremendous symbol of Street's principle that Gothic could be modern as well as medieval.

As well as offering a new vision of modern architecture, the building was to teach its visitors how to understand the natural world. Its architecture, therefore, was to be an act of interpretation founded on science. For the opponents of Gothic, the very notion of a Gothic science museum was absurd. Those who were not content merely to champion classical aesthetics for their own sake couched their arguments in terms of association. When *The Athenaeum* reviewed the different entries for the architectural competition in November 1854, the reviewer attacked the dominance of the 'dreary asceticism of the first and middle pointed school of ecclesiology', which he accused of 'emulating, in an entire absence of grace, the *naïve* transcripts of nature, so painfully elaborated by the votaries of pre-Raffaelism'. Instead, he proposed, 'a building destined as a tribute to the dignity of Experimental Philosophy' should be built in the Tudor style 'loved and honoured by men such as Bacon and Erasmus'.[61] James Fergusson, an inveterate loather of Gothic, satirised the museum mercilessly for its medievalism in his *History of the Modern Styles of Architecture* in 1862. No age was less scientific than the Middle Ages, so no architecture was less appropriate to a science museum than Gothic. 'Nothing', it appeared, 'was to betray the hated and hateful nineteenth century, to the cultivation of whose sciences it was to be dedicated.' The bookcases in the library were 'monkish', the instruments in the laboratory ornamented in such a way as to suggest that they belonged 'to an age long anterior to that when those impostors Torcelli, or Galileo, or Newton, are said to have invented these things'. The laboratory itself (fig. 62), modelled on the Abbot's Kitchen at Glastonbury, led him to imagine that the Professor of Chemistry would be 'practising alchemy' within. It was a surprise, Fergusson remarked, that the dons had not 'reprinted Lyell's "Geology," or Darwin's "Origin of Species," in black letter, and illuminated them, like the building, in the style of the thirteenth century'. The whole thing, he declared, was a 'great forgery'.[62] Fergusson's critique, though funny, is patently unfair. He entirely ignores the modernity of the museum's roof, for

emphatic words, '*precisely the same relative place* that it did in God's own Museum, *the Physical Universe in which it lived and moved and had its being*'.[77] For Phillips, nature was equally emphatically 'the expression of a DIVINE IDEA, the harmonious whole of this world of matter and life'.[78]

Acland refined this idealist version of natural theology, recasting the God of natural theology not as a designer in the functional sense but as a creative and imaginative artist. As he went on to explain in the Harveian Oration which he gave at the Royal College of Physicians in 1865, nature as a whole could not be adequately understood as 'mere mechanical adaptation', but only through 'considering the operation of the Artist's mind'. As 'Art in its highest sense is Creation', so creation is art.[79] In his 1858 lecture, he had indicated how this same conception of creation as art was to be expressed in the museum building itself. Where science allows us to glimpse 'traces . . . of the unexpressed Art of the great Artificer', the arts in their turn preserve the sparks of 'true light from the Beauty and the subtle Law which stamp the meanest work of the Everliving, Everworking, Artist'.[80] For Acland, science and art come together at the Oxford museum in the discovery and representation of the divine artistry of nature. Grounded in science and holding itself to scientific standards, Pre-Raphaelitism was the perfect model for yet another new kind of sacred art that would celebrate not the incarnation, in this case, but the creation.

Students and visitors are inducted into this Christian vision of nature as they enter the museum. The Gothic windows and arches, along with the cloister of the central court, imply that nature is God's work. The arch in the centre of the façade over the main doors makes the same claim boldly and directly. The surviving stages of the design for this arch reveal successive refinements and containments of its message. Woodward initially hoped that the museum would have a projecting porch. Several of his designs survive.[81] They vary in their precise forms, but none is sufficiently detailed to reveal more than the broad ecclesiastical analogy suggested by the use of Gothic itself. When the porch was deemed too expensive by Convocation in May 1858, exposed brickwork was left on the façade which needed cutting away and facing with stone. With an eye to carving what would have been the dominant decorative element of the façade, the spandrels above the entrance arch were faced with the large blocks of pale Portland stone, which Woodward had used extensively at Trinity College both on the façade and for the interior decorations.[82]

Woolner was commissioned to produce a design for relief sculpture to decorate these spandrels. He proposed the subject of the Fall of Man in a letter to Woodward in September 1858.[83] His design (fig. 67), completed by June 1860 at the latest, survives in two copies, one in the British Museum and a smaller copy seemingly done for Thomas Combe which now belongs to the Oxford University Museum itself.[84] It is modelled on the same scene as it appears on the ceiling of the Sistine Chapel (fig. 68). Like Michelangelo, Woolner uses the Tree of Knowledge to divide his pictorial space into two halves. On the left, Eve takes the apple from the serpent and in turn tempts Adam. On the right, Adam and Eve are banished from Eden by an angel armed with a sword, perhaps deriving from Masaccio's mural *The Expulsion from the Garden of Eden* in the Brancacci Chapel in Santa Maria del Carmine in Florence. The composition is highly charged. Woolner takes advantage of the need to fit his figures into the spandrels above the arch to emphasise Eve's erotic desire and her desirability. In Michelangelo's painting these qualities are suggested by her flushed face and its proximity to Adam's genitals. But Michelangelo's Eve has yet to bite the fruit that the serpentine devil is passing her, and she is not yet aware of her own sexuality. Woolner's Eve too has yet to bite the apple, but she is already looking up into Adam's eyes, taking him by the hand, and moving to pull him down to join her on the ground beneath the tree. As she sits down, her breasts are thrust forward out of the plane of the picture, while the curve of her left leg emphasises her thigh and bottom. These effects would have been still more pronounced in the relief sculpture that Woolner was envisaging. For his part, Adam looks stern, yet the grip of his left hand on his right wrist, and the awkward tension this gives to his well-muscled upper body, hint that it is a struggle for him to resist Eve and his own virile desire.

In taking the apple from the snake, Woolner's Eve inadvertently puts her finger in its mouth, while one tip of its forked tongue reaches round the fruit to lick her thumb. These details are at once grotesque and suggestive, implying (*top*) 67 Thomas Woolner, copy of design for spandrels above entrance to Oxford University Museum, *c.*1858–60. Oxford University Museum.

(*bottom*) 68 Michelangelo Buonarotti, *Original Sin and the Expulsion from Paradise*, 1508–12. Sistine Chapel, Vatican Museums and Galleries, Vatican City.

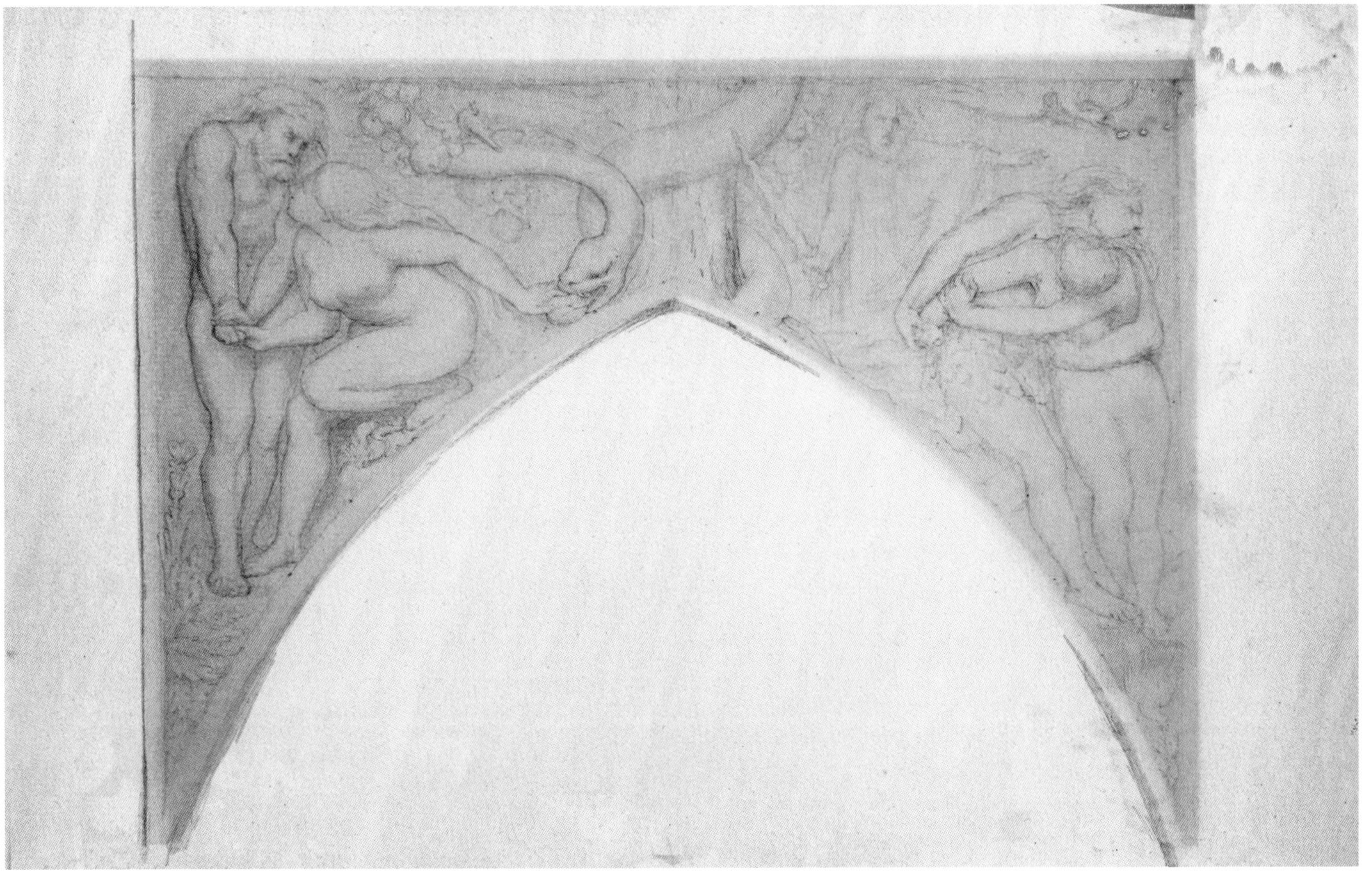

an instantaneous transfer of the erotic charge from the serpent through Eve to Adam. The serpent itself is striking for its naturalism. Where medieval and early Renaissance artists including Michelangelo typically represent the serpent as a demon combining reptilian and human features, Woolner

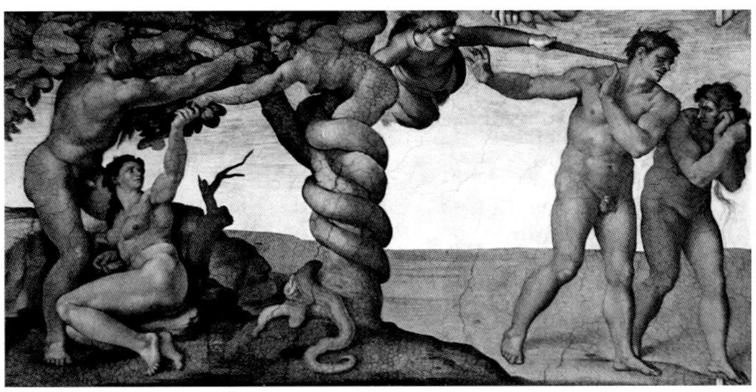

draws a straightforward snake of massive proportions. Its head reveals its malignity through features that are those of an actual snake – the vertical slit of its eye, the fangs and the forked tongue. Its raised nostrils and overhanging brows too are not unusual in snakes, although their massiveness in Woolner's design lends them a mammalian quality which enhances this particular serpent's bestial presence.

The Pre-Raphaelite naturalism of Woolner's snake opens up an equally Pre-Raphaelite richness of interpretation. The judgement passed by God on Eve and Adam is clear within the composition as a whole, but it is not so clear that we should acquiesce in it. The crucial question is whether sexuality should be considered to be evil or natural or both. As well as Adam and Eve, the left-hand panel includes a bird singing lustily in the apple tree and a butterfly landing on a thistle flower. Birdsong and the pollination of flowers

69 John Hungerford Pollen, design for entrance to Oxford University Museum, *c.*1860–61. Oxford University Museum.

are natural processes of sexual reproduction, offering an alternative interpretation of Eve and Adam's desire from the myth. In the right-hand panel the bird chases the butterfly down into the bottom corner of the spandrel, symbolising the introduction of violence and predation into nature with the Fall. But again the naturalistic snake implies that predatory impulses were already present in nature from the outset. Perhaps all we are seeing is two sides of nature – reproduction and predation – not two stages in a narrative. The depiction of Adam and Eve themselves in this second panel is striking too. Where Michelangelo and Masaccio show them each absorbed in their own suffering, Woolner shows Adam protectively embracing Eve as she buries her face in her arm in shame or sorrow. The Fall has brought them together, not driven them apart. Although Eve's face is hidden and Adam's indistinct, it is clear too that Woolner has not followed his early Italian models in cursing the once beautiful couple with a hideousness that is the objective correlative of their moral collapse. The unmoved angel is a figure of moral rectitude, yet Woolner's design implicitly questions whether Adam and Eve are truly at fault, and whether they have not gained as much morally as they have lost.

Woolner's design at once depicts the legend of the Fall and interrogates it, biologically and morally. Woodward approved his design, but it was never carved, probably in part, as O'Dwyer suggests, because of its eroticised nudity, perhaps too because of the subtlety of its interrogation of the biblical narrative.[85] It is surely also likely that it was rejected because it was hopelessly inappropriate. A legend that insists that the pursuit of natural knowledge is fundamentally sinful is hardly fitting for a museum dedicated to teaching natural knowledge, even within a design that subtly calls that moral into question. It was the Tree of Knowledge specifically that was unsuitable, however, not Adam and Eve themselves. In 1834, only twenty years before the museum was commissioned, *A Preliminary Discourse on the Study of Natural History* was published in the same *Cyclopedia* as Herschel's *Preliminary Discourse on the Study of Natural Philosophy*. Its author, the ornithologist William Swainson, could still comfortably take Genesis as a source for his science. In his account of the 'Rise and Progress of Zoology', Swainson accepts that the early history of this science 'must ever remain more a matter of conjecture and of theory than of positive fact'. The one thing he can affirm with confidence, however, is that 'Some acquaintance with the properties of animals was certainly possessed by our first parents, who were enabled, by the Divine agency, to assign names to the beasts of the field, and to distinguish such as were adapted to their wants.'[86] Twenty years later in May 1854, just after the competition to build the Oxford museum had been launched, Richard Owen, the country's foremost comparative anatomist and the curator of the Hunterian Museum at the Royal College of Surgeons in London, affirmed in a lecture on humans and apes that 'in reference to the unity of man's species, and the fact of his being the latest, as he is the highest, of all animal forms upon our planet, the interpretations of God's works coincide with what has been revealed to us of our own nature and origin in his Word.'[87] Whether taken as a literal account of early human history, as in Swainson's textbook, or as figurative but accurate in its essentials, as in Owen's lecture, the account of the creation of human beings in the Bible remained a point of reference for mid-Victorian natural history. As long as the evidence for human origins uncovered by science corresponded in broad terms with the account in Genesis, then Genesis itself could be taken to lend weight to the project of natural theology.

Viewed in these terms, it was apposite, not inappropriate, to represent Adam and Eve themselves on the front of a natural history museum. Of course, it underscored too Oxford's commitment to natural theology. It is not clear exactly how or when the job of designing the arch passed from Woolner to Pollen, or even whether they may have worked on it together, as O'Dwyer suspects.[88] Either way, Pollen's subsequent arch incorporates Adam and Eve. In his design (fig. 69) – which O'Dwyer identifies as the drawing Pollen exhibited at the Royal Academy in 1861, and which was carved in part the same year – the spandrels are again divided in two by a tree growing from the top of the arch.[89] This is no longer the Tree of Knowledge, however, but rather the Tree of Life. Its fruit has been replaced with leaves and flowers. Though pictorially it remains in the same position, structurally its function is different, as it no longer provides a narrative break between two consecutive frames. Instead, the two frames mirror one another as the left-hand panel

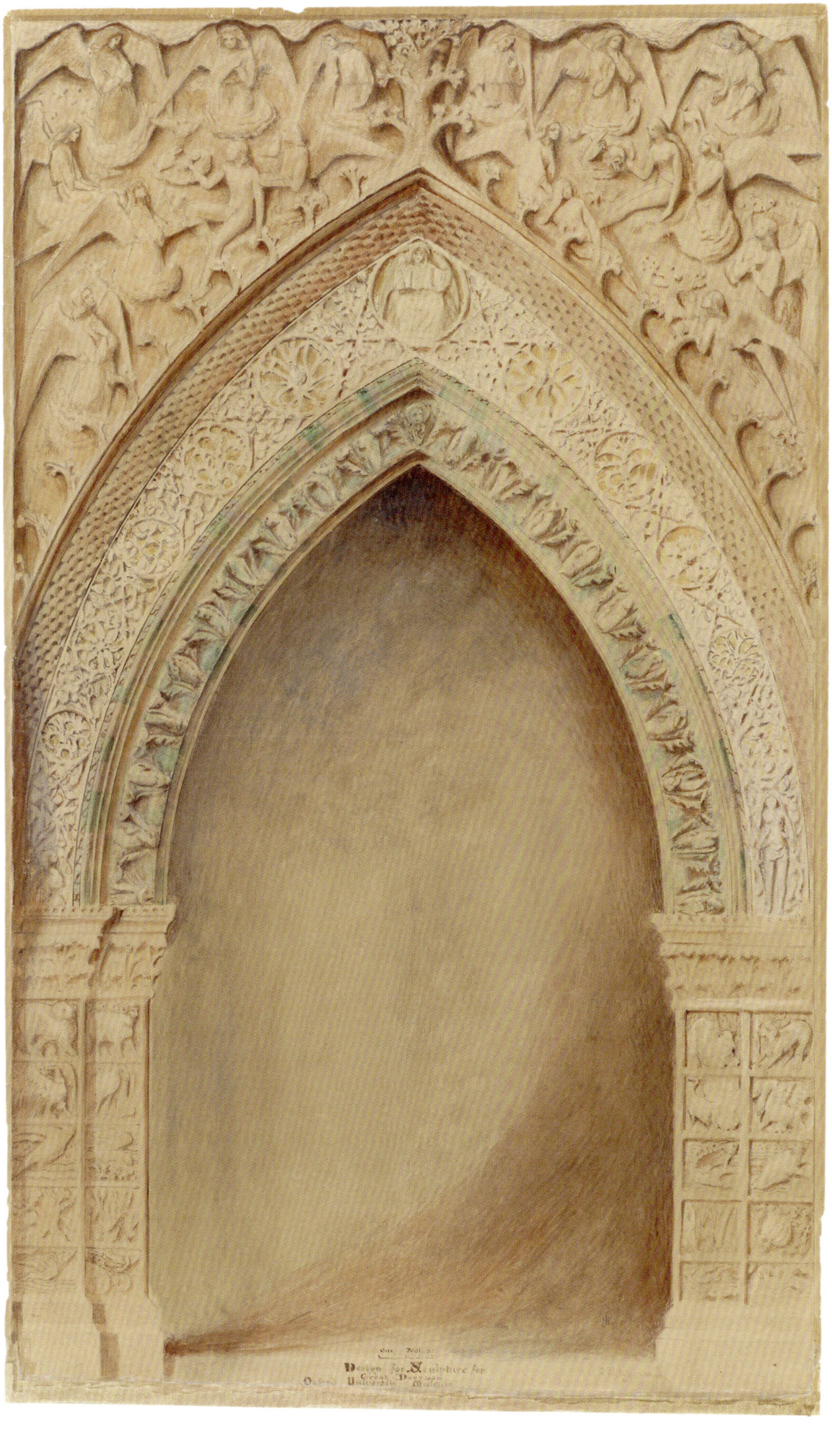
Design for Sculpture for
Great Doorway
Oxford University Museum

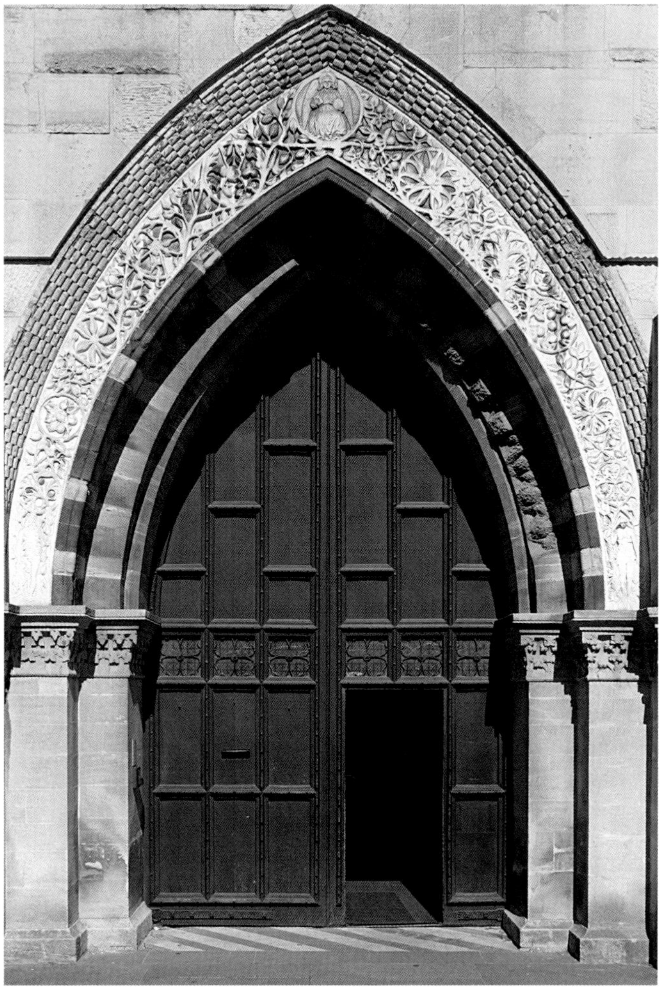

70 Entrance to Oxford University Museum.

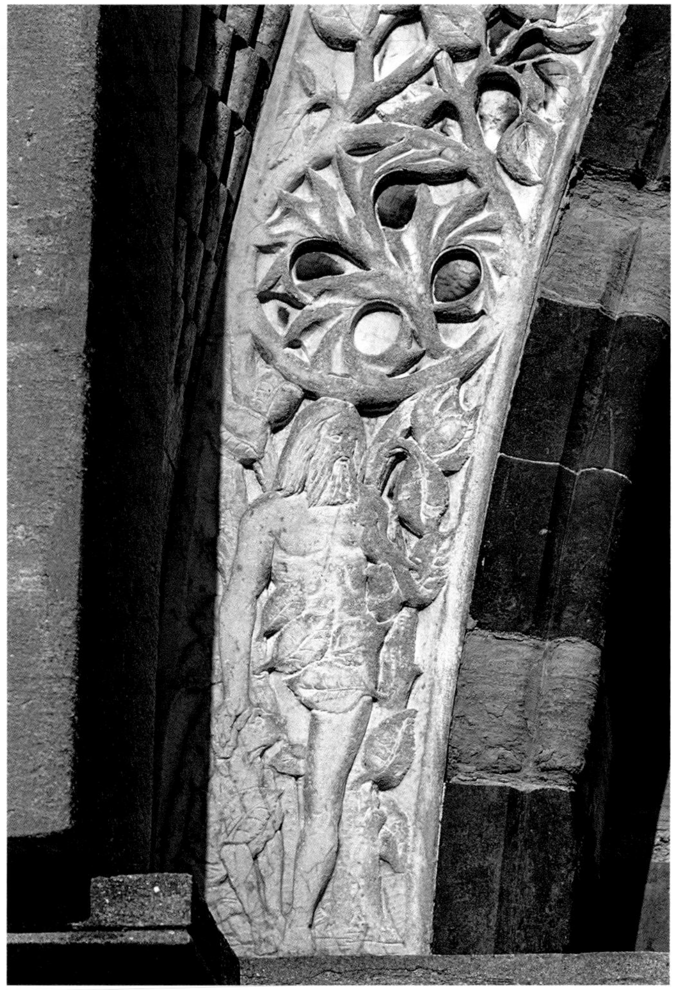

71 Entrance to Oxford University Museum, detail of Adam.

shows the resurrection of a man, the right-hand panel that of one or perhaps two women. Each of these figures is surrounded by angels. Another angel sits in a circle or tondo at the apex of the arch, above the face of Christ holding his hand up in blessing. Where Woolner's design probes the Christian mythology it rehearses, Pollen's is unequivocal in its optimistic affirmation of Christian belief. It includes the Fall, but its significance is muted, as Eve and the serpent are marginalised within the composition. The form of the arch itself pulls the eye upwards away from the Fall to Christ, the central angel and the vision of heaven. Here, above Eve's head, we see woman forgiven and resurrected, the Fall reversed by Christ's incarnation and blessing.

Curiously, Adam does not appear at all in Pollen's design, but he was reinstated in the carving on the arch itself (fig. 70). In what is apparently a unique piece of iconography, he is shown with his hand patting the head of a bloodhound (fig. 71). His other hand, now damaged, held a bird. For her part, Eve stands impassive as if in a trance, a winged serpent at her feet. When Acland glossed this design over thirty years later, he took both figures to be 'symbolic representations of obedience or disobedience to the spiritual law', with the hound as a 'type of material ferocity and agent of material death, restrained until sin had broken loose upon man'.[90] On this reading, Adam is as yet unfallen and Eve just about to fall. But the relationship between Adam and the animals

72 John Hungerford Pollen, design for entrance to Oxford University Museum, *c.*1860–61, detail of the angel (see fig. 69). Oxford University Museum.

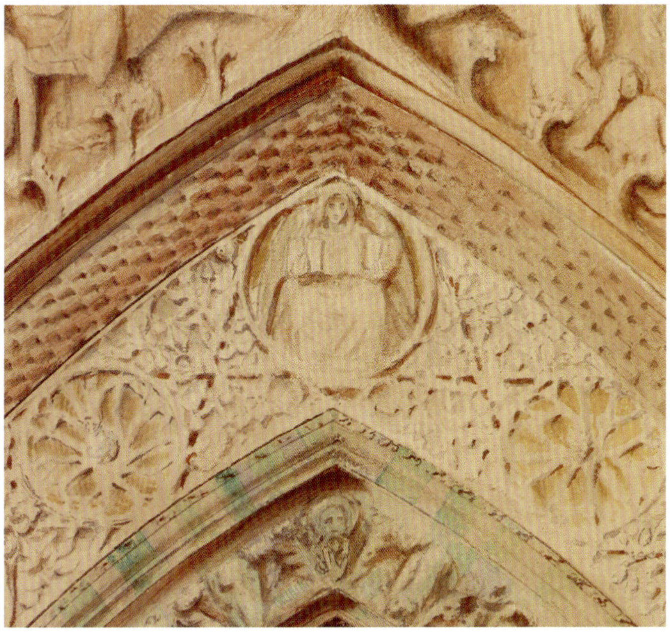

73 Entrance to Oxford University Museum, detail of the angel.

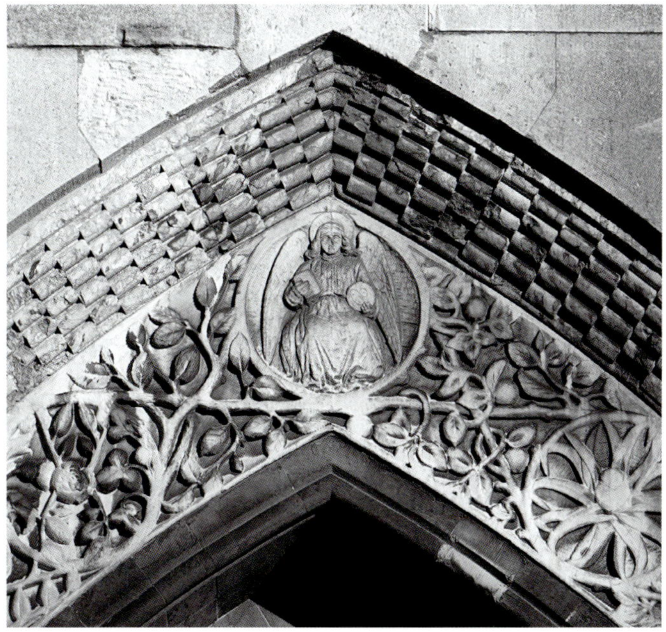

also suggests a trust between human beings and other creatures that is lost through the Fall but partly regained through the companionship of domestication and the attentive study of natural history. In line with Swainson's textbook, Pollen's iconography implies that Adam's divinely appointed understanding of the natural world forms a model for the science of zoology as taught in the museum. Under God's direction, science itself can bring about a reversal of the Fall, as Deverell had suggested through the figure of Newton in his poem 'The Garden'.

Pollen's design for the spandrels was never carved. Only the middle band of the arch was completed. Even so, the angel above the door is enough to affirm that God is the sponsor of the museum's scientific project. Here too Pollen's original design was subtly altered in carving. In Pollen's drawing, the angel holds two books: the Bible and the Book of Nature (fig. 72). In the carving, the angel holds one book and, to quote Acland, 'a Disc, on which are carved the dividing cells and nuclei of *Schwann* and of *Schleiden*' (fig. 73). Again, Acland interpreted this disc long after the fact as a symbol of the cycle of life, death and rebirth.[91] More pertinent here is how this change to Pollen's design recasts the science of the museum. The metaphor of the Book of Nature is traditional. It affirms directly that nature is God's work, giving it an equivalence to the Bible. At the same time, it implies that the hermeneutic strategies of reading can be carried over into science. By symbolising nature not through a book but through nucleating cells, Acland and his colleagues affirmed that observation and experiment, not hermeneutics, were the means to understand the material processes of God's creation. At the same time, they confirmed that their reverence for the Bible was matched by a commitment to modern science, as only around twenty years before had Matthias Schleiden and Theodor Schwann demonstrated that plants and animals were made up of cells, and that cells themselves had nuclei which were essential to their reproduction and therefore to growth. In the image of the cell, Acland brought together the several different sciences studied and taught at the museum, including zoology, botany, the chemistry to which cell theory ultimately subordinated biology, and his own discipline of medicine, then in the process of being revolutionised by Rudolf Virchow's new textbook *Cellular Pathology*, published in 1858 and translated into English in 1860, the year the museum opened.[92]

The angel who presides over the scientific enterprise at the Oxford museum has one more story to tell. As well as sponsoring science, he symbolises one of its key findings, as some at least of the Oxford scientists understood it. For Daubeny, science had failed to give the lie to the evidence for miracles.[93] Strickland had gone further, arguing that geology confirmed the direct intervention of God on earth, by revealing the successive extinctions of living forms and the creation *ex nihilo* of new types. As he explained in a pamphlet on the study of geology at Oxford, where natural theology could 'confute the Atheist', only modern geology could confound the deist, by demonstrating 'what Revelation declares in another language, that Miracles are *not* contrary to Experience, or at least *not* contrary to Fact'.[94] Pollen's design for the entrance to the museum affirmed that science was not merely accommodating to Christianity but in step with it, revealing at once God's artistry in the forms and laws of nature, and his power to suspend or transcend those laws and create new forms at will.[95]

Passing through the entrance to the museum, you enter a small stone lobby. Initially this room was undecorated, until one of the O'Shea brothers was recalled in 1879 to carve life-sized plants and animals, including waterbirds, owls, parrots and a snake, on the capitals.[96] A short flight of steps takes you to the two doors to the museum itself. You step first into one of the lower galleries that run around the central court and then into the central court itself. Here the distinct structural elements of the building combine to form a rich and complex representation of nature, combining zoological, botanical and geological metaphors. As Trevor Garnham has pointed out, the frame of the iron roof supports suggests a gigantic skeleton, echoing and enclosing the skeletons on display in the collection.[97] This impression is enhanced by the close visual similarity between the Gothic arches and the jawbone of a whale that early photographs show to have been on display at the museum since at least the 1870s, as well as other recurrent forms such as elephant tusks and antelope horns. At the same time, the central iron columns create what Stephens disparagingly called a '*quasi*-forest' – it was the only feature of the museum he actively disliked – as the light from the sky shines through Skidmore's iron spandrels wrought into the forms of the leaves and fruit of different trees.[98] Plants are also represented on the wrought-iron capitals attached at balcony level to the supporting iron columns. These are typically ground or water plants – bulrushes, for example – so what is superficially a canopy does service as a symbol of geology, with the inorganic strata of minerals supporting the life on the earth's surface. This idea is enhanced by the two rows of stone columns around the central court chosen and positioned by Phillips. As the capital of each column is carved to form a plant, they repeat the example of the iron columns in the middle of the court itself. But as one row of columns runs above the other around the perimeter, this hints too at stratification. Reading the building's overall structure in these multiple terms, the specimens in the central court acquire added significance as examples of the interdependence of animal life, of the ecological relationships between creatures and their habitat, and of the hidden treasures to be found as fossils deep in the earth.

In a letter read by Acland to the Architectural Congress in 1858, Ruskin argued that Gothic art was most effective when it formed 'part of a connected system'.[99] The connected system formed by the Oxford museum's materials and decorations is a rich, symbolic model of nature as a complex but unified whole. The central court is where this system coheres, fulfilling Acland's promise that the museum would convey the divine artistry of nature, as affirmed by Pollen's archway and the religious resonances of Woodward's Gothic style.

ECOLOGY AND EVOLUTION IN STONE: THE PRE-RAPHAELITE ART OF THE O'SHEAS

Turning from this unity to the diversity that comprises it, the cracks within mid-nineteenth-century natural theology begin to show. The artists who represented individual animals and plants with the greatest precision and attentiveness at the museum were the O'Shea brothers and their nephew Whelan. The O'Sheas have often been seen as, in Mordaunt Crook's words, 'the embodiment of the hypothetical Ruskinian artisan', reincarnations of the men who carved the decorations on Europe's Gothic cathedrals in the Middle Ages.[100] Following Acland's account of the decoration of the museum, and marking the transition from no less detailed or animated but less scientific carvings in Dublin, it is equally

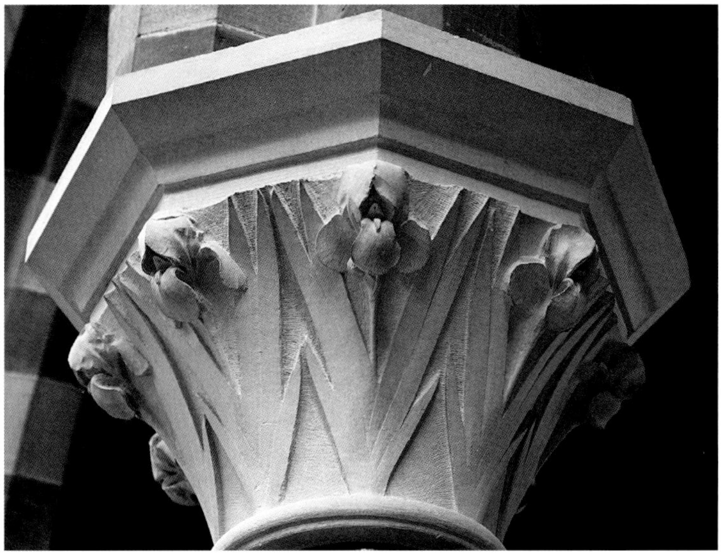

74 John O'Shea, German irises, 1860. Oxford University Museum.

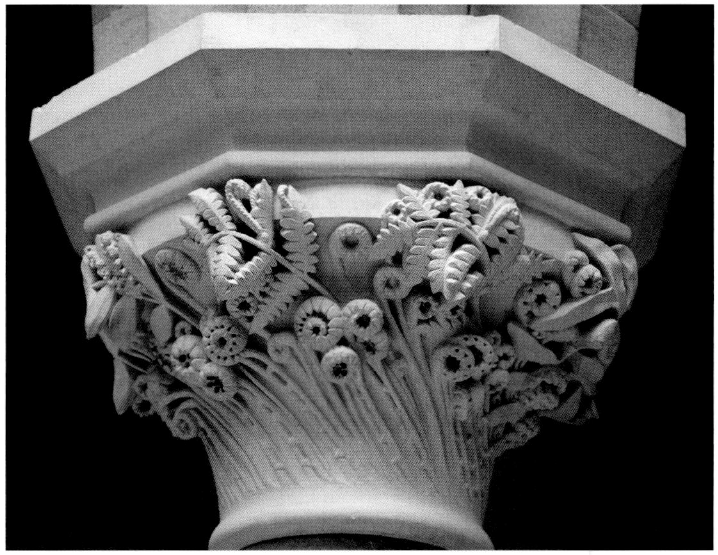

75 James O'Shea, golden leather ferns, 1860. Oxford University Museum.

possible to see their carvings in Oxford as a sustained programme of Pre-Raphaelite natural history sculpture.

Collectively, the O'Sheas' work captures a remarkable ecological vision of the natural world. There are, however, contrasts in emphasis between James and John O'Shea's work, in particular, which were recognised at the time and which reveal tensions between alternative interpretations of nature within both Pre-Raphaelite art and mid-Victorian science.[101] Building on Ruskin's arguments, Street claimed in his pamphlet on the museum that the figures carved by medieval craftsmen were 'not conventionalized so far as to be unlike nature; but just so far as to stereotype, as it were, nature's most perfect mood'.[102] Stephens found this satisfying degree of stereotyping in John O'Shea's work. In comparison with those of his brother and nephew, his carvings were 'simpler in the general forms, chastened and more severe,' with 'less luxuriance' to them, suggesting that he had seen 'deeper into the needs of architectural art'.[103] Stephens singles out his carving of German irises on the south side of the court as exemplary (fig. 74). As Blair Gilbert has remarked in the only close analysis of the O'Sheas' carvings as art since Stephens's own, in this work 'only the simplified and very stylized flower is rendered with any definition and prominence', while 'all else is merely suggested and traced on to the face of the capital.'[104] Stephens's remark about the requirements of architectural art tallies too with the symmetry of this capital, as the forms repeat themselves with little variation on each of its eight faces.

John O'Shea's carvings tend to recall what Lorraine Daston and Peter Galison have characterised as the normative truth-to-nature of botanical illustration.[105] They are in line with Acland's and Phillips's ambition that the capitals should provide a taxonomic catalogue of 'the vegetable creation' in order 'to satisfy the botanist as well as the artist', as the plant specimens themselves were to remain at the University's Botanic Gardens.[106] By contrast, Stephens faulted James O'Shea's carvings for not being sufficiently conventionalised, yet at the same time he praised them as 'exquisite for faithful rendering in stone'. Although he argued in principle that 'mere imitation and reproduction, however exquisite, is not the end of art,' in practice Stephens was carried away by the 'soft-rounded forms, rich diversified outlines, and almost merry look' of James O'Shea's carvings. 'Nothing', he wrote, 'can be more intelligently faithful than these works.'[107] Stephens had in mind specifically the mallow and ferns carved in the last bay on the north side of the museum, but his description applies at least as well to the astonishing group of golden leather ferns on the capital in the previous bay (fig. 75).[108] Where John pares the iris down to a stereotype, James gives us an exuberant growth of vegetation. Each separate frond seems alive, uncurling before our eyes. Far from being muted, the detail is intense and precise. These are individual plants first, types only as an afterthought, or more precisely after we have done our own work as botanists, describing and classifying them – work that John O'Shea does for us in his carvings.

James O'Shea's close attention to plants as individual living organisms recalls Rosa Brett's *Thistles* and Tupper's insistence on the gain to be had from knowing a single daisy. His brother's tendency to depict plants as types, on the other hand, is pre-emptively critiqued in Dante Gabriel Rossetti's well-known poem 'The Woodspurge', which, according to his brother, he wrote in 1856, while the construction of the museum was underway:

> The wind flapped loose, the wind was still,
> Shaken out dead from tree and hill:
> I had walked on at the wind's will, –
> I sat now, for the wind was still.
>
> Between my knees my forehead was, –
> My lips, drawn in, said not Alas!
> My hair was over in the grass,
> My naked ears heard the day pass.
>
> My eyes, wide open, had the run
> Of some ten weeds to fix upon;
> Among those few, out of the sun,
> The woodspurge flowered, three cups in one.
>
> From perfect grief there need not be
> Wisdom or even memory:
> One thing then learnt remains to me, –
> The woodspurge has a cup of three.[109]

Rossetti's poem combines precise itemisation of physical details with the close study of a mental state. Indeed, the one embodies the other. The speaker's mental condition is one of paralysis and alienation brought about by despair. In the first stanza, his motivations bear only a circumstantial relation to their supposed cause in the strength of the wind. In the second, the parts of his own body become dissociated from his will, as if they too are only his by circumstance. The very listing itself which makes up these first two stanzas, with each line end-stopped, suggests his inability to break out of the constrained state of 'perfect grief', in part because he cannot give that grief itself voice, as he acknowledges in the sixth line.

This syntactical and structural paralysis is broken by an act of natural history. The speaker recognises one particular weed, the woodspurge or euphorbia, growing on the ground 'out of the sun', perhaps because euphorbia tends to grow in shade, or because it is shaded beneath his own overhanging hair. He is already familiar with the plant, but it is only then that he observes one of its diagnostic features. The modified leaf structures or bracts which encase its flowers tend to grow in clusters of three, with two cups growing on thin stems up from a somewhat flatter cup or disk. The Trinitarian symbolism of the three cups in one is too obvious to miss, yet it is steadfastly ignored within the poem itself. Instead, the speaker reiterates this taxonomic detail itself as the only thing remaining to him of his sorrow. The subjective experience of perfect grief itself is beyond recall – hence again its objectification in his bodily posture, described in the first person but as if from a third-person perspective. The poem affirms the psychological benefit of natural history in a similar way to Woolner's 'Of my Lady. In Death', as it is his attention to nature which enables the speaker to move beyond his abject condition. Yet at the same time, as well as denying the plant itself any symbolic significance, it explicitly denies that the 'thing' the speaker has learnt about it gives him 'wisdom' or, in the previous draft of the poem, 'knowledge'.[110] Taxonomy for Rossetti may be memorable, but it is the fact of memory itself that is significant – that is, the importance of the individual plant to the individual person. Divorced from these interactions, botanical details themselves do not signify.

As art, the O'Sheas' carvings encourage this kind of reflexive engagement with other living things. This engagement links their work to the Pre-Raphaelitism typified by William Rossetti's 'Fancies at Leisure' and Hunt's and Millais's early paintings. Following Pre-Raphaelite procedures in their art, and in the context of a museum dedicated to the science of natural history, they trace the complex interactions in nature between individuals of different species. In his published letter to Acland, Phillips proposes to vary the carvings of plants around the court by including 'animals geographically

76 James O'Shea, date palm, 1860.
Oxford University Museum.

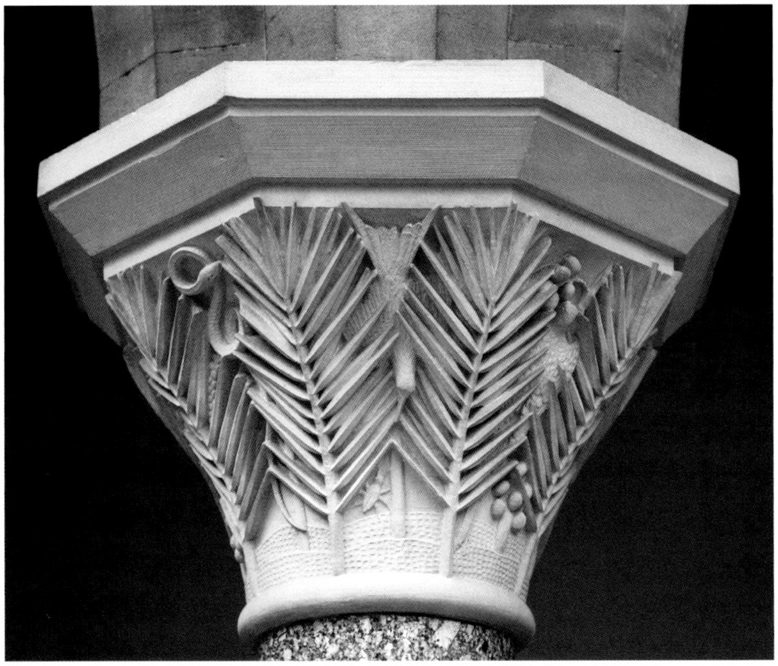

and naturally associated with them'.[111] Three capitals carved by James O'Shea on themes initially sketched out by Pauline Trevelyan follow this principle closely.[112] The first represents water plantain, among whose stems move a frog, a snipe and a crayfish. The second shows the leaves and flowers of a patch of rush. One small bird makes as if to fly away across the court, while another had landed on some flowers just ahead of a dragonfly that is almost as big as it is itself. The third shows the fronds of a date palm (fig. 76). Here a more exotic bird, perhaps a roller, is hunting a beetle. To the right, another bird feeds on a ripe date, while to the left a snake slithers up on the other side of the same group of leaves. In each case, we see animals appropriate to the environment in which the plants are growing. The plants themselves form part of their habitat, while animals feed on them and predate on one another.

The sense that what we are looking at is a living ecosystem in these carvings is so strong that Stephens concluded the carvers themselves had explored and studied these environments directly in situ, seeking 'in the lush recesses of the neighbouring river-bank for crisp water plants, or those that love the margins and whip the stream'.[113] In keeping with Acland's commitment to Pre-Raphaelite principles, the O'Sheas seemingly did work from life, although rather than go into the field themselves they were supplied with specimens from the Botanic Gardens by Phillips and Daubeny. As William Tuckwell, who was headmaster of New College School at the time, recalled decades later, 'every morning came the handsome red-bearded Irish brothers Shea [*sic*], bearing plants from the Botanic Garden, to reappear under their chisels in the rough-hewn capitals of the pillars.'[114]

The O'Sheas' ecological vision is so convincing not only because it is founded on this close observation of the plants themselves, but also because it incorporates information provided by the scientists about the animals that share their habitats. This is most apparent in the fireplace in the entomology museum upstairs, carved by Whelan (fig. 77). Whelan's carving turns this fixture within the room itself into an illustration of insect ecology. On the left, stag beetles move about in their favoured habitat, an oak tree. Alongside a large male, Whelan includes a female in flight, a pupa and a larva. On the right, a caterpillar is feeding on some potato leaves. Nearby another caterpillar is undergoing its metamorphosis in its chrysalis. The two death's head hawkmoths above show what it will emerge as. With both the beetle and the moth, Whelan matches the insect to its habitat and shows us successive stages of its life cycle. As J. M. Edmonds, curator of the geological collection at the museum from 1955 to 1976, wrote in an unfinished history of the museum, 'it is clear that the sculptor was guided by an expert,' probably, O'Dwyer suggests, John Obadiah Westwood, the first professor of entomology at the museum.[115]

The choice of subjects on the fireplace hints at the relationship between insects and humans through plants we both depend on. This integration of animals, plants and humans in an overarching ecology is explored in more depth in the capital and corbels of a bay on the north side of the museum, again attributed to Whelan.[116] This bay forms a single composition. Where Millais shows humanity's role within the ecology of an English woodland directly in *The Woodman's Daughter*, requirements of scale led Whelan to probe this theme indirectly, as indeed Hunt does in *Our English Coasts*. The flat stone of the fireplace limits both Whelan's scope to provide depth to his composition and

77 Edward Whelan, fireplace, 1860. Hope Entomology Museum, Westwood Room, Oxford University Museum.

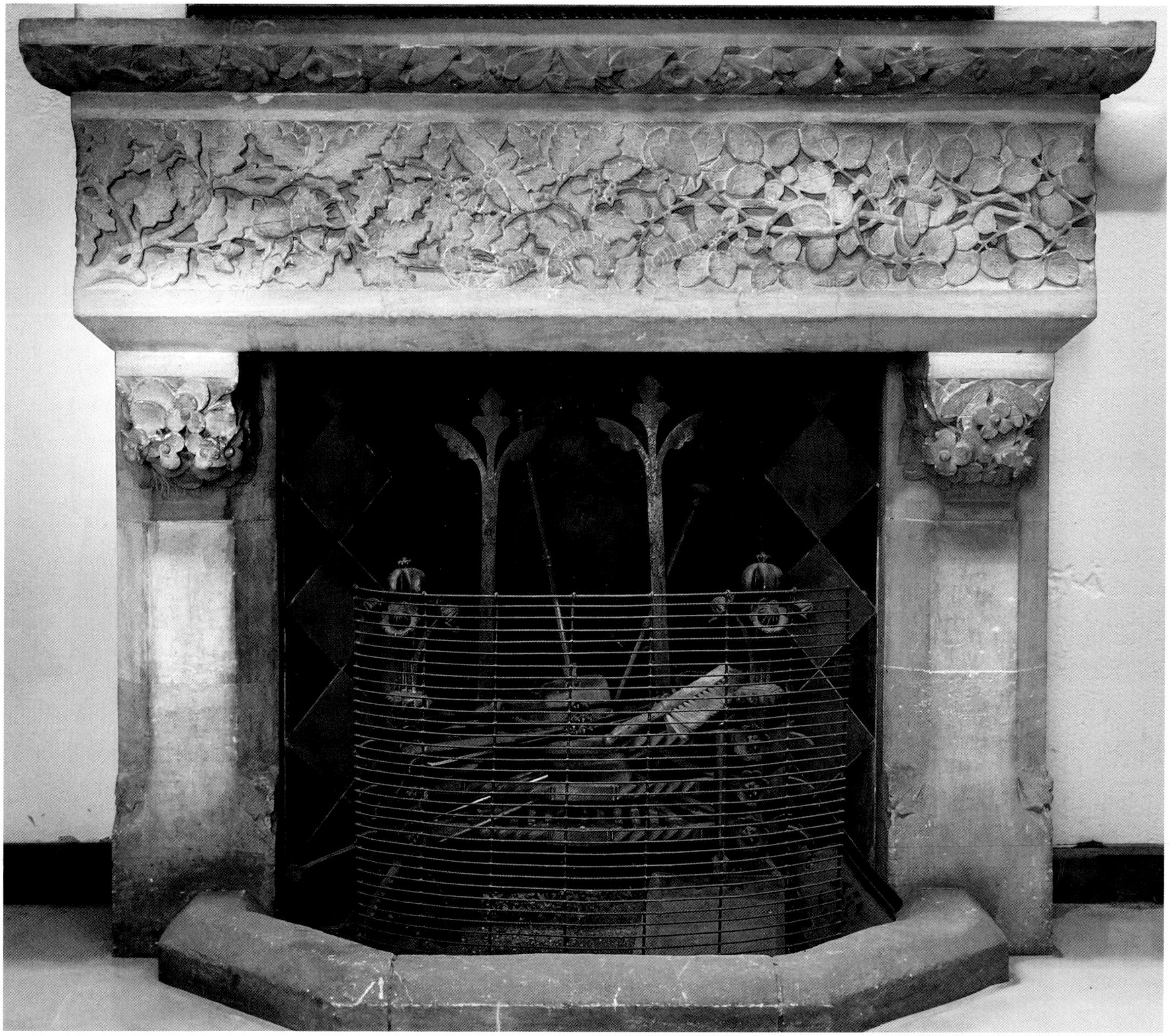

ours to experience it in the round. The carvings around this bay, on the other hand, can each be seen from several angles, enhancing the impression that we are looking at a real habitat existing in three dimensions. Yet in fact this series of scenes is as much symbolic as real. Approaching from the entrance to the museum, the first corbel represents the wild grass brome, with birds eating its seeds (fig. 78a). The capital depicts a wide range of cultivated grasses, including wheat, barley, oats and sugar cane (fig. 78b). A flock of sparrows has landed and is feasting on the crops. The second corbel continues this same theme among rice and canary grass (fig. 78c). While several birds eat the grains, a quail is moving through the grasses. Another small bird has made a nest and laid eggs in among the stems. Unbeknownst to this nesting

78 Edward Whelan, capital and corbels depicting grasses, 1860. Oxford University Museum.

a (*left*) Bromus.
b (*right*) Wheat, barley, oats, Indian corn and sugar cane.
c (*below right*) Rice and canary-grass.

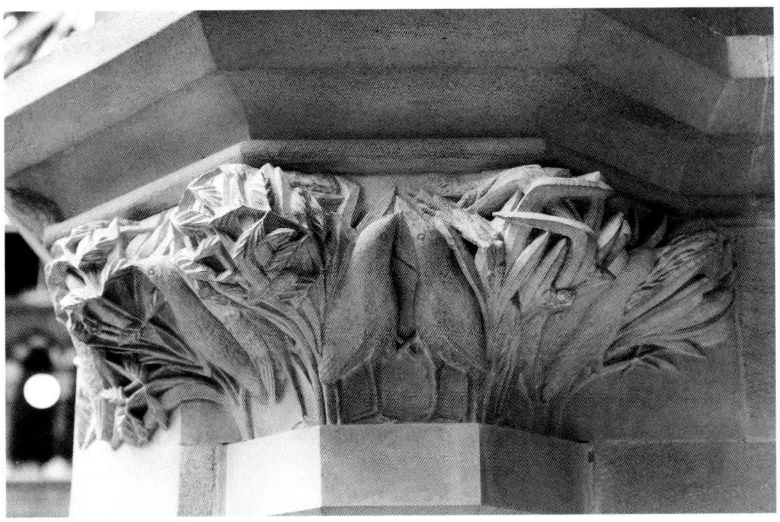

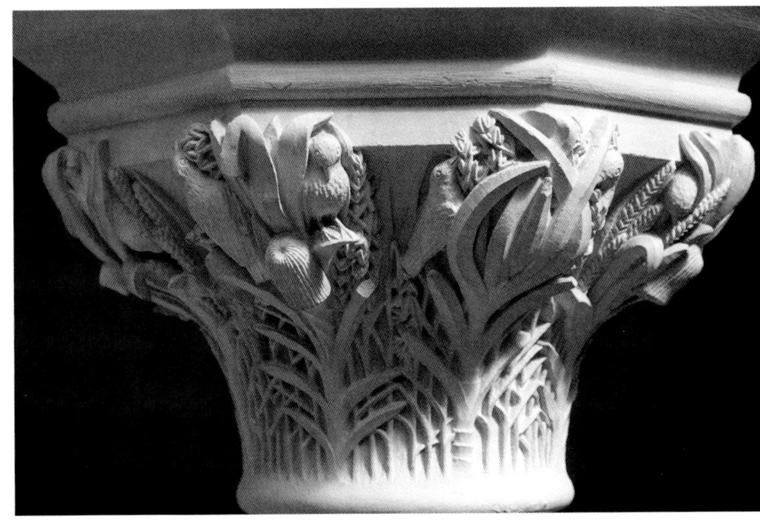

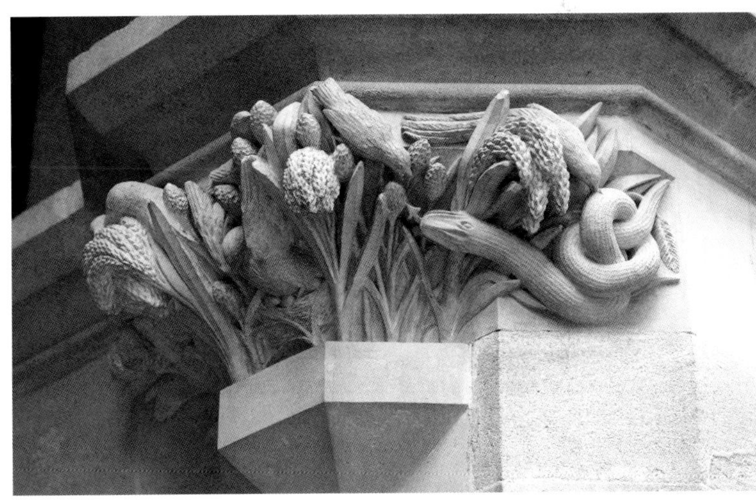

bird, a snake is approaching. Two other small birds have seen the snake, however, and one has stabbed it with its beak in an effort to scare it away. This last detail may be implausible, but otherwise the ecology of the scene as a whole is convincing and compelling.

Together the three carvings tell a narrative of ecological development from a natural landscape to a cultivated one, and from opportunistic feeding to settlement and the formation of complex ecosystems involving many different species. Where James O'Shea represents real ecosystems made up of plants and animals that live alongside one another, Whelan is less concerned here with geographical accuracy than with the principle that we too participate in ecology. The plants are crops, and diverse crops that one would not expect to grow together. In consequence, ecology in this bay becomes linked to economics. Human beings are not represented in Whelan's sculpture, but we are implicated in them. Farming transforms landscapes, as our wheat and sugar and rice provide new habitats for birds and other animals. Contemplating these new environments, our sympathies may lie with the nesting bird. From an economic perspective, however, the snake is keeping down the populations of the birds who steal our crops.

The O'Sheas and Whelan may have been tutored in natural history by Phillips, Daubeny and Westwood, but their ecological vision of life raises questions which these old-school natural theologians may have found unsettling. The systematic side of ecology – the study of what would become known as ecosystems – is not necessarily incompatible with natural theology, which views the whole of nature itself as a single, divinely ordained system. Woodward's architecture as a whole embodies this system. But the cluster of birds among Whelan's crops seems more haphazard and opportunistic than systematic. Then there is the problem of predation. Paley's breezy insistence that this is a happy world may be supported by the pleasure we take in nature through the O'Sheas' art, but it is if not confounded then at least

83 James O'Shea posed carving the cat window, *c.*1859. Oxford University Museum.

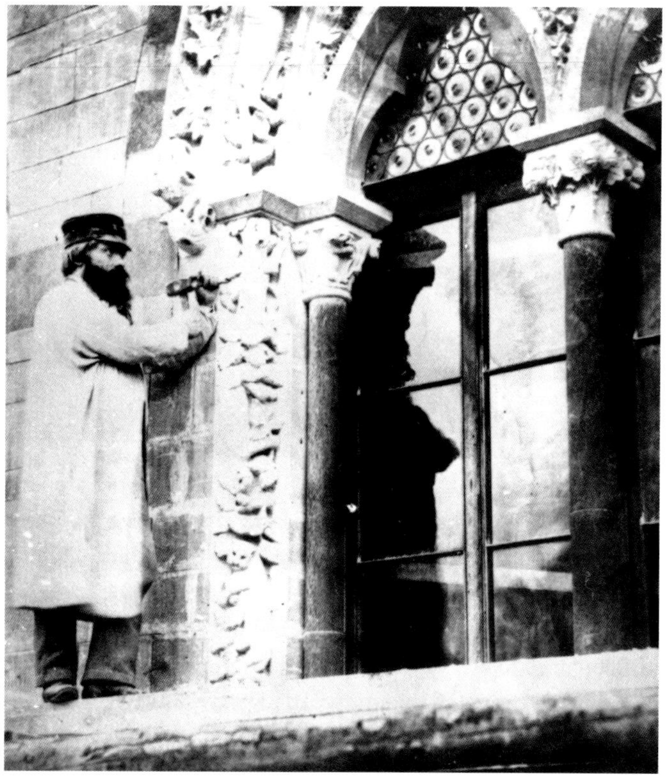

to have been at the least unconventional. Certainly, the debacle over the cat window had nothing to do with Darwin directly, as the window was carved in October 1859, a few weeks before *On the Origin of Species* was published. O'Shea's cats are not the only monkeys at the museum, however. In the upper corners of one of the bays on the east side there are two small monkeys looking across at one another (figs 84a and b). One is peeking out from behind its arm and holding the end of its tail; the other is chewing its thumb and scratching its bottom. The trees here are coniferous, so the monkeys seem somewhat out of place. But the capital in the next bay round, carved by John O'Shea, is a monkey puzzle tree.

The history of these two monkeys is obscure, but they seem to be part of an elaborate joke or riddle, slyly teasing the museum's visitors with the question of humanity's primate origins. This same question had notoriously been put to Darwin's close ally and champion T. H. Huxley by the Bishop of Oxford, Samuel Wilberforce, at the meeting of the British Association for the Advancement of Science hosted by the museum in June 1860. Wilberforce reputedly asked Huxley if he considered himself to be descended from an ape on his grandfather's or his grandmother's side. Huxley supposedly replied with the put-down that he would rather be descended from an ape than a man of ability and influence who diminished a serious scientific issue with ridicule. Huxley was supported by Darwin's closest friend, the botanist Joseph Hooker, who by most accounts gave a fuller refutation of the bishop's arguments against Darwin, while Wilberforce himself was believed to have been briefed by Owen.[122] Among the Oxford dons, Greswell sided with Wilberforce, stressing the distance between humans and apes.[123] Two days earlier Daubeny had made the same point in his own talk, denying that it was possible to trace 'the higher faculties of the Human Race' from 'the glimmering Intelligence perceived in the Brute'.[124] For their part, the monkeys are unsure where they stand. In keeping with the comic play with nature which the O'Sheas introduced into the museum, seemingly with Acland's blessing, the monkeys pose a puzzle even as they look puzzled themselves.

The most compelling indication that Acland and O'Shea had Darwin in mind while they were working on the museum is a little-noticed carving on the equivalent window to the cat window on the left-hand side of the façade. After carving the cat window, O'Shea first carved the ground-floor window directly beneath it to a design drawn by Ruskin in 1855. He then collaborated with Ruskin again on the first-floor window immediately to the left of the central tower, no longer copying Ruskin's drawing directly but incorporating aspects of several of his other early designs into fuller, richer and more vibrant carvings. O'Shea was working on this window in April 1860.[125] In May Ruskin left for the Alps, and in early June Woodward, who had not been on site since before O'Shea started work on the cat window, returned for the first of several supervisorial visits.[126] In their absence, O'Shea worked on a third first-floor window, most likely reporting to Acland as he had when he was working on the cat window.

For the most part, the carving on this window is comprised of foliage of various kinds, with small birds perched among it in O'Shea's characteristic ecological style. But the capitals of the pillars at the sides of the window do not fit in with

84a and b Monkeys, *c.*1860. Oxford University Museum.

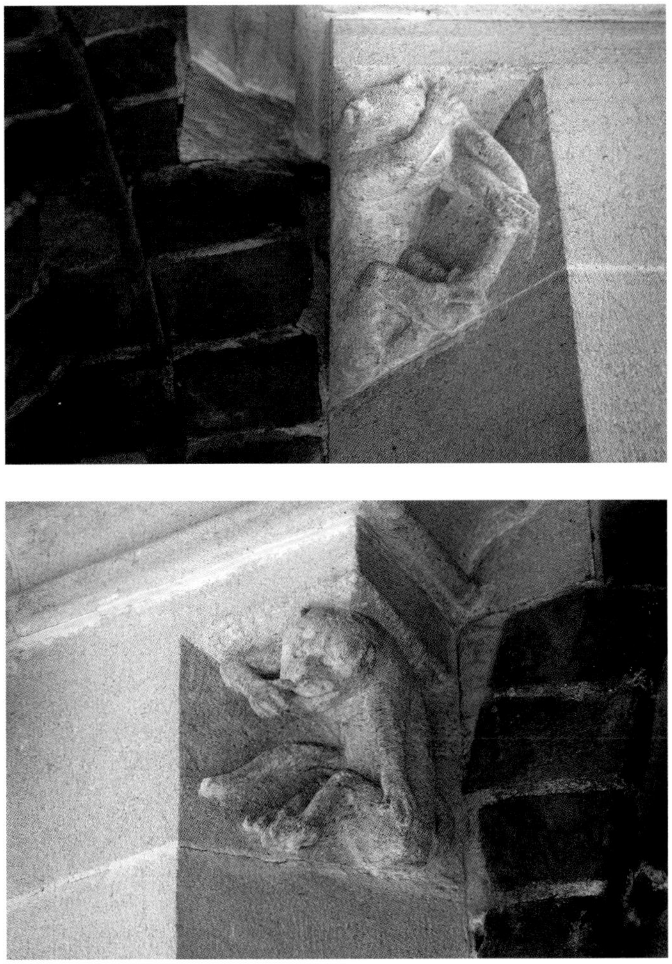

85a and b James O'Shea, Bear-whales, 1860. Oxford University Museum.

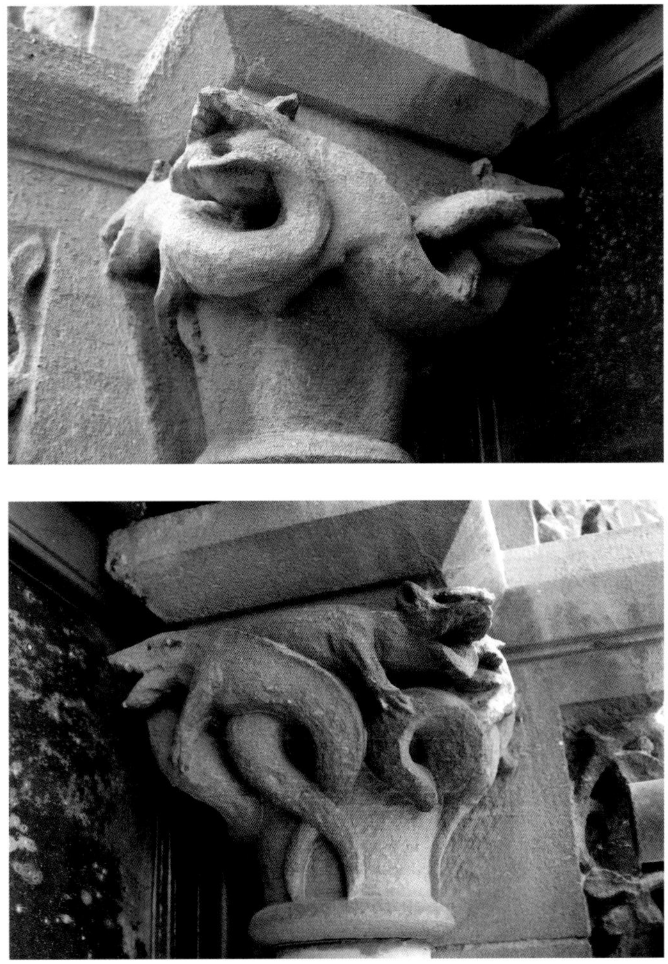

this overall composition. In fact, they do not fit in with the decorative scheme of the museum at all, as alone of all the O'Sheas' carvings there, aside from the innocuous eared snake, they depict animals that do not exist. Their front ends look like heavily built carnivorous quadrupeds, with their mouths wide open. But their back halves are long tails, like those of sea-monsters. Given when they were carved, and the function of the museum itself, it seems certain that they must allude to a well-known passage in *On the Origin of Species* where Darwin uses a thought experiment to explain how evolution could come about through natural selection:

> In North America the black bear was seen by Hearne swimming for hours with widely open mouth, thus catching, like a whale, insects in the water. Even in so extreme a case as this, if the supply of insects were constant, and if better adapted competitors did not already exist in the country, I can see no difficulty in a race of bears being rendered, by natural selection, more and more aquatic in their structure and habits, with larger and larger mouths, till a creature was produced as monstrous as a whale.[127]

In his sea-monsters O'Shea imagines the result of this evolutionary change. They are bear-whales (figs 85a and b), their mouths so grotesquely distended that they have ended up catching not only insects but each other's eel-like tails.[128]

The bear-whales are the surest sign that Darwin's ideas were incorporated directly into the decoration of the Oxford museum. It may have been Acland who directed O'Shea to this passage from Darwin's book, or perhaps it was Daubeny. Towards the end of his career, Daubeny began gathering together a book of *Fugitive Poems Connected with Natural History and Physical Science*. Two of the poems home in on Darwin's image of a bear evolving into a whale in order to expose his arguments as ludicrously implausible. One of them – 'The Origin of Species. A New Song', by Lord Neaves – was culled from *Blackwood's Magazine* in May 1861, a year after O'Shea had carved his window. The other, 'Darwin on Species', is anonymous and undated. After a lengthy disquisition on the evolution of the greyhound and the hare from dogs and rabbits, it ends with this general principle:

> Thus does *selection's* power elaborate
> Great things from little, little things from great,
> To reach the wants of each peculiar state.
> In million ages lions grew from cats,
> In million ages seals fined down to sprats;
> And black bears dabbling in the sea for play,
> Lapsed into whales, and grandly swam away.[129]

As the punchline of this poem, Darwin's bear-whales become an image of his fancy getting carried away by his theories. For this comic poet, the bear-whales are a *reductio ad absurdum* of the theory of evolution by natural selection. O'Shea's bear-whales are also comic grotesques which at once enjoy and ridicule the science they allude to. Yet as well as extrapolating from Darwin's imagined example of evolution for comic effect, O'Shea mythologises it. The bear-whales look more like the product of a medieval than a modern imagination, a pre-scientific legend like the mermaid. But they are also science-fiction creatures, projections into the future of what may yet evolve according to Darwin's theories.

Acland and O'Shea enter imaginatively into Darwin's evolutionary narrative both to test its validity and to amuse each other and us with its possibilities. In the process, they introduce another unruly element into the museum's vision of nature in stone, iron and glass. The O'Sheas' independent Pre-Raphaelite study of the natural world, informed but not shackled by the views of the scientists who employed them, reveals the same interconnections, fluctuations and violence as Darwin's own researches. The open plan of Woodward's museum can accommodate this ecological idea of the natural world alongside the more orderly patterns of natural theology. The O'Sheas share too Darwin's relish at the energy and beauty of life itself. He famously concluded his book by asserting, 'There is grandeur in this view of life.'[130] The museum's sheer visual brilliance and originality, incorporating the O'Sheas' sculpture, enable it to celebrate this grandeur as much as it does the work of Acland's 'Everliving, Everworking, Artist'. Yet in its architecture and decoration, the museum remains ambivalent about the theory of evolution itself, tied as it is by Pollen's entrance arch to natural theology, and responding to Darwin only in hints and jokes. Even so, these jokes, like Tupper's comic play with anatomy, evolution and astronomy in his poems in *The Germ*, depend upon a shared familiarity with science and with the questions that it asks. Carved in stone, even hints of evolution will not go away. Instead, they remain to tease and provoke.

Phillips, Daubeny and Acland all argued that, ultimately, Darwin's theories were compatible with natural theology. But for Phillips, in his book *Life on the Earth*, published in the same year as the Oxford debate, this was the last line of defence in a sustained attempt to hold the threat of Darwinism at bay. He rejected Darwin's claim that the fossil record was inadequate and dismissed as unfounded his argument that it was possible to extrapolate unlimited and undirected change from the limited and directed change that occurs with domestication. Only then did he suggest that, even if natural selection were an actual force, it would be nothing other than providence under another name, 'continually acting for the good of its subject, – encouraging it, or rather compelling it to continual advancement'.[131] For Daubeny too, Darwin's theories had some local explanatory power and were valid to that degree, but he did not think that they could account for such radical shifts as the emergence of vertebrates from invertebrates. Even if they could, however, he agreed with Phillips that this would not undermine 'the cause of Natural Theology'.[132]

For Acland, in contrast, Darwin's ideas were increasingly convincing on something closer to their own terms. Acland's embrace of evolution was partly a matter of scientific integrity. If Darwin's science was compelling, it would be dishonest as well as fruitless to oppose it. As he remarked to Charles

Longley, who was archbishop first of York then Canterbury, 'to pervert facts so as to square with our temporary notions can lead to no good.'[133] But Acland also saw Darwin's theory as compatible with his own view of God as a divine artist working not arbitrarily but through carefully devised patterns and principles. Far from announcing the 'law of higgledy-piggledy', as John Herschel had dismissively remarked, in Acland's view Darwin had made a 'remarkable attempt to reduce the doctrine of what used to be called Chance to an orderly philosophical system'. In keeping with his own notion of God as an artist, Acland insisted in his Harveian Oration that the theory of evolution by natural selection 'in no way derogates from the splendour of the instrument, but only attempts to account for the work of its construction'.[134]

The third edition of Acland's guidebook, published a year after his oration in 1866, incorporates a detailed description of the collections prepared by James Parker. Its interpretation of the exhibits includes elements of both evolutionism and anthropocentric natural theology. The museum's fossils were arranged chronologically according to the different geological epochs, not to present these as successive creations, as Strickland had insisted, but rather to demonstrate that 'the present system of nature is connected with the past by a chain of continuous life, stretching far back into the supra-cretaceous period.'[135] This continuity was demonstrated by the discovery of mammals in the Jurassic rocks in Stonesfield in Oxfordshire and of 'a singular feathered creature, perhaps more bird than reptile' – the archaeopteryx – from the same period in Solenhofen in Bavaria, as well as of the more recent remains of humans alongside extinct mammals in caverns in England and France.[136] At the same time, we are told that it was only very recently that 'the creatures specially serviceable to man' appeared, while in spite of Acland's reservations the display of human beings themselves still followed Owen's separation of Bimana from Quadrumana – that is, of humans from apes and monkeys, as separate mammalian orders – rather than Huxley's more recent reinstatement of Linnaeus's original order Primates.[137]

Under the leadership of Acland, who held the Regius Professorship in Medicine from 1857 right through to 1894, and George Rolleston, who served as Linacre Professor of Physiology from 1860 when the museum opened to 1881, Oxford ultimately came to accept that a Christian university could and should accept evolutionism as the true account of God's plan of action. Acland even argued retrospectively that the museum building itself endorsed it. In 1894, in a pamphlet occasioned by the unveiling of the statue of Thomas Sydenham by the University's new Chancellor Lord Salisbury, he went so far as to reinterpret Pollen's expressly creationist design for the main entrance in evolutionary terms. The plants weaving their way up either side of the arch in semi-geometric patterns, he said, represented 'a perpetual upward growth of life and evolution till the complex web of organization has reached the feet of the Angel of Life seated on the summit of all'. Adam and Eve he relegated to mere moral symbols, while he even made a virtue of the fact that the spandrels above the arch had never been carved, reading their blank stone as a reminder of the hidden 'spiritual revelations of a life beyond'.[138] Acland's reading of Pollen's symbolism combines Christian metaphysics with materialist science, but the firm certainties of mid-century natural theology have been sacrificed for the mysteries of a theistic evolutionism in which science and revelation, rather than complementing one another, occupy separate spheres. As an interpretation of Pollen's art, it shows how even the museum's most emphatic symbols of natural theology could be recast for the age of evolution, but also how far they would have to be distorted in the process. Ultimately, though, the Oxford museum's very inconsistency has been a part of its success. It remains a comprehensive vision of nature precisely because it contains conflicting views of what the science of natural history entails and what it reveals, views which come together within the space of the building and yet refuse to be collapsed into one another.

EMBODYING SCIENCE: THE PRE-RAPHAELITE STATUES AT THE OXFORD MUSEUM

If the O'Sheas' carvings make up one programme of Pre-Raphaelite sculpture at the Oxford museum, the statues of scientists surrounding the court comprise another. Where the architecture and decoration of the museum represent natural history specifically, these statues represent the full

86 Thomas Woolner, *Francis Bacon* (detail), 1856–7. Oxford University Museum.

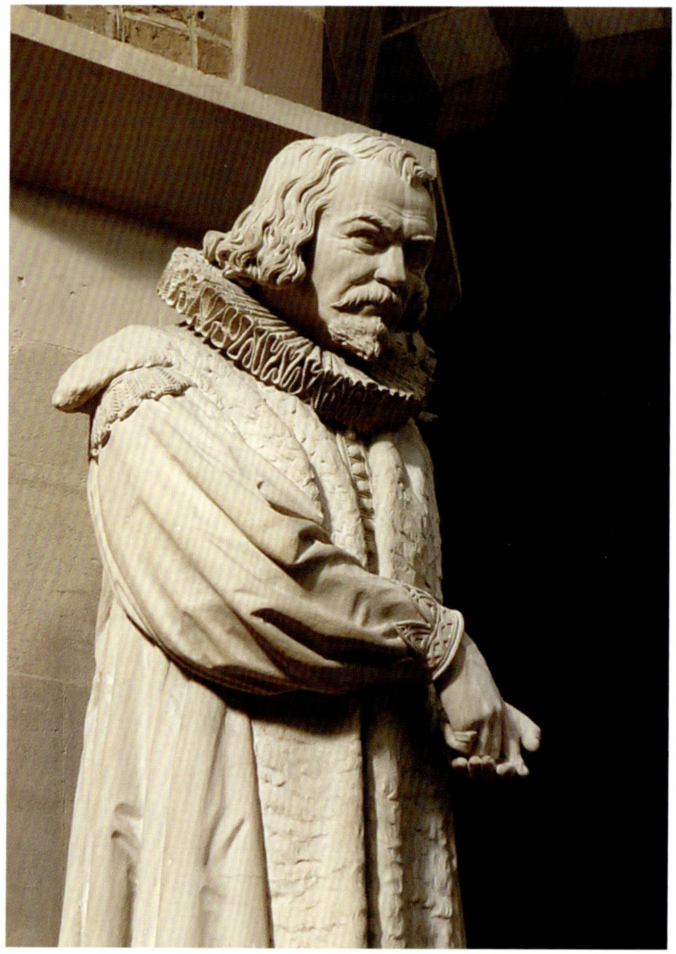

range of the sciences, encouraging scientists from all different fields to come together within the single shared space of the court. The original list of statues the University proposed was deliberately inclusive, and not only in terms of scientific disciplines. In a pamphlet on scientific education at Oxford published in 1848, Acland argued that Oxford should 'add some study of the fundamental arrangements of the Natural world to the general education' it offered, for fear that 'if we do not *add* it, we may live to see . . . such knowledge *substituted* for our present system.'[139] Acland was an accomplished diplomatist. Whether his fear was genuine or merely a device for winning over recalcitrant classicists and clergymen, his ideal of science as complementing rather than supplanting the traditional Oxford curriculum was neatly symbolised in the selection of scientists to be included among the statues in the court.

When the University voted its initial grant of £30,000 for building the museum, it made it clear that this would not be sufficient to cover the interior decorations. Instead, it launched an appeal for donations. As well as the marble pillars and carved capitals – and a series of up to a hundred 'Illustrative Inscriptions', which were never carved – the University proposed seventeen statues of 'the great Founders and Improvers of Natural Knowledge'. In addition to eleven 'Moderns', this included six 'Ancients', firmly establishing science as stemming from the same root as Oxford's classical curriculum.[140] The queen herself took a lead in sponsoring the statues of modern scientists, stumping up £350 to fund the five the University had identified as in the top rank: Bacon, Galileo, Newton, Leibnitz and the Danish physicist Hans Christian Oersted, who had discovered electromagnetism in 1820. The founder of the Department of Entomology at Oxford, Frederick Hope, sponsored Linnaeus, while Ruskin's father went for Hippocrates. Oxford's undergraduates, true to form and their education, clubbed together to put down the £70 asked for the statue of Aristotle.[141] Aristotle and Bacon were placed side by side, flanking the main entrance to the court, as the two men 'who have laid the deepest and widest the foundations of science', as Acland put it in the second and subsequent editions of his guidebook.[142] By giving Aristotle pride of place, Acland affirmed that science and classics were allies, not opponents. At the same time, although only one more of the ancients besides Aristotle and Hippocrates – Euclid – was carved, their presence alongside seventeenth-, eighteenth- and nineteenth-century natural philosophers and scientists gave the overall impression that science, like the humanities, was firmly grounded in its own history.

If the choice of statues puts forward a particular interpretation of science, so do the statues themselves. For one thing, Woolner, Tupper and Munro carried the rigour of Pre-Raphaelite painting over to sculpture. The curators of the 'Sculpture Victorious' exhibition, held at the Yale Center for British Art and Tate Britain in 2014 and 2015, discuss in their catalogue the complex relationship between nineteenth-century sculptors and their assistants, which has led

87 John Lucas Tupper, *Linnaeus*, 1859. Oxford University Museum.

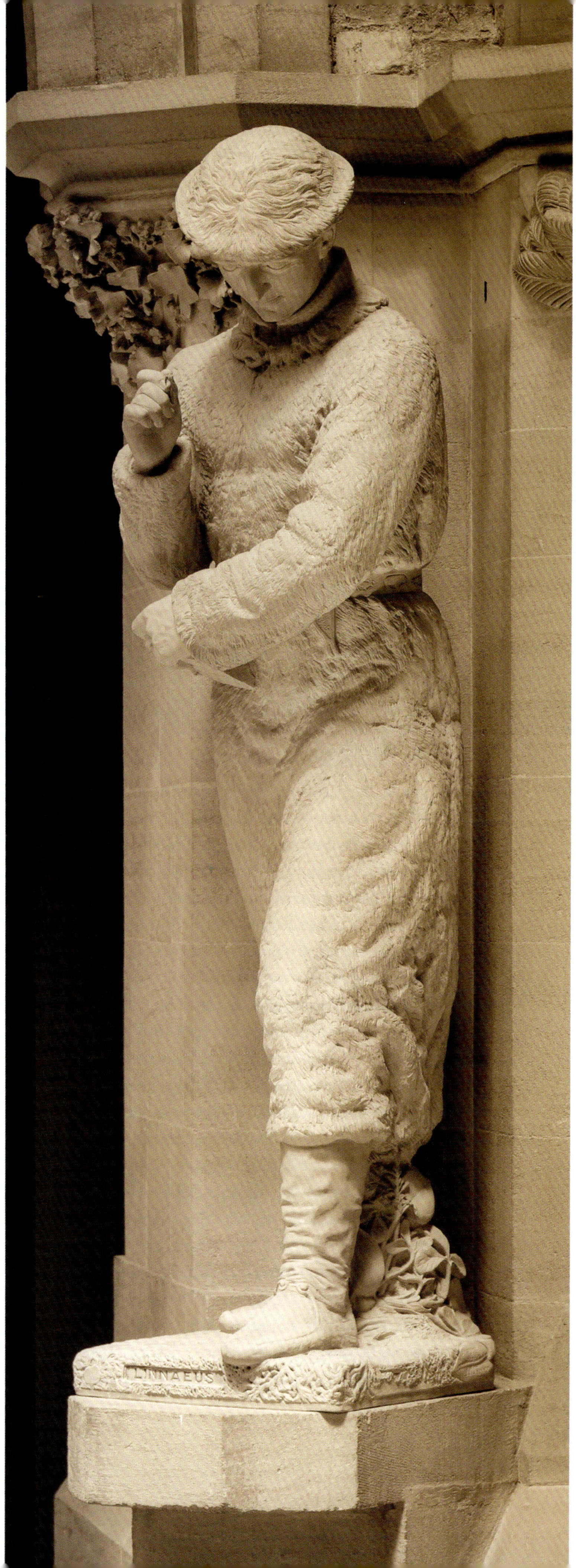

to the perception that sculptors were in effect designers who did not carve their own statues. They suggest that this is at once unfair in itself and unduly prejudiced in favour of the idealised and rarely real figure of the artist as lone genius.[143] It is striking nonetheless that the Pre-Raphaelite sculptors retained an express commitment to carving the details of their statuary themselves. In the case of Woolner's *Bacon* (fig. 86), this was because of the inadequate job done by the stone-carver who transferred his design.[144] But for both Munro and Tupper it was a point of principle, encapsulated, as Joanna Barnes and Alexander Kader point out, in Tupper's signature on his statue of *Linnaeus*, which reads 'Designed and Carved by J. L Tupper'.[145]

Through this act of hand-carving, Woolner and Tupper in particular were able to bring out a really astonishing degree of detail in their stonework, as the O'Sheas do in their natural history carvings. In his account of Woolner's work in 'The Two Pre-Raphaelitisms', Stephens notes how successfully he captures the 'heavy softness' of Bacon's velvet robes.[146] The crisp curls and folds of his ruff are if anything even more impressive. In an article on Woolner for the art magazine *The Portfolio*, Tupper praises him as 'one of the few living sculptors who . . . interprets each nicest variety of form truthfully, because not literally; by equivalents, not by equals'. When Woolner carves hair and other accessories, Tupper notes, he does so in a manner that is 'commensurate with the impression which these leave upon the eye'. This practice, Tupper argues, is the same as Hunt's in *The Hireling Shepherd*.[147] Tupper's comments on Woolner show how the defining method of Pre-Raphaelite painting – to represent as closely as possible what can be seen in the circumstances, not what we know is there but cannot see: that is, to be true to the visible phenomena – could be carried over to sculpture. In the case of his own statue, he follows the tradition of representing Linnaeus in the traditional Lapland costume he wore on his field trips. Tupper's *Linnaeus* (fig. 87) is clothed in perhaps the most meticulously rendered fur suit and hat of any statue ever carved, accompanied by boots (fig. 88) which, as Prettejohn has commented, 'seem malleable in the representation of the leather'.[148] Like Woolner, Tupper bestows extraordinary care on creating the impression that we can see every hair of Linnaeus's coat,

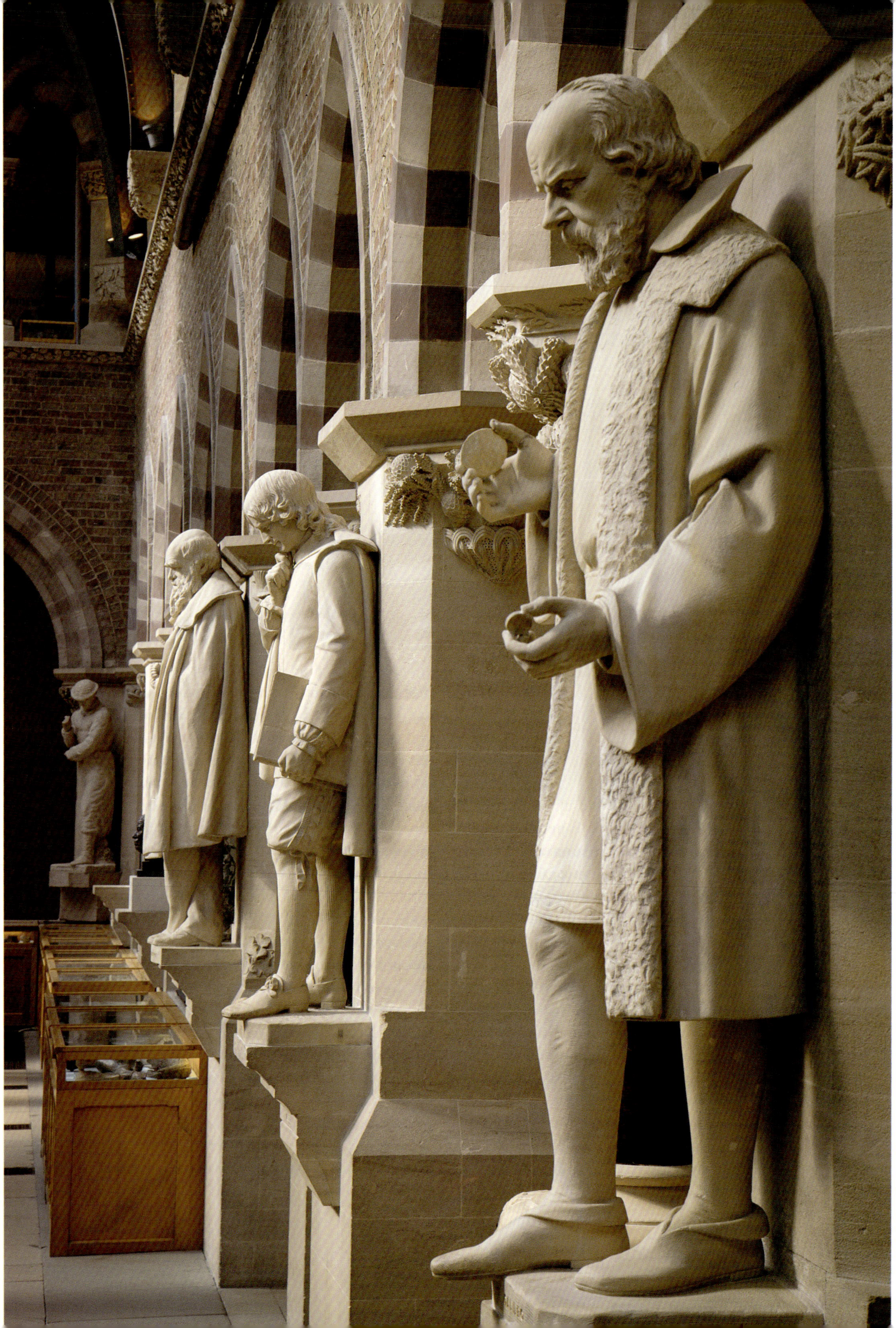

(*facing page*) 93 Oxford University Museum statues, including (from foreground) Alexander Munro, *Galileo* and *Newton*, *c.*1857–60, and Henry Hope-Pinker, *Darwin*, 1899. Oxford University Museum.

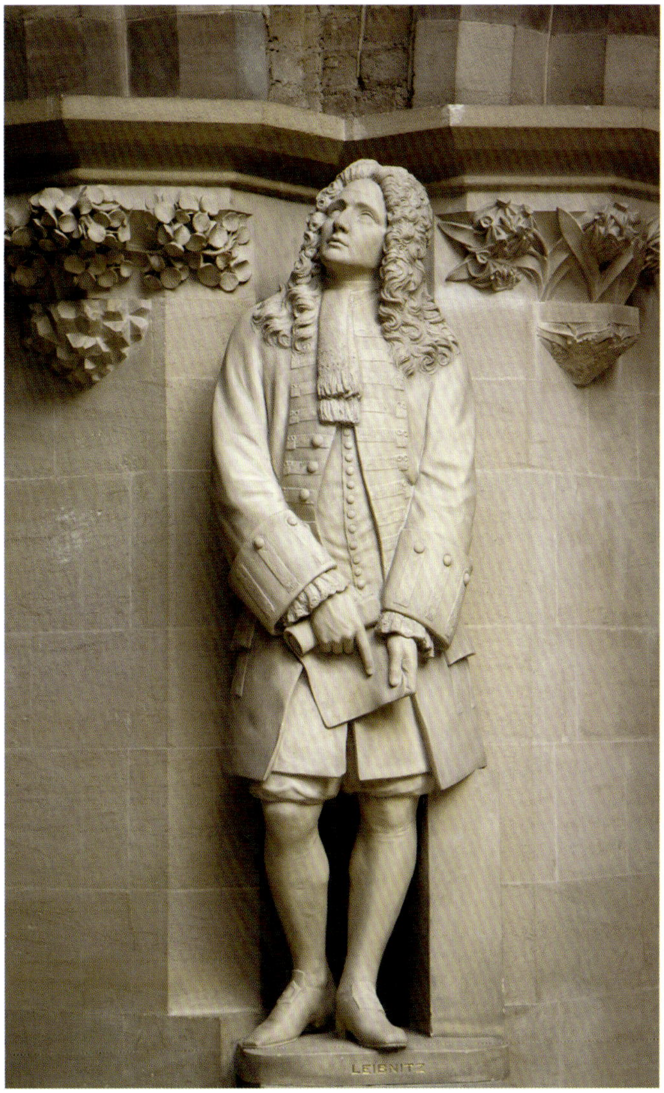

94 Alexander Munro, *Leibnitz*, *c.*1857–60. Oxford University Museum.

upon the other'.[157] His face is alert, authoritative, compelling. While designing his statue, Woolner consulted Spedding, who was preparing his edition of Bacon's works at the time.[158] Woolner shows Bacon to be an acute thinker, an assured teacher and a persuasive controversialist – precisely the qualities needed to lead what the Victorian scientists and the Pre-Raphaelites both saw as an intellectual revolution.

Munro's statues may lack the remarkable precision of texture of Tupper's and Woolner's, but they capture the same individuality and animation. The most engaging are *Galileo* and *Leibnitz*. Galileo is shown not with an emblematic telescope, but holding two lenses of different sizes, his brow furrowed as he contemplates the best way to assemble them at the correct distance apart (fig. 93). Munro's depiction of Leibnitz is more puzzling but no less effective (fig. 94). According to Acland, Leibnitz was included among the statues as the pre-eminent modern mathematician.[159] Munro represents him looking up, presumably at the night sky, to check what he sees against a roll of parchment he holds in his hands. The allusion is obscure, though it may hint at Leibnitz's controversies with Newton over the nature of space. Perhaps he is confirming his own calculations through observation, or checking someone else's data. The composition is perhaps not strictly believable, given that the telescope which Munro's Galileo is in the process of inventing had been in use among astronomers for several decades by the time Leibnitz was disputing with Newton. But it is a vivid depiction all the same of a scientist doing science.

Munro's *Leibnitz* contrasts effectively with his *Newton* too, who stands on the far side of Galileo from him (see fig. 93). Where Leibnitz is looking up, Newton is staring down at the famous apple, much as Deverell imagines him in 'The Garden'. Munro's Newton is a young man, as he was when he supposedly saw the apple fall. At the same time, he is shown proleptically with a copy of the *Principia Mathematica* under his arm. Like Galileo and Leibnitz, he is directly engaged in thought, 'lost in meditation' as William Whewell described him twenty years earlier in his *History of the Inductive Sciences*, with 'the utmost devotion of thought, energy of effort, and steadiness of will' as he embarks on the speculations and analyses that would take him from the apple to the publication of his magnum opus twenty years later.[160] This downward gaze is repeated across many of the other statues, including *Linnaeus*, *Galileo*, *Aristotle* and *Hippocrates*, until it becomes almost a motif. It has another advantage besides casting the scientist as absorbed in thought. As the statues are placed on plinths around the court, it ensures that we as viewers can exchange thoughts with them ourselves along a direct line of sight.

The characterisation of the scientist as a profound and active thinker continues throughout the original series. Like Woolner's *Bacon*, Stephens's *Priestley* is explaining a theory

95 Henry Hugh Armstead, *Aristotle*, *c.*1858–60. Oxford University Museum.

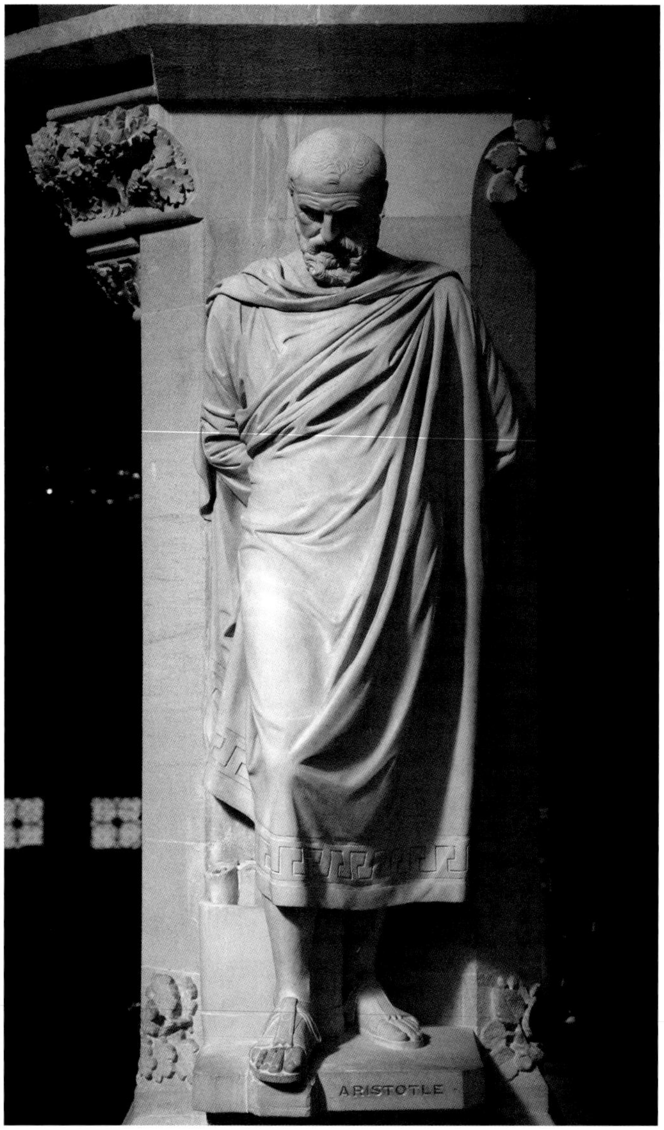

or phenomenon, the notes from his experiments clasped tight in his hand, unconsulted, as he knows so well the conclusions his science has led him to and the process by which it has reached them. Aristotle was traditionally referred to as the Peripatetic because of his habit of walking back and forth while lecturing. Armstead shows him starting off on a new train of thought, his manuscripts similarly neglected in a box at his feet (fig. 95). Woolner objected to the way Armstead had put both Aristotle's hands behind his back. He commented to Woodward that he preferred to see the hands 'illustrating the character of the subject treated'.[161] Yet taking Aristotle's posture as a whole, the fact that his hands are behind his back is suggestive. He is recognisably in a posture of thought, and active thought at that, the kind of thought that prevents the thinker from keeping still. Munro's *Hippocrates* is also shown thinking, reaching towards a diagnosis in his head, while his *James Watt* looks set to begin testing a model of a new piece of equipment. Only the last of Munro's statues, *Humphry Davy* – copied closely from a portrait by Thomas Lawrence – is placid, the public man conventionally posed with emblems of his greatness, in this case two books and a relief sculpture of a Davy lamp.

The contrast between these statues and more conventional depictions of scientists and other public men is realised most vividly and thoroughly in Tupper's *Linnaeus* (see fig. 87). In the other portraits in the Hoffman tradition, Linnaeus's costume is merely an emblem of his research, while in more formal representations, such as the statue on the façade of 6 Burlington Gardens in Mayfair, carved by Patrick McDowell in the late 1860s, even the emblems are condensed, leaving a portrait of Linnaeus as a successful eighteenth-century gentleman whose specific achievements and profession are revealed only by the quill pen in his right hand and the book and plant in his left. But in Tupper's statue we see Linnaeus in the middle of his fieldwork. He has collected a specimen of his trademark flower – the *Linnaea borealis* or twinflower – which he holds in his left hand, while one finger is keeping his place in his notebook. An early description of the statue records that his right hand originally held a marsh arum, plucked from the ground where more specimens can still be seen growing.[162] Something else on the ground appears to have caught his attention, and in just a moment he will bend down to pick that up too.

This moment is realised in yet more detail, and as an exemplary instant in Linnaeus's life and science, in a poem entitled 'A Vision of Linnaeus' that Tupper wrote at the same time as he was carving the statue:

> I saw a youth walking upon the hills
> In the breme Lapland morning, while the sun
> Now swerving upward (as a swallow turns

That has not rested on the earth) emblazed
The close fur wrapping him with gold that rippled
I' the flying wind: what time I certified
His cap of fox-skin, and his coat of deer:
And, as he walked, how he would stay his step,
Against the unconquered wind to scrutinize
The ground with flowers and rare growths mottled
  o'er
In that high region; and the rocks and pools
Sucked there by spongy herbage – not as a girl
Culling wild flowers, who looks for these alone,
But taking with a wide glance all that was
As each a limb of one great animal.
For whether it were moss or flower or fern,
Or fungus growth of rottenness, the bare
Bleached jaw-bone of some stag, or wind-bleached
  rock,
Or raven's wing in rocky cleft, or foot
Of hare the eagle-owl left, nesting close:
Each sang keen notes of one great anthem still,
Of which the dominant (man, in health, disease,
Or death) rang joyous, with a cry that rent
The harmony up through sunny air to heaven.
Grandly he walked, or grander stood, the wind
Passing, and great thoughts passing on more swift
Within him, what the world had been and was;
While in his hand the flower, held listlessly,
I saw he saw not, for his soul was rapt –
As one who has fasted feels a lightness go
Throughout his frame, conversing more with air
Than solid earth, and running seems to fly.
I saw him hovering about that hill
Like an alighted eagle, staring round
A strange world with a glory in his gaze:
A visitant who momently we fear
Even while we gaze may find his task complete,
And merge into the skies in mystery.[163]

Together Tupper's statue and his poem form a double work of art. The double work, uniting poetry and painting, is a well-known innovation of Pre-Raphaelitism, particularly in the work of Dante Gabriel Rossetti. In all such double works, there is an ambiguity as to the relationship between the two component pieces, each treating the same theme but in different media. 'A Vision of Linnaeus' was not published during Tupper's lifetime, and we do not know its precise compositional history. It can be read as a verbal cartoon for the statue or as an exhibition note for the finished piece. Returning to the statue on reading the poem, however, we can see how Tupper adds colour to the texture of Linnaeus's clothes by identifying which animals the fur comes from, and how he incorporates into the statue and/or glosses in the poem several of the same details, such as the stag's jaw-bone embedded among the marshy and mossy plants at his feet (see fig. 88). The poem invites us to imagine the statue as a man moving across a landscape, with that motion only momentarily arrested in our gaze.

'A Vision of Linnaeus' casts the naturalist as an inspired yet attentive visionary. As such, he exemplifies the scientist as depicted across these Pre-Raphaelite statues. In their animation, these statues capture the energy and intense thought that go into pioneering science. There is, of course, much that the Pre-Raphaelites leave out of their vision of science. Each of these great men stands alone, without collaborators, a solitary genius, for all that they converge on this one court. Some of them are taking measurements or initiating experiments. Linnaeus and Leibnitz are making observations; Galileo and Watt are making or testing specialist instruments. But the others are just thinking intensely, undertaking what Dante Gabriel Rossetti would later call 'fundamental brainwork' in the context of the arts.[164] Indeed, it would be hard to tell that they were scientists at all without a knowledge of their biographies and of their relationship to the museum. The most powerful scientific instrument, the statues suggest, is the imagination. Notwithstanding their own Baconian ideals and those of the scientists who commissioned the museum, the Pre-Raphaelite sculptors cast the scientist not as dogged and respectable, but as challenging and dynamic. In their practice as painters, poets and sculptors, the Pre-Raphaelites aspired to make the artist into a scientist. In these statues, they embody a complementary vision of the scientist as an inspired artist or poet, gifted and brilliant, creating a new world-view through intense concentration and the exercise of his own profound imagination.

As well as forming a gloss on Tupper's statue, 'A Vision of Linnaeus' is a prompt for how to read the museum as a whole in its relation to the inspired artist, poet or scientist. This becomes clear when we compare the poem to its likely source: Wordsworth's much longer didactic poem *The Excursion*, first published in 1814. Initially disparaged by the younger Romantics as marking Wordsworth's defection from the radical cause, *The Excursion* became required reading for the early Victorians, and remained a key point of reference for Acland when writing about the museum as late as 1893.[165] The central character in the poem is the Wanderer, a retired pedlar of sixty who roams freely across the remote hills and valleys of the Lake District. Having been taught 'the rudiments / Of science', he is attentive to the natural world around him, taking 'delight / To measure the altitude of some tall crag / That is the eagle's birthplace' or to trace the geology of 'some peak . . . that shows / Inscribed upon its visionary sides, / The history of many a winter storm, / Or obscure records of the path of fire.'[166] In imagining Linnaeus, Tupper takes these passing instances of the Wanderer's scientific eye and develops them into a more detailed and comprehensive itemisation of the signs and kinds of living nature in a still more rugged landscape. In the process, he marries a Wordsworthian style and rhythms to the precision identified by the Pre-Raphaelites with science.

While the Wanderer is one model for Linnaeus as Tupper imagines him, another is Wordsworth himself. In a passage from his perpetual work-in-progress *The Recluse*, quoted in the Preface to *The Excursion*, Wordsworth characterises his poetic undertaking in terms which anticipate Tupper's closing account of Linnaeus's scientific enterprise. 'For I must tread on shadowy ground,' Wordsworth writes, 'must sink / Deep – and, aloft ascending, breathe in worlds / To which the heaven of heavens is but a veil.'[167] Wordsworth as poet and Linnaeus as scientist are each cast as angelic figures, capable of descending to earth but liable too to ascend back to heaven. Tupper's characterisation of Linnaeus as a 'visitant' here is especially revealing, as Wordsworth's Wanderer uses precisely the same word to characterise science itself. In Book IV of *The Excursion* the Wanderer anticipates a time when science will be 'a precious visitant'.[168] The condition that must be met for this to be the case, he says, is that it must learn in studying nature to uncover the moral messages for humanity latent in the natural world:

> Happy is he who lives to understand,
> Not human nature only, but explores
> All natures, – to the end that he may find
> The law that governs each; and where begins
> The union, the partition where, that makes
> Kind and degree, among all visible Beings;
> The constitutions, powers, and faculties,
> Which they inherit, – cannot step beyond, –
> And cannot fall beneath; that do assign
> To every class its station and its office,
> Through all the mighty commonwealth of things;
> Up from the creeping plant to sovereign Man.
> Such converse, if directed by a meek,
> Sincere, and humble spirit, teaches love.[169]

Wordsworth describes a Linnaean exercise in taxonomy imbued with a spirit of natural piety, much as Tupper characterises Linnaeus's own work in his poem. It is no accident that the notebook Tupper's Linnaeus carries is blazoned with a cross.

Both Tupper and Wordsworth see true science as natural theology, as God's creation is revealed to the scientist in the natural world. One of the major tenets of such a position is that human beings are a race apart, in accordance with their unique standing within the Bible and Christian tradition. Both *The Excursion* and 'A Vision of Linnaeus' share with William Rossetti's poems a sense of the value of non-human life, and of the moral benefits which the close study of other living things can bring. When Rossetti asks at the end of 'A Quiet Place' how a visitor to the riverbank could see 'no cause to thank / For the grass growing here so green and rank', he too hints at the arguments of natural theology. Through an unobtrusive pun, grounds for thanks become themselves a testimony to the divine 'cause' to whom those thanks are owed. But Tupper's vision, like Wordsworth's, is more overtly anthropocentric than Rossetti's. Yet his conception of man as the dominant note in the anthem of nature is both more original and more complex than Wordsworth's rehearsal of the familiar hierarchy of the chain of being. As the dominant note, humanity gives structure and meaning to nature as a whole – as its culmination and, through science, its interpreter. At the same time, our note is dominant in a very different sense, tearing apart the harmony of the rest of nature by virtue of our own exceptionalism. For Tupper, we are both in nature and outside it, both of it and beyond it, as we alone are made in God's image, with immortal souls and the potential for new understanding and moral growth.

The specimens that Linnaeus finds in Tupper's statue and poem – the twinflower, the deer's jaw-bone, the raven's wing, the hare's foot, the rock eroded by the wind and so on – are precisely the kinds of objects on display in the museum itself. As the founder of modern scientific taxonomy and the premier biologist among the original set of statues, Linnaeus sets the standard for how to interpret them. In *The Excursion* Wordsworth is keen to affirm the 'union' in nature, yet his description of taxonomy persistently emphasises the 'partition' that separates one kind from another. For Tupper, Linnaeus's subdivision of the natural world enables him to hear more clearly the 'one great anthem' of nature. Tupper's metaphor implies the reunification of different strains into one complete and complex whole, like Linnaeus's own economy of nature. As Stephens recognised in his article on the museum, this 'principle of diversity in unity' is carried over from the science of nature to the art of the building itself.[170] Where Tupper imagines this unity as music, Woodward and his collaborators realise it as architecture and design. The iron ribs of the glass roof encourage us, like Tupper's Linnaeus, to take a 'wide glance' and to see 'all that was / As each a limb of one great animal'. At the same time, like the O'Sheas and indeed like Collinson in 'The Child Jesus', Tupper does not shy away from the predation and suffering in nature. The remnants of the stag, raven and hare that Linnaeus finds all bespeak death and dismemberment. As Tupper puts it in the startlingly bleak poem 'What the Sun Sees', written a couple of years before, the sun sees 'Death everywhere, or pain!'[171] Even so, for Tupper human exceptionalism means that ultimately this pain and death do not countervail over the joy of nature as a whole, as we alone suffer profoundly, and we are saved.

Acland and his colleagues took a marked risk in commissioning the Pre-Raphaelite sculptors, with their lack of experience in public sculpture, to design and carve the statues around the court at their new museum. It was a risk that paid off tremendously. The animation of these statues is wholly unlike the staid pomposity of so much Victorian public sculpture. It captures with precision an idea of the scientist as gifted with an intense and incisive imagination, capable of perceiving 'the unexpressed Art of the great Artificer', in Acland's words. As the collection of statues grew from ten in 1860 to seventeen by the end of the century, the Pre-Raphaelites' concentration on scientists caught in the act of imagination and apprehension was diluted somewhat, though the four statues completed by Henry Hope-Pinker from the 1880s onwards matched it with an appealingly relaxed informality equally unusual in Victorian public sculpture.

Hope-Pinker's statue of Charles Darwin (see fig. 93) was unveiled in June 1899 by the Vice-Chancellor of the University after a speech by Hooker.[172] It memorialised a scientific theory which was alien to Tupper's vision of nature, although it had been hinted at in the O'Sheas' carvings. By the 1890s, Oxford had become one of the most determinedly Darwinian centres for the study of natural history in the country. In 1883 the evolutionary anthropologist E. B. Tylor had taken over from the mathematician Henry Smith as Keeper of the museum.[173] The following year General Augustus Pitt Rivers gifted his anthropological collection to the University, and an extension was built on the back of the museum by Thomas Deane, the son of Woodward's partner, to house and display this collection according to Pitt Rivers's Darwinian principles.[174] This became the Pitt Rivers Museum, under the curatorship of a third evolutionary anthropologist, Henry Balfour. In 1891 E. Ray Lankester arrived as Linacre Professor of Comparative Anatomy. He was joined by the equally staunch Darwinian E. B. Poulton, appointed Hope Professor of Zoology in 1893. Yet another Oxford scientist, the Waynflete Professor of Physiology John Scott Burdon Sanderson – soon to be Acland's successor as Regius Professor of Medicine – endorsed ecology as a science for the first time in England as President of the British Association in 1893, declaring that it was as important a branch of biology as morphology and his own field physiology.[175] Finally, in 1899 the Darwinian statistical geneticist W. F. R. Weldon was appointed Linacre Professor of Comparative Anatomy.

96 Henry Hope-Pinker, *Roger Bacon* (detail), 1914. Oxford University Museum.

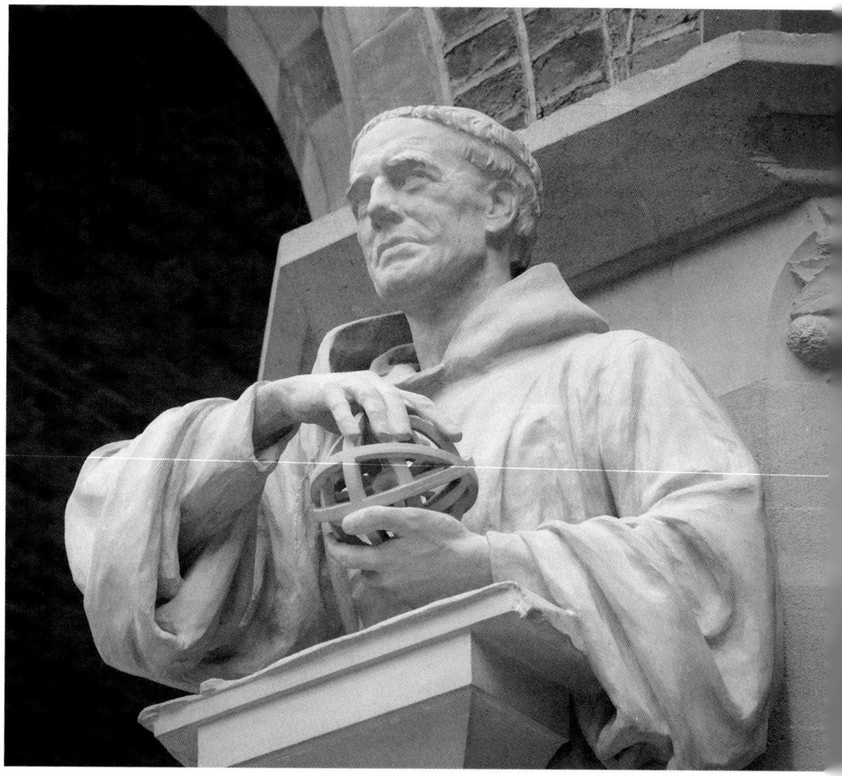

It was Poulton who sponsored Darwin's addition to the museum's pantheon of eminent scientists. His eventual place in that pantheon was as revealing as his joining it. At the museum's founding, the undergraduates of the University had pledged the money for a statue of Georges Cuvier alongside that of Aristotle.[176] Cuvier was the great French comparative anatomist who had dominated natural history in the early decades of the nineteenth century. He was a determined opponent of evolution. But the money and the statue were never forthcoming. With Cuvier absent, Darwin's only rival as the paramount modern biologist in the museum's pantheon was Linnaeus, and it was easy to construct a narrative of scientific progress that could accommodate them both.

Darwin's seniority was reflected in the location of his statue. Two alternative proposals for positioning it were put forward by a committee in February 1897. One was to place Darwin on the other side of the entrance to the eastern gallery from Newton. The other was to move Aristotle and Bacon into the small entrance hall and replace them with Darwin and Newton.[177] This would have signified a deliberate change in the museum's loyalties, relegating the classics and empiricism to mere heralds of science, while declaring Newton and Darwin to be the twin – and English – heads of modern science. Under this scheme, Darwin would have commanded the biological sciences as Newton did the physical sciences. This more radical proposal was set aside, however, and Darwin was placed where he stands today, next to Newton on the far side of the court from the museum's entrance. But while the committee implied that either proposal would place Darwin on an equal footing with the great physicist, this is not quite the case. In fact, Newton remains the central statue on the eastern side of the court, diametrically opposite the main entrance and flanked by Darwin and Galileo. The resemblance between these two flanking statues – both bald, full-bearded, in long coats – enhances the sense that they are more equivalent to one another than they are to the youthful Newton, for all that Hope-Pinker's statue stands half a head taller than Munro's ones.

Darwin was eventually welcomed in to the fold of the Oxford museum, signalling that the biology taught there was now Darwinian biology. The move realigned the museum's decorative scheme, encouraging visitors to see the ecological vision that had always been there, thanks to the O'Sheas' carvings on Pre-Raphaelite principles. But although Darwin's place in the museum was a prominent one, he remained a latecomer. On Darwin's death, Owen had characterised his arch-rival as the Copernicus of biology. It was a back-handed compliment, as it presupposed, as Owen had no qualms saying, that the science had yet to be 'blessed with its Galileo, its Kepler and finally its Newton'.[178] Oxford's final estimate of Darwin at the end of the century was a notch higher than Owen's, placing him on a level with Galileo, but he was still not quite the equal of Aristotle, Bacon or Newton.

After Darwin, one last statue was added to the pantheon in 1914. Again carved by Hope-Pinker, this was of the medieval philosopher, mathematician and experimentalist Roger Bacon (fig. 96). Bacon was a Franciscan friar, and the statue was erected in honour of his 700th anniversary. His presence in the court adds a new element to the museum's history of science, acknowledging that science was not wholly eclipsed in Christian Europe during the supposedly dark ages between the dusk of Greco-Roman culture and the dawn of Renaissance empiricism represented by Francis Bacon and Galileo. But, as his monastic robes assure us, his natural philosophy was sanctioned by theology. An Aristotelian trained at Oxford, Roger Bacon is another reminder of a long tradition of science grounded in the classics and at the service of faith. If you walk across the court from his statue and past Darwin into the east gallery, turn and look behind you, you will see the two small monkeys. Darwin's statue answered the riddle set by the O'Sheas forty years before. But in preserving Newton's primacy and in admitting Roger Bacon to the court, Oxford insisted too that it was possible to embrace evolution while retaining a firm belief in a divinely ordered universe discoverable through science.

Following pages Detail of fig. 128.

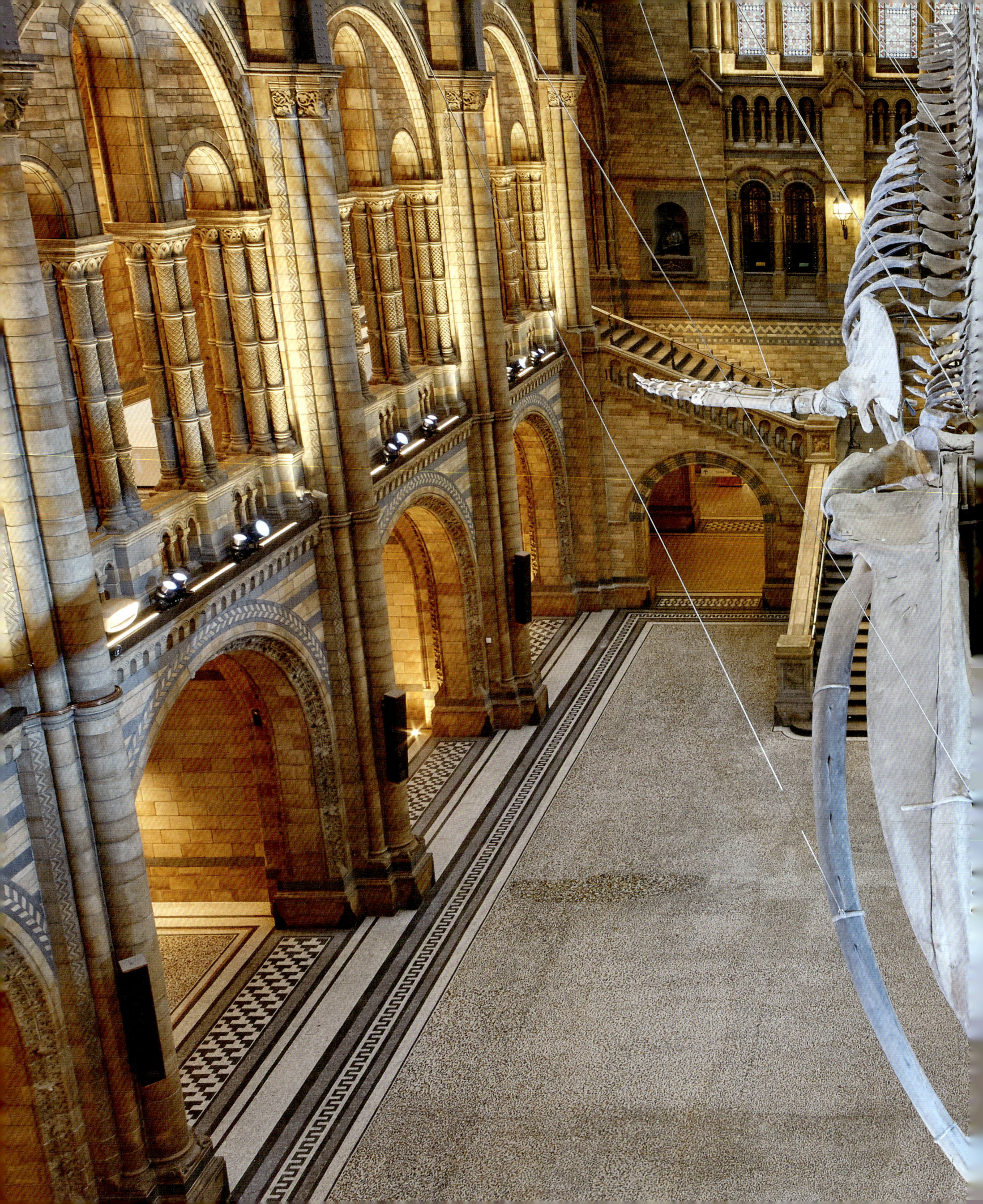

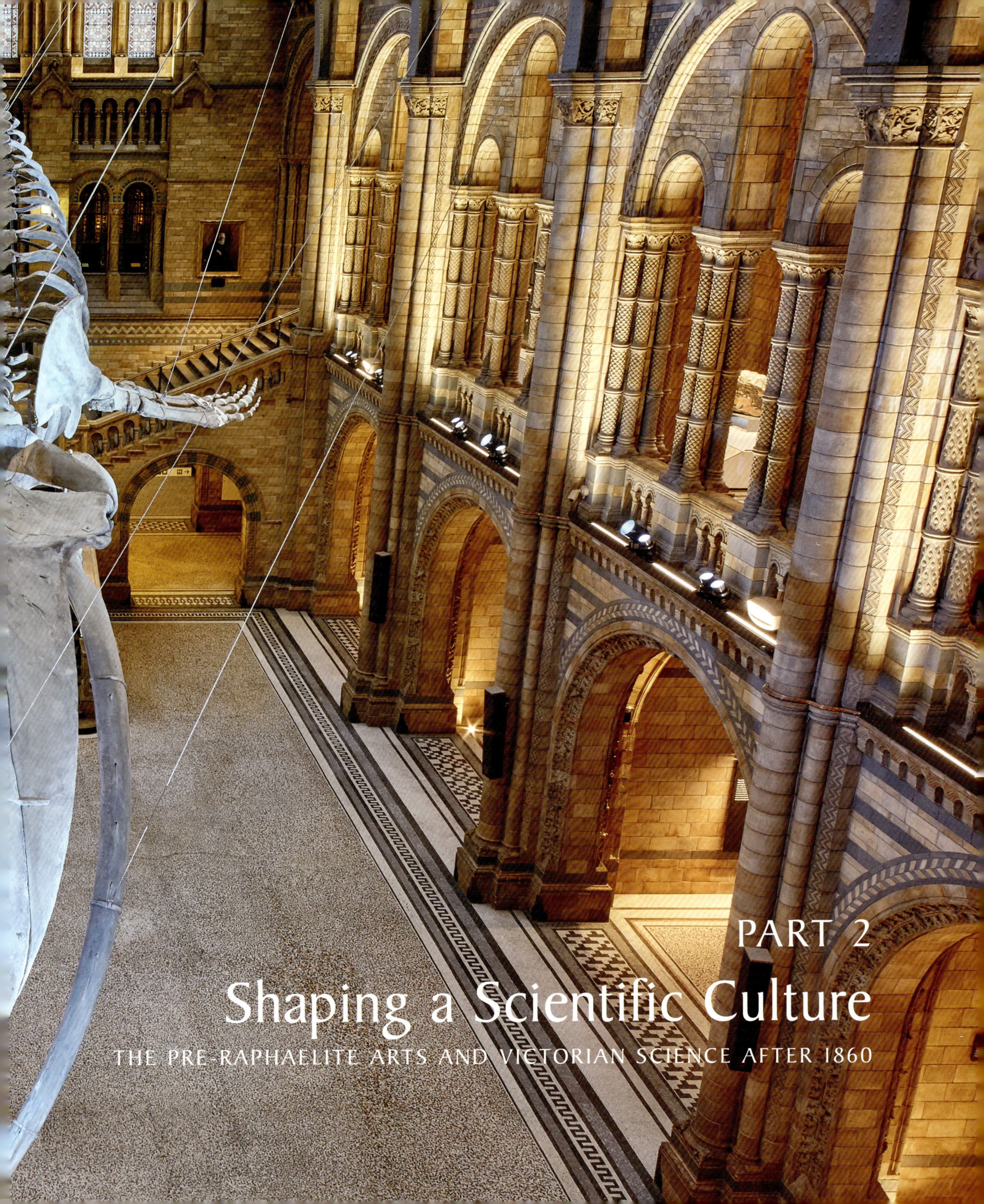

PART 2

Shaping a Scientific Culture

THE PRE-RAPHAELITE ARTS AND VICTORIAN SCIENCE AFTER 1860

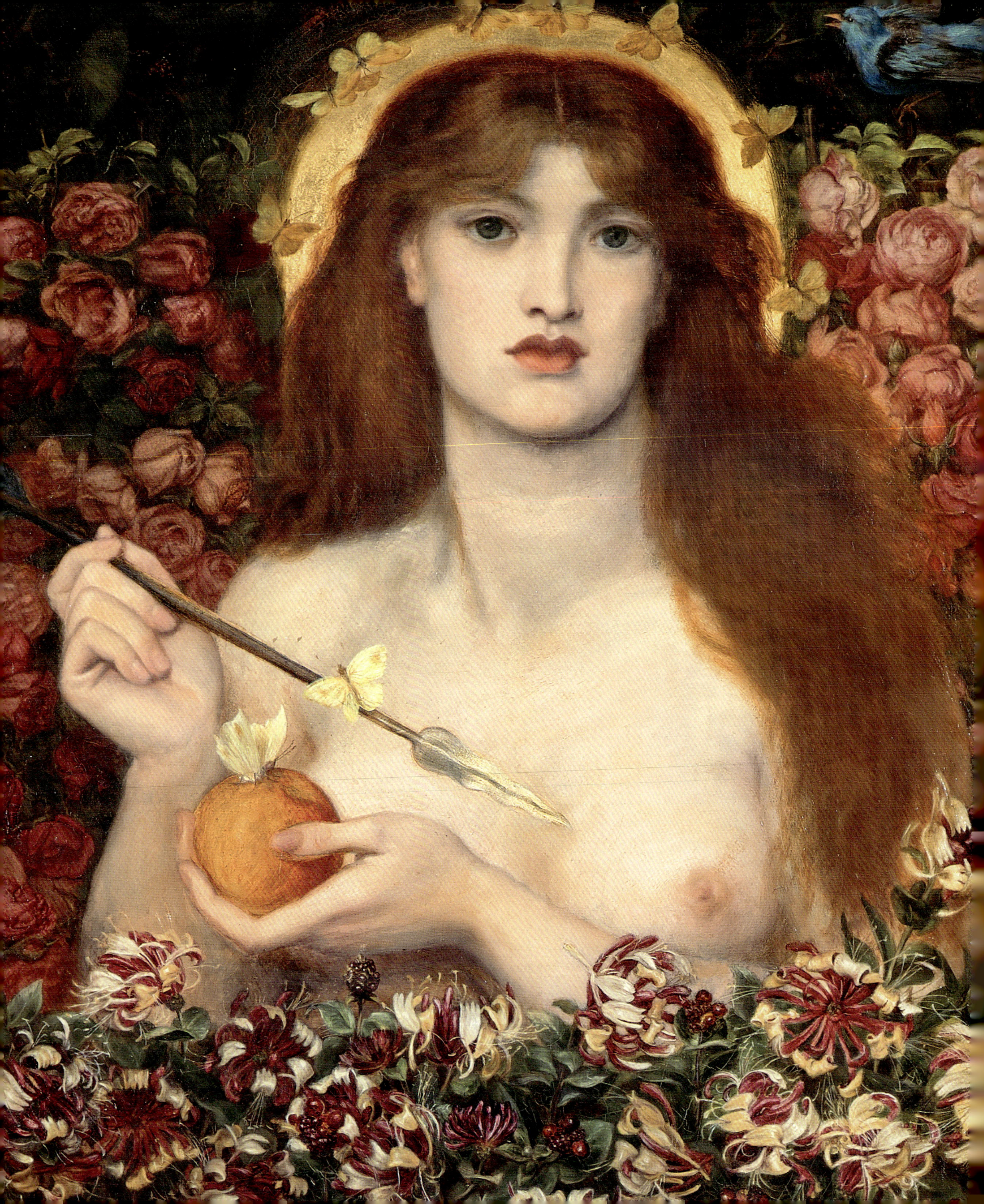

6 · Fault Lines

PRE-RAPHAELITISM AND SCIENCE AFTER 1859

On 17 April 1860 Holman Hunt's sacred painting *The Finding of the Saviour in the Temple* was exhibited at the German Gallery in London.[1] In June the same year, Oxford University opened its new museum to host the annual gathering of the British Association for the Advancement of Science, with Hunt and Woolner among the delegates.[2] Together these two events mark the culmination of the Pre-Raphaelite project to revitalise art through a rigorous new engagement with science. The Pre-Raphaelites had turned to science as a model for their art and the scientists had repaid the compliment, taking Pre-Raphaelitism as a model for how to communicate conceptions of nature and of science itself in plastic form. But even at its very moment of triumph the original Pre-Raphaelite project was effectively over. On the one hand, the publication of *On the Origin of Species* in November 1859 had opened a fissure in science, epitomised in the fraught debate over human evolution between Huxley, Wilberforce and others which took place at the Oxford meeting of the British Association. On the other hand, Pre-Raphaelitism itself was fragmenting. In the same month that Darwin's book was published, Rossetti wrote to William Bell Scott to tell him that he had recently painted a 'little half-figure in oil'.[3] This was *Bocca Baciata* (fig. 97), which would be on display at the Hogarth Club's exhibition when Hunt unveiled *The Finding of the Saviour*.[4] *Bocca Baciata* signalled a new departure in Rossetti's art, a turn away from the genre of history painting that the Pre-Raphaelites had set out to reform. From now on he would devote himself to portraits of women, abstracted away from their supposed or actual biographies to become vehicles of sensual experience, on the one hand, and the focus of elusive symbolisms, often private to the artist and to each viewer in turn, on the other.

Facing page Detail of fig. 101.

97 Dante Gabriel Rossetti, *Bocca Baciata*, 1859. Museum of Fine Arts, Boston.

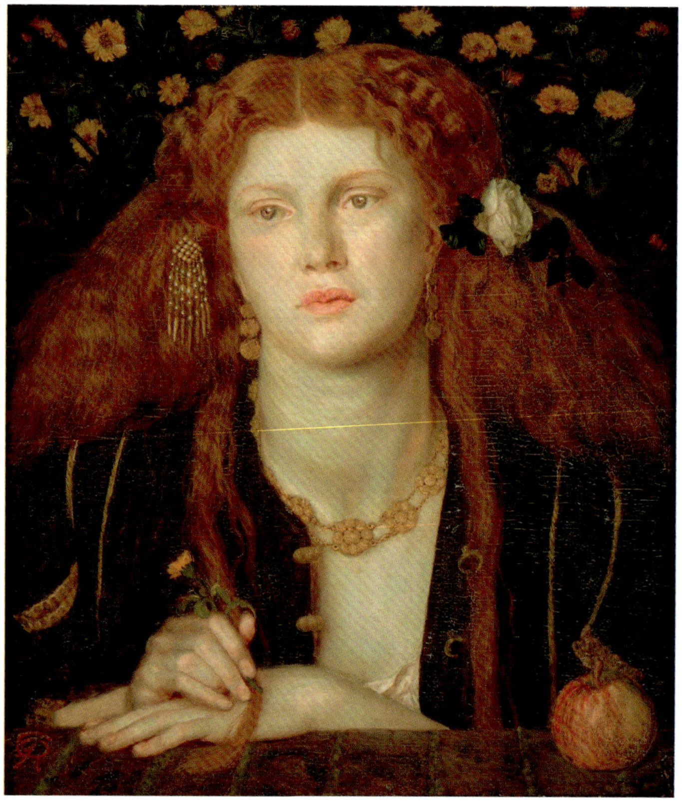

Rossetti and Darwin both died in April 1882, only ten days apart. After their deaths, Rossetti's friend Theodore Watts paired them together, writing in an article for the *Nineteenth Century* that 'Rossetti's reflected influence throughout the entire world of English taste has been as potent almost as the influence of Darwin throughout the entire world of English thought.'[5] Like *The Origin of Species*, *Bocca Baciata* marks a decisive moment in that influence, signalling a turn among Rossetti's closest collaborators towards the aestheticism and symbolism which would come to define late Victorian avant-garde art.[6]

The fault lines in both Victorian science and Pre-Raphaelitism were beginning to appear in the mid-1850s. From the 1860s onwards, neither could any longer claim to be a unified and coherent project. In the second part of this book, I want to explore the relationship between science and Pre-Raphaelite art in this newly fractured intellectual and cultural landscape. In particular, I want to examine the part that Pre-Raphaelite art in its various forms played in shaping the scientific culture of later Victorian England. There have been several studies of how individual painters, poets and critics associated with the aesthetic movement engaged with specific scientific disciplines or discoveries.[7] The three chapters which follow this one each take instead a particular Pre-Raphaelite art form – poetry, architecture and painting – to examine its function within and contribution to Victorian scientific culture as a whole. The relationship of art to science is different in each case, and as with the Pre-Raphaelite project of the 1840s and 1850s, the conception – or rather conceptions – of science itself will be as central to the discussion as any given scientific theory or finding. Pre-Raphaelite poetry, it transpires, played an important role in giving meaning to a new culture of scientific materialism. By contrast, Pre-Raphaelite architecture, in the shape of the Natural History Museum in London, was marshalled in the service of natural theology, while Pre-Raphaelite painting turned its attention to recording and interpreting the history of science unfolding before it. Before we move on to each of these developments in detail, however, we need to establish two things: what part the original aesthetic programme of Pre-Raphaelitism was able to play in the new scientific context of the 1860s and after, as the old natural theology was pitted against a new methodological materialism in science, and what the place of science was within the new aestheticism and symbolism that came to define the art of the second generation of Pre-Raphaelites.

NATURAL THEOLOGY VERSUS SCIENTIFIC NATURALISM

On the Origin of Species marks a watershed in the history of Victorian science. It is best remembered as the book that laid the foundations for modern evolutionary biology. Evolution itself was not a new idea. In the late 1700s savants such as the Comte de Buffon and Erasmus Darwin had posited that the history of life on earth was one of species transforming over time, not of successive acts of creation. As a scientific hypothesis, transmutation became firmly established though not predominant in France early in the nineteenth century through the work of Jean-Baptiste Lamarck and Etienne Geoffroy Saint-Hilaire. It remained marginal in England until the 1840s, when successive editions of *Vestiges of the Natural History of Creation* won evolutionary ideas a wide hearing.[8] Even so, evolutionism still had little scientific legitimacy, being condemned equally by the Baconian old guard and by rising men like Huxley who would begin to mount their challenge to the old world-view in the later 1850s. For British scientists, it was Erasmus Darwin's grandson Charles who turned evolution from a hypothesis into a theory. Darwin established two crucial new concepts: the mechanism of natural selection and the principle of descent with modification, which gave that mechanism something to work on. At the same time he amassed an immense and diverse body of evidence to support his theory – evidence which, he argued, other theories simply could not account for. Meanwhile, concurrent research by archaeologists and anatomists demonstrated with increasing certainty that human beings themselves had a long prehistory and were closely related to the great apes. These findings were summed up in two more landmark books that appeared four years after the *Origin* in 1863, both written by friends and supporters of Darwin: Lyell's *Antiquity of Man* and Huxley's *Man's Place in Nature*.[9] Opposition to the principle of evolution or descent with modification itself was short-lived among Victorian scientists. By contrast, the exact nature and force of the mechanisms that might drive that process – whether natural selection was a sufficient cause on its own or only one of many, whether evolution happened gradually or in fits and starts, whether it was in some sense predetermined – were hotly debated for decades, as was the exact status of humanity within this process.[10]

Darwin's arguments struck more deeply into the heart of the mid-Victorian scientific establishment than this relatively easy victory implies, however. His book challenged two of its most fundamental tenets, both closely associated with the original Pre-Raphaelite project: that the fundamental method of science was Baconian empiricism, and that what it revealed to us was a world that was recognisably God's creation. In his autobiography, Darwin cast himself as having 'worked on true Baconian principles'.[11] In the early 1850s he had cemented his reputation as one of the leading naturalists of the day through the unimpeachably Baconian endeavour of minutely distinguishing, describing and cataloguing all known species of barnacle.[12] Yet as James Moore has explained, Darwin's challenge to the Baconians was to shift the aim of scientific argument from establishing the truth to weighing up probabilities.[13] In *On the Origin of Species* he sought to show that descent with modification, shaped by natural selection, was an altogether more likely and therefore more satisfying explanation for the facts of nature than the special creation of each individual species. But he could not prove that this is what happened because the processes or events under discussion – the origins of living species – took place in the distant past. So his method of argument is to test the competing hypotheses against the evidence, rather than to amass the evidence first and then describe the patterns that emerge from it.

Darwin meets the Baconians on their own ground, arguing his case by assembling fact upon fact which cannot be accounted for under the theory of special creation except by tautologically affirming that things are the way they are because they were made that way. But the use to which he puts this data is openly probabilistic. How unorthodox this was as science becomes clear when the *Origin* is compared to one of its own key intellectual sources, Lyell's *Principles of Geology*, which Darwin read eagerly and assiduously during the voyage of the *Beagle*. Lyell's argument for uniformitarian geology is founded on the premise that the entire geological history of earth can be accounted for through phenomena that can be observed in action today, from sedimentation to earthquakes and volcanoes. Darwin's argument for evolution, by contrast, works by reverse engineering current living forms to reveal long biological histories which of necessity cannot be observed.

E. Ray Lankester put the case for Darwin's new kind of science in his book *Degeneration: A Chapter in Darwinism*, published in 1880. Then Professor of Zoology at University College London, Lankester was one of Darwin's most committed and influential supporters. He would go on to be in succession Professor of Physiology at Oxford, Director of the Natural History Museum, and one of the leading scientific popularisers of the early twentieth century through his *Daily Telegraph* column 'Science from an Easy Chair'.[14]

98 Thomas Woolner, *Charles Darwin* (detail), 1868–9. Herbarium, Cambridge University, Cambridge.

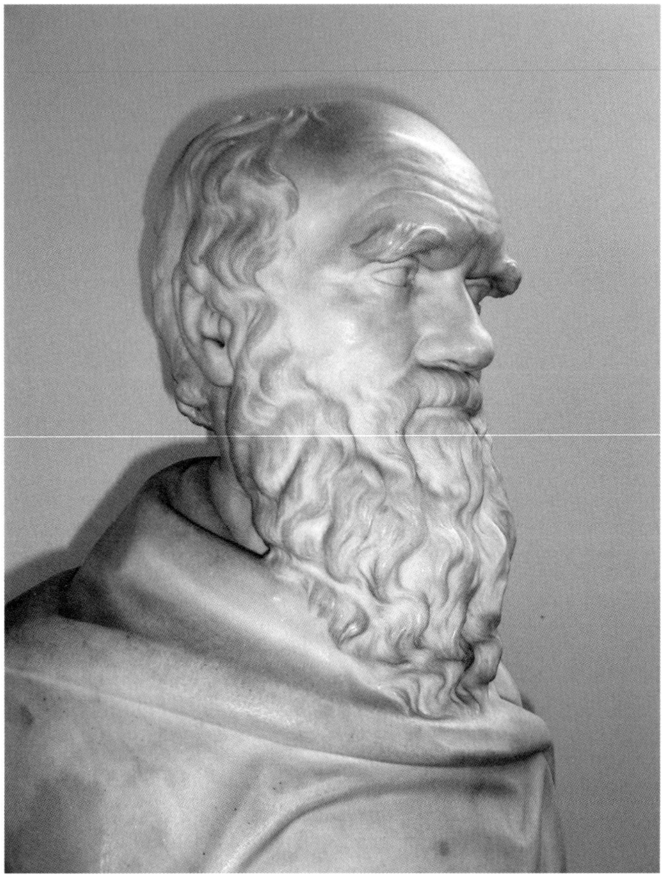

Science, Lankester argues, is concerned with the causes of things, not with 'the mere inspection and cataloguing of natural objects'. 'It is a common mistake', he argues, 'to consider all knowledge of raw products, of living objects and other natural objects, as necessarily "science." The truth is, that a man may have great knowledge of these things as so many facts, and yet be devoid of "science".' This was where the mid-Victorian natural historians went wrong:

> Fact upon fact was observed and recorded by students of plants and animals, but having no hypothesis as to the causes of what they were studying, the naturalists of twenty years ago, and before that day, though they collected facts, made slow progress and some strange blunders. Suddenly one of those great guesses which occasionally appear in the history of science, was given to the science of biology by the imaginative insight of that greatest of living naturalists – I would say that greatest of living men – Charles Darwin.[15]

Lankester's description of Darwin as an inspired visionary is anticipated in the portrait bust that Woolner carved of him in the late 1860s. Woolner's *Darwin* is one of the defining images of the scientist as sage (fig. 98). His bald, high-domed head and cascading beard give him the appearance of an ancient philosopher, while his blank stone eyes seem to stare into the middle distance with, to borrow Lankester's words, 'imaginative insight' into the nature of things. Woolner remains subtly true to the modernity and the specificity of his subject. Darwin's jacket marks him out as a man of the nineteenth century, his blanket or shawl as an invalid. At the same time, the sweep of the shawl reinforces the impression that we are looking at a classical bust of a man elevated beyond his own time to become part of the pantheon of great thinkers.

Lankester's account of Darwin chimes with Woolner's bust, as it does with the Pre-Raphaelites' ideal of the scientist as poet embodied in their statues at the Oxford museum. But his argument lays waste the Baconian model of science that they sought to follow in their own art. On Lankester's terms, the 'adherence to fact, to experiment and not theory, – to begin at the beginning and not fly to the end' that Stephens took to be characteristic of science and an example to the arts is not scientific at all. Darwin's second great challenge to the scientific establishment lay in insisting that science should seek only material explanations for material phenomena. Throughout the century, natural theologians from Paley to Phillips and Strickland had argued that natural phenomena betokened a divine cause. For all that it was denounced as materialist, even *Vestiges* routinely assigns this role to God, whether as 'First Cause', 'Creative Power', 'Divine Author' or 'Great Father'.[16] Wielding Ockham's razor, Darwin made it a principle of biology and of science in general that supernatural causes should not be inferred where material causes were sufficient to sway the balance of probabilities. In writing that all life on earth had probably 'descended from some one primordial form, into which life was first breathed', he tacitly admitted that there were as yet no known natural causes which could explain the origin of the first living beings.[17] But his theory made away with all reference to God as a designer or as the miraculous creator of successive living forms. Instead, Darwin identified random variation, the brutal struggle to survive and sexual desire as the engines driving creation. Each of these was unpalatable to natural theology, which held the natural world to be the orderly product of God's intrinsic goodness, in accordance with the pacific benevolence and modesty of received Christian morality.

Natural theology stood and fell on the correlation between the findings of science and the precepts of religion. Once Darwin's ideas had taken hold, there was much to be said for keeping science and faith apart, rather than trusting that the two would continue to be mutually supportive. The crucial disagreement dividing Darwin and his followers from the natural theologians was not principally over religion itself, but over its the relationship to science. The Darwinians espoused a strict materialism in their methodology, yet many of them retained their faith in the Christian God. Even Huxley and his equally pugnacious friend the chemist John Tyndall, who were frequently charged with espousing materialism in the full sense of believing that the world consists exclusively of matter, identified themselves rather as 'agnostics' or 'scientific naturalists'. Their point was that, as transcendental questions were intrinsically unanswerable, they should be set aside for all practical purposes, whether in science, education or politics.[18]

In the eyes of the natural theologians, however, this stance was disingenuous. To propose that science and religion each be allowed their own discrete domains – what Stephen Jay Gould would call their 'non-overlapping magisteria' or 'NOMA' in a re-run of the same debate in America at the end of the twentieth century – was to abandon the intellectual primacy of Christianity, to concede science to a secularism that was in effect atheism, and to leave faith no more than a matter of private conscience.[19] That this was indeed the direction of travel seemed to be confirmed in 1871 when Darwin published *The Descent of Man*. In 1859 he had tactfully – or perhaps menacingly – limited his intervention on human evolution to two short sentences: 'Psychology will be based on a new foundation, that of the necessary acquirement of each mental power and capacity by gradation. Light will be thrown on the origin of man and his history.'[20] The second sentence is the famous one. For all Darwin's restraint, all his readers and commentators realised immediately that his book had implications for our understanding of human origins. But the first sentence is ultimately the more charged. What Darwin hinted at here, and would go on to argue in detail in *The Descent of Man* and its sequel, *The Expressions of the Emotions in Man and Animals*, published the following year, was that human mental capabilities had been acquired through evolution.

This, again, was a marked departure from the conclusions drawn by his evolutionary precursors. Arguing that 'we have faculties in full force and activity which the animals either possess not at all, or in so low and obscure a form as to be equivalent to non-existence,' Chambers had concluded in *Vestiges* that 'these parts of mind are those which connect us with the things that are not of this world.'[21] For Darwin, there are no grounds to presume such exceptionalism on our own behalf. Our feelings and thought processes could be placed on a graduated scale with those of other animals. There really was nothing unique or special about human beings, beyond that ubiquitous kind of uniqueness that defines any species as distinct from others. For many Victorians who had by this point accepted that our bodies were descended from those of ape-like primates, the divine gift of a soul remained the condition of our humanity and the guarantee of our immortality. To accept Darwin's arguments in *The Descent of Man* would be to give up the things that are not of this world, to yield even the soul to science, not merely as an object of study, as implied in the term 'psychology' – 'psyche' meaning 'soul' in Greek – but as a final concession to a secular world-view. For the natural theologians, above all, it would be quite literally to give up the ghost.

The crisis in science in the 1860s and 1870s pitted the old Baconian natural theologians against the new and often younger Darwinian scientific naturalists. It divided the Pre-Raphaelites and their circle too. Tupper, Hunt and Ruskin held firmly by their belief in natural theology. Brett, Stephens and the Rossetti brothers aligned themselves to varying degrees with scientific naturalism. Yet if the Pre-Raphaelites themselves had to choose where to stand, Pre-Raphaelite aesthetics could be pressed into service on both sides. Even if Baconianism was obsolete, Pre-Raphaelite rigour in observation was still exemplary, and the principle that truthful art could help to establish the truth of science held good regardless of what that truth turned out to be.

That Pre-Raphaelitism was equally serviceable to natural theology and the scientific naturalism of its opponents can be seen in essays published by Tupper, Brett and the astronomer Norman Lockyer in the late 1860s and 1870s. In 1869 Tupper published a book entitled *Hiatus* calling for Pre-Raphaelite principles in art education. The following year Brett published an article on 'Natural Science at the Royal Academy' in the new scientific journal *Nature*. They both argued, just as Stephens had done in *The Germ*, that artists needed to understand science in order to know how to judge the accuracy of their own drawings. They both extended this argument too, claiming that science itself could learn from the precision of Pre-Raphaelite art. As Brett put it, 'scientific men pay too little attention to the broader aspects of the visible world; while the artists on their part pass by the clear fountain of natural beauty, and content themselves with dreamily sipping lukewarm water from the corroded vessels of their forefathers.'[22] This suggestion that art modelled on science could benefit science in its turn was new within Pre-Raphaelite art theory, although it had already been implemented at the Oxford museum, where, according to the third edition of Acland's guidebook, published in 1866, an artist had been engaged 'for the purpose of making original drawings of scientific objects, and of instructing Natural History students in drawing from Nature'.[23]

In *Hiatus* the argument that training in drawing and in science are indispensible for one another underpins Tupper's rearguard defence of pre-Darwinian biology. Tupper was unequivocal about the implications of Darwinism. As he wrote to Hunt in 1872, 'Darwinism is *practical atheism*.'[24] It was also bad science, on Baconian criteria, as he explained in *Hiatus*: 'To observe a gradient of structural resemblance from the lowest vertebrate up to man is scientific; but upon such observed facts to found a theory of actual development of each vertebrate into the next above it, each to be produced in its totality from nothing but that one below it, help of time, and material conditions, is unscientific, till structural and functional progress are observed to be coextensive.'[25] In Tupper's view, Darwin's inferences were not warranted by the evidence in front of him and his use of data was prejudicial. In his letter to Hunt, he outlined an unpublished and now lost satire that he was writing against Darwin. The thrust was that Darwin's theory would serve just as well to explain a wholly different set of natural facts from those we are actually faced with, such as 'Why our commonest butterfly has the under surface of the wing blue and the upper green', 'Why there are no white moths', 'Why owls are always black', 'Why the bat's wing is more richly ornamented than that of birds', 'Why peacocks have melodious voices', and so on.[26] If Darwinism could account for everything, whether or not it matched the reality of the natural world, then it could not explain any given phenomenon at all.

Where Lankester attributes a preternatural vision to Darwin, for Tupper Darwin's fatal weakness as a scientist is precisely that he cannot see. The solution is to learn to draw. Tupper's case is that Pre-Raphaelite practice can help to underpin natural theology because if a person studies nature closely, he will see that it is designed. 'Let him draw his leaf veins hap-hazard without design,' by contrast, 'and nature must be, *to him*, hap-hazard without design.'[27] As in Tupper's earlier writings, the close study of anatomy is foundational here, but it is not sufficient in itself. Indeed, the evolutionists' mistakes are to lay too much stress on anatomy in determining kinship and to equate structure too exclusively with function. Why, for example, should the closer structural similarity to a human of an ape than a beaver outweigh the 'loss of house-building faculty' which 'removes it so much further from human function'?[28]

Tupper's most intriguing arguments along these lines concern beauty as a category of evidence for affinity. Drawing, he suggests, leads us to recognise not only patterns in form but also similarities in our own aesthetic responses to which Darwin and his allies are oblivious: 'For had they studied with eye and drawn with hand the forms of birds, flowers, palms, and panthers, and the ineffably fairer form of man and woman; they had found out that beauty has also a significance: that, if likeness is the evidence of kindred, an ape, which is less beautiful than a panther or a horse, is, so far, less like man and less akin to him than is horse or panther.'[29] Tupper rehearses an unapologetically anthropocentric argument made by the Duke of Argyll at the Royal Society of Edinburgh a few years earlier. Summarising Argyll's argument in an address to the Devonshire Association for the Advancement of Science, Literature and Art, Daubeny had concluded that the beauty of a flower can 'only be said to minister to the gratification of the higher instincts and faculties of our nature'.[30] Building on this same logic, Tupper argues that our own response to beauty constitutes data in itself. To ignore that data is again bad science:

> when rejecting as absurd the idea that certain colours of birds and insects were designed to give pleasure to man, and maintaining that they are only the remains of ancestral structure once useful but now useless to the possessor, you are going on the last baseless assumptions; you are denying that they were designed for a function which you know and feel they possess, that of affecting you with pleasure and beauty, rejecting this certain evidence for a series of suppositions grounded upon a primary supposition.[31]

'You, who profess to take evidence impartially,' he concludes triumphantly, 'allow the uncertain to supersede the certain, in your eagerness to dispense with design.'[32]

Tupper demands a recalibration of science to take account of aesthetic responses as evidence in their own right. Like Acland, he moves from a mechanical principle of design to an artistic one. Darwin's tree of life, he argues, is only one way of conceiving of the relationship between different animals. It prioritises adaptationist interpretations of structural similarities over other kinds of evidence which might bear on the question, including our own aesthetic responses. To use the terms of Tupper's own argument from twenty years earlier in *The Germ*, we need to be electricians sensitive to physiological influences and not merely geometers fixated on form. But to appreciate this we need to be able to look closely, both literally at nature and reflexively at ourselves.

Tupper published his book under the pseudonym Outis, which means 'nobody' in Greek. This is the name Odysseus adopts to trick the Cyclops before he blinds him in Homer's *Odyssey*. Tupper may be a nobody – a mere schoolteacher who had failed to make a career for himself in art – but he is also the scourge of the one-eyed in science as in art. Without the meticulous observation of nature through drawing – without, in effect, Pre-Raphaelitism – not only art but science will go astray. Yet he is also a trickster. It is hard to gauge how far Tupper means us to take his argument for a taxonomy predicated on beauty seriously. His primary objective in science is to undermine the case for evolution as Darwin presented it. His argument that aesthetic responses amount to data in themselves is both determinedly Baconian and radically subjective, a brilliantly bold argument for an idealist biology and a *reductio ad absurdum* of the idea that you can make a case for a scientific theory on a selective interpretation of the facts.

In *Hiatus* Tupper marshalled Pre-Raphaelitism against materialism. Brett's article, by contrast, appeared in a journal committed to methodological materialism in science and leaning decidedly towards scientific naturalism in its world-view. Brett first met the founding editor of *Nature*, Norman Lockyer, in the late 1860s.[33] He joined Lockyer as the artist in residence on an expedition to Sicily to observe the solar eclipse at the end of 1870. With Lockyer's backing, Brett was elected a fellow of the Royal Astronomical Society and went on to publish papers on the transit of Venus, as well as reporting regularly on astronomical meetings for *Nature*.[34] As Ruth Barton has shown, Lockyer was keen that *Nature* should be a site of scientific argument, not overly attached to one party or another.[35] Indeed, Tupper wrote a couple of reviews of books on geometry for *Nature* himself.[36] But he and Hunt both recognised that Lockyer was an ally of the Darwinians and would not be likely to welcome his satirical swipe at evolutionary theorising.[37] Lockyer had connections with Huxley and Tyndall going back to an earlier magazine called *The Reader* which he had edited in the mid-1860s. *Nature* itself became 'Darwin's publication of choice', in Melinda Baldwin's words, and that of his younger followers including Lankester and G. J. Romanes.[38] Alongside his scientific connections, Lockyer had several links to networks of Pre-Raphaelitism. As well as working closely with Brett, he was a friend of Hunt, knew Woolner, and had employed William Rossetti as a contributor to *The Reader*.[39] Lockyer's own commitment to Pre-Raphaelitism can be seen in the extended articles he wrote for *Nature* on science for artists on successive occasions in 1878, 1887 and 1891.[40]

Brett's essay set the keynote for *Nature*'s art criticism. As a theorist, Brett lacks Tupper's sophistication or mercurial wit. But that is part of his point. Most writing on art by artists, he suggests, is mere '"gas,"... to use an expressive americanism'. In its place, Brett offers precepts for judging art to 'satisfy the scientific student', written in the 'grammatical and reasonable speech' typical of 'scientific men'. Like Tupper, he stresses the importance of beauty as the 'central motive of fine art'. Again, like Tupper, he proposes a Baconian approach to establishing what that beauty is. Rather than defining it, he urges his readers to 'go to the Academy to seek for it', though he cautions them that they will only find it 'about in the same proportion as pure gold to the pebbles in the bed of an African river, or sense in a railway novel'.[41]

Where Tupper uses the artist's sense of beauty to correct modern science, Brett adopts the pose of a level-headed scientist correcting conventional artists' all too common violations of natural law. But he also sees scientists as in need of correction in their turn. The painter whose canvas at the Royal Academy reveals to *Nature*'s readers 'that they are positively living on a planet which enjoys the advantage of a solid lemon-yellow sky' is at fault, but so is the 'scientific man' who argues on the theory of complementary colours that it is because the sun looks orange that the sky looks blue.[42] Both need to pay better attention to the phenomena before them, to look and notice what they see, and to judge their aesthetic prejudices and scientific preconceptions by this same light. Ideally, each should 'go to school with the other', to strengthen and refine the capabilities of both.[43] The tone of Brett's essay suggests that he is a man talking to men, a scientist and artist talking to artists and scientists. But his proposal, like Tupper's in *Hiatus*, is also an educational programme to improve the teaching of both science and art by adopting Pre-Raphaelite principles.

Lockyer's articles, like Brett's, rehearse the principles articulated by Stephens in 'Modern Giants'. In his six-part article on 'Physical Science for Artists' published in *Nature* in 1878, Lockyer argued: 'In art, as in science, imagination must have a basis to work upon, and the surer the basis the more will the imaginative effort which transcends it be in sympathy with those hidden powers of the mind and those hidden feelings which it is the function of art to bring into play.'[44] A solid grounding in science was therefore the best foundation for art. Most of his article is given over to detailed scientific explanations of key phenomena in his own field of optics which, he argues, artists need to study alongside perspectival geometry and anatomy to ensure the accuracy of their art. But he also weighs up and passes his judgement on specific paintings from the year's Royal Academy exhibition. Lockyer judges Brett's *Cornish Lions* (now lost) to be a 'remarkable picture: the colours and the atmospheric absorption, and therefore transformation of the colour, perfect'. W. H. W. Foster's *After the Rain*, by contrast, is marred by an 'unnatural sunset' with the 'colour and distribution of light wrong', while Cecil Lawson's *An Autumn Sunrise*, he sardonically remarks, is 'interesting as a foretaste of the future when the sun shall have cooled'. His verdicts on other paintings are more nuanced. Tristram Ellis's *Sunrise Gun, Castle Cornet, Guernsey* has a 'good' sky colour, but unfortunately the colour of the water is 'impossible' under the 'sky conditions given'. George Hering's *Struyve Rocks, Coast of Arran* is 'nearly perfect' as a sunset, though 'if the yellow had faded into green it would have been better.'[45]

In his articles on art for *Nature*, Lockyer takes on the role of scientific arbiter projected by Stephens and Brett in their essays. The resemblance between his interventions in art criticism and Pre-Raphaelitism are not incidental. Rather, they are a direct continuation of the Pre-Raphaelite project by a leading scientist writing in the leading scientific journal of the moment. Lockyer admits that, although he is no artist himself, he has discussed these judgements with 'two artist friends of great distinction', who are presumably among the few modern painters who, as he writes, 'enter into the study of physical science not only for its own sake, but in order to embrace it in their art'.[46] He does not name his artistic consultants, though given their close working relationship it is highly likely that one of them was Brett himself. Another may have been Hunt, who had recently returned from another long stay in Jerusalem, and who for his part consulted Lockyer on questions of science.[47] Lockyer remarks that his own scientific investigations have also benefited greatly from the opportunity to discuss 'questions bearing on art' which have arisen from his researches with professional artists, though

99 William Holman Hunt, *The Triumph of the Innocents*, 1883–4. Tate Gallery, London.

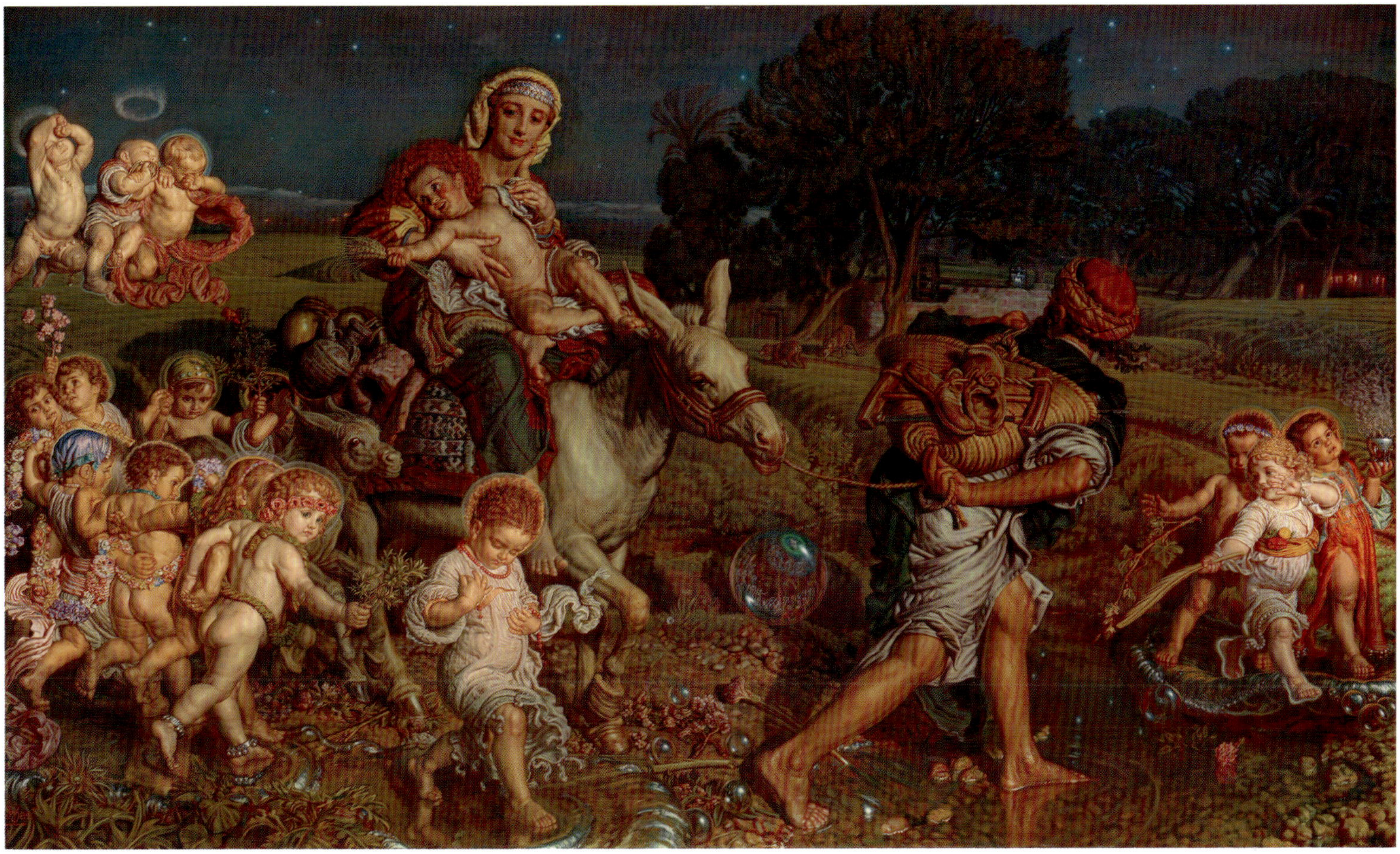

he does not specify how these consultations fed into his findings.[48] Like Tupper, Brett and the scientists at the Oxford museum, he implies that scientists – and therefore science itself – can benefit from holding themselves to Pre-Raphaelite principles of scrupulous observation and even from direct collaboration with artists working along these lines.

Tupper's *Hiatus* and Brett's and Lockyer's essays in *Nature* show that the original Pre-Raphaelite project to ground art in science could be put at the service of natural theology, or it could equally well be squared with scientific naturalism. This was no easy neutrality, however, as Tupper's argument in particular reveals. The terms of the Pre-Raphaelite project were at issue as much as those of science. One flashpoint was the exhibition of Hunt's late sacred painting *The Triumph of the Innocents* in London in 1885 (fig. 99). For Ruskin, who saw an earlier version of the painting in Hunt's studio before he had completed it, it promised to be 'the greatest religious painting of our time'.[49] For Stephens, it was incoherent.

Of the original Pre-Raphaelites, Hunt had remained most committed to the principle of independent investigation on the basis of fact in his paintings. *The Triumph of the Innocents* takes the flight of the Holy Family into Egypt as its subject. As with his earlier biblical history paintings *The Scapegoat*, *The Finding of the Saviour in the Temple* and *The Shadow of Death*, Hunt was scrupulous in his historical and geographical accuracy. He based the landscape on studies he had made in Gaza and Ramleh, the first of these carefully situated on the route between Bethlehem and Egypt. He modelled the ass and its foal on the Mecca breed of donkey specifically and Joseph's tools on local examples.[50] Yet alongside these precise material details, he paints the immaterial souls of the innocents massacred by Herod. These infant souls, risen into eternal life, travel in company with their infant saviour. Even by the standards of Pre-Raphaelitism, their physicality is intense, as they are brilliantly lit while the scene as a whole takes place in darkness.

Hunt's visual depiction of these invisible souls seems like an obvious departure from the rigorous study of the material facts of the case in his earlier paintings from the Holy Land. Yet the vivid, almost tactile presence of the babies within *The Triumph of the Innocents* is a sign that for Hunt their life after death is itself a literal fact. Hunt seeks to turn an invisible reality into a visible one. In doing so he combines the literal depiction of the infants themselves with a complex pattern of symbolism which leaves such literalism behind but which is itself no less materially present. The children are bathed in a bright heavenly light which marks them out from the Holy Family themselves, who are currently earthbound and lit only by the stars. Unsatisfied with the 'unearthly light' alone, Hunt also represents the infants as stepping along what he calls in his exhibition notes for the painting the 'living fountains of water – the streams of eternal life'. In depicting this metaphorical stream, Hunt deliberately departs from nature, showing it 'as ever rolling onward and breaking – where it might if real water be dissipated in vapour – into magnified globes'.[51] Through these unreal globes, Hunt is able to give the souls of the dead babies a concrete presence and thereby to represent the heavenly life which for him is a fundamental truth of the Christian revelation.

In Ruskin's eyes, Hunt's painting triumphantly confirmed the 'immutable historical fact that all the beautiful work, and all the happy existence of mankind' depended upon 'the doctrine of the Resurrection'.[52] It combined the two attributes which, he suggested, elevated Hunt above Rossetti as an artist: his 'respect for physical and material truth' and his conviction that 'the story of the New Testament' was 'not merely a Reality, not merely the greatest of Realities, but the only Reality'.[53] Throughout the 1870s Ruskin had sustained a heroic if doomed campaign for a science inflected with art and rejecting mechanism and materialism, much as Tupper had done in *Hiatus*.[54] In a series of lectures at Oxford in 1872 on the relationship between science and art, later published as *The Eagle's Nest*, he even tried to hold back the tide of materialism by redefining science itself. Ruskin argued, 'It is not the arrangement of new systems, nor the discovery of new facts, which constitutes a man of science; but the submission to an eternal system; and the proper grasp of facts already known.' In redefining science as 'not the knowledge of general laws, but of existent facts', Ruskin goes beyond reaffirming Baconianism in the face of Darwinism to committing science to a return to the state of knowledge before the Baconian project itself was overturned.[55] Hunt's painting of the literal truth of the salvation of the infants' souls is a vivid expression of the subordination of material science in submission to the eternal system of Christian theology.

Stephens reviewed Hunt's painting for *The Portfolio* in 1885. Unlike Ruskin, he found his painting of Christian mythology as material fact to be untenable. The painting may have been 'magnificent in almost all respects of conception, design, and execution', but its 'attempt to represent the unseen by substantial means and all too faithful methods' was nonetheless 'self-contradictory'.[56] As Stephens notes, the Pre-Raphaelite physicality of the children and the stream is an endeavour 'to represent spiritual essences with substantial appearances'.[57] Pre-Raphaelitism demands the close and faithful study of material phenomena, but in carrying the same aesthetic across to the representation of immaterial ideas and beings Hunt fails to represent their very immateriality.

Stephens mocks Hunt's 'living fountains' as 'apparently gelid sheets and curving planes of an unknown ice-like substance' and his 'glorified Innocents' as 'self-radiant, though massive'.[58] This quasi-scientific diction parodies Hunt's own scientific pretensions. Stephens closes his essay on a more positive note, praising Hunt's technical mastery and complementing him on his Innocents in particular. But the terms in which he does so again reveal his sense that Hunt has lost his claim to be a scientific artist: 'Dismissing the supernatural and accepting all that remains of the noble beauty of these resplendent human creatures, there is more than enough to reward the artist's pains, the long delay in his labours, the fine and lovely thoughts he has expressed. In all art I recall no company of children so gloriously beautiful as these.'[59] Stephens can acknowledge and appreciate the glorious beauty of Hunt's Innocents, but only after 'dismissing the supernatural', in effect rejecting the premise of the painting itself. For Hunt and Ruskin, the immortal soul is a fact which can be represented as literally present without flouting the principles of Pre-Raphaelitism, just as natural theology can accommodate revelation without violating its claim to scientific rigour. For Stephens, by contrast, natural theology is no longer science, and neither science nor Pre-Raphaelitism can presume to judge what lies beyond the limits of perception. Like Tupper and Brett, they take opposite sides in the controversy over science's methods and its findings. All five remain true to the Pre-Raphaelite project, but in doing so they reveal how far the rift in science had made a fissure in that project itself.

SECOND-GENERATION PRE-RAPHAELITISM AND SCIENCE

As we shall see in Chapter 8, the disagreements over the implications of Pre-Raphaelitism in an age of materialist science were played out in the building of the Natural History Museum in South Kensington and its reception in *Nature*. For the Pre-Raphaelites themselves, however, a more fundamental rift opened within the movement itself in the late 1850s. When Stephens published the first instalments of his series of essays 'The Two Pre-Raphaelitisms' in *The Crayon* in 1856, his account of modern Pre-Raphaelitism was of a movement committed to grounding art in science, just as the first Pre-Raphaelites – the artists before Raphael – had been committed to the close observation of nature. Stephens reprinted his appeal from *The Germ* for a modern art that could be judged by the standards of science, and praised Millais's 'transcripts of nature', Hunt's 'thoughtful power and deep scientific observation' and Dante Gabriel Rossetti's insight into 'psychological problems'.[60] But by the time he resumed his series nearly a year later, it was no longer possible to claim that all the Pre-Raphaelites still shared these scientific aims. In his original series in 1856, Stephens had limited his coverage of Rossetti's work to two short paragraphs on *The Girlhood of Mary Virgin* and *Ecce Ancilla Domini!*. In 1857 he extends it by an entire long article in two parts. Here Stephens transforms Rossetti from an addendum to Millais and Hunt into the most brilliant of the three artists. He judges his Dantesque and Arthurian watercolour drawings to be 'the highest of high Art, the most subtle in feeling, the most splendid and delicate in color, the most learned in composition'. For Stephens, Rossetti's drawings are 'like a mystical dream, some like the progress of a gorgeous romance, and some like the singing of a song of highest glory to God'.[61] The contrast with the factual, scientific realism of the original Pre-Raphaelite project could hardly be more pronounced. Now there really were two quite distinct Pre-Raphaelitisms.

Of all the close Pre-Raphaelite circle, Rossetti had always been the most ambivalent towards science. As he admitted in a doggerel letter to Tupper as early as 1850, 'as to NATURE, Jack, / (Poor dear old hack!) / Touching sky, sun, stone, stick, and stack, / I guess I'm half a quack.'[62] After his death, his brother remarked, 'Science and metaphysics were totally out of Rossetti's ken', while Hunt recast the whole history of Pre-Raphaelitism as a struggle between himself, convinced that science was 'of the greatest poetic and pictorial importance for modern art', and Rossetti, who 'despised such inquiries'.[63] And yet in the late 1840s Rossetti had taken a lead in devising and refining a rigorous approach to psychology through poetry and painting, even if he had always been too lazy or impatient for the rigours of painting minutely in situ and, like Collinson, had remained casual about factual details. The psychology of 'A Last Confession' remains startlingly acute and closely studied, even if the arresting opening lines – 'Our Lombard country-girls along the coast / Wear daggers in their garters' – are undercut by William Rossetti's wry remark, 'Every one except Dante Rossetti knows that Lombardy has no coast.'[64]

Rossetti's ambivalence towards science can be seen as early as *The Germ*. His brother and Hunt both recalled that he led the drive to bring out a Pre-Raphaelite magazine.[65] As well as contributing his own poetry and fiction, Rossetti came up with the magazine's subtitle, 'Thoughts towards Nature', and drafted the endpaper, which set out its commitment 'to encourage and enforce an entire adherence to the simplicity of nature' in art.[66] These descriptions of the magazine's remit tally with the aesthetic agenda set out by

Stephens, Tupper, Brown and William Rossetti in their critical essays. As *The Germ*'s leading champion, Rossetti was presumably in sympathy with this agenda. Yet he also sponsored 'A Dialogue on Art', sent in by the dying painter John Orchard and published posthumously in the fourth issue. Alone of the essays in *The Germ*, Orchard's 'Dialogue' challenged the consensus that the arts should look to the example of the sciences, insisting that to put too much weight on the merely physical facts of science is to risk creating 'mere natural history paintings from the animal side of man'.[67] The way Rossetti incorporates astronomy into 'The Blessed Damozel' is revealing too. Hunt quotes Rossetti as asking dismissively, 'What could it matter whether the earth moved round the sun, or the sun travelled about the earth?'[68] But the image in this poem of the earth spinning 'like a fretful midge' is precisely Copernican, capturing with remarkable precision the difference between the Damozel's sped-up perception of time from the outskirts of heaven and her bereaved lover's drawn out sense of it on earth. Rossetti imagines the Damozel herself surrounded by 'ether', the medium which Victorian physicists posited for the transmission of light waves across the universe. In the poem, as in the scientific hypothesis, the ether conducts light but not sound, leading to 'The peace of utter light / And silence' where the Damozel stands on the edge of space.[69] Yet the fact that space has an edge shows that this modern scientific physics is superimposed onto a medieval cosmology.

As we have seen, Rossetti took part in the original Pre-Raphaelite project of grounding art in science across his poetry and painting in the late 1840s and early 1850s. But he did so on his own terms. In 'The Blessed Damozel', for instance, he is more interested in the imaginative possibilities of science than in its methods, and he was always more concerned with the human mind and its impressions than with the material reality beyond it. Rossetti's art developed over the 1850s through his close creative partnership with Elizabeth Siddal, whose poetry and paintings are similarly preoccupied with medieval subjects and forms.[70] By the late 1850s, he had all but disengaged from the scientific aspect of Pre-Raphaelitism. One of the new paintings described by Stephens in 'The Two Pre-Raphaelitisms' illustrates this change within Rossetti's work particularly well. It is a delicate

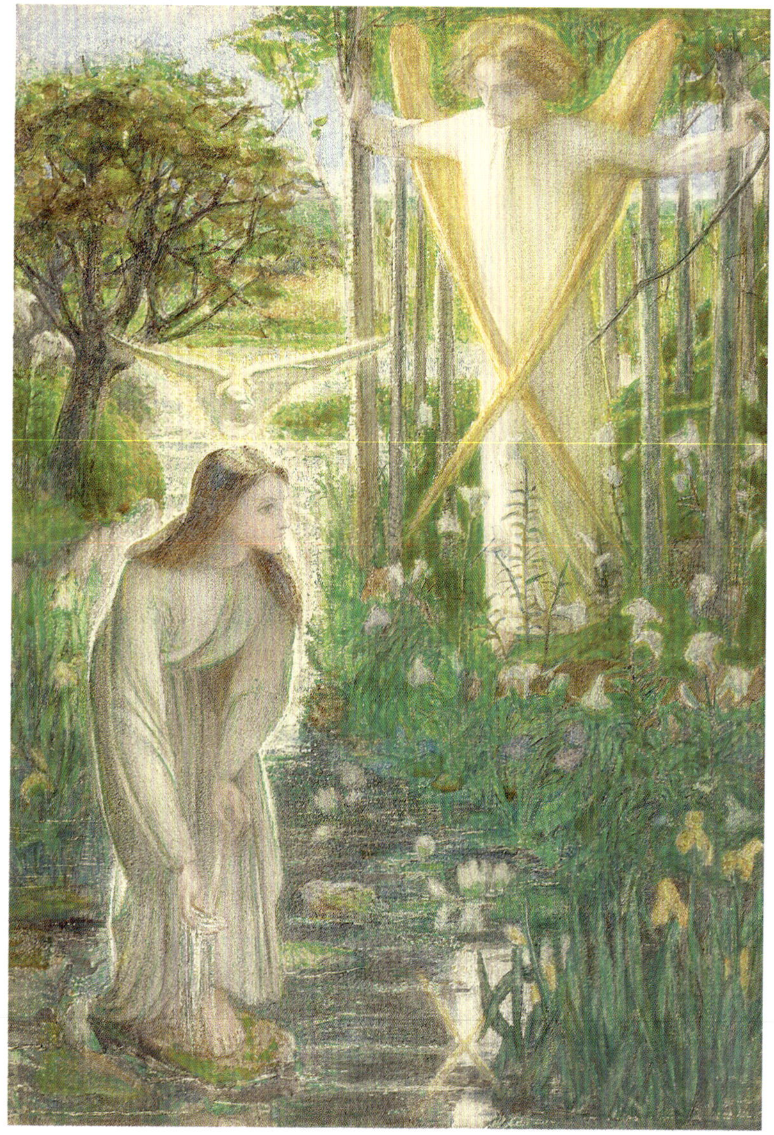

100 Dante Gabriel Rossetti, *The Annunciation*, 1855. Private collection.

Annunciation which he painted in watercolour for Boyce in 1855 (fig. 100). This new version of the scene makes a stark contrast with *Ecce Ancilla Domini!* (see fig. 42). In place of the blank interior of the earlier painting, Rossetti places Mary within a summer landscape characterised by abundant flowers amid woods and meadowland. Mary herself stands on a rock in a shallow, barely moving stream, washing her hands. The colours are bright, the sense of life rich, but the detail is uneven and at times little more than indicative. The tall thin trees in the wood by the stream, for example, are virtually featureless. Nor does Rossetti show much interest any longer in probing Mary's psychology, or in how it might feel to experience such a profound personal revelation. Instead, his concern seems to be predominantly with mood. The angel's appearance is sudden and his looming presence potentially threatening, but Mary's response shows that she does not feel it in this way. Her composure and the serenity of the landscape as a whole convey a stillness to the picture which is not compromised by the water running through her fingers. Where the power of *Ecce Ancilla Domini!* resides in its psychological challenge to our preconceptions about the Annunciation, the power of this second painting resides in the preternatural quality given to it by this combination of brightness and tranquillity, as the event itself is transposed out of history into a mystical, dreamlike landscape.

Beyond the partnership of Rossetti and Siddal, the splitting of Pre-Raphaelitism into two distinct schools of art was signalled in the contrast between the decorative schemes of the two buildings built by Benjamin Woodward in Oxford in the 1850s. For all Woodward's encouragement, Rossetti never contributed anything concrete to the decoration of the Oxford museum, which as we have seen became the fullest expression of the Pre-Raphaelite project of grounding art in science. In 1857, however, he persuaded Woodward to let him and a group of his friends and fellow artists paint a series of murals based on Arthurian legends on the walls of the new debating chamber (now the library) of the Oxford Union Society.[71] Although a number of artists worked on both buildings, including Munro, Pollen and Swan as well as Woodward himself, the impetus behind the Union Society murals came from Rossetti and two young artists who shared his enthusiasm for medieval art for its own sake: William Morris and Edward Jones, later Burne-Jones. Like Rossetti's watercolours, the murals make no pretence to be transcripts of the world before us. Instead, they match their medieval subject matter with a much more direct invocation of the flattened, two-dimensional forms of medieval art than was ever intended within the original conception of Pre-Raphaelitism. Along with the poet Algernon Charles Swinburne, who would befriend them later the same year, Rossetti, Morris and Burne-Jones would be the driving force behind what has become known as second-generation Pre-Raphaelitism.

Over the next forty years, they would carry forward across the full range of the arts the symbolist and aestheticist project heralded by Rossetti in 'The Blessed Damozel'. In painting, the luminescent medieval aesthetic of Rossetti's watercolours and the Union Society murals would be taken up and developed by Burne-Jones, Simeon Solomon, Marie Spartali, John Roddam Spencer Stanhope – another of the Union Society painters – and Evelyn De Morgan. The frank eroticism of *Bocca Baciata* would resonate throughout the oil paintings of female models that dominate the second half of Rossetti's own career as a painter. So too would its formal properties – the intense proximity of the subject to the viewer, the rich colours of the hair, clothes and surrounding flowers, the Pre-Raphaelite precision in the observation of detail. The forms of Rossetti's paintings and their concentrated sexuality would be echoed in the work of Frederick and Emma Sandys and, with a homoerotic inflexion, Simeon Solomon. Together these various strands of second-generation Pre-Raphaelitism would develop a distinctive English tradition of symbolist art, while their aestheticism would become a leitmotif of later Victorian art more widely, manifest in many different forms from Frederic Leighton's classical nudes to Whistler's nocturnes. In poetry, Rossetti, Morris and Swinburne would become recognised as the leaders of a new movement, notoriously attacked as the 'Fleshly School of Poetry' by Robert Buchanan.[72] Younger poets who would consciously follow their lead included Arthur O'Shaughnessy, Edmund Gosse, Oscar Wilde and even John Masefield.[73] In the decorative arts, second-generation Pre-Raphaelitism would have a profound influence through the work of the design company Morris, Marshall, Faulkner & Co., established in 1861 with Morris at the helm and Rossetti, Burne-Jones, Ford Madox Brown and Philip Webb among the partners.

As Stephens's essay implied, the new Pre-Raphaelitism had a very different relationship to science from its predecessor. In their visual art, the second-generation Pre-Raphaelites retained their commitment to the careful observation of natural forms. Fiona MacCarthy's biography of Burne-Jones, for example, recounts his spending 'whole days in Bagley Woods making detailed studies of foliage and flowers' as a

101 Dante Gabriel Rossetti, *Venus Verticordia*, *c.*1863–8. Russell-Cotes Art Gallery and Museum, Bournemouth.

(*facing page*) 102 Dante Gabriel Rossetti, *La Ghirlandata*, 1873. Guildhall Art Gallery, London.

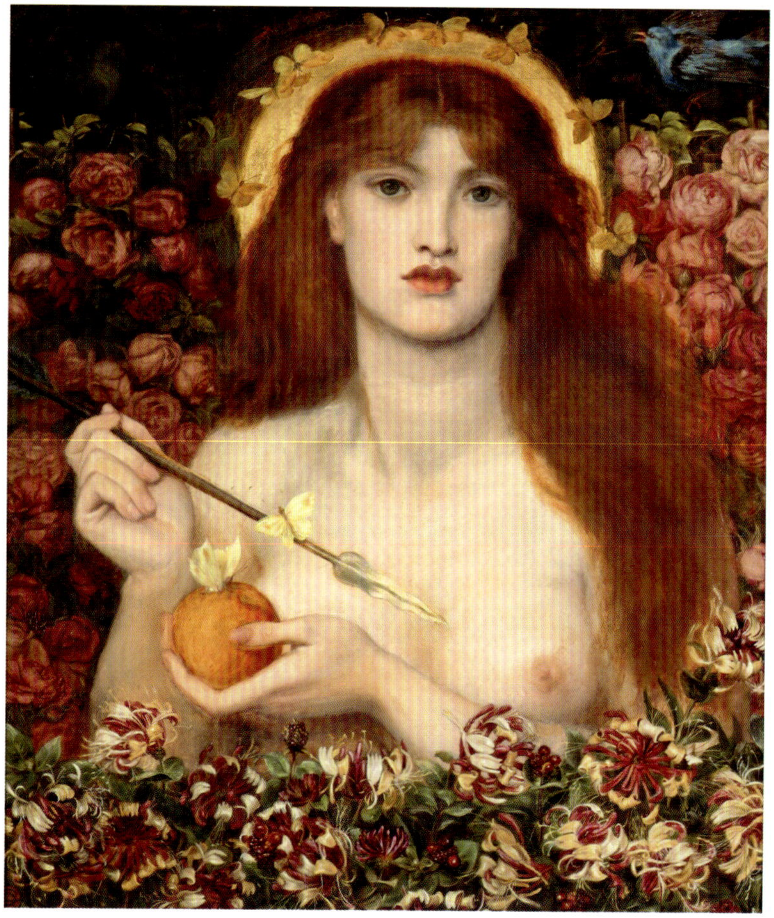

student, and still copying tangled roses for the *Briar Rose* series from specimens as late as 1890.[74] But they no longer identified their practice with science or applied it in a scientific way. In her compelling study *Art for Art's Sake*, Prettejohn notes Swinburne's insistence in his study of *William Blake*, published in 1868, that art was incompatible with science.[75] Defending himself from Buchanan in 1871, Rossetti wrote that 'the motive powers of art reverse the requirement of science, and demand first of all an *inner* standing-point.'[76] The practice he describes – entering wholeheartedly into the mind of a human subject within a work of art in order to comprehend that subject – bears a close resemblance to the approach outlined by Brown in 'On the Mechanism of a Historical Picture' and Stephens in 'Modern Giants' twenty years earlier, and followed by Rossetti himself in 'The Bride's Prelude' and 'A Last Confession'. But where they conceived of this process by analogy with science – specifically anatomy – Rossetti now saw it rather as its opposite.

Two of Rossetti's compositions from the later 1860s and early 1870s, *Venus Verticordia* (fig. 101) and *La Ghirlandata* (fig. 102), are subtly representative of this new attitude to science. By the mid-1860s, according to his friend the Irish poet and diarist William Allingham, Rossetti cared 'nothing about natural history, or science in any form or degree'.[77] Rossetti took great pains to ensure that his roses and honeysuckles for these paintings were of particular varieties and in perfect condition, reproducing each one with hyperrealist fidelity.[78] Yet this horticultural precision was matched with botanical carelessness. According to his brother, he included a detailed study of larkspur in the foreground of *La Ghirlandata* because he mistook it for the more sinister and poisonous monkshood.[79] But the blue bird with black markings which occurs in both these paintings most epitomises the move away from science towards aestheticism in Pre-Raphaelite art. Rossetti's drawing and brushwork in both pictures is precise enough that the species can be identified as a Blue Dacnis, a common but strikingly beautiful forest tanager from South America. But this taxonomy only demonstrates the bird's lack of connection to either the mythology or the ecology of Rossetti's paintings.

In describing Rossetti's medievalist watercolour *The Blue Closet*, Prettejohn remarks that 'the musical instrument, comprised of bells, strings and keyboards,' which forms the central focus of the painting, 'is not authentic to any historical period,' and that 'the numerous mystic symbols that decorate its surfaces and hangings have no intelligible message.'[80] The contrast with Hunt's scrupulous recreation of the physical and technological landscape of the Holy Land at the time of Christ is immediately apparent. The details of *Bocca Baciata* likewise 'project visual gorgeousness rather than intelligible symbolic meaning', in contrast, as Prettejohn shows, with those of *The Awakening Conscience*.[81] The Blue Dacnis fits exactly into this new aesthetic. Neither its scientific classification nor its natural history provides any clue as to its symbolic significance. Critics and curators have begun to explore connections between Darwin's theory of sexual selection and aestheticism.[82] But Rossetti associated the

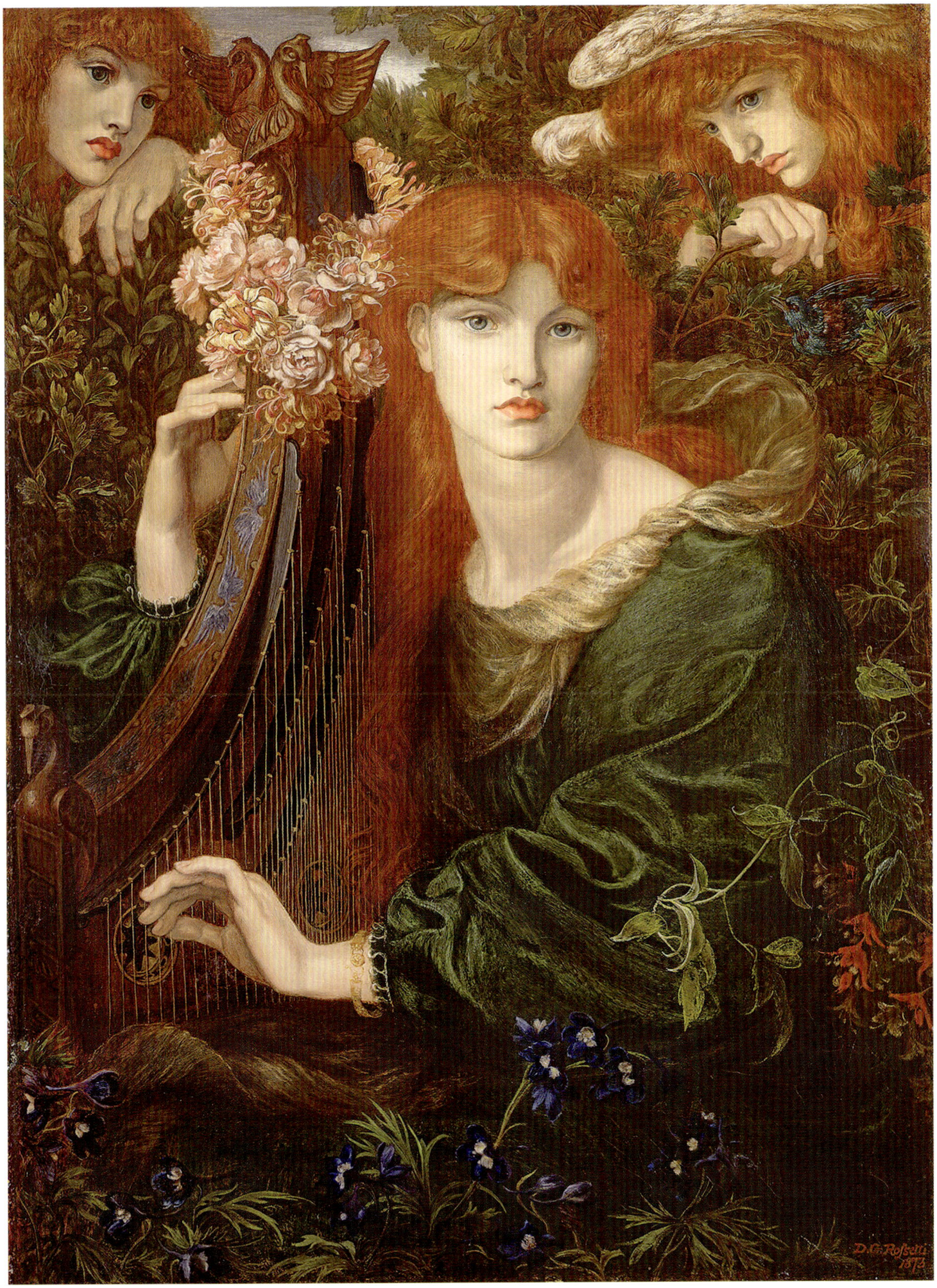
D.G.Rossetti
1873

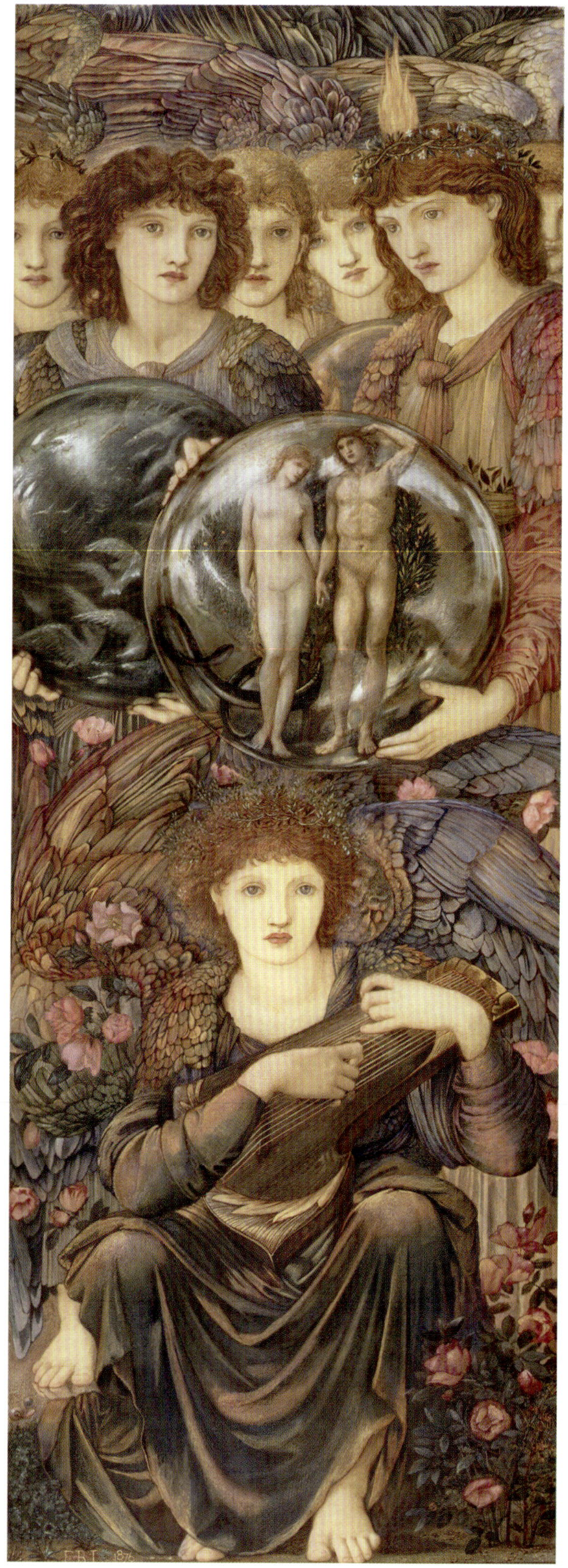

103 Edward Burne-Jones, *The Sixth Day of Creation*, 1870–76. Fogg Art Museum, Harvard University, Cambridge, Mass.

Dacnis with Venus well before Darwin put forward his detailed exposition of sexual selection in birds in *The Descent of Man and Selection in Relation to Sex*, and even then there would be no specific scientific reason for his choosing this particular bird over others. Its significance lies rather in its sheer beauty and in the exoticism of that beauty in Victorian London.

In aesthetic terms, Rossetti's practice in these two paintings is in line with the theory of art formulated by Edward Poynter in an essay for the *Fortnightly Review* in 1871. Poynter was Burne-Jones's brother-in-law and a future President of the Royal Academy and Director of the National Gallery. In his article on 'Beauty and Realism, in Construction and Decoration', he argues that an artist needs to pursue 'the closest and most continual observation' in order to move beyond the 'broad external facts of Nature' that are obvious to everyone and to make new discoveries. But the object of these discoveries is not new knowledge or understanding, whether physical or moral, but rather beauty itself. Like Stephens writing in *The Germ* twenty years earlier, Poynter identifies 'truth' as an object in art, but he does so because truth is 'the essential of beauty', not because it is in itself the artist's goal. For Poynter, 'the highest beauty is attained by the highest application of the realistic or imitative faculty' because this enables the artist to realise 'the beauties of Nature to the utmost'.[83] Rossetti's richly detailed treatment of plants and animals in his later paintings operates through just such an intense gaze. He discovers and shares new beauties that would escape the casual observer by fixing in precise lines and brilliant colour the forms of each individual flower or bird. But his aim, like Poynter's, is to capture and create beauty for its own sake.

Once the Dacnis has been identified with Venus by Rossetti in *Venus Verticordia*, it acquires a new significance beyond its beauty. This new symbolism is revived by Rossetti in *La Ghirlandata*, which was commissioned by William Graham, who owned a replica of *Venus Verticordia* in watercolour.[84] But it remains a private significance, shared between Rossetti and his clients, not publicly recognised as part of the iconography of Venus or the symbolic language of love. Such symbolism – whether actually or implicitly private – is characteristic of later Pre-Raphaelite art, in contrast with the readily interpretable (though by no means unambiguous or simplistic)

104 Edward Burne-Jones, *The Wheel of Fortune*, 1875–83. Musée d'Orsay, Paris.

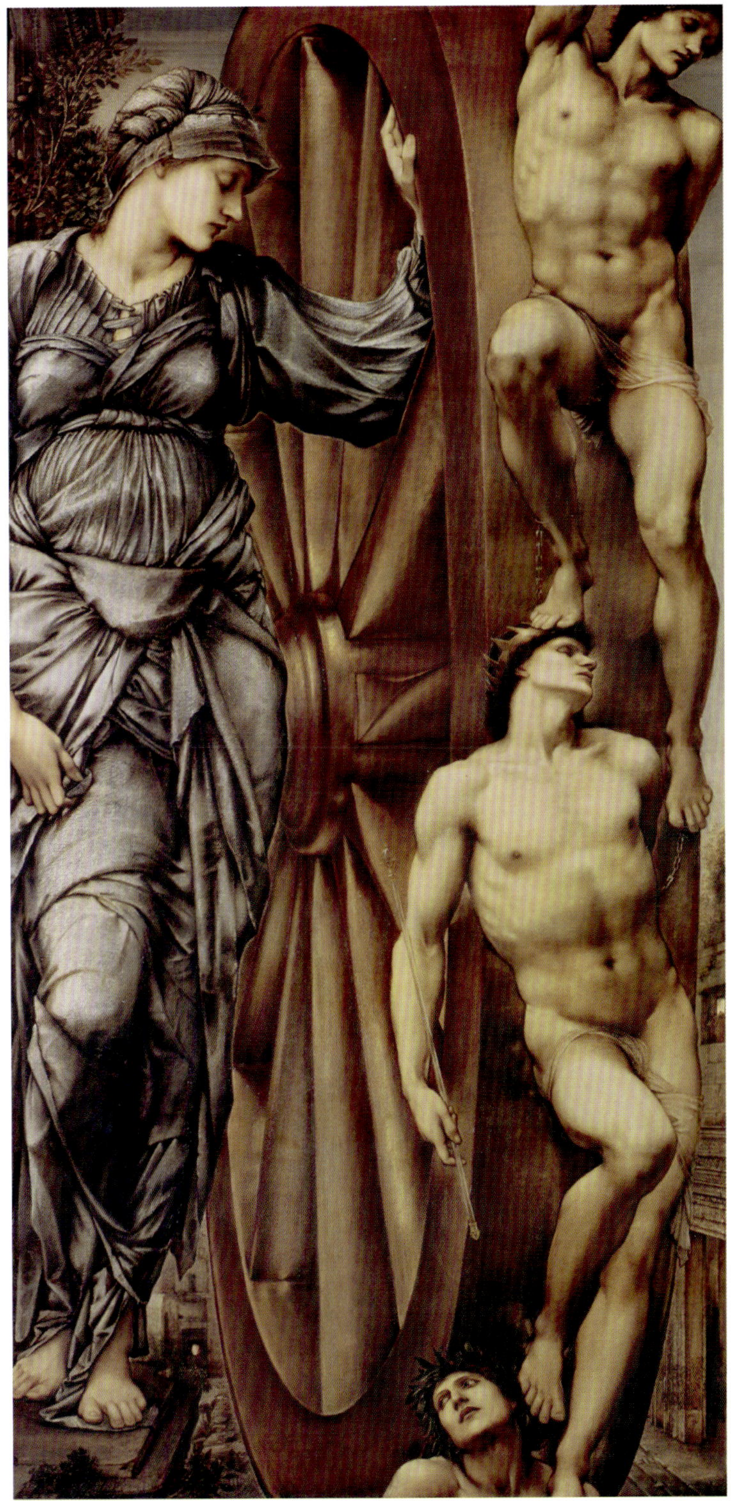

visual codes of Hunt's paintings. It hints at an occult significance, as though the painter and the viewer are initiates in the same mysteries. Both can apprehend them, but neither can translate them into the 'reasonable speech' of 'scientific men', to use Brett's idiom. In weaving the Dacnis into his private symbolism, Rossetti thus distances his meticulous study of a particular species of bird still further from the public discourse of scientific knowledge.

In its aestheticism and its symbolism, and both in theory and in practice, second-generation Pre-Raphaelite art appeared to define itself in opposition to science. Yet the divorce between the later Pre-Raphaelites and science was not as total as this suggests. A number of them as individuals took an active interest in science. O'Shaughnessy was employed as an assistant in the natural history department of the British Museum; Swinburne was a Fellow of the Anthropological Society of London; both these affiliations bear on their poetry.[85] Piers Hale, Stephen Eisenman and others have shown that Morris engaged with the political implications of different strands of evolutionism as a socialist and a theorist of design.[86] For his part, Ruskin still saw the practices of late Pre-Raphaelite art as analogous to those of science. He based his claim that Burne-Jones's 'grotesque and wild forms of imagination', far from being a reaction against science, were 'a part of that science itself' on a continuity, as he saw it, between Burne-Jones's practice in paintings such as his *Days of Creation* (fig. 103) or *The Wheel of Fortune* (fig. 104) and the scientific standards of early Pre-Raphaelitism. Such paintings, Ruskin declared, were 'the result of infinitely more accurate scholarship, of infinitely more detective examination, of infinitely more just and scrupulous integrity of thought, than was possible to any artist during the two preceding centuries'. For Ruskin, Burne-Jones was a rigorous investigator, just like the Pre-Raphaelite artists of the 1850s as William Rossetti had described them. The subject of his investigations was mythology itself. Through meticulous and scrupulous research, in sympathy with science and meeting its standards, he could discover and reveal to us 'the meaning, in what a fable said'.[87]

By the 1880s Ruskin was viewed as a maverick within scientific circles. Much like his endorsement in the same lecture of Hunt's *Triumph of the Innocents*, his account of Burne-Jones's mythological paintings as grounded in science would be unlikely to have matched the scientists' definitions of their own practices or aims in the way that Stephens's Baconian account of the Pre-Raphaelite project had done in the 1850s. A more telling association between late Pre-Raphaelite art and science was intimated by Walter Pater, the leading theorist of aestheticism, in his classic of aesthetic criticism *Studies in the History of the Renaissance*, published in 1873.[88] Pater cast both the artist and the art critic as scientists. Works of art, he claims in the preface to his book, 'are valuable for their virtues, as we say in speaking of a herb, a wine, a gem'. The 'function of the aesthetic critic', therefore, is 'to distinguish, analyse, and separate from its adjuncts, the virtue by which a picture, a landscape, a fair personality in life or in a book, produces this special impression of beauty or pleasure, to indicate what the source of that impression is, and under what conditions it is experienced. His end is reached when he has disengaged that virtue, and noted it, as a chemist notes some natural element, for himself and others.'[89] Ultimately, the aesthetic critic sets out 'to define beauty not in the most abstract, but in the most concrete terms possible, not to find a universal formula for it, but the formula which expresses most adequately this or that special manifestation of it'.[90] The role of the artist in turn is to discover each new beauty or pleasure in creating the work of art and to embody it within it. The Renaissance men from the generation before Raphael who form the subject of Pater's *Studies* – Pico della Mirandola, Botticelli, Leonardo and Michelangelo – are all investigators. Like the Pre-Raphaelites themselves, their lives and work combine art and science through various forms of scholarship, observation, experimentation and theorisation.

Pater's account of the aesthetic critic and the Renaissance artist draws on three distinct models of science, each of which bears on second-generation Pre-Raphaelitism in a different way. First, Pater's conception of the aesthetic critic as a chemist isolating the elements of beauty remains resolutely Baconian. It chimes with Brett's invitation to scientists to look for concrete examples of beauty at the Royal Academy, Tupper's insistence that the only way to understand beauty is to look at it and draw, and Poynter's encouragement to artists to realise beauty through studying natural forms. All four of them cast art and art criticism as forms of discovery on scientific lines. But the model of science that they follow, Baconian induction, had been superseded within science itself. On Lankester's terms, the best this method can give us is 'knowledge', not science. Even so, its survival as a popular conception of science helped keep Pre-Raphaelite art and science connected in the public mind.

The second model of science that Pater invokes is still more comprehensively removed from the reality of late Victorian scientific practice and theory. Writing about Leonardo's 'philosophy', Pater observes that 'much of the spirit of the older alchemy still hangs about it.' Pater imagines Leonardo 'poring over his crucibles, making experiments with colour, trying by a strange variation of the alchemist's dream to discover the secret, not of an elixir to make man's natural life immortal, but rather of giving immortality to the subtlest and most delicate effects of painting'.[91] This portrait of the artist as an alchemist stands as an analogue to Pater's account of the aesthetic critic as a chemist in his preface. But the artist's discoveries amount to an esoteric and uncorroborated science. Botticelli too was no 'mere naturalist', though he had the skill to be. Instead, he was a visionary whose genius 'usurps the data before it as the exponents of ideas, moods, visions of its own'. Pater's Botticelli, playing 'fast and loose with those data, rejecting some and isolating others, and always combining them anew', is another variant on the chemist distinguishing and analysing new elements.[92] At the same time, he is wilfully subjective, irresponsible before the true facts of nature, and preoccupied with his own visions. The parallel with Rossetti's combination of acute observation and private symbolism in *Venus Verticordia* and *La Ghirlandata*, or with Ruskin's account of Burne-Jones revealing the meanings behind myths, is close. On these terms, second-generation Pre-Raphaelitism may hint at arcane knowledge, but it is not science in the modern sense.

There is, however, a third model of science in play in Pater's work which brings Pre-Raphaelite art back into company, if not into contact, with modern science. In his conclusion to *Studies in the History of the Renaissance*, Pater returns to chemistry. 'What is the whole physical life', he asks, 'but a combination of natural elements to which science gives their names?' That physical life includes 'the passage of the blood, the wasting and repairing of the lenses of the eye, the modification of the tissues of the brain by every ray of light and sound – processes which science reduces to simpler and more elementary forces'.[93] We are also composed of elements that science can identify and which interact in processes that science can explain. Like Darwin, Pater does not pause at the human mind. His reference here to 'the tissues of the brain' signals his thoroughgoing commitment to scientific naturalism, allowing no place for the soul, in sharp contrast with Ruskin and Hunt. An idea, such as those that Botticelli explores in his art, remains an object for Pater, but not because it has its own independent existence. Rather, it exists as 'a fact in the brain', as the poet George Meredith – another member of Rossetti's circle – put it in 1862 in his powerful Romantic 'Ode to the Spirit of Earth in Autumn'.[94]

For Pater, scientific naturalism is the method of modern science, materialism its implication. The strongest bond between second-generation Pre-Raphaelitism and scientific naturalism lay in their shared rejection of the received truth of Christian theology – the 'eternal system' that seemed to hold good up to the 1850s and that Hunt and Ruskin still espoused. As we shall see in the next chapter, second-generation Pre-Raphaelite poetry spoke to those who felt that modern science had made their old faith unsupportable, yet for whom both the Baconianism and the aesthetic mysteries of Pre-Raphaelitism had a lingering appeal. Through their acts of reading, even Rossetti, who had parted company with science seemingly for good, would be brought back into its cultural orbit.

105 Dante Gabriel Rossetti, *The Question*, 1875.
Birmingham Museums and Art Gallery.

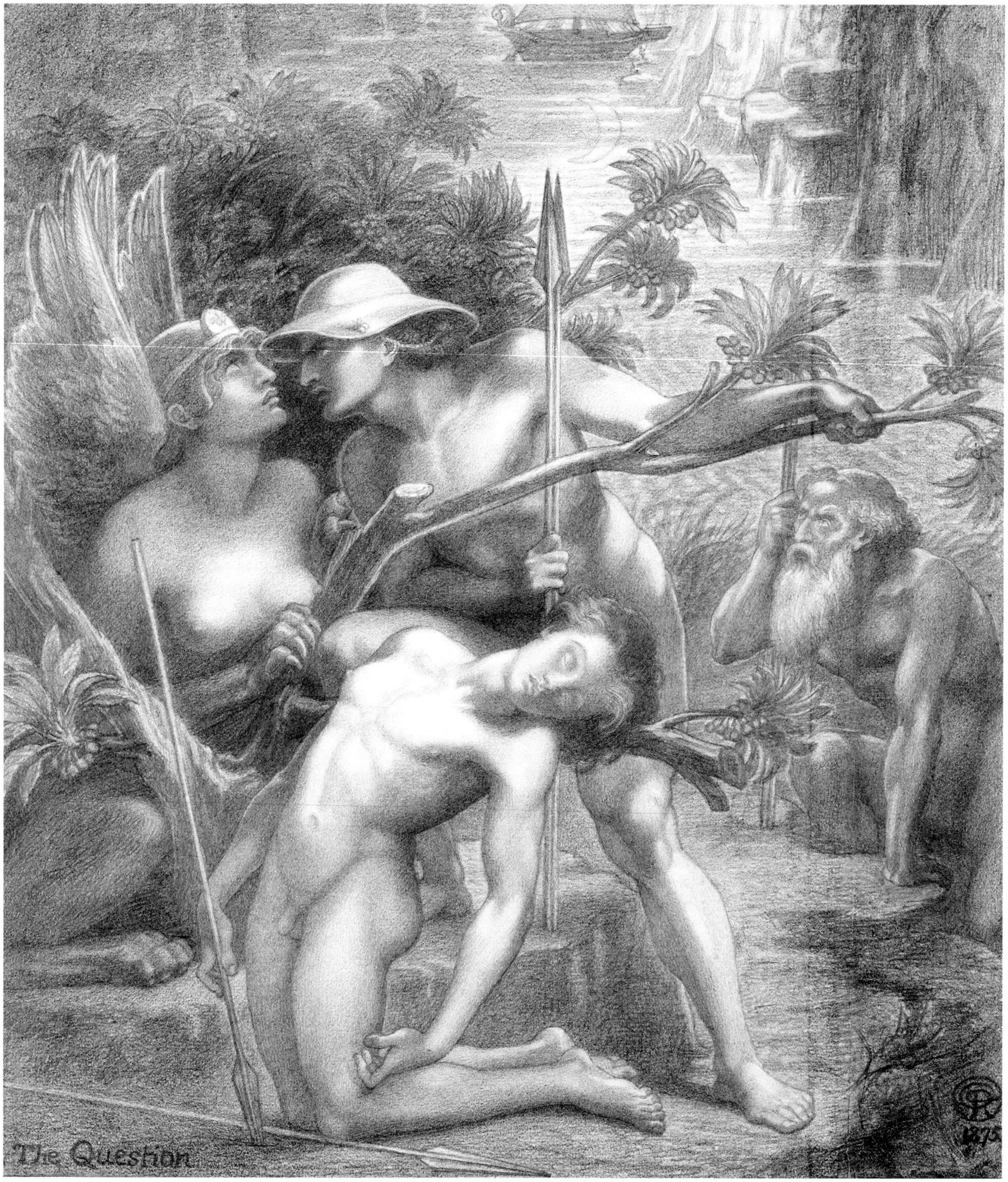

7 · The Facts of the Case

SCIENTIFIC NATURALISM AND PRE-RAPHAELITE POETRY

Dante Gabriel Rossetti took little interest in science after the 1850s, no longer aspiring to be even eclectically scientific in his own art. Writing for the art magazine *The Portfolio* as early as 1870, Sidney Colvin suggested that Morris and Burne-Jones too asserted their 'freedom from science' in seeking to 'create for their art an unscientific and uncritical No-man's-land of beauty and enchantment'.[1] Indeed, of the four artists at the centre of the second generation of Pre-Raphaelites, only Swinburne actively sided with science, piquing himself on a 'habit of mind . . . not (I hope) unscientific', as he remarked to Theodore Watts in 1874.[2] Nonetheless, the second-generation Pre-Raphaelites as a group found themselves feted by the advocates of scientific naturalism. As Gowan Dawson has shown, aestheticism was equated with scientific naturalism by their mutual enemies, who saw them as alternative expressions of the same materialism and moral irresponsibility.[3] This was not merely an attempt by their opponents to tar each with the other's brush, as the Pre-Raphaelites also enjoyed active support from the supporters of scientific naturalism in the pages of liberal and open-minded periodicals such as the *Nineteenth Century* the *Fortnightly* and *Westminster Reviews*.

Pre-Raphaelite poetry especially resonated with scientific naturalism. To see why, and to understand the significance of that poetry for the scientific culture of its time, we need to return the poems to their original context. This chapter explores what it was that drew not necessarily scientists themselves, but the men of letters who embraced and championed scientific naturalism, to the poetry that Rossetti, Morris, Swinburne and their circle were writing in the late 1860s and early 1870s. Why, for example, did John Morley choose to publish the Pre-Raphaelite poets specifically in the pages the *Fortnightly Review*, a leading liberal journal with marked materialist leanings? By reading their poems alongside the *Fortnightly*'s prose articles from the same period, we can begin to see how they enter into dialogue with science, playing an important role in defining the journal's intellectual outlook as a whole. In a second case study I will turn from the publication of poetry to its reception, looking at how and why reviews of the most popular as well as the most monumental Pre-Raphaelite poem of the period – Morris's *Earthly Paradise* – identified it with science. These reviews shed new light on the poem itself, enabling us to trace scientific subtexts in a poem that appears on the face of it unconcerned or even at odds with science. Both the *Fortnightly*'s prose and the reviews of *The Earthly Paradise* show how Pre-Raphaelite poetry resonated with readers for whom scientific naturalism had fundamentally changed how they viewed this world and the next. In his final poem for the *Fortnightly*, however, and in a relatively little-known double work of art which stemmed from it, Rossetti would call this project and the hopes of science itself into question. The final section of this chapter, then, addresses Rossetti's doubts about science, while revealing how even these doubts show more of an affinity with the original Pre-Raphaelite project than we might expect.

READING THE PRE-RAPHAELITES IN THE *FORTNIGHTLY REVIEW*

In 1865 a new heavyweight periodical was launched in Victorian London. The *Fortnightly Review* was conceived by the novelist Anthony Trollope as a non-partisan journal which would be unique in encouraging intellectually serious contributors of all stamps, but also in requiring them to sign their names to their articles and reviews. Trollope's main aim was to break the divisive practice of anonymous reviewing, which enabled critics to assume the authority of a major journal while hitting out at their rivals and log-rolling for their friends. He persuaded the prominent intellectual George Henry Lewes to take on the editorship. While Lewes remained true to the principle that contributions should be signed, his own political liberalism soon began to be stamped on the pages of the *Fortnightly*, for all that he kept it open to contributors with other views.[4] Lewes was a close student of physiology and natural history, who would go on to become one of the leading theorists of embodied psychology. As editor, he brought in an impressive cast of scientists and scientific theorists to write for the journal, including Herschel, Huxley, Tyndall, Spencer and Tylor. Like Lewes himself, many of them were freethinkers and staunch scientific naturalists who rigorously excluded any form of supernatural interventions from their science. In this too a party line began to take shape.

After two years, Lewes resigned as editor due to ill health. He was replaced by the young lawyer and journalist John Morley, who would later serve as a cabinet minister in successive Liberal governments from Gladstone's short-lived administration of 1886 up until the outbreak of the First World War. Under Morley's editorship, the *Fortnightly*'s commitments to radical and liberal politics and to scientific naturalism, though never exclusive, became deeper and more explicit. In several of his own articles Morley sought to put the discipline of history on a scientific footing, while Lewes argued that religion should be grounded in science and Francis Turner Palgrave even set about prescribing a method for the scientific study of poetry.[5] At the same time, Huxley and Tyndall pushed their own ideas closer towards out and out materialism in essays such as 'On the Physical Basis of Life' and 'Climbing in Search of the Sky'.[6]

While its political and scientific commitments, though clarified and deepened, remained broadly consistent, the *Fortnightly*'s poetry was transformed under its new editor. Under Lewes, the review's most prominent poet and literary commentator had been Robert Buchanan, whose poems on London poverty chimed with its broad political sentiments. But although Buchanan was a radical and a freethinker, he was also a fastidious prude who had attacked immorality in literature in the pages of the *Fortnightly* in September 1866.[7] As editor, Morley had little time for what he called the 'moral hypochondriac', insisting a few months later in the same paper: 'Morality is not the aim and goal of fine art, any more than it is an aim or measure of cobbling or of the art of physic.'[8] With Morley in charge, Buchanan's access to the *Fortnightly* was gone. He was supplanted by Swinburne, his bête noire, whose *Poems and Ballads* Buchanan – and as it happens Morley himself, against his own later, better judgement – had savaged in print on moral grounds only months before.[9]

Swinburne had initially been invited to contribute to the *Fortnightly* by Lewes, but it was not until January 1867 that his first poem, 'Child's Song in Winter', appeared in the magazine.[10] Over the next decade, he would contribute a total of fifteen poems and twelve critical essays. Swinburne encouraged William Bell Scott to contribute, and Scott in turn introduced Morley to Dante Gabriel Rossetti.[11] Each of them had poetry published in the *Fortnightly*, while Rossetti also contributed a critical review and a revised version of his story 'Hand and Soul', which had previously appeared in both *The Germ* and *The Crayon*. This lineage itself marks the *Fortnightly Review* out as a Pre-Raphaelite periodical. Morris also contributed four poems and co-authored a translation of an Icelandic saga. The only other poet who published repeatedly with the *Fortnightly* during the same period – and the only one to have established himself as a contributor under Lewes – was George Meredith, who published nine poems over the same period as Swinburne's fifteen. Like Scott, Meredith was a friend and fellow-traveller of the Pre-Raphaelites. His first collection of poems had been warmly praised by William Rossetti in *The Critic* back in 1851, while Swinburne had championed his second collection, *Modern Love and Poems of the English Roadside*, against a moralistic attack from *The Spectator* in 1862.[12] Meredith shared the Pre-Raphaelites' sensitivity to ecology, their commitment to the minute examination of embodied psychology, a similar refusal to be hamstrung by prudery, and the love of the ballad as an intense form of poetic narrative. He was also attentive to developments in science, particularly evolutionary theory, scrutinising its implications in his poetry.[13]

For ten years from 1867, then, the Pre-Raphaelites made up the house school of poetry in the *Fortnightly Review*. Swinburne and Meredith contributed throughout this period, Scott, Rossetti and Morris over the five years from 1868 to 1872.[14] Given that Rossetti and Morris distanced their art from science, why was Morley, with his overtly scientific outlook, so keen to publish them? There are on the face of it several fairly straightforward answers to this question. Meredith had been something of a mentor to Morley. Morley himself had a growing rapport with Swinburne and appreciated what he called the 'exquisite loveliness' of Rossetti's poems.[15] He was also determined to take a stand against moral hypochondria and what the mid-twentieth-century historian of the *Fortnightly*, Edwin Mallard Everett, called 'the inept didacticism of Victorian respectability'.[16] As Everett observed, Morley was also in sympathy politically with Swinburne's and Meredith's cosmopolitan radicalism, as voiced in the *Fortnightly* in their respective odes on the insurrection in Candia in Crete in 1867 and on the condition of France in 1870 during the Franco-Prussian War, and with Morris's support for the working classes against aristocratic privilege and abuse in his poem 'The God of the Poor', published in the *Fortnightly* in 1868.[17]

But Morley was aware too of the close if submerged connection between Pre-Raphaelitism and science. Colvin identified 'the scientific spirit' as the 'very essence' of Pre-Raphaelitism in an article for the *Fortnightly* on 'English Painters and Painting in 1867', naming Rossetti as 'perhaps the only one' of the original Pre-Raphaelites who still combined 'the passion for beauty with intellectual subtlety and executive mastery'.[18] Morley picked up and twisted this same thread in an essay sparked by Pater's *Studies in the History of the Renaissance* which he published in the *Fortnightly* in 1873. Pater's own essays on Leonardo, Botticelli, Pico and Michelangelo had all been published in the *Fortnightly*.[19] In his essay, Morley drew a parallel between the Oxford Movement of the 1830s and 1840s and an aestheticism which he traced running from Ruskin and the Pre-Raphaelite painters through Swinburne and Morris to Pater. The crucial distinction between the two, he argues, is that since the Oxford Movement 'science has come'. Science was for Morley a key precondition for the growth of aestheticism, which emerges in his essay as the synthesis of a dialectic in which science is the antithesis of sacramentalism. In its turn, aestheticism becomes an important counterweight to what would now be called scientism or, in Morley's words, the dominance of science over 'other than scientific manifestations of intellectual activity and fruitfulness'.[20]

Morley's analysis bears directly on the role of Pre-Raphaelite poetry within the *Fortnightly Review* under his editorship.

His first issue had included a powerful essay on 'The Significance of Ritualism' by James Cotter Morison. It was through Morison's influence that the editorship passed to Morley when Lewes resigned, and his views in his contributions to the *Fortnightly* closely resemble Morley's own. Morison was one of several writers in Lewes's and Morley's stable who were spokesmen for the philosophical position known as positivism, alongside Frederic Harrison, John Henry Bridges and Richard Congreve. Originating in the works of Auguste Comte, positivism was rigorously empirical, arguing that science was the only source of knowledge about the external world. Like the scientific naturalists, the positivists denied that theology had any authority or access to truth. However, they were more determined to recapture the aesthetic and symbolic resonances which had been vested in religion but which for freethinkers such as themselves had been irretrievably severed from it.[21] In his essay, Morison sets up his own dialectic in which ritualism had been an emotional revolt against what he calls the 'cold brilliance of science', unable to 'fill the void within'. But ritualism itself was only 'a gorgeous palace, built upon a foundation which is crumbling beneath it every day'. 'The problem of the future', Morison writes, 'is to satisfy the heart without resorting to the narcotics and anodynes of an incredible and preposterous faith.'[22]

Morley's dialectic reverses Morison's terms, placing science as the reaction against ritualism rather than the other way round, but the problem remains essentially the same. He finds a solution to this problem in Pre-Raphaelite painting, the poetry of Swinburne and Morris, and Pater's essays. Several of the poems by Pre-Raphaelite poets published in the *Fortnightly* are attempts to meet the need for spiritual satisfaction and consolation which positivists such as Morison and scientific naturalists such as Morley found wanting in science itself. Both Swinburne in 'Child's Song in Winter' and Meredith in his lyric sequence 'In the Woods', published in 1870, encourage their readers to see their own lives as part of larger cycles which are powerfully elemental but strictly natural. For Meredith, the central symbol is a tree growing as part of a forest. The woodland in his poem is both an allegorical landscape, symbolising our journeys through life with no clear sense of direction or end, and an ecological metaphor, as we ourselves are equated first with individual trees 'Rooted in earth' and then, on an even smaller level, with the fruit they bear, as in the closing lyric of the sequence:

> A wind sways the pines,
> And below
> Not a breath of wild air:
> All still as the mosses that glow
> On the flooring and over the lines
> Of the roots here and there.
> The pine-tree drops its dead:
> They are quiet as under the sea.
> Overhead, overhead,
> Rushes life in a race,
> As the clouds the clouds chase:
> And we go,
> And we drop like the fruits of the tree,
> Even we,
> Even so.[23]

Meredith's imagery contrasts the powerful elemental forces of the wind and the sea with the steady continuities of growth, death and rebirth. By identifying our individual deaths with the fall of fruit, he implies that our lives also bring about new growth even after we ourselves are dead. There is nothing in this poem to imply personal immortality, but in its place Meredith offers us an enduring legacy and a place within an effectively unending and perpetually resonant cycle of nature.[24]

Where Meredith takes the tree as his symbol, Swinburne takes a rose. The contrast is indicative. Meredith is forever roaming forests and the wider countryside in his poems, while Swinburne returns time and again to gardens. But each is an emblem of regrowth. In 'Child's Song in Winter', the rose itself survives blanketed in snow though the flower-heads die. But while Meredith's imagery suggests the birth of new life through the fruit, Swinburne's suggests rather the revival of the same plant in new flowers. In his essay on ritualism, published in the same issue as Swinburne's poem, Morison sees that the strength of the Oxford Movement lay in its 'prayer, and frankincense, and flowers' and in the 'solemn or spirit-stirring psalmody' with which it voiced the 'mysteries, and beauties, and terrors' it borrowed from Roman Catholicism.[25] Swinburne's poetry looks to take the same emotional hold through natural equivalents of these symbols, forms and suggestions. The rebirth of the rose is a form of resurrection, though Swinburne like Meredith is careful not to equate this with life after death. The rose flower itself has been laid to rest by fairies who 'In red leaves wound her, / With dead leaves bound her / Dead brows, and round her / A death-knell rang.'[26] The insistent repetition of 'dead' and 'death' three times in as many lines underscores the mortality of the individual flower. But by casting winter as death and spring as a resurrection, Swinburne hints that the cycle of the seasons can be grasped as a form of immortality in its own right.

This implication is reinforced but also complicated in the second half of the poem. Here the same structure of resurrection in which vitality is first suspended in winter and then revived in spring is applied not to a rose, but to one of the elemental forces itself. In the poem's northern winter the streams, personified in the figure of a water fairy, are frozen, only to thaw in spring:

But winds will vary,
And wise and wary
The patient fairy
Of water waits;
All shrunk and wizen,
In iron prison,
Till spring re-risen
Unbar the gates;
Till, as with clamour
Of axe and hammer,
Chained streams that stammer
And struggle in straits
Burst bonds that shiver,
And thaws deliver
The roaring river in stormy spates.

In fierce March weather
White waves break tether,
And whirled together
At either hand,
Like weeds uplifted,
The tree-trunks rifted
In spars are drifted,
Like foam or sand,
Past swamp and sallow
And reed-beds callow,
Through pool and shallow,
To wind and lee,
Till, no more tongue-tied,
Full flood and young tide
Roar down the rapids and storm the sea.[27]

The river here is at once newborn and immensely powerful, flooding with a force strong enough to uproot trees. The martial undertones of wielding axes and storming the sea are literalised in the succeeding stanza, where the spring storms on this northern coast are likened to the 'swift sea-kings', the Vikings who ravaged this coast ten centuries before.[28] Where the resurrection of the rose transposes the central mystery of Christianity, this elemental power is equivalent to the terror of divine power Morison finds in ritualism, a point Swinburne again underscores by closing his poem with a prayer to God 'whose hand is the strength of the sea'.[29] The poem is putatively spoken by or for a child, so this invocation of a personal deity takes its place alongside its flower and water fairies as part of a juvenile mythology of nature akin to the 'animation hypothesis' characterised in a later series of *Fortnightly* essays by the anthropologist J. F. McLennan, another of Rossetti's circle of friends, as the initial form of religious faith across all human cultures.[30] At the same time, by characterising an elemental deity in terms that are familiar from the Christian God – associating him with angels, suggesting that he 'made our wonder to work his will', and declaring 'Praise God, the giver / Of tongues to praise' – Swinburne redirects awe and worship towards a very different idea of God as a personification of natural forces.[31]

In praising 'psalmody', Morison implicitly calls for the formal qualities of verse to be put at the service of a religion grounded not in theology but in science. Meredith and Swinburne were both acutely sensitive to the possibilities of rhythm and metre in poetry. In his closing lyric from 'In the Woods', Meredith uses a broadly anapaestic measure,

Six shall stumble with wounded shins;
For the rood is stiff whoever kneels,
And God never stops his chariot wheels,
Nor looks out of his narrow window,
Over the drifts and steeps of snow;
But Satan for a thousand years
Has gotten a lease of our hopes and fears –
To catch men's souls by the eyes and ears.
Let us every one beware
Little faith or overcaring,
Pride of heart or overdaring,
Lest we come within his snare.[40]

The poem is spoken at some unspecified time after the events it relates are supposed to have occurred. Its sporadic archaisms and its irregular four-beat rhythms, derived from the quasi-medieval metre of Coleridge's 'Christabel' – another poem of demonic intervention – enhance the impression formed by Scott's imaginative embrace of superstition that the narrator is voicing an anti-Enlightenment perspective. From within this world-view, he charts as a decline the move away from superstition which for the *Fortnightly* signified progress. This narrator does not represent Scott's own authorial voice, and the poem's readers are not expected to share his idealisation of early Christianity. Even so, through him Scott raises both the question asked by Cobbe in an earlier essay for the *Fortnightly* – 'What is Progress, and Are We Progressing?' – and the problem of evil in a world in which received religious teachings can no longer assert an automatic claim to truth.[41] Scott's poem hints at the same conclusions Cobbe reaches in her essays: that material and intellectual progress are insufficient without moral progress, and that evil is no less a problem merely because we are emancipated from the old myths and prejudices. But where Cobbe deconstructs and dismisses the Devil, Scott casts God himself as divorced from his own creation, leaving the lease of it to a Devil whose nonexistence does not prevent him from effectively symbolising the temptations of pride and evil persisting within a society that can no longer claim access to God.

Like Scott, Rossetti probes the implications of materialism in a series of sixteen sonnets published under the comprehensive title 'Of Life, Love, and Death' in March 1869. These are the most moving and suggestive of all the Pre-Raphaelite poems published in the *Fortnightly*. They are also those with the most subtle and complex, but also the most stark, relation to the journal's scientific methods and values. Morley's comments on Rossetti's poems themselves are less helpful in looking to understand this relationship than his favourable reviews for the *Fortnightly* of two other poets admired by Rossetti. For the same issue which carried Rossetti's sonnets Morley wrote a review of *The Ring and the Book*, Browning's immense cycle of dramatic monologues centred on a murder in seventeenth-century Rome. The previous June he had reviewed the first instalment of Morris's even more vast cycle of mythological and medieval tales, *The Earthly Paradise*.

Morley sees Browning as a model of a 'scientific attitude' in poetry.[42] As Stephens had argued in his early manifestos for Pre-Raphaelitism in *The Germ*, this attitude shows itself not in the themes of his poetry but in its approach. It is telling that the poet Morley implicitly disparages by contrast with Browning and indeed Morris is Tennyson, who engaged more directly with scientific theories and findings in his poetry than either of them. In characterising Browning's poetry, Morley writes:

> the master is he who presents the natural shape, the curves, the thews of men, and does not labour and seek praise for faithful reproduction of the mere moral drapery of the hour, this or another; who gives you Hercules at strife with Antæus, Laocoon wrestling in the coils of the divine serpents, the wrestle with circumstance or passion, with outward destiny or inner character, in the free outlines of nature and reality, and not in the outlines of a dress-coat, either of Victorian or Arthurian time.[43]

The reference to Arthur marks Tennyson out as the butt of Morley's critique. Like Stephens, Morley casts Browning as an expert on human anatomy; like Tupper, he recognises the same grasp of embodied psychology within antique sculpture. But where Browning gives us a scientific study of human nature, all Tennyson does in his own ongoing cycle of poems *The Idylls of the King*, as far as Morley is concerned, is rehearse the dominant values of his time. As he remarks: 'we have this long while been so debilitated by pastorals, by graceful presentation of the Arthurian legend for drawing-rooms, by idylls, not robust and Theocritean, but such little pictures as might adorn a ladies' school, by verse directly didactic, that a rude inburst of air from the outside welter of human realities is apt to spread a shock, which might show in what simpleton's paradise we have been living.'[44] Morley's charge that the poetry of Tennyson and his fashionable followers is intellectually inadequate is anticipated from another angle in his review of Morris. He praises Morris for eschewing 'the little doubtings, and little believings, and little wonderings, whose thin wail sounds in a conventional manner through so much of our current writing, whether in prose or verse, weakening life and distorting art'.[45] Like the academic artists the Pre-Raphaelites defined themselves against in the 1840s and 1850s, Tennyson is hamstrung in Morley's view by his inability to break with convention. Because he cannot emancipate himself from received wisdom, the beliefs and speculations articulated in his poetry – and here Morley would seem to have *In Memoriam* in mind as much as the *Idylls* – remain trivial and unimpressive.

Morley does not mention Rossetti in either of these reviews, but his friend and contributor Colvin later recalled that this same dissatisfaction with Tennyson made his generation enthusiastic for Rossetti.[46] Morley's admiration for Rossetti suggests that he saw Rossetti as allied to Browning and Morris – themselves curious bedfellows – rather than Tennyson. His reviews provide a frame for considering the sequence 'Of Life, Love, and Death'. On the face of it, these sonnets would appear to have more in common with *In Memoriam* than with either *The Ring and the Book* or *The Earthly Paradise*. Like Tennyson's elegy, they constitute a lyric sequence meditating principally on bereavement: the widowhood symbolised in the first four sonnets through the allegorical landscape of Willowwood. They are preoccupied with mortality, and their focus is almost entirely on the interior of one man's mind, implicitly inviting readers to generalise this out into 'the cry of the whole human race', as Tennyson characterised his own poem.[47] The 'outside welter of human realities' do not impinge here, at least not in any socio-political sense, and there is none of the attention to embodied psychology found in Browning, antique sculpture, or earlier Pre-Raphaelite poetry and painting.

Yet there are several crucial differences between *In Memoriam* and 'Of Life, Love, and Death' which may help to explain why Morley was keen to publish Rossetti's sonnets in the *Fortnightly* even as he denigrated Tennyson's poetry in its pages. In one crucial sense the structure of Rossetti's sequence resembles Browning's and Morris's massive narratives more closely than it does Tennyson's sustained lyric. *In Memoriam* is a poem which charts a clear if halting progress through grief. Tennyson works through several iterations of the same basic problems – the intransigence of loss, the resulting sense of alienation from oneself and those around us, the existential shock at the prospect our own extinction – but over the course of the poem he manages to surmount his grief and fear. While Rossetti's fundamental concerns are the same, there is no such progress in 'Of Life, Love, and Death'. Instead, like the monologues in *The Ring and the Book* and the tales in *The Earthly Paradise*, each sonnet returns afresh to the same problems, exploring them from a new angle of vision, but not getting any closer to a final understanding or state of ease.

The implication of Browning's, Morris's and Rossetti's poems is that there is no such final answer – that any truth can only ever be provisional, and must always be subjected to further scrutiny. In this sense, although Tennyson grapples with the implications of science, their poems are more thoroughly scientific in their outlook. For a reader such as Morley, there is something factitious in the logic of Tennyson's poem, as the consolations he works his way towards – Christian faith, personal immortality, the institution of marriage – are the same as the ideals he set out with, and clearly conventional. Tennyson may protest that 'There lives more faith in honest doubt, / Believe me, than in half the creeds', but his doubts do not prevent him from avowing at the end of the poem his faith in 'That God, which ever lives and loves, / One God, one law, one element, / And one far-off divine event, / To which the whole creation moves.'[48] Even the teleological evolutionism which was Tennyson's new solution to the problem of extinction when he wrote the poem in the 1830s and 1840s had become conventional, at least in its theistic guise, as Lewes pointed out in his review of the Duke of Argyll's *The Reign of Law* in the *Fortnightly* in July 1867.[49] The onset of Darwinism failed to shift Tennyson's comforting vision of nature as purposive, even if he saw that it menaced it.[50] All told, from Morley's perspective, Tennyson's 'doubtings' had shown themselves to be 'little', personally painful but unable to seriously alter his 'believings', which were themselves diminished in the process.

Rossetti included a self-consciously factitious sonnet in his sequence which exposes precisely this kind of self-deluding sense that there is a direction or transcendental purpose to our existence. This is 'The Landmark', first written, like Scott's 'Anthony', back in the 1850s:

> Was *that* the landmark? What, – the foolish well
> Whose wave, low down, I did not stoop to drink,
> But sat and flung the pebbles from its brink
> In sport to send its imaged skies pell-mell,
> (And mine own image, had I noted well!) –
> Was that my point of turning? – I had thought
> The stations of my course should rise unsought,
> As altar-stone or ensigned citadel.
> But lo! the path is missed, I must go back,
> And thirst to drink when next I reach the spring
> Which once I stained, which since may have grown
> black.
> Yet though no light be left nor bird now sing
> As here I turn, I'll thank God, hastening,
> That the same goal is still on the same track.[51]

This early sonnet conveys the sense of a particular person in a particular landscape that is characteristic of William Rossetti's 'Fancies at Leisure', but here the woodland well is allegorical as well as literal. Like Meredith in 'In the Woods', Rossetti borrows from Dante the motif of life as a journey through dark woods. The speaker has lost his way through not realising the significance of one apparently insignificant moment on that journey. He turns back, confident that he can still regain the correct path, but we have to doubt this, not only because we notice the fading of the light, but because by this point in the sequence we have read nine other sonnets, all of which affirm the finality of events in the past. There is no turning back in Rossetti's poem, or at least no hope if you do turn back that you will be able to reach the same goal on the same track. There is no God, either, at least not one who will help you. The truth of the matter, the nine sonnets before this one and the six after all imply, is that the spring has indeed been indelibly blackened.

In Memoriam is a sustained exercise in reassuring its doubting author and readers that their belief in personal immortality is true. Like Meredith in 'In the Woods', Rossetti allows himself and his readers no such prospect. But where Meredith steers us towards accepting our own short-lived existence as contributing to the life of a natural world that will outlast us, Rossetti's resolute focus on individual consciousness and his reiteration of the fundamental problem of loss call this consolation into question too. Time in Rossetti's poem is not a cycle but an inexorable movement in one direction only. There may, he suggests in the seventh sonnet, 'Run and Won', be a vantage point from which it is possible to comprehend a life or even life itself as a complete whole, but even that knowledge is fruitless, and even the knower must succumb in the end:

> Around the vase of Life at your slow pace
> He has not crept, but turned it with his hands,
> And all its sides already understands.
> There, girt, one breathes alert for some great race;
> Whose road runs far by sands and fruitful space;
> Who laughs, yet through the jolly throng has pass'd;
> Who weeps, nor stays for weeping; who at last,
> A youth, stands somewhere still, with silent face.
> And he has filled this vase with wine for blood,
> With blood for tears, with spice for burning vow,
> With watered flowers for buried love most fit;
> And would have cast it shattered to the flood,
> Yet in Fate's name has kept it whole; which now
> Stands empty till his ashes fall in it.[52]

In the context of the *Fortnightly Review*, the man who turns the 'vase of Life' appears as an expert whose knowledge carries with it the authority of a detached, meticulous, scientific gaze. As readers, we see the runner's life unfold on the vase, but we do not comprehend it. We know he weeps, but we do not know why. We do not know where it is that he is standing in the final frame, nor why he is silent, though we do know that he is still only a youth in spite of his experience of grief. The turner of the vase, at least by his own estimate, knows all this, yet he has not transcended his own subjectivity all the same. Like Morison, he craves the comfort of ritual, and cannot do without it, and yet he knows that it brings no comfort. The only ritual that remains at the end of the sonnet is the bleak and empty one of his own interment in the vase itself. The 'vase of Life' has shown itself little more than a funerary urn.

If the turner of the vase resembles a scientist in his claim to objective knowledge, he resembles the poet too. Throughout the sequence Rossetti examines life from different sides, describing and interpreting allegorical figures like the runner and their movement through physical landscapes representing the passage of time. The first four sonnets, grouped together as 'Willowwood', stage an encounter between the speaker and visions of his past self and his dead lover beside a woodside well – a sign that does not bode well for the speaker in 'The Landmark'. Across succeeding sonnets, these revenants reappear in new forms. In 'Lost on Both Sides', they are personifications of extinguished hopes. In 'Known in Vain', they are Work and Will, cast as lovers who 'awake too late, to gaze / After their life sailed by'.[53] In 'Lost Days', they are the very days of his life, encountered at the instant of his own death as 'Each one a murdered self'.[54] As testimonies to an intensely felt and closely examined subjectivity, these sonnets offer precisely the complementary corrective to purely 'scientific manifestations of intellectual activity' demanded by Morley. As the positivist Harrison argued in an essay for the *Fortnightly*, 'nothing is true knowledge that is not relative to the human nature in its complex whole.' To achieve knowledge therefore required a synthesis of positive knowledge – science – and subjective self-scrutiny.[55] This 'subjective synthesis', as Harrison called it, parallels the synthesis of science with aestheticism – itself a synthesis of ritualism and science – which Morley was to propose in his essay on Pater. Through publishing Rossetti's 'Of Life, Love, and Death', together with other Pre-Raphaelite poems examining similar existential questions, including Meredith's 'In the Woods' and 'A Mark in Time', Swinburne's 'A Lost Vigil' and 'Ave Atque Vale', Morris's 'The Dark Wood', and even Scott's 'Anthony', Morley was able to create precisely such a synthesis within the pages of the *Fortnightly Review* itself.

The structure of *In Memoriam* is ultimately that of comedy. It ends with a marriage and the avowal of Arthur Hallam's immortality. 'Of Life, Love, and Death', by contrast, ends with the only prospect any life can lead to within the logic of Rossetti's sonnets and Morley's materialism. The speaker faces his demise in a pair of sonnets entitled 'Newborn Death':

To-day Death seems to me an infant child
  Which her worn mother Life upon my knee
  Has set to grow my friend and play with me;
If haply so my heart might be beguil'd
To find no terrors in a face so mild, –
  If haply so my weary heart might be
  Unto the newborn milky eyes of thee,
O Death, before resentment reconcil'd.
How long, O Death? And shall thy feet depart
  Still a young child's with mine, or wilt thou stand
Fullgrown the helpful daughter of my heart,
  What time with thee indeed I reach the strand
Of the pale wave which knows thee what thou art,
  And drink it in the hollow of thy hand?

And thou, O Life, the lady of all bliss,
  With whom, when our first heart beat full and fast,
  I wandered till the haunts of men were pass'd,
And in fair places found all bowers amiss
Till only woods and waves might hear our kiss,
  While to the winds all thought of Death we cast: –
  Ah! Life, and must I have from thee at last
No smile to greet me and no babe but this?
Lo! Love, the child once ours; and Song, whose hair
  Blew like a flame and blossomed like a wreath;
And Art, whose eyes were worlds by God found fair;
  These o'er the book of Nature mixed their breath
With neck-twined arms, as oft we watched them there: –
  And did these die that thou might'st bear me Death?

Like the other sonnets in the series, Rossetti's shockingly original vision of Death personified as a newborn baby owes a great deal to his own life. In 1860 Rossetti married Elizabeth Siddal after a protracted and unsteady courtship. The following year their only child was stillborn, and less than a year later

Siddal died, almost certainly a suicide. These bleak biographical details are the root of Rossetti's conceit in this sonnet, yet the sonnet itself transcends them to become a meditation on the inevitability of every death as the final end of every life, the lastborn child which erases all others for the individual concerned.

In the figure of Death as an infant child, the first sonnet holds out the possibility that there may yet be something to cherish in this ending, for all that we do not know how long we might have to live with the knowledge that we are dying, or whether death will be easeful when it comes. But the second sonnet returns us once again to the theme of loss. With the preceding sonnets still fresh in our mind, Life is both a personification and the lost lover herself. But it is their children who stand out in the context of the *Fortnightly Review*, and indeed of the Pre-Raphaelite project as a whole. That Rossetti should cast Song and Art – poetry and painting – among the children lost to him through death is not itself remarkable. What is striking is that he watches them as they read 'the book of Nature' together. Rossetti's reference here to the traditional image of the field of scientific enquiry is a reminder of the original object and methods of Pre-Raphaelite art. It hints too at the synthesis of art and science proposed by Morley. As with the contrast between objective knowledge and subjective experience in 'Run and Won', however, the end of all such endeavours remains for each individual the same.

The closing sonnet of Rossetti's short sequence ends on a question, but its power lies in the fact that it presupposes the answer. I. A. Richards's praise for Thomas Hardy as 'the poet who has most steadily refused to be comforted in an age in which the temptation to seek comfort has been greatest' is equally apt to Rossetti in these sonnets.[56] As the art critic Harry Quilter remarked, reflecting on Rossetti for the *Contemporary Review* after his death, 'He never paltered with the facts of the case, no matter how terrible; but in the life of others, as well as in his own, cut down to the truth'.[57] Quilter returns to the analogy between art and anatomy proposed by Stephens thirty years before in *The Germ*. Rossetti's determination to leave himself and his readers undeceived – to cut down to the bone – was why Morley so admired him and his sonnets. In place of Tennyson's 'little doubtings, and little believings', Rossetti offers us nothing to believe in and no doubt whatsoever, only 'the facts of the case'.

READING *THE EARTHLY PARADISE* WITH SCIENTIFIC MEN

Like Rossetti's sonnets, Morris's *Earthly Paradise* offered Morley a counterpoint to the thin conventional wail of mid-Victorian poetry and prose. From the opening line of its prologue – 'Forget six counties overhung with smoke' – *The Earthly Paradise* seems the very antithesis of the paraded modernity of science, with its new knowledge and technological accomplishments.[58] Morris's reader is charged to step back five centuries to the Middle Ages, before being whisked still further away from the real, modern world by a voyage to unknown lands in the West as a group of wanderers flee the plague in search of a fabled land of immortality. They end up at last settled as guests on an island in the Atlantic, a marooned remnant of a classical civilisation. Here old men from the two societies – the wanderers and their hosts – regale one another with tales from their own cultures, retold by Morris from a diverse range of ancient, medieval and modern sources.[59] One tale from each tradition is told each month, beginning in March and ending with a year completed in February.

It would be easy to brand this as determined escapism, stretched as it is over twenty-five long verse tales in total. As such, the poem would be an extreme example of Morley's ideal of aesthetic poetry as a counterweight to science, and a distraction at least from the doubtings and believings which so irritated him. But it is harder to see how it could be a product of science too, as Morley's argument in his essay on Pater would suggest, or how, if Morris's proffered earthly paradise were mere escapism, it could help to dispel the other 'simpleton's paradise' of the more conventional Victorians. Yet Morley was not alone among scientifically minded intellectuals in his regard for Morris's poem. Lewes and his partner, George Eliot, relished *The Earthly Paradise*.[60] Huxley also read it.[61] A reviewer in the *Westminster Review* specifically remarked: 'To our personal knowledge, political economists and scientific men to whom Shelley is a mystery and Tennyson a vexation of spirit, read the *Earthly Paradise* with admiration.'[62] These glimpses of a scientific readership for Morris's poem, though only that, are nonetheless intriguing. Unlike both Shelley and Tennyson, Morris did not write poetry which engaged directly with science, so clearly this admiration was not predicated on a desire for poetry to reflect on scientific theories or discoveries. In seeking 'to explain this phenomenon' of Morris's popularity with 'scientific men', the *Westminster* reviewer takes them to be a subset of 'those, who as a rule, do not care to read any poetry', suggesting that the appeal of *The Earthly Paradise* to such readers rests on its 'easy flow and simple construction', its unforced rhymes, its 'bright and vivid' pictorial quality – its unchallenging accessibility, in effect.[63] While these virtues surely do account in part for the poem's remarkable popularity, especially given its sheer bulk, there is a thread within the reviews which suggests that it may have appealed specifically to advocates of modern science because the distinctive qualities of Morris's poetry bore a relation to science itself.

In his review for the *Fortnightly*, Morley identified 'a vigorous and healthy objectivity' as Morris's 'central quality' as a poet.[64] This emphasis on the objectivity of Morris's poems recurs in review after review. The American critic Charles Eliot Norton detected it in his previous long poem, *The Life and Death of Jason*, commenting that Morris told his story 'objectively, with simple regard to its own development, and with no direct expression of the emotions or thoughts its course awoke in himself'.[65] Reviewing the final instalment of *The Earthly Paradise* for the periodical *The Academy*, Colvin remarked that Morris's landscapes encapsulated 'the precise and delicate record of observation by perfectly alert senses free from all preoccupation, trouble, or impediment'.[66] Morley also remarked on the 'effective precision and truthfulness' of his observations, while an anonymous reviewer for the *New Englander* characterised him as 'a deliberate and accurate observer of details, little disturbed by a consciousness of anything more in what he sees than can be apprehended by the senses'.[67] All of these critics explicitly or implicitly identify Morris's attention to detail as painterly. As the *New Englander* reviewer remarks, 'He has a Pre-Raphaelite's longing to give us everything – everything external.'[68] But their accounts of his practice also mark it out as analogous to science in its objectivity, empiricism and precision.

Like the earlier Pre-Raphaelites, Morris directs his empiricism towards the natural world, on the one hand, and the human body and mind, on the other. In looking at landscapes, plants and animals, Morris sets aside his own preoccupations and prejudices to record what he can apprehend about nature through observation. Two brief passages from 'The Story of Cupid and Psyche', the third of the twelve classical tales retold in *The Earthly Paradise*, demonstrate the characteristic detail of his descriptions of landscapes. In the first, Psyche, who has just been ejected from Cupid's palace for daring to look upon his body, awakens beside a river:

> There, at the head of a green pool and deep,
> She stood so long that she forgot to weep,
> And the wild things about the water-side
> From such a silent thing cared not to hide;
> The dace pushed 'gainst the stream, the dragon-fly,
> With its green-painted wing, went flickering by;
> The water-hen, the lustred kingfisher,
> Went on their ways and took no heed of her;
> The little reed birds never ceased to sing,
> And still the eddy, like a living thing,
> Broke into sudden gurgles at her feet.
> But 'midst these fair things, on that morning sweet,
> How could she, weary creature, find a place?[69]

This scene marks the beginning of a long period of homeless wandering for Psyche within the poem. Some unspecified time after, she finds her way to a landscape of cornfields which, it transpires, surrounds a temple sacred to Ceres, the goddess of agriculture:

> Now, midst her wanderings, on a hot noontide,
> Psyche passed down a road, where, on each side
> The yellow cornfields lay, although as yet
> Unto the stalks no sickle had been set;
> The lark sung over them, the butterfly
> Flickered from ear to ear distractedly,
> The kestrel hung above, the weasel peered
> From out the wheat stalks on her unafeard,
> Along the road the trembling poppies shed

On the burnt grass their crumpled leaves and red;
Most lonely was it, nothing Psyche knew
Unto what land of all the world she drew.[70]

Like Millais's and Hunt's early paintings, William Rossetti's poems and the O'Sheas' carvings, these passages set down particular ecologies in detail. We are introduced first to the landscape – the pool or the cornfields – described plainly in terms of its colour, spatial relations and temperature. The pool is 'deep' and 'green'; the fields, which line the road 'on each side', are 'yellow'. When Psyche arrives at the cornfields it is 'a hot noontide'. Though the temperature of the pool itself is not directly stated, the word itself hints at its rhyme-word 'cool', a hint that is realised just after the second of these two passages when Psyche, reaching another stream, 'slipped adown into the shaded pool, / And with the pleasure of the water cool / Soothed her tired limbs awhile'.[71]

Once the environment has been established, Morris populates it with wildlife. As the 'wild things' are undisturbed by the passive Psyche, he is able to record them going about their daily lives. Morris combines different sensory impressions in his accounts of the animals and plants which comprise these different ecologies. He sees colours and shapes, and hears the songs of birds and the sounds of the river. Above all, he captures movement within the landscape and the degrees of intent behind it. While insects flicker, the small dace swim more purposefully against the current of the stream, and the moorhen and kingfisher go 'on their way', implying that they have ways they mean to go. Even the static hanging of the kestrel and peering of the weasel bespeak close observation on their part as well as on his. By focussing on the actions and experiences of other animals within the landscape, Morris downplays both his own responses, as Norton observes in his review of *Jason*, and Psyche's. Her despondency noticeably fails to colour his account of the natural life around her. Instead, he contrasts the vitality of nature and the communities of living things with her own enervation and loneliness.

The *Westminster* reviewer identified 'the beauty and freshness with which all natural objects are described' in *The Earthly Paradise* as among its most winning qualities.[72] Like Pre-Raphaelite landscape painting, Morris's detailed observation in the early tales of *The Earthly Paradise* affirms the value of the natural world for their readers. The poem's later tales increasingly tip the balance away from nature and romance, leaving 'psychology unexpectedly in the ascendant', as a reviewer for the *Pall Mall Budget* remarked.[73] Morris's presentation of psychology, like that of the earlier Pre-Raphaelites, concentrates on the visible signs of mental states. In this sense, his method is again objective. In 'The Lovers of Gudrun', a retelling of the Icelandic *Laxdaela Saga*, Morris shows us the collapse of the relationship between Gudrun, her lover Kiartan and his foster-brother Bodli, occasioned by absence and torn loyalties. Following his source, he charts this process with the meticulousness of a realist novel, but where George Eliot, for example, focalises her narratives through her characters' own interpretations of events, Morris suggests their emotions, which they themselves often do not fully comprehend, through their bodily experience and expression of them.

As in Millais's *Isabella* (see fig. 36), Hunt's *Valentine Rescuing Sylvia from Proteus* (see fig. 38) or Collinson's *Answering the Emigrant's Letter* (see fig. 44), the most powerful expressions of embodied psychology in 'The Lovers of Gudrun' occur in scenes where characters are shown interacting with one another. These scenes can seem remarkably cinematic. After drawing his sword on the King of Norway for fear that he and his fellow Icelanders will lose their freedom, the hero Kiartan reacts to the king's promise that they may go in peace. At this point Morris gives us a close-up of his face:

To Kiartan's face, pale erst with death, there rose
A sudden flush, and then his lips, set close,
And knitted brow, grew soft, and in his eyes
There came at first a look of great surprise,
Then kind they grew, and with shamefaced smile
He looked upon the king a little while,
Then slowly sank his sword, and, taking it
By the sharp point, to where the king did sit
He made his way.[74]

Where a painting could only record one of this sequence of expressions, Morris's poetry can describe each in turn, tracking the gradual change from one to another. We see Kiartan's fear and then his elation as they unfold first in the colour of his face, then in the relaxation of his facial muscles and the dilation of his eyes. The change to kindness in his eyes reflects that he is regaining control of his situation, while the embarrassment in his smile – hinted at too in his flushed cheeks – shows his realisation that this control is only partial as the king retains the upper hand. Morris traces these complex emotions on Kiartan's face not from within his mind, communicating them as an actor would, much as Brown had recommended in his essay 'On the Mechanism of a Historical Picture' in *The Germ* twenty years before. Kiartan's feelings are also revealed in his speech immediately after this passage, where he not only pledges his fealty to the king by offering him his sword but accepts his invitation to convert to Christianity. It is because he has been emotionally disarmed that he is so ready to disarm himself, and to acquiesce to a religion that itself purports to disarm the old gods and to lay down its own arms as well.

Morris traces how emotions are experienced as bodily sensations by those who feel them. When Gudrun first realises that Kiartan not only likes but desires her, she feels a sudden and disorienting rush of desire herself:

> But, turning round,
> Kiartan upon the other hand she found,
> Gazing upon her with wild hungry eyes
> And parted lips; then did strange joy surprise
> Her listless heart, and changed her old world was;
> Ere she had time to think, all woe did pass
> Away from her, and all her life grew sweet,
> And scarce she felt the ground beneath her feet,
> Or knew who stood around, or in what place
> Of heaven or earth she was; soft grew her face;
> In tears that fell not yet, her eyes did swim,
> As, trembling, she reached forth her hand to him,
> And with the shame of love her smooth cheeks
> burned,
> And her lips quivered, as if sore they yearned
> For words they had not learned, and might not know
> Till night and loneliness their form should show.[75]

Again Morris gives us specific and successive stages in the unfolding of an emotion, here correlating Gudrun's own interior feelings with their expression on her face. He captures the immediacy of awakening to new desire and the intensity of the transformation this brings with it. This experience as Morris describes it happens too quickly for Gudrun to grasp it fully herself. Her body's responses to desire – lightheadedness, blushing, trembling, smiling, the welling-up of tears at the realisation of a previously subconscious hope – are all inadvertent, and yet they express Gudrun's feelings more directly and powerfully than she can articulate them. Indeed, they are her emotions. At the end of this passage, with rare third rhymes placed in the middle of two successive lines, Morris affirms that her burning cheeks bespeak her yearning, even though her lips have not the learning to describe it, and that only in the loneliness of the night will her emotions show themselves to her in full and will she come to know them. That learning is at once beyond the reach of the preceding couplet and superfluous to it, as Gudrun's body itself expresses her feelings with great lucidity. This lucidity is matched by Morris's poetry, which for its own part does not seek to move beyond describing her bodily experience into any more abstract taxonomy of emotions or analysis of their narrative significance or moral worth.

The Earthly Paradise recalls the scientific standards of the earlier Pre-Raphaelites, not because Morris sets out to model his art on science as they did, but because his practice is grounded in their precepts and example. This legacy is identified as a determined objectivity within the reviews. Morris's poem as a whole revives the ecological imperative of Pre-Raphaelitism, while 'The Lovers of Gudrun' marks a return in verse to the careful study of embodied psychology in early Pre-Raphaelite painting. But the analogue between Morris's poem and the Pre-Raphaelite project of grounding art in science goes further than this. Like William Rossetti's 'Fancies at Leisure', Morris's poetry intimates that there is a personal moral benefit to observation itself. This is also noticed within the early reviews. Reviewing *The Life and Death of Jason* for the *North American Review*, Henry James commented: 'To the jaded intellects of the present moment, distracted with the strife of creeds and the conflict of theories, it opens a glimpse into a world where they will be called upon neither to choose, to criticise, nor to believe, but simply to feel, to look, and to listen.'[76] For *The Spectator*, too, *The Earthly Paradise* is 'an effectual antidote for the overwrought self-consciousness of this generation', not least because it eschews 'the complex questionings that vex our modern poetry' and 'hardly ever touches on matters of speculation'.[77] Like Morley, these reviewers find in Morris's detailed attention to material things a tonic for an age suffering from intense intellectual anxieties.

Science is typically seen as a major factor causing these anxieties, as it is for Tennyson in *In Memoriam*, while poetry such as Tennyson's is recognised as a powerful and subtle literary form for working through them. But the reviews of *The Earthly Paradise* suggest a different role for both science and poetry, as practices which, through a shared objectivity, can take people away from their subjective troubles. For the *New Englander*'s reviewer:

> Where Wordsworth notices that 'every flower enjoys the air it breathes,' Mr. Morris remarks that it is small and as 'red as blood,' or that it 'flames' in 'grey light,' or that it has a black or a yellow centre and takes a certain curve in bending. To know such facts as these about flowers, and similar facts about all natural objects, is of more importance to his purpose than to be impressed with a belief that they are alive and enjoying the life they feel.[78]

In characterising Morris as determined to observe, know and report 'facts', this reviewer casts him in the same Baconian role as Stephens and Rossetti cast the Pre-Raphaelite painters in the 1850s. Morley, James and *The Spectator* reinforce this interpretation of his poetry by rehearsing Stephens's opposition between facts and theories. While theorisation was increasingly accepted as an essential component of scientific practice, their reviews suggest that even for advocates of scientific naturalism such as Morley the ideal of Baconian empiricism provided a haven from the 'conflict of theories', allowing scientists to concentrate on observation in its own right. *The Earthly Paradise* offers readers a realm in which this ideal still holds good. 'Scientific men', whether scientists or not, prized *The Earthly Paradise* in part because it enacted a form of scientific enquiry in its own approach to nature and narrative. Baconian empiricism was at once prior to and simpler than modern science. It affirmed the value of the careful study of nature while providing a respite from the controversies and complexities of science in the aftermath of Darwin.

To identify *The Earthly Paradise* as scientific in its methods is, of course, to ignore the fact that the worlds of the tales – though not that of the frame narrative – are characterised by magic, from the manifestations of pagan gods and goddesses and the slaying of monsters in the classical tales, to the shape-shifting of fairies and journeys to lands beyond time in the medieval ones. As a reviewer remarked in the *Pall Mall Budget* in 1869, after the appearance of the second instalment of the poem, Morris takes us into a world 'where nature exists but has left her laws behind her'.[79] This reviewer – Peter Faulkner tentatively identifies him as Colvin – proposes another dialectic to resolve this paradox. There are, he suggests, two main schools of poetry: romanticism, which affirms the freedom of the imagination, be it 'never so irregular, never so unscientific, never so fantastic', and its unnamed opposite which prefers 'to deal with clear conceptions testable by rules of reason and canons of propriety, semi-scientific conceptions, not too remote from sight and hearing'. Although Morris represents the former school, he brings to it some of the values of its opposite. Thus, while he shows us 'a remote romantic world', that world is 'conceived and set forth with the inexhaustible detail of a loving eye-witness'. His world is 'heterocosmic' – that is, it is another universe, apart from and unlike our own – yet Morris gives it 'the clear definition, the intelligible form and beauty that can be grasped and tested'.[80] By grounding his presentation of romance so firmly in observable details of the real world, Morris transforms romance itself from an 'unscientific' mode to a 'semi-scientific' one.

The most striking of these claims is that the form and beauty of Morris's world can be not only 'grasped' but 'tested'. In making this claim, the reviewer draws an analogy between reading poetry and the experimental method. The precise detail of Morris's observations of physical and living nature enables readers to visualise his world for themselves, and so encourages them to test how far it conforms to or departs from their own perceptions. In this sense his poem creates its own community of naturalists, as William Rossetti does across his 'Fancies at Leisure'. But the implications of the reviewer's interpretation reach beyond this. It is not only forms that can be tested in the poem but beauty itself. By implication, even the aesthetic values of *The Earthly Paradise* are somehow verifiable or at least disprovable, whether by 'rules of reason', 'canons of propriety', or tests against the empirical data of experience. Beauty for this reviewer is a property that can be identified and calibrated by Baconian methods, as it is for Pater, Poynter, Tupper and Brett.

One sound basis for viewing the poem in this light may be that it enacts this very process of testing through its own structures. The poem can be read as a series of imaginative experiments to test the proposition that it might be possible to achieve a state akin to paradise on earth. In the prologue, wanderers from fourteenth-century Europe encounter six different societies in different regions of the yet-to-be-discovered New World in their quest to find an earthly paradise. The closest approximation they find to it – though it is a slight substitute for the immortality they had hoped to find, and that Morris, like Morley and Rossetti, will not allow them – is the surviving remnant of the classical world on the island where they end up, where the social relations between the classes, generations and sexes appear remarkably harmonious. Like the prologue, each of the twenty-four tales told on the island investigates from one angle or another the possibility that its heroes, heroines or societies might themselves attain some approximation to paradise. Several consider the institution of kingship, for example, such as 'The Man Born to be King' and 'The Love of Alcestis', asking how and how far a king may be able to foster the happiness of his subjects, and examining the tendency of monarchy to corrupt monarchs and society as a whole. Others play out quest narratives which raise the possibility of personal fulfilment through adventure, wealth or love.

Many of Morris's tales hold out the prospect of attaining happiness through sexual love, typically prompted by beauty. This is the possibility and the risk that is tested most fully, not least through recurrent interventions by the goddess Venus. As the embodiment of sexual desire, Venus has different attributes in different tales. In 'The Story of Cupid and Psyche', she seems kind – 'What depths of love from out her grey eyes beamed!' – yet she is jealous and petty.[81] In 'Atalanta's Race' and 'Pygmalion and the Image', she is the benevolent sponsor of male lovers, bringing about Milanion's victory over Atalanta through trickery, ensuring that she must marry him, and kissing Pygmalion's statue into life to be his lover. In both cases, the female characters are happy in the end to comply, though they have little choice, suggesting the coercive nature of many sexual contracts, as well as the compulsion of the will by erotic desire itself.

In the last two medieval tales, both told in the poem by the Swabian priest and alchemist Laurence, Venus persists as a telling survival from the pagan world into the world of medieval Christianity. In 'The Ring Given to Venus', both the coercive and the compulsive aspects of sexual desire are realised as Venus forces herself on a bridegroom, also called Laurence, who has foolishly placed his wedding ring on the finger of her brass statue to avoid it being damaged in a game:

> Then round about him closed the mist;
> It was as though his lips were kissed,
> His body by soft arms embraced,
> His fingers lovingly enlaced
> By other fingers; until he
> Midst darkness his own ring did see.
>
> Nought else awhile; then back there came
> New vision: as amidst white flame,
> The flower-girt goddess wavered there,
> Nor knew he now where they twain were
> Midst wild desire that nigh did rend
> His changed heart; then there came an end
> Of all that light and ecstasy;
> His soul grew blind, his eyes could see;
> And, moaning from an empty heart,
> He saw the hangings blown apart
> By the night wind, the lights flare red
> In the white light the high moon shed
> O'er all the place he knew so well,
> And senseless on the floor he fell.[82]

In this passage, Morris describes the experience of intense sexual desire, regardless of the object of that desire. Rather than describing Venus principally through the sense of sight as he does in the other tales, he concentrates instead on Laurence's bodily sensations, first of the touch of her body to his, and then of his orgasm and the subsequent sense of void as his 'wild desire' evaporates with the goddess herself. While Venus compels Laurence to experience this erotic transport, she does so not because she desires him but because he has jokingly mocked her statue.

Sex for Venus in this tale is an exercise of power, not of love. Even for Laurence himself, this sexual satisfaction brings shame rather than pleasure, as he is effectively raped by sexual desire itself. And yet in the last tale, 'The Hill of Venus', when she becomes the lover of a young man called Walter – a version of Tannhäuser – Venus reappears rather as the embodiment of his desire, manifest to him as a fulfilment and confirmation of his solipsism. In spite of her passivity, Walter persuades himself that their loves are mutual and that, as he exclaims, 'For this, for this, / God made the world, that I might feel thy kiss!'[83] Both of these tales are sympathetic studies in erotic fantasy, as are 'Atalanta's Race' and 'Pygmalion and the Image', but the fantasy itself, like desire, varies from one to another. The results too are different. Pygmalion is fulfilled as his fantasy of his statue coming to life is realised. Milanion wins his prize as Atalanta yields willingly to the new pleasure of sexual love and a release from the impulse to compete and destroy. Laurence only finds happiness in marriage once he has atoned for his frivolity, however, while Walter leaves Venus to atone through pilgrimage yet returns to her at last.

The overarching structure of *The Earthly Paradise* submits these tests of erotic and other paradises to further scrutiny. Where happiness is attained, whether in sexual love and companionship as in these five tales and several others, or in exemplary kingship as in 'The Love of Alcestis', it is typically through magic or divine intervention. Even in the cases of Perseus and Andromeda or Bellerophon and Philonoë, where the happy union of lovers is not brought about directly by magic, it is the hero's superhuman achievements which enable him to marry his beloved. Morris keeps the prospect of paradise on earth alive in our imaginations, but we are aware that, in the poem at least, these paradisal states are dependent on magic and are only found in the world of romance. While fulfilment is achieved in many tales, there are several in which it is not, with an equivalent depth of loss in its place. In 'The Watching of the Falcon', for example, the king attains transcendent sexual gratification and later great power, but only on the precondition that he will lose them both, which he does as his realm collapses into war. As well as setting out their different results, the tales themselves are evaluated by their tellers, their auditors and the overarching narrative voice. A community of different assessments is thus built up within the poem to which readers can add their own judgements. Overall, the poem's findings suggest that it is possible to approximate to paradise on earth only by lowering one's expectations and abandoning that very quest itself, as the wanderers are forced to do. Paradise, it implies, is a state of being that may if we are lucky be realisable in a given moment, not a condition that can be achieved once and for all.

In its very doubts, Morris's poem bears traces of the changes within science since the inception of the Pre-Raphaelite project twenty years before. His practice may be Baconian, but the results of his investigations into nature are strictly materialist. In spite of the pervasiveness of magic, neither nature itself in *The Earthly Paradise* nor the tales that unfold within it reveal any divine order or purpose to life. Unlike the fairies in Swinburne's 'Child's Song', the fays in tales such as 'The Watching of the Falcon', 'Ogier the Dane' and 'The Land East of the Sun and West of the Moon', though actively enchanting, are not personifications that intimate that flowers or rivers have purposes or feelings. For their part, the pagan gods Venus, Minerva and Apollo are mythical and at times symbolic beings operating within the world he describes, not manifestations of an omnipotent god capable of determining its laws and development.

For the *New Englander*'s reviewer, this is a limitation in Morris's poetry. It is 'not seeing things deeply', for all that it is 'seeing them with surprising accuracy and clearness'.[84] But for Morley it is a sign that Morris, like Rossetti, is undeceived. Time and again in the poem Morris reminds the reader that life and happiness are transitory. These reminders typically occur at moments when one might expect to be allowed to bask a little longer in the paradises attained by his characters. At the end of 'Pygmalion and the Image', we are left with an idyllic scene as the harvest is celebrated under an August moon. But as we move into the preamble to the next tale we are told that, although 'August had not gone by', in the wind 'The fruit-hung branches moved, and suddenly / The trembling apples smote the dewless grass, / And all the year to autumn-tide did pass.' The abruptness of this transition is captured in the moment of moving from one line to the next, its irreversibility heard in the rhyme on passing. To underscore his point, Morris adds that it was the kind of day when young men 'whose eyes can see not death at all, / To thoughts of stirring deeds and pleasure fall,' recalling the same association between dropping fruit and dying that Meredith makes in 'In the Woods'.[85]

At the close of the penultimate tale, 'Bellerophon in Lycia', the teller of the tale makes change an even more immediate counterweight to happiness as he meditates on the wedding of Bellerophon and Philonoë:

And even as a man new made a god,
When first he sets his foot upon the sod
Of Paradise, and like a living flame
Joy wraps him round, he felt, as now she came,
Clear won at last, the thing of all the earth
That made his fleeting life a little worth.

My heart faints now, my lips that tell the tale
Falter to think that such a life should fail;
That use, and long days dropping one by one,
As the wan water frets away the stone,
Should change desires of men, and what they bring,
E'en while their hearts with sickening longing cling
Unto the thought that they are still the same,
When all they were is grown an empty name.

O Death-in-life, O sure pursuer, Change,
Be kind, be kind, and touch me not, till strange,
Changed too, thy face shows, when thy fellow Death
Delays no more to freeze my faltering breath![86]

This is the only time in the whole poem when Morris explicitly identifies a loving marriage with paradise. Yet even before this sentence is finished he has reminded the reader of Bellerophon's chronic depression, caused by his accidentally having killed his own brother. Letting this reminder take hold in the penultimate word of the last line of a paragraph, Morris undercuts the value of the marriage, implied in the rhyme-word 'worth', just at the moment when we might hope it would be affirmed. The implication is that the marriage will not cure Bellerophon's misery, even if his initial feeling is one of joy – a feeling that, characteristically, Morris expresses through his physical experience of intense emotion.

The last line of this first paragraph undermines the hope of marriage – or of any other earthly paradise – by reminding us that even a heroic life is 'fleeting'. But the prospect of death is, Morris suggests, at least more welcome than a long life cursed with change. For the narrator of this tale, change in human affections is as steady and inevitable as erosion. The desperate but futile attempt to 'cling' to an unchanged image of oneself is captured in internal echoes which reverberate yet slip away. Like Rossetti, Morris insists that there is no stopping time or recovering what we have lost, but where Rossetti's sonnets take their pattern from his widowhood, the close of this tale is a testimony to the state of Morris's own irreparable marriage. Because sex and the beauty of youth are so central to the promise of love throughout *The Earthly Paradise*, the prospect that desire between two lovers may fade or be diverted is profoundly distressing. Yet to cling to the pretence that this does not happen is to foster a longing that is itself 'sickening' because its goal cannot be secure. Morris's projection of the failure of desire here is deeply moving for the steadiness of its gaze, the forensic objectivity with which he faces up to the slow death of his marriage as a consequence of the implacable movement of time.

Of all Morris's reviewers, it was Pater who saw most clearly the role of time within his poetry, and by extension the promise that the poem offered readers in an age of scientific naturalism which relegated the supernatural, and with it the afterlife, to the realm of fancy. Pater wrote a retrospective on Morris as a poet for the *Westminster Review* in 1868, soon after the publication of *The Life and Death of Jason* and the first two parts of *The Earthly Paradise*. In his essay, he expressly linked 'the desire of beauty quickened by the sense of death' within Morris's poems to 'modern philosophy', which, he suggests, reveals a world comprised of material elements in perpetual motion, where all perceptions and therefore all phenomena as such are evanescent.[87] As a 'counted number of pulses only is given to us of a variegated, dramatic life', our best chance to wring the most out of that life is to experience these moments as intensely as we can. For this reason, the best that can be done to help the human spirit, Pater argues, is 'to startle it into a sharp life and eager observation'.[88] *The Earthly Paradise* at once stimulates and models this act of eager observation in its treatment of the natural world and human emotion. At the same time, it offers us new and vivid experiences in itself. In Pater's epicurean ethics, 'Not the fruit of experience but experience itself is the end.'[89] Morris's poetry is thus for Pater exemplary of an art which can deliver that intensity of experience and thereby 'a quickened, multiplied consciousness'.[90] Like the desire of beauty, this consciousness is 'quickened', brought to life, by the 'sense' or present knowledge of death.

Although Pater refers to modern philosophy in his essay, he makes it clear that it is 'science' specifically that has made us more and more aware of the forces acting on and through 'our physical life', intensifying in turn this sense of death, as we are forced to realise that all sensations are 'unstable, flickering, inconsistent' and, like our own lives, will be 'extinguished with our consciousness of them'.[91] Pater reused this passage from his essay on Morris as the 'Conclusion' to his *Studies in the History of the Renaissance*, the book Morley reviewed so favourably in the *Fortnightly* in 1873. In reprinting the concluding section of his essay on Morris, Pater gave it a new epigraph from Plato referring to the pre-Socratic Greek philosopher Heraclitus's adage that all things are always in motion.[92] He went on to quote a version of the same saying – 'Παντα ρει, all things fleet away' – in the first of his course of lectures on *Plato and Platonism*, published twenty years later in 1893.[93]

This thread of allusion leads back to modern science, and expressly to Darwinian evolution, which Pater characterises in his lecture as 'old Heracliteanism awake once more in a new world, and grown to full proportions'.[94] For Pater, Darwinism was 'the very law of change', Heraclitus's insight reified into a truth of science, which 'every month is adding to its evidence'.[95] As the science of our physical life, Darwinism compels us to accept the reality of perpetual change within and around us. At the same time, it enables us to see in this change a new vitality. In *Plato and Platonism*, Pater recognises in a typically acute insight that in Darwinism the species or '"type" itself properly *is* not but is only always *becoming*'.[96] Looking back to his essay on Morris, this new understanding of life and nature becomes the model for how to live one's own life. 'Failure is to form habits,' Pater writes, 'for habit is relative to a stereotyped world', a world to which Darwin's science has given the lie. If we can resist this stereotyping, we can achieve 'success in life', experiencing it with the continual intensity that Pater captures in his famous injunction to 'burn always with this hard gem-like flame'.[97] Morris tests out different experiences of love and beauty across *The Earthly Paradise*. As Prettejohn remarks in an incisive reading of Pater's response to Morris and its place in his aesthetic theory, for its readers the 'test now is a subjective one' as 'we must ask ourselves whether the work of art does indeed yield us "quickened, multiplied consciousness"' as Pater promises it does.[98]

The deep admiration that Pater, Morley, Colvin and other intellectuals who accepted scientific naturalism felt for Morris's poem was predicated on a realisation that art is not an escape from life but a heightened experience within it. Across his poem as a whole, Morris no more lets us forget the inevitability of change and death than Rossetti does in his sonnets or Pater does in his essays. Nor does he dangle before us any hope of compensation or relief in an afterlife. What he offers us instead is the pleasure of the poem itself, a pleasure that extends for over forty thousand lines. As Jeffrey Skoblow notes, *The Earthly Paradise* itself is 'a site for the enactment of desire, a site to *be with* rather than (as with Dickens, say) to get to, get through, get, have, want'.[99] In Pater's more economical and vivid phrase, the poem is 'water to bathe and swim in'.[100] Morris reminds us throughout the poem that the only paradises available to us are temporary states of intense feeling, fortunate interludes of intimacy and social cohesion, and forgetting ourselves altogether in the observation of nature. In showing us such moments in his poem, he invites us to experience their beauty vicariously; to foster them through the act of reading the poem aloud, as Lewes and Eliot did, and as he did himself among his own friends; and ultimately to recognise and value them in our own lives. The poem itself becomes the thing its prologue shows us cannot be found. For as long as it lasts, it is an earthly paradise in itself, as, like the tellers and listeners within the poem, we are transported for a time into a clearer and more intense world.

ROSSETTI'S RADICAL IGNORANCE: POSING THE QUESTION OF SCIENCE

Pre-Raphaelite poems such as Rossetti's 'Of Life, Love, and Death' and *The Earthly Paradise* gave the advocates of scientific naturalism an aesthetic satisfaction that was not predicated on what they saw to be false hopes and outworn truths. Yet for their own part Rossetti and Morris remained sceptical of science. Morris's doubts about science were increasingly political. In 1877, in a lecture on 'The Decorative Arts' which later became famous under the title 'The Lesser Arts', he criticised science for being ideologically bound up with capital, 'in the pay of the counting-house . . . and the drill-sergeant', and implicated in the very ugliness and utilitarianism that his art sets out to counteract.[101] Stephens had embraced science's modernity for its own sake, but Morris repudiated such an unguarded love of science as politically naive, in part because of science's failure to address the social and environmental problems of industrialisation.

Rossetti's doubts about science were more fundamental. They concerned its very claim to knowledge, which he interrogates in the last of the poems he published in the *Fortnightly Review*, entitled enigmatically 'The Cloud Confines'. Each of the poem's twelve-line stanzas juxtaposes a stern gaze at the existential uncertainties of life with a refrain which appears to assert our potential to achieve enlightenment after all:

> Still we say as we go, –
> 'Strange to think by the way,
> Whatever there is to know,
> That shall we know one day.'[102]

The jaunty banality of this refrain jars with the pitiless vision of life which runs through its verses, hinting that it may express only the 'simpleton's paradise' Morley equated with conventional wisdom. On this reading, this apparent confidence in future knowledge is Tennyson's confidence that illumination awaits us beyond the veil of death. But Rossetti's refrain also suggests the overconfidence of late Victorian science itself. In the final stanza, he rewrites this refrain as a series of rhetorical questions which undermines any such glib self-assurance:

> What word's to say as we go?
> What thought's to think by the way?
> What truth may there be to know,
> And shall we know it one day?[103]

Depending on how these lines are spoken, either the emphatic openings of each line of the refrain are set back behind unstressed, uncertain questionings ('What *word's* to *say* as we *go*?'), or the stress falls on the questioning itself ('*What* word's to *say* as we *go*?'). Either way, confidence is replaced with uncertainty. Where the rhetorical question that closes 'Newborn Death' presumes a particular answer – that Death is indeed the last child of Life for whom all the others have to die – these questions open only onto a radical ignorance which seems to call into doubt any hope of productive enquiry or investigation.

Rossetti's final double work of art, *The Question* (fig. 105, p. 182), encapsulates his intuition that the pursuit of knowledge or understanding is ultimately futile. Rossetti completed the design in pencil in 1875, describing it to Brown as a visual equivalent of 'The Cloud Confines'.[104] Although he never transferred it to canvas, on his deathbed he dictated a pair of sonnets interpreting the drawing to Hall Caine:

> This sea, deep furrowed as the face of Time,
> Mirrors the ghost of the removed moon;
> The peaks stand bristling round the waste lagune,
> While up the difficult summit steeply climb
> Youth, Manhood, Age, one triple labouring mime;
> And to the measure of some mystic rune
> Hark how the restless waters importune
> These echoing steeps with chime and counterchime.
> What seek they? Lo, upreared against the rock
> The Sphinx Time's visible silence, frontleted
> With Psyche wings, with eagle plumes arched o'er.
> Ah, when those everlasting lips unlock
> And the old riddle of the world is read,
> What shall man find? or seeks he Evermore?
>
> Lo, the three seekers! Youth has sprung the first
> To question the Unknown: but see! he sinks
> Prone to the earth, – becomes himself a sphinx, –
> A riddle of early death no love may burst.
> Sorely anhungered, heavily athirst
> For knowledge, Manhood next to reach the Truth
> Peers in those eyes; till, haggard and uncouth
> Weak Eld renews that question long rehearsed.
> Oh! and what answer? From the sad sea brim

The eyes of o' the Sphinx stare through the
midnight spell,
Unwavering, – man's eternal quest to quell:
While round the rock-steps of her throne doth swim
Through the wind-serried wave the moon's faint rim,
Sole answer from the heaven invisible.[105]

These paired sonnets and 'The Cloud Confines' can each be seen as the pendant to Rossetti's design, making for not just a double but a triple work of art.

In both its forms, *The Question* depicts three men approaching the Theban Sphinx by moonlight. As Robert Upstone has pointed out, Rossetti reverses the traditional legend, so that rather than the Sphinx posing them a riddle, each of them in turn asks her to explain what Rossetti calls 'the old riddle of the world'.[106] In the sonnets, the Sphinx is equated with 'the Unknown' and 'Time's visible silence'. Rossetti presents time itself as a series of enigmas in 'The Cloud Confines', declaring 'Our past is clean forgot, / Our present is and is not, / Our future's a sealed seedplot.'[107] This same structure is repeated in *The Question*, with the pilgrimage of Youth, Manhood and Age or Eld representing the past – the youth dies as we watch – the present and the future. In the earlier poem, Rossetti accuses Time of having a 'heart of hate' and 'the pitiless eyes of Fate'.[108] This hatefulness is represented in *The Question* not only in the youth's death – another 'riddle', according to the poem – but in the Sphinx's own blank eyes, which 'stare through the midnight spell, / Unwavering, – man's eternal quest to quell'. The stoniness of her gaze is enhanced in the design by Rossetti's monochrome pencil, recalling the Sphinx's Egyptian as well as her Greek antecedent.

The Sphinx's particular monstrosity in Rossetti's design hints at the animal origins of humanity discerned by Darwin and others behind the opacity of Time. Her body is partly obscured by a branch, but it appears to blend human and animal features rather than to switch abruptly from one to another. Notably her arms begin with human shoulders but end in a lion's tendons and claws. Her posture is also upright even though she squats on feline haunches. According to the sonnets, her wings are 'Psyche's wings, with eagle plumes arched o'er'. As the Sphinx's body represents the animal within the human, so her wings symbolically associate the soul and the mind – Psyche and the psyche – with violence and predation. In 'The Cloud Confines', the 'Red strife from the furthest prime' of the natural world – Tennyson's 'Nature red in tooth and claw' – and 'War that shatters her slain' are both attributes of Time.[109] In her lion's arms and eagle's wings, the Sphinx similarly brings together the violence of the natural world with the violence inherent in regal and imperial power. Sex becomes associated with both violence and enigma, both in her 'everlasting lips', on which Manhood seems about to plant an aggressive kiss, but which show no signs of opening either to receive his kiss or to answer his question, and in her naked human breasts which catch the moonlight and mirror the eroticised nudity of the dying youth.

In his depiction of the Sphinx in *The Question*, Rossetti implies that the truths that lie out of reach, forever lodged in the past or the future, may bear a resemblance to the claims of modern Darwinian science. But it is only a resemblance, not a clear likeness, and no further investigation will ever be able to confirm the actual truth. The 'Sole answer' offered by Rossetti in the poem which accompanies the design comes in the reflection of the moon in its closing lines. But as in Rossetti's earlier series of sonnets, there is no sign here of anything beyond material phenomena. Heaven is literally 'invisible' in the design, where the artist's vantage point obscures any view of a horizon, let alone the sky. In the sonnets, this literal invisibility takes on a symbolic significance, suggesting that any divine presence is likewise imperceptible. Instead, all we can see is an astronomical body – the moon – glimpsed through an optical process – reflection. But if we look back to the design it is clear that science is not going to tell us anything meaningful about this phenomenon either. In the design the moon's faint rim is implausibly clear and crisp, suggesting the kind of definite outline that we find in early Pre-Raphaelite art, but transposed onto a rippled waterscape which in reality would blur it. Yet within that outline there is no detail other than the ripples themselves.

The moon is, in effect, empty, while the water too resists our attempts to peer beneath it. As with the Sphinx, whose body hints at what her mouth will not say, the only answers they can offer us come not as knowledge but rather as a suggestive symbolism which in the final reckoning appears to deny even the possibility of knowledge. Rossetti answers his own question not through science but through an art which refuses to defer to it. Yet not even art can free us from the perpetual cycle of questioning. Indeed, Rossetti himself was still framing the question in the act of dying. So while *The Question* seems to repudiate his brother's insistence from thirty years before that the Pre-Raphaelite artist should investigate the world, it is in its own right a profoundly sceptical investigation of the nature of knowledge which sets down an unflinching avowal of its results. As such, it is at once a repudiation and a fulfilment of the original Pre-Raphaelite project. At the same time, it confirms Morley's dialectic, showing how aestheticism, born of science, could serve to remind it of its limitations and to counter the hubris of its more totalising ambitions.

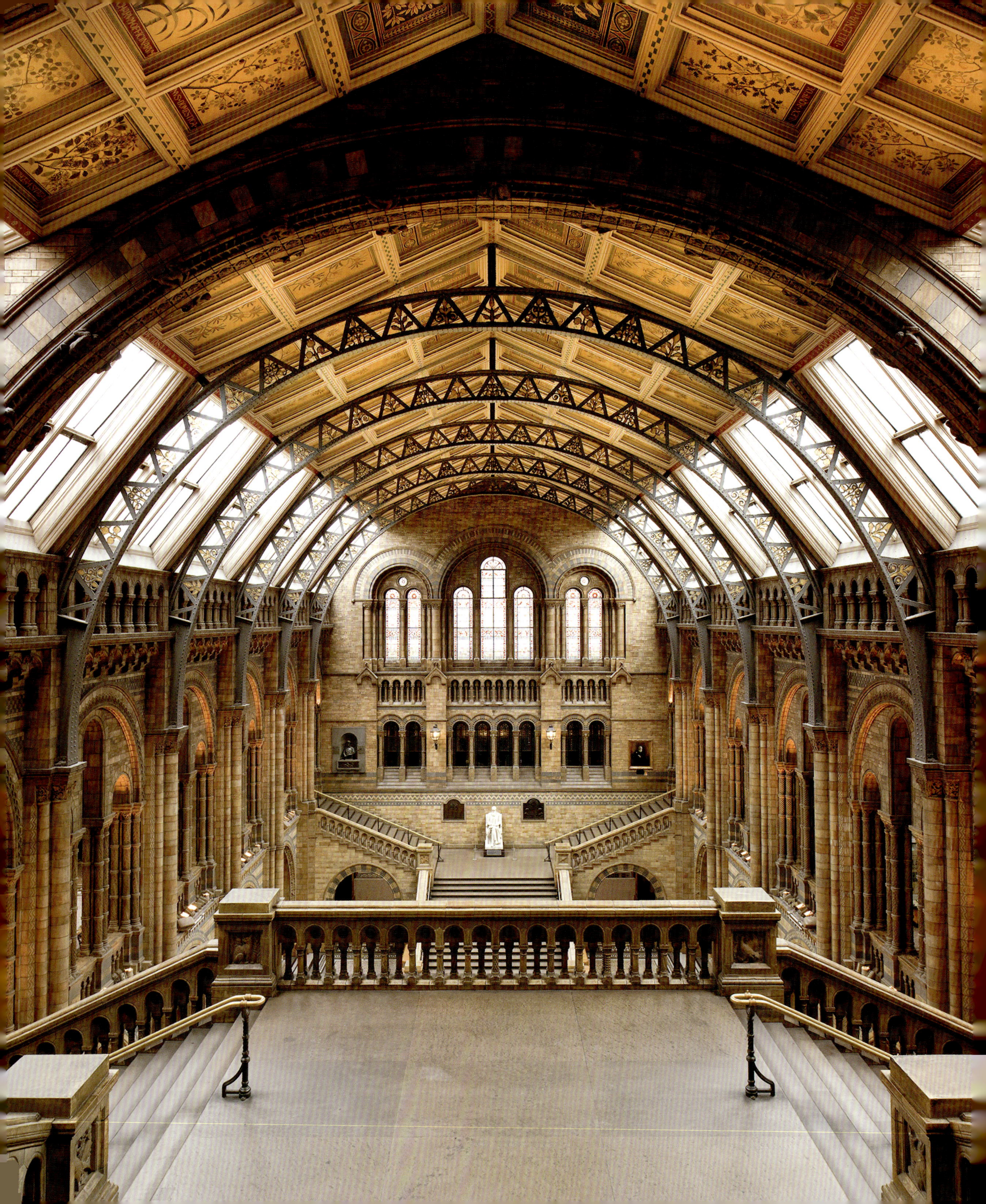

8 · Pre-Raphaelite Architecture and Scientific Controversy after Darwin

THE NATURAL HISTORY MUSEUM VERSUS *NATURE*

The publication of Rossetti, Swinburne and their circle in the *Fortnightly Review* and the popularity of *The Earthly Paradise* among 'scientific men' reveal the role that later Pre-Raphaelite poetry played within the wider discourse of scientific naturalism. For free-thinkers like Morley and Pater, Pre-Raphaelite poetry, like scientific naturalism, set aside the false consolations of the supernatural, including divine benevolence and the promise of an afterlife, offering instead an undeceived encounter with the facts of the case. We have only fleeting glimpses of actual scientists reading Pre-Raphaelite poetry. Huxley drew a comic analogy between himself and Morris's 'The Man Born to be King' in conversation, while his friend the brilliant and outspoken young mathematician William Kingdon Clifford quoted Swinburne's radical poetry in essays published in the *Fortnightly* and the *Nineteenth Century*. The anthropologist M. J. Walhouse cited Swinburne's classical tragedy *Erechtheus* in a paper on child sacrifice.[1] Other scientists continued to follow the visual precepts of Pre-Raphaelitism, adapting them to their own ends as Acland and Phillips had done so brilliantly at Oxford. In this chapter I want to turn to another of the Pre-Raphaelite arts – architecture – to examine its place within the scientific culture of later Victorian England. Tupper's *Hiatus* and the articles in *Nature* by Brett and Lockyer show that the original Pre-Raphaelite principles in art could be squared equally well with natural theology and scientific naturalism. The conflict between these two scientific methods and world-views was played out in the design and reception of a second natural history museum, on a far grander scale than the one in Oxford. As at Oxford, the aesthetic principles of Pre-Raphaelitism would be instrumental to the design of the Natural History Museum, built in South Kensington in the 1870s

Facing page Detail of fig. 112.

to house the natural history collections of the British Museum. They would shape its reception too, as the museum was judged by the same Pre-Raphaelite standards in the pages of *Nature*. Where the Natural History Museum was a belated reaffirmation of natural theology, *Nature* upheld its own commitment to scientific naturalism. Both laid claim to the Pre-Raphaelite principle of truth to nature, but their truths themselves were, in the end, incompatible.

A SECOND PRE-RAPHAELITE MUSEUM

The Natural History Museum took even longer to get off the ground than its Oxford predecessor. Richard Owen (see fig. 133) first floated the idea of a national natural history museum in the late 1840s. In 1856 he moved from the Hunterian Museum to become Superintendent of the natural history collections at the British Museum. Soon after, he began a steady and determined campaign to move these collections from their cramped conditions in Bloomsbury to a new site. In 1863 Parliament agreed to purchase a plot of land for the new museum in South Kensington on what was then the outskirts of London. In 1864 the competition to build it was won by Francis Fowke, the architect of the South Kensington Museum, now part of the Victoria and Albert. Fowke died in 1865. The following year a young architect from Manchester, Alfred Waterhouse, was asked to execute his plans. After a further seven years of vacillation, including a hiatus when the site was provisionally moved to the Thames embankment, building work finally started to a new design by Waterhouse. The building itself took eight more years. The Natural History Museum, as it was immediately if unofficially known, was finally opened to the public in 1881, though even then many exhibits were still to be transferred from the main British Museum itself.[2]

The expression of the Natural History Museum's subject through its plans, forms, materials and decoration is very much in line with the Pre-Raphaelite architectural principles pioneered by Street and Woodward in the 1850s. Like Acland and Lockyer, Owen was closely connected to the Pre-Raphaelites. Although he was an ambitious, autocratic and at times divisive figure within science, Owen was well liked by writers and artists. His friends included Dickens and Tennyson, and when he died the President of the Royal Academy, Sir Frederic Leighton, contributed a generous five guineas to the fund for a memorial statue.[3] Broderip, the naturalist who bought *The Hireling Shepherd*, introduced Hunt to Owen in 1852, explaining to him, in Hunt's self-satisfied words, 'that the great Professor had been one of my stoutest champions throughout'.[4] It was presumably through Hunt that Owen became an honorary member of the Hogarth Club and befriended several other Pre-Raphaelites.[5] The surviving details of these friendships are remarkably intimate. Millais's son John Guille Millais describes Owen as 'a friend of the whole family', recalling that he used to take him and his brothers to the British Museum to show them round the natural history collections.[6] This experience was clearly a formative one, as the younger Millais went on to become an accomplished natural history artist and author, and a specimen collector for the British Museum.[7] Owen had an influence on Tupper's career too. He may well have written one of the testimonials that got Tupper his post at Rugby School – Woolner certainly suggested that Tupper ask him – and Owen helped him with enquiries over scientific literature.[8] To the Millais family, he was 'dear old Owen'.[9] To Tupper, writing to Stephens, he was 'the dear, big, blue-eyed-boy-like Owen'.[10] The affection between the Pre-Raphaelites and Owen was clearly genuine, as were his respect for their art and theirs for his science.

Waterhouse had his own ties to Pre-Raphaelitism. After visiting the Oxford museum, he employed Woolner and the O'Sheas to carve the statues and decorations at the Manchester Assize Courts.[11] This building made his name and earned him the contract to build the Natural History Museum. Later he would commission Brown to paint twelve murals for Manchester Town Hall, his second great civic building in that city, built concurrently with the Natural History Museum in the 1870s. Following Street, with whom he often found himself in competition, Waterhouse became one of the leading designers of modern Gothic architecture. But where Street's attachment to Gothic was ideological as much as practical, Waterhouse was altogether more flexible, adapting his forms freely to suit the requirements of different clients and buildings. This flexibility, combined with personal charm, imagination and a flair that marks out all his diverse buildings as his own, makes him, alongside Gilbert Scott, one of the most commercially successful architects of the Victorian age. The sheer range of Waterhouse's well over 400 commissions – from middle-class homes to aristocratic palaces, from churches, university colleges and museums to town halls and the National Liberal Club – makes him hard to pigeonhole even as a Gothic architect. Similarly, although he collaborated on occasion with Pre-Raphaelite artists, unlike Woodward he did not work closely with them and unlike Street he did not identify himself as a Pre-Raphaelite by another name. Nevertheless, the Natural History Museum fulfils Street's ideal of Pre-Raphaelite principles in architecture, with representational art integrated into a building which is itself representational in its structure and which aspires to be true to its subject.

Where the Oxford University Museum is a superlative example of collaboration between several scientists and artists, guided but not controlled by the architect, the design of the Natural History Museum is virtually entirely the work of just two men. At Oxford, the O'Sheas carved their own designs. The role of the lead sculptor at South Kensington, M. Dujardin, was to realise in terracotta the designs supplied to him by Waterhouse. That he did this with great skill, but also with attentive fidelity, is clear from the two large portfolios of Waterhouse's drawings held at the museum.[12] Several more portfolios of drawings for the building itself survive in the Royal Institute of British Architects Library, across Exhibition Road from the museum at the Victoria and Albert. These reveal Waterhouse's remarkable attention to every detail of the building, down to the door-handles.[13] This meticulous and direct involvement by the architect in every aspect of the building gives the Natural History Museum a uniformity of style which Woodward's Ruskinian practice at the Oxford Museum expressly rejected. Its scientific vision is also self-consistent, with Owen in effect in sole charge, where Acland, Phillips and other Oxford scientists – not to mention the artists themselves, studying animals and plants on Pre-Raphaelite lines – had all made their own contributions to the scientific interpretation of nature through art at Oxford.

The collaboration between Owen and Waterhouse was close and harmonious. Looking at the different features of the museum as it was built, it is possible to trace who was responsible for which aspects of the design in broad terms. Where the Oxford museum was designed to be accretive, Owen's earliest sketch (fig. 106) for a national natural history museum and the plan he worked out with the surveyor Henry Hunt, a friend, ahead of the competition to design it were strictly symmetrical.[14] The sketch includes a central hall displaying British natural history and typical specimens of each main group. These became the themes of the two halls in the finished museum. From the plan, it is clear that the only public staircase in the building was to lead up from the central hall, as it does, or did until further external staircases were added more recently. The broad layout of the galleries, stretching back from the two wings of the museum, was in Owen's early drawings too.

Although the early plans do not expressly require it, the new museum retained the pre-existing practice within the British Museum of housing the palaeontological and zoological collections separately. This distinction was mirrored in the decorations, with the east wing decorated with extinct creatures and the west wing with extant ones. The suggestion that the museum should be ornamented with animals and plants was again Owen's.[15] The *Building News* reported in January 1878 that the extinct animals were 'modelled from actual fossils in the British Museum'.[16] While this is possible, it is certain that in November 1873, as building work on the museum was getting underway, Owen supplied Waterhouse with more than fifty illustrations of extinct species on which to base his designs.[17] A surviving letter from October 1874 reveals that, once the designs had been drawn and modelled, Waterhouse again consulted Owen for approval before they were cast in terracotta.[18] Carla Yanni has shown that the palaeotherium and the pterodactyl enclosed in roundels on the gables of the main façade, for example, are derived respectively from Owen's *History of British Fossil Mammals and Birds* and an educational wall poster by Benjamin Waterhouse Hawkins, who had worked with Owen on the life-sized dinosaurs in the park at the Crystal Palace in South London.[19] Owen also seems to have exercised his influence over the choice of the extinct animals. Many were creatures he had worked on. Although he is best remembered now as the man who grouped the dinosaurs together under one

106 Richard Owen, 'Idea of a Museum of Natural History', 1859. Natural History Museum, London.

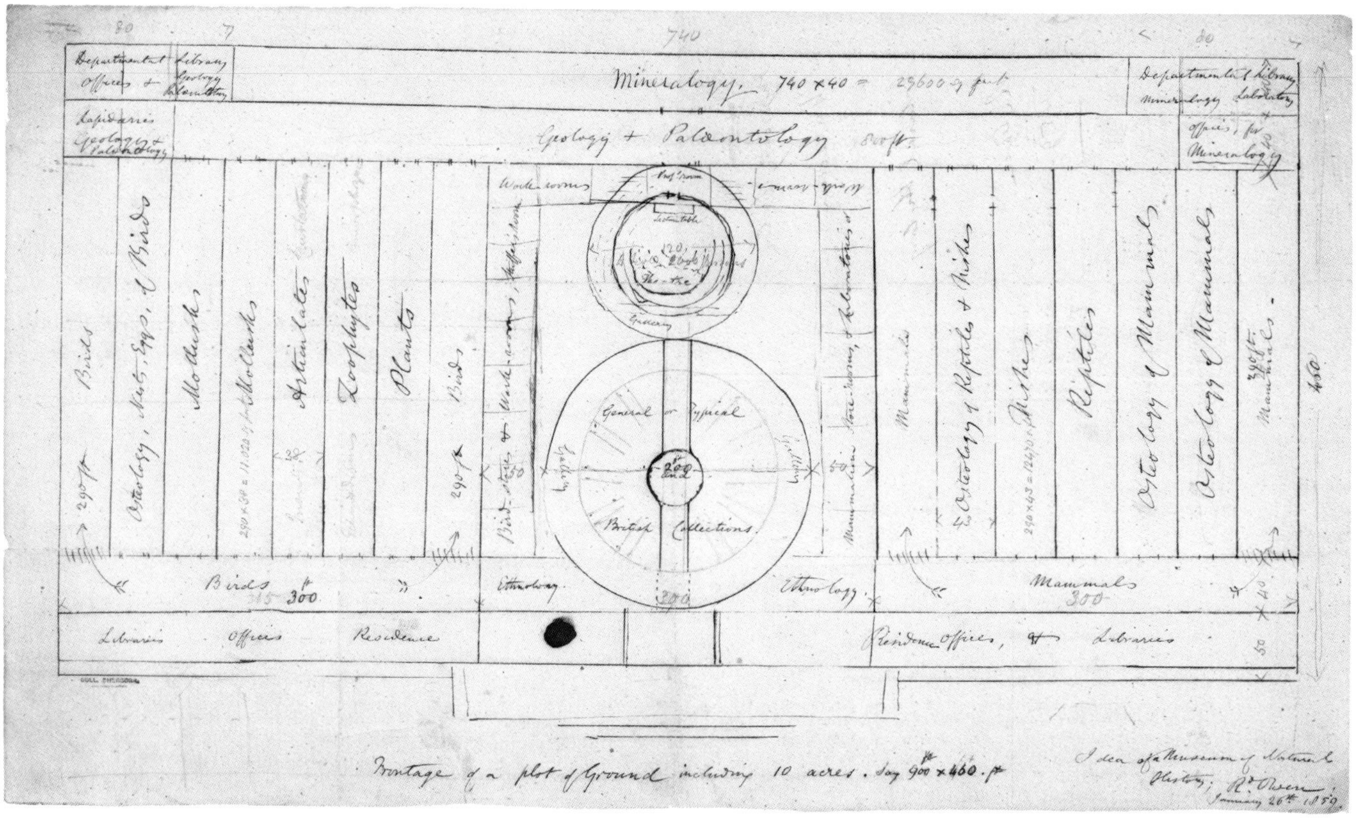

name, no dinosaurs are included in the decorations on the museum, most likely for reasons of scale. But Owen was also an expert on prehistoric mammals, and the three animals monumentalised as colossal statues on the parapet of the east wing – palaeotherium, the ground sloth mylodon and the scimitar-toothed lion *Machairodus* – were all animals he himself had described.[20] Though it was only recently extinct, the dodo took pride of place in a half roundel in the interior of the east gallery. This too was one of Owen's animals, as he had published a *Memoir on the Dodo* in 1866, incorporating several earlier accounts by his friend Broderip, who had died a few years before.[21]

Colin Cunningham speculates that the relief sculptures of fish on the piers in the main south galleries were 'directly copied from a book', as they are conventionally drawn, with no attempt to show turning or motion.[22] A glance across a number of palaeontology textbooks current in the 1870s confirms this suspicion. The collection of Waterhouse's drawings at the Natural History Museum includes two sheets of extinct fish designed for the piers of the south gallery in the east wing (fig. 107).[23] Several genera, including *Dipterus*, *Diplacanthus* and *Smerdis*, are copied directly from illustrations in Owen's own book *Palaeontology*, first published in 1860. The dorsal fin from a portrait line drawing of a fossil of *Semiophorus* in the same book is bent back to fit Waterhouse's landscape orientation, while to give his design more animation *Platysomus gibbosus* is shown in the act of swallowing a smaller fish.[24] Waterhouse's *Holoptychius*, an extinct relative of the coelacanth, bears a clear resemblance to the picture of the same fish incorporated into the fifth edition of A. C. Ramsay's *Physical Geology and Geography of Great Britain*, published in 1878, but as illustrations for textbooks were borrowed and recycled without compunction by Victorian publishers, it is equally likely that he had seen the same drawing or a similar one in another book (figs 108 and 109).[25] One illustration it seems certain that he

107 Alfred Waterhouse, designs for extinct fish, *c.*1877–8. Natural History Museum, London.

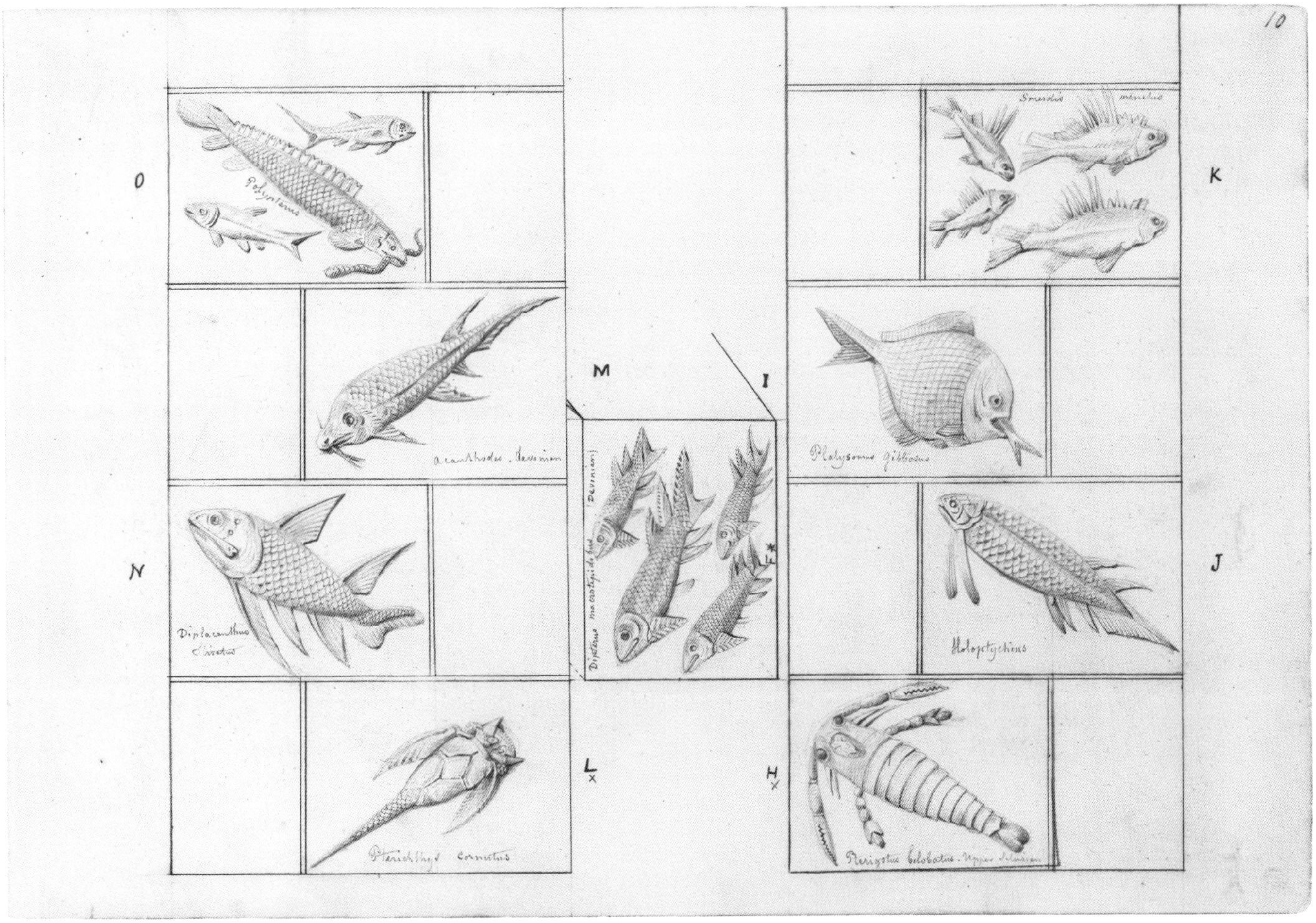

must have used directly is to H. Alleyne Nicholson's *Ancient Life-History of the Earth* of 1877. It shows two of the fish included in Waterhouse's designs: *Osteolepsis*, described as 'a Devonian Ganoid', and *Polypterus*, 'a recent Ganoid fish' (fig. 110). In the text Nicholson explains that the Polypteri are living fish allied to the earlier Ganoids found today in the Nile and Senegal rivers.[26] The fact that Waterhouse includes a *Polypterus* among his extinct fish strongly suggests that he used this illustration (which brings them together) without reading the text which should by rights have set them apart again. Ramsay's book was presumably Waterhouse's direct source in this case, as, although his fish designs are undated, he was still working on designs for the capitals for the same gallery in 1878.[27]

It is harder to trace Waterhouse's sources for his decorations for the west wing, modelled on living animals, as far more illustrations would have been available to him and he does not seem to have asked Owen to point him to particular sources. In some cases he may have worked directly from specimens, in others from illustrations.[28] But where the decorations at the earlier museum had typically been carved from living specimens on strict Pre-Raphaelite principles, Waterhouse's designs reflect a more attenuated Pre-Raphaelitism, retaining the commitment to scientific accuracy and the right to artistic interpretation, but jettisoning the requirement that the artist study the natural object for himself. His use of mechanical reproduction in the creation of the terracotta sculptures likewise breaks faith with Ruskin's

108 *Holoptychius nobilisimus*, from A. C. Ramsay, *The Physical Geology and Geography of Great Britain*, 5th edn, 1878.

109 *Platysomus gibbosus*, from Richard Owen, *Palaeontology*, 2nd edn, 1861.

110 *Polypterus* and *Osteolepsis*, from H. Alleyne Nicholson, *The Ancient Life-History of the Earth*, 1877.

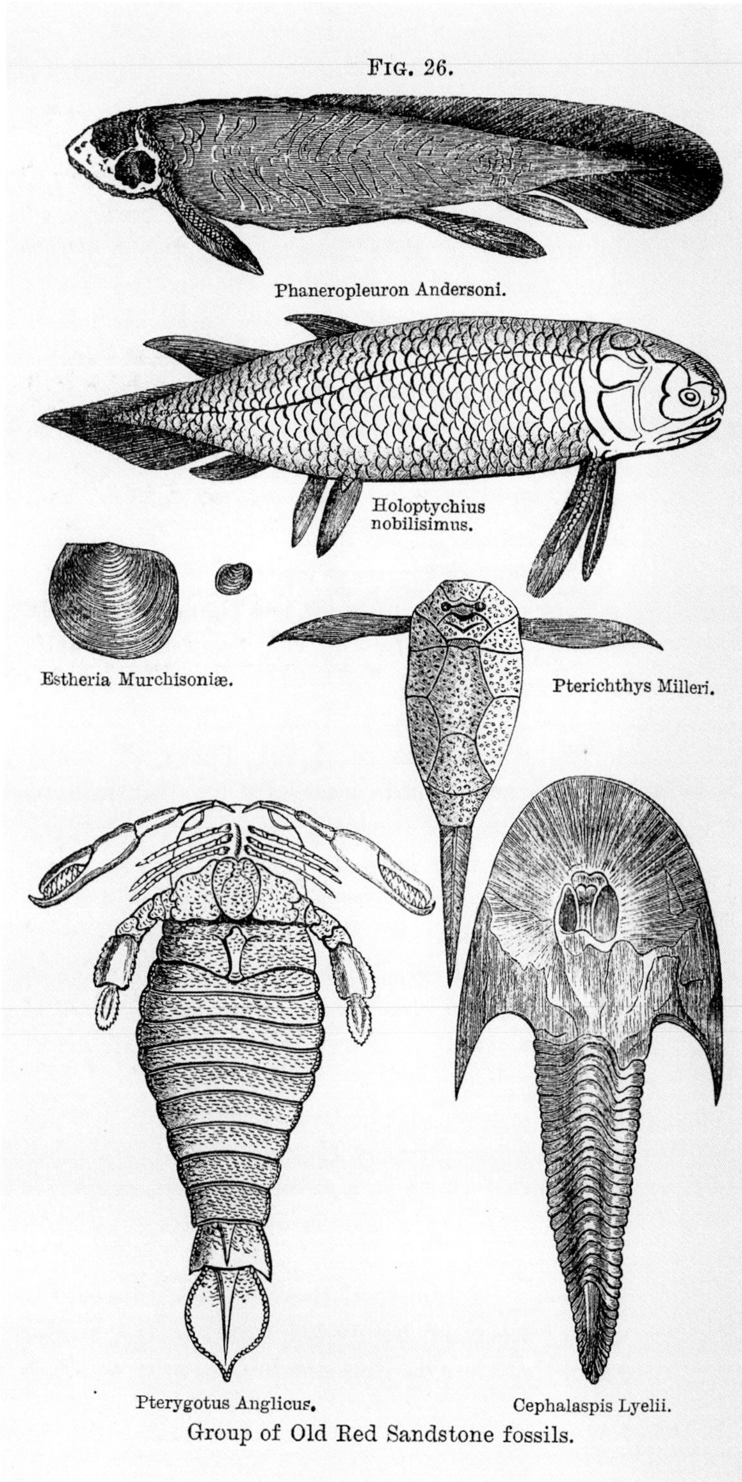

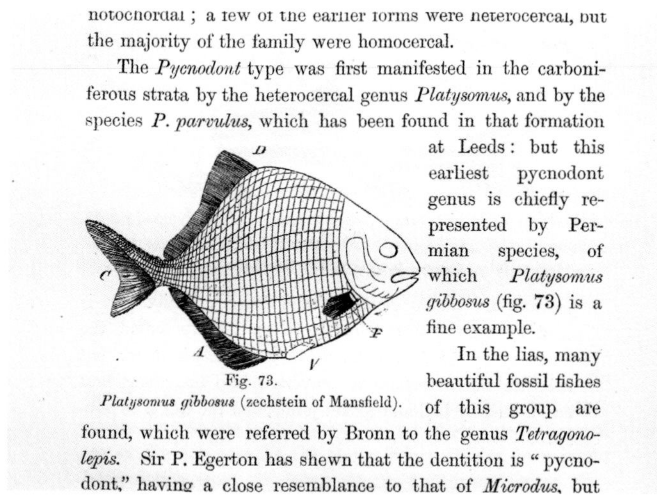
notochordal; a few of the earlier forms were heterocercal, but the majority of the family were homocercal.

The *Pycnodont* type was first manifested in the carboniferous strata by the heterocercal genus *Platysomus*, and by the species *P. parvulus*, which has been found in that formation at Leeds: but this earliest pycnodont genus is chiefly represented by Permian species, of which *Platysomus gibbosus* (fig. **73**) is a fine example.

In the lias, many beautiful fossil fishes of this group are found, which were referred by Bronn to the genus *Tetragonolepis*. Sir P. Egerton has shewn that the dentition is "pycnodont," having a close resemblance to that of *Microdus*, but

Fig. 73. *Platysomus gibbosus* (zechstein of Mansfield).

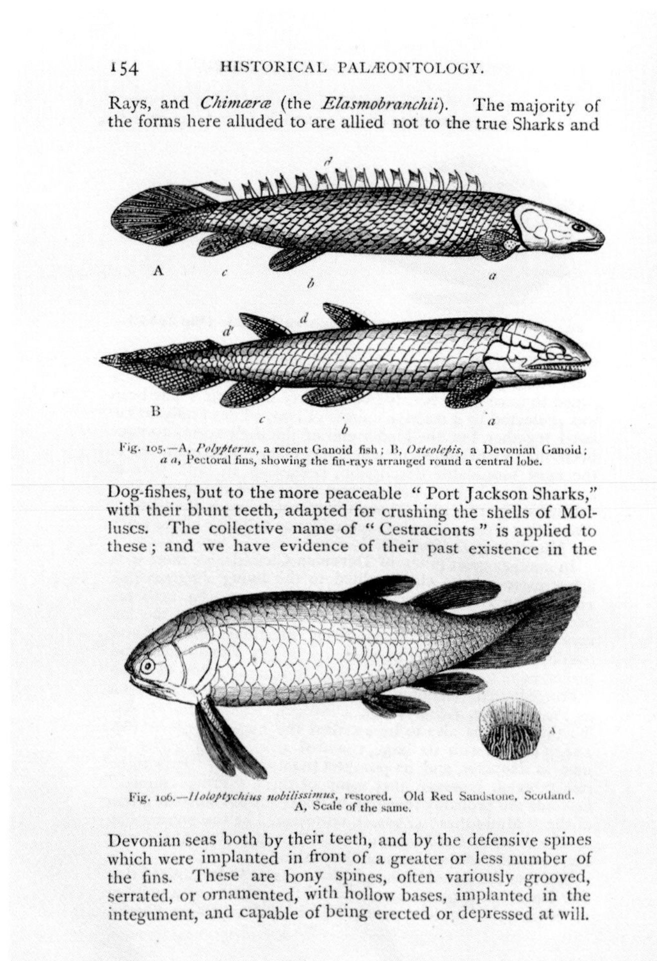
154 HISTORICAL PALÆONTOLOGY.

Rays, and *Chimæræ* (the *Elasmobranchii*). The majority of the forms here alluded to are allied not to the true Sharks and Dog-fishes, but to the more peaceable "Port Jackson Sharks," with their blunt teeth, adapted for crushing the shells of Molluscs. The collective name of "Cestracionts" is applied to these; and we have evidence of their past existence in the Devonian seas both by their teeth, and by the defensive spines which were implanted in front of a greater or less number of the fins. These are bony spines, often variously grooved, serrated, or ornamented, with hollow bases, implanted in the integument, and capable of being erected or depressed at will.

Fig. 105.—A, *Polypterus*, a recent Ganoid fish; B, *Osteolepis*, a Devonian Ganoid; *a a*, Pectoral fins, showing the fin-rays arranged round a central lobe.

Fig. 106.—*Holoptychius nobilissimus*, restored. Old Red Sandstone, Scotland. A, Scale of the same.

insistence on carving by hand according to the carver's own designs, while the aesthetic of his designs is markedly closer to the stylisation of nature in the tapestries and wallpapers of William Morris – another of his occasional contractors, like Woolner and the O'Sheas – than to the minute observation of early Pre-Raphaelite painting. On the other hand, he was careful to be as scientifically accurate as he could, using the best scientific sources for his illustrations and confirming with Owen that he had got them right. The overall effect was a building which, with more determination and coherence than the Oxford museum if less complexity or fine detail, put in stone a comparable vision of the natural world.

ARCHITECTURE AND THE ARCHETYPE: NATURAL THEOLOGY AT THE NATURAL HISTORY MUSEUM

Owen had been a mentor to both Greswell and Acland, and Waterhouse's building gives spatial and visual form to an idea of nature closely related to their notions of God's own museum and of divine artistry.[29] Yet the forms, plans, materials and decoration of the Natural History Museum all tend towards a subtly but consistently different inflexion of this idea from those of the Oxford University Museum. Owen originated and to some degree oversaw the decorative scheme for the Natural History Museum, but he paid tribute to the 'supreme mastery' with which Waterhouse himself had selected and combined its different elements.[30] He also acknowledged that it was Waterhouse who proposed that Romanesque, the round-arched precursor to Gothic in medieval architecture, be adopted as the overall style for the building. Owen had originally opposed the adoption of Gothic in Oxford on the same associational grounds as Fergusson, remarking in a letter to Acland: 'The sciences were not born or nursed where that style originated.'[31] But Waterhouse made a strong case for Romanesque, telling the First Commissioner of Works, Lord John Manners, in 1868 that it would 'afford both the grandeur and simplicity which should characterise a building of this description' and convincing Owen that it would be the best style for the decorative scheme he had in mind.[32] Fowke had originally proposed a French Renaissance design, which he had intended should combine 'a certain amount of Gothic freedom with the classical grandeur of proportion and unity which are so essential in public buildings of such magnitude and importance'.[33] In proposing Romanesque, Waterhouse moved the building a step closer to Gothic while preserving the rounded arches of the original design.

Waterhouse flanked the main entrance to the museum with two high towers reminiscent of the Romanesque cathedrals of Germany. In the process he intensified the ecclesiastical resonances of the Oxford museum which, as Mark Girouard argues, had already been introduced into the new museum in Fowke's design through allusions to the alternative cathedral architecture of Renaissance Rome.[34] The façade (fig. 111) resembles a Christian 'Temple of Nature', to quote *The Times*'s celebratory coverage of the opening of the new museum.[35] Its grandeur is married to the simplicity, proportion and unity promised by Fowke and Waterhouse, in contrast with the intricacies and idiosyncrasies of the Oxford University Museum. The prevailing impression of nature that is conveyed by the façade of the Natural History Museum is of a divinely ordained order. The two central towers, the two wings, the two pavilions all balance one another, matching the symmetry of Owen's initial plans. The rounded arches balance one another as well, while each individually alludes to the perfect form of a circle. Comparing the terracotta shells with which Waterhouse decorates his museum – often arranged into roundels on the rounded arches – with the illustrations of bivalves that run throughout Ramsay's and Nicholson's books, these too have been given an artificial symmetry. As many of the arches are nestled within one another in perfect proportion, so they suggest an orderly, hierarchical pattern within the natural world.

The impression that nature has been planned in perfect proportion by God is reinforced by the spaces and the plan of the interior. Even before it had been built, Waterhouse himself likened its Central Hall (fig. 112) to 'the Nave of a Cathedral'.[36] This was exactly how it struck the reviewer for *Building News* when he visited the building in 1878. The reviewer remarked: 'We might fancy ourselves in the nave of a Romanesque church of great width.'[37] Another reviewer, this time for *Nature*, made the same remark three years later,

111 Alfred Waterhouse, Natural History Museum, London, 1873–81, photographed in 1902.

(facing page) 112 Alfred Waterhouse, Central Hall (now Hintze Hall), Natural History Museum, London, 1873–81.

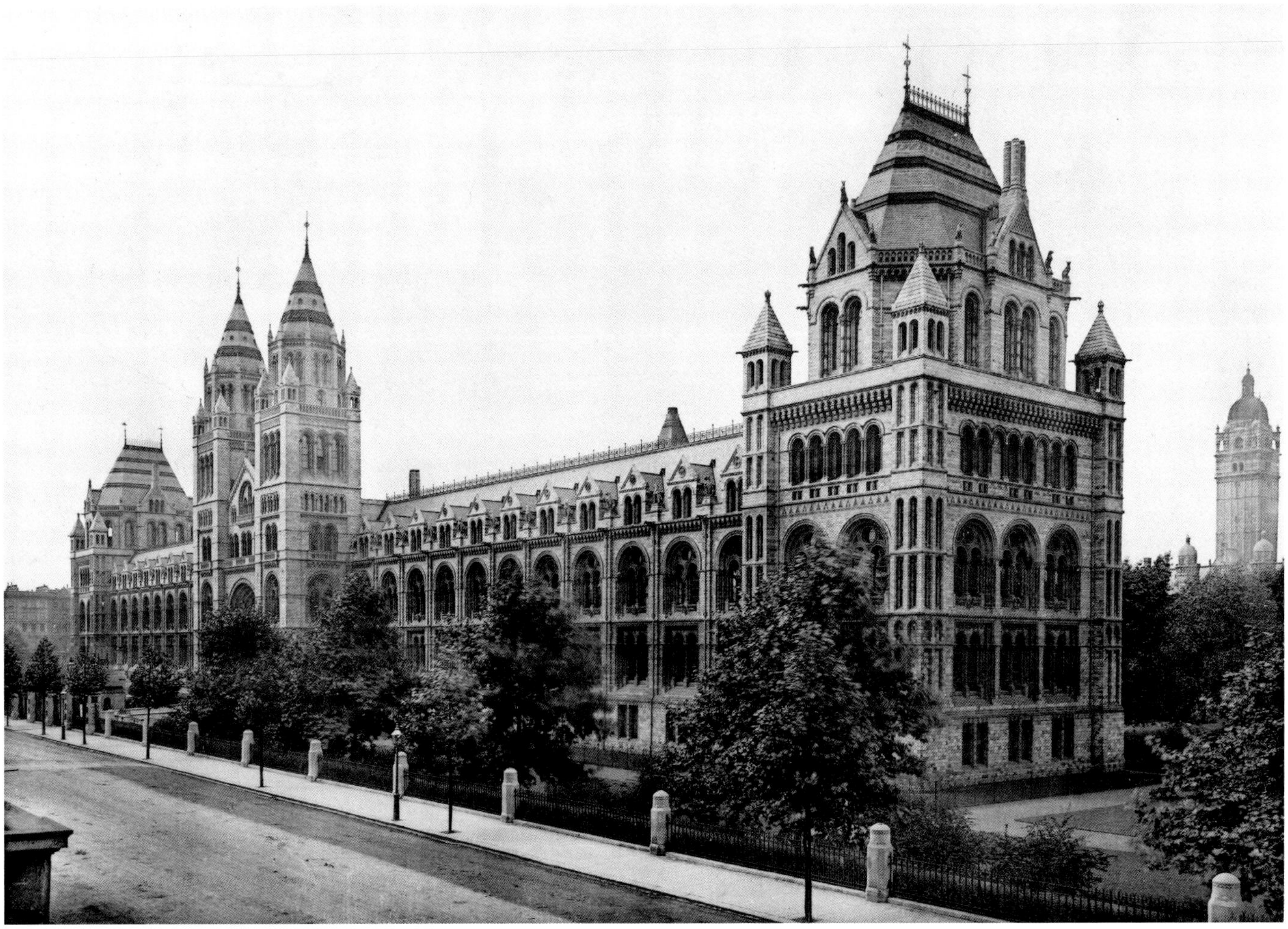

commenting that it 'has the general appearance of a cathedral'.[38] If the Central Hall itself is the nave, the arches along its sides lead into individual chapels, each originally dedicated to one particular group of organisms within what Owen conceived of as his Index Museum – an epitome of all life on earth which visitors could then follow up over a wider range of specimens in the different galleries.

As at Oxford, these strong ecclesiastical echoes in the architecture are combined with other associations. As a national building, the Natural History Museum is an emblem of the importance of science and nature to British culture and history. This symbolism is expressed primarily through the ceiling of the Central Hall. Like its predecessor, albeit with less audacity, the partly exposed roof integrates the 'railway materials' iron and glass into the structure of the hall, giving it natural light and tensile strength. This reminder of Britain's industrial modernity is matched with paintings of plants, primarily crops, from across the globe, suggesting the economic reach of the British Empire. Like the terracotta decorations, these were designed as cartoons by Waterhouse himself, and then executed by an employed artist, the painter Charles James Lea. Where Owen advised on the designs for the terracottas, the consensus is that it was probably the Keeper of Botany, William Carruthers, who advised on these ceiling paintings, although there is no direct evidence to demonstrate this.[39]

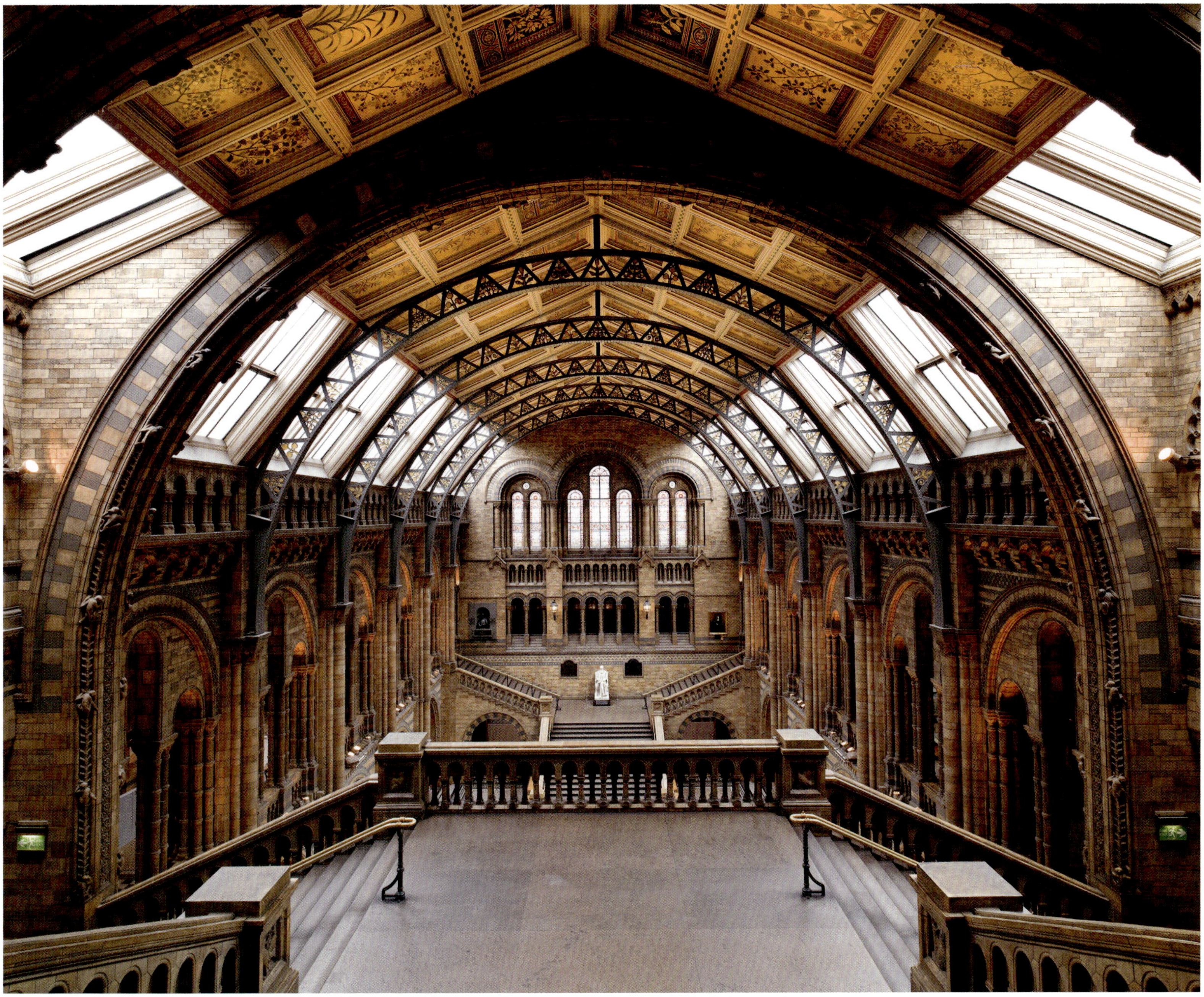

The scale, ecclesiastical design, iron and glasswork, and intricate ceiling decorations all combine to make the Central Hall an imposing and engaging space. But Girouard is surely right to remark: 'What turns the Central Hall from an impressive space into an extraordinary one is the staircase which has been inserted into the cathedral nave.'[40] The broad staircase leads up from the far end of the hall to a platform before splitting to either side to carry you up to the colonnaded balconies. You then have to walk back along the length of the hall to reach the first-floor galleries along the front of the building. But even as you reach them a magnificent vaulted staircase (fig. 113) invites you to climb up to a balcony with a view commanding the whole of the hall. From there you climb up a further flight over another vault to what were the second-floor galleries and libraries, though these are not currently open to the public. Like Woodward's Gothic

113 Central Hall of the Natural History Museum, London, from *A General Guide to the British Museum (Natural History)*, 1891.

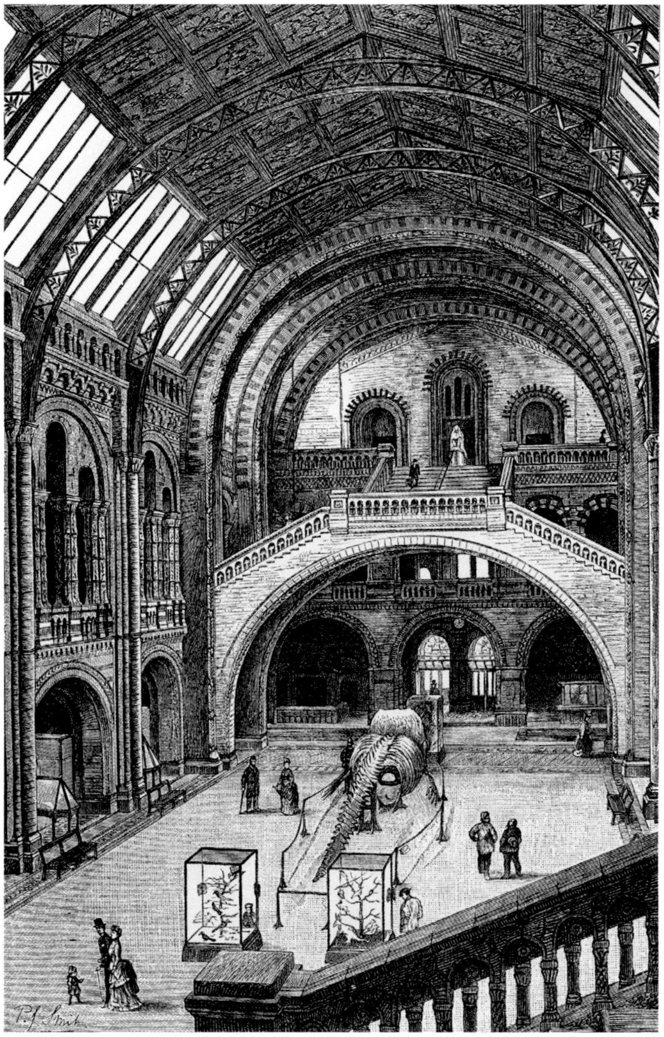

iron and glass ceiling at Oxford, Waterhouse's staircase proclaims his architectural brilliance, transforming a space that is heavily allusive into one that is also original and unique. Strikingly, Waterhouse's transformation of his model of the Romanesque nave would find its way back into ecclesiastical architecture itself in the vaulted staircase that cuts across the nave of the Anglican cathedral in Liverpool. This was to be the last great building of the Gothic Revival in England, designed by Gilbert Scott's grandson Giles Gilbert Scott, begun in 1904 and finally completed in 1978.

Waterhouse's staircase transforms the Central Hall into an aesthetic masterpiece, but it serves a directive and controversial function within the museum's architecture as well. It is not the staircase itself that has been faulted, but the fact that it was the only public staircase in the building. As the plan of the building shows, the museum is divided into three further sections radiating out from the Central Hall (fig. 114). Geology and botany were in the east wing, with geology and palaeontology on the ground floor, mineralogy on the first floor, and botany on the second floor. The west wing was devoted to zoology, the North Hall to British natural history. To get from any one of these spaces to another you had to pass through the Central Hall, and because this was where the only staircase was you also had to pass through the Central Hall to get from the galleries on the ground floor to those on the upper floors, even if you were staying in the same wing or moving from east to west to study the same kind of creature in its ancient and modern forms. Girouard calculates that a visitor wanting to get from the modern fish gallery to the extinct fish gallery might have had to walk over 600 feet, even though the physical distance between the two spaces was less than a quarter of that.[41] Not only must this have been 'infuriating', as he remarks, but as Yanni notes it also 'did not lend itself to linear processions, or to the unfolding of knowledge that might accompany such a procession'.[42]

As museum studies has grown as a discipline, scholars have become increasingly attentive to how the layout of museums, including natural history museums, impose interpretations on their visitors not just through the display of exhibits, but through the narratives implicit in their architectural structures.[43] The plan that Greswell proposed for the Oxford museum in 1853 (fig. 115) shows how the layout of a building could steer visitors towards the conception of nature as God's creation, structured as a great chain of being rising up to humankind. Greswell located his museum on the first floor of a typical collegiate quadrangle, above the laboratories, lecture rooms and dissecting rooms. As a visitor to Greswell's museum, you would have been able to reach it by any one of the four staircases at the corners of the quadrangle. But once you were there you could only move either up or down the chain of being reaching from rocks and plants through invertebrates, fish, amphibians and reptiles, onto birds and mammals and finally to humanity. Greswell

114 Alfred Waterhouse, plan for the ground floor of the Natural History Museum, London, 1872. Natural History Museum, London.

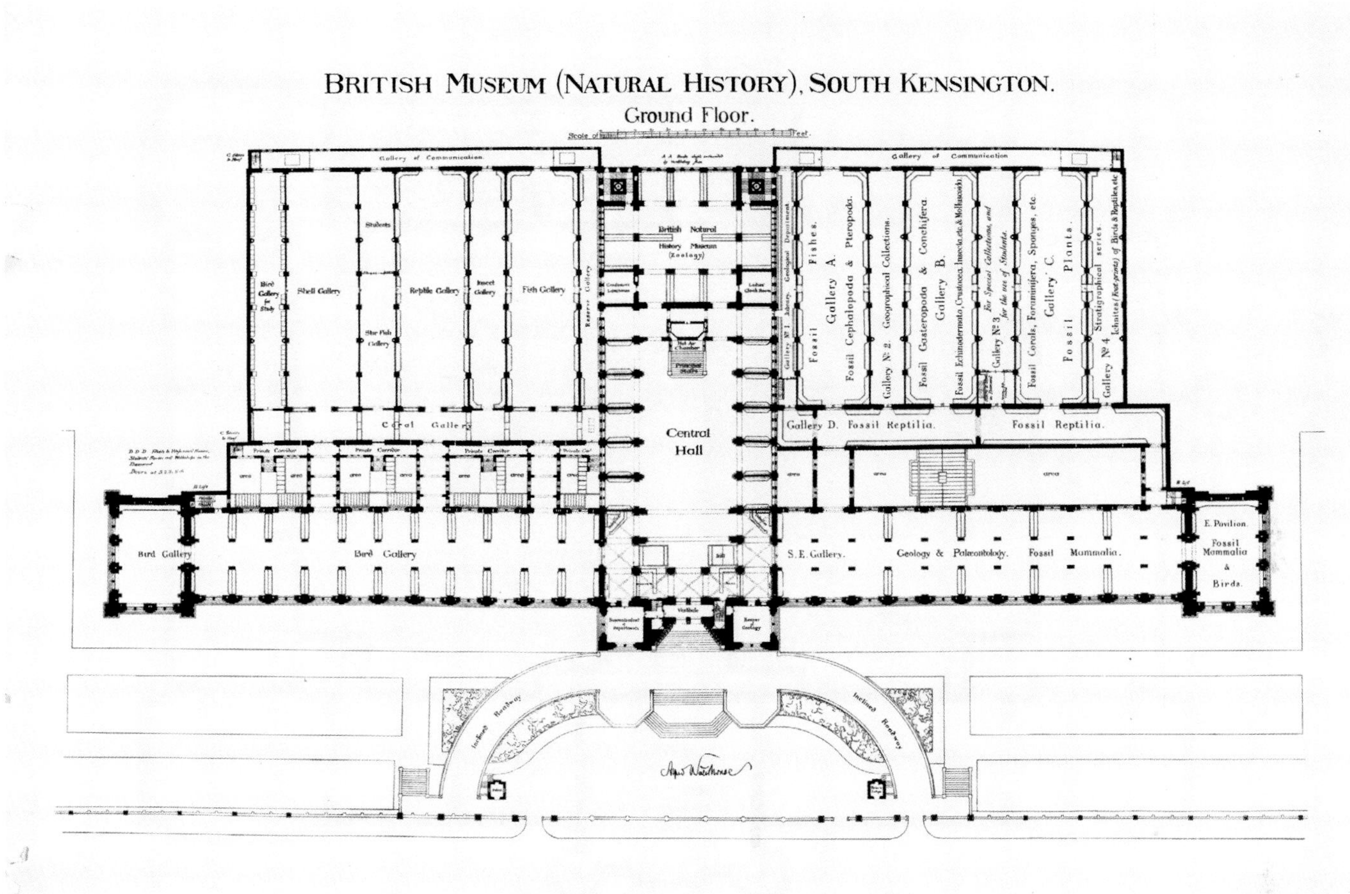

achieved this by the simple expedient of placing a solid wall at the end of the third room devoted to human beings, the Pathological Museum, where every other wall on this first floor is left partly open. If you had walked leftwards round the museum to reach the galleries of Human and Comparative Anatomy and Physiology, you would also be enlightened as to the logic of your journey, as displays on the opposite wall explained the philosophy of life and organisation.

In the pamphlet in which he set out most clearly his proposals for a national museum of natural history, Owen toyed with a similar progressive structure, suggesting that the layout of the galleries should allow visitors 'to see how the Mammalian type is progressively modified and raised from the form of the fish or lizard to that of man'.[44] When it came to it, however, the design of the museum itself staunchly resisted this kind of organisation. This may have been partly due to institutional resistance. Owen was Superintendent, but the different departments of the museum each had their own administrations, so redeploying their collections would have been hard to arrange. The narrative structures which Greswell envisaged as reinforcing natural theology were also now increasingly apt to imply evolution. Owen was not opposed to evolution per se. Indeed, in his monograph on the dodo he expressly sides with the early French evolutionists Buffon and Lamarck against those who insist on the necessary perfection of the creation.[45] But he was set against the naturalism that he saw as implicit in Darwin's particular version of the theory. If evolution occurred, then for Owen it occurred through marked steps in the unfolding of a divine plan, not as a result of gradual

117 Charles James Lea, *Aloe succotrina*, *c.*1878, designed by Alfred Waterhouse. Natural History Museum, London.

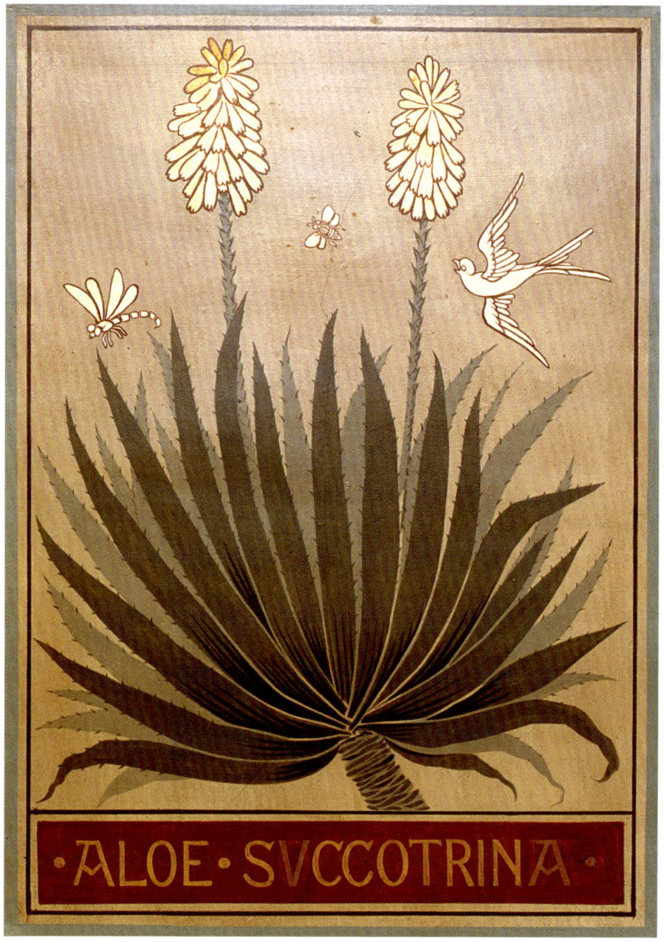

Central Hall are paralleled in his terracotta animals throughout the museum. The set-piece bas-reliefs of birds and mammals on the staircases and balconies in the Central Hall are again bounded with frames, as if to affirm that each species is distinct. This also creates a sense that the animals themselves are representatives of the type, illustrated in a typical habitat, not individual creatures living within a complex shared ecology. Waterhouse's looser, less detailed style of sculptural design contributes to this impression. He had increasingly come to feel that the O'Sheas' work 'looked much better when half worked', as he told a meeting of the Royal Institute of British Architects in 1865.[54] In his own designs, he emphasised the overall form of a given species of animal or plant rather than the precise detail of a given specimen.

The impression that the animals which adorn the walls of the Natural History Museum represent types, not individuals, is enhanced by the material from which they are made. There were several practical and aesthetic reasons for Waterhouse to have chosen to face his building, inside and out, with terracotta. It was resistant to the acid smog of Victorian London, could be easily cleaned, and protected the iron and concrete pillars that formed the skeleton of the front galleries from fire.[55] It enabled him too to embrace Ruskinian polychromy at the same time as modernity and fashionable taste. As Cunningham notes, because it was integral to the structure of the building, there was less risk that the funding would run out as it had at Oxford and that the decoration be left incomplete. And it gave Waterhouse more control, as he could design all the sculptures himself, approve Dujardin's models, and then have them mechanically reproduced.[56]

But Waterhouse's use of terracotta also matched Owen's view of nature. As the sculptures done by the O'Sheas at Oxford were hand-carved, each is necessarily unique, even John O'Shea's more conventionalised pieces. The terracottas at the Natural History Museum, though designed by hand, are machine-made. In principle, they are infinitely replicable, and many of the same forms recur around the museum. Where each bird or insect carved by the O'Sheas is an individual animal pursuing its own needs or desires in life, the birds around the arches that lead off the Central Hall are repeated at regular intervals, with the same two designs alternating on each arch. Where the Oxford birds nest and feed among the leaves of individual plants, Waterhouse's birds perch on matching pairs of maple leaves growing on either side of a generic stem which leads down to generic roots. The bird and insects who fly around the aloe in Lea's painting are similarly stereotyped, drawn from stencils which pattern the ceiling of the Central Hall balconies. Even the individual framed panels on the stairs are repeated. The overall effect is to show nature comprised of kinds, not of individuals, and of established forms, not evolving populations or emergent ecosystems.

In addition to affirming a fixed Linnaean order of nature, Waterhouse's decorations embody and playfully explore one of Owen's most distinctive scientific theses: the theory of

118 Alfred Waterhouse, designs for panels for the Central Hall, 1877–8. Natural History Museum, London. (a) the hare; (b) the cavy; (c) the fennec fox

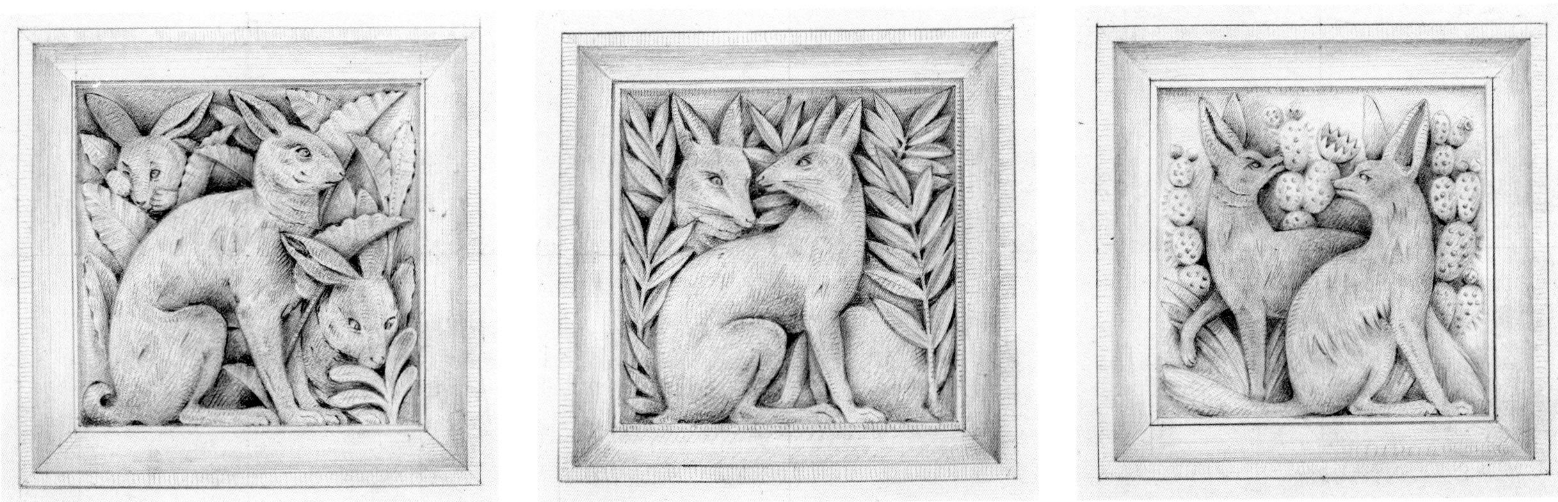

the vertebrate archetype. Owen's biographer Nicholaas Rupke explains that this conception underwent several transformations within his thinking, originating in German idealist morphology as a force or tendency for recurrence within living forms, but emerging under the influence of Owen's Oxbridge patrons as a Platonic conception in the divine mind, a pre-existing blueprint for all different vertebrate designs.[57] As a non-adaptive alternative to the argument from design within natural theology, it was a significant influence on Acland's conception of nature as God's art.

Owen's conception of the archetype bespeaks a divine artist with a preoccupation with patterns and forms. The same interest in recurrent forms in nature can be seen across Waterhouse's apparently haphazard terracotta menagerie. Several of the different relief sculptures on the main staircase echo one another visually. In the illustrations of the golden pheasant and the black grouse, Waterhouse has arranged the pairs of each bird in the same position relative to one another. This enables the viewer to see how these two gallinaceous species vary from one another, each modifying the same basic structure. In this case, what we see is a homology, whether we trace it to common ancestry or a shared blueprint. In the cases of the hare (fig. 118 a), the cavy (fig. 118 b) and the fennec fox (fig. 118 c), we see rather a common anatomical feature reappearing across several different lineages. What draws Waterhouse to these creatures is their long ears, as he delights in the endearing and even whimsical appearance it gives to the animals themselves. In Darwinian terms, these are instances of convergent adaptive evolution. Extrapolating from Owen's idea of the archetype, what we see instead is the same structure taken from the stock of variations latent in the archetype and redeployed by the creative mind of God in different creatures. As the historian of evolution Peter Bowler has noted, according to those late Victorian biologists who followed Owen in seeing predetermined trends rather than random variation and natural selection as the key to evolution, it was perfectly possible for the same inherent tendency to lead to the same feature or even near-identical creatures in different lines of descent.[58] Form itself is what matters here, not its function, and what it reveals is not common descent but a shared impulse.

Form is what matters for Waterhouse as well. Where the carving and ironwork at the Oxford University Museum suggest individual animals or plants, the capitals and lines of the Natural History Museum are often made up of parts of animals that cannot be imagined as actually present or alive. The doorway to the mineralogy gallery, for example, is made up of snakes (fig. 119), not hunting as those at Oxford are, but plaited together in a purely decorative rhythm.[59] The animal heads that surmount the piers in the main south galleries peer through foliage, but there is no space for bodies behind them. The sheep, cattle and mule

119 M. Dujardin, interwoven snakes, designed by Alfred Waterhouse in 1874. Natural History Museum, London.

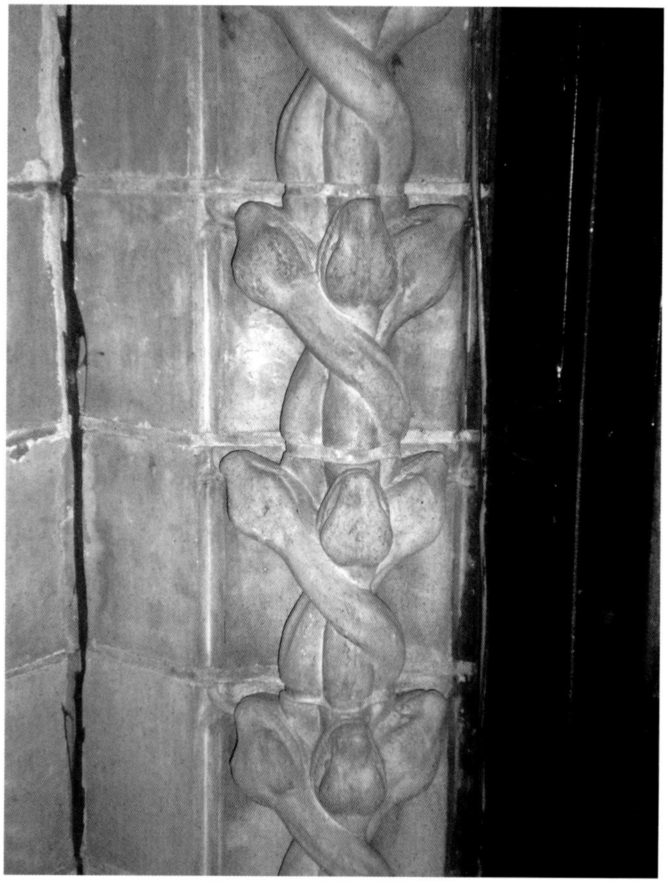

120 Alfred Waterhouse, designs for capitals, *c.*1875–8. Natural History Museum, London. (a) rams' heads; (b) ammonites

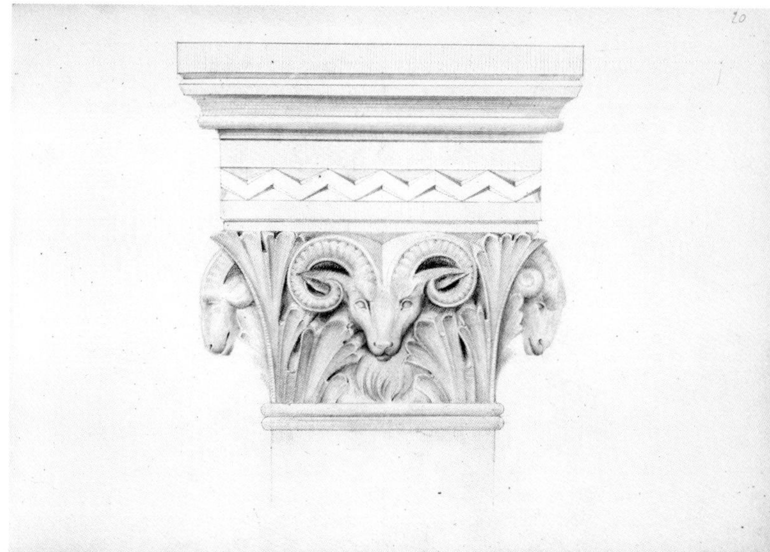

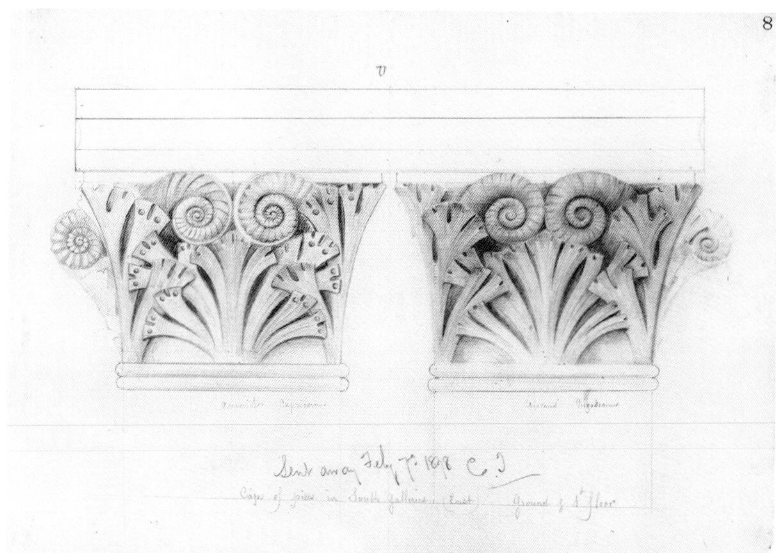

heads in the west wing are matched by ancient ungulates in the east wing.

Other pairings are suggestive in other ways. The rams' heads (fig. 120a), in particular, make a witty allusion to the ionic order of columns in classical architecture, reminding us how natural forms are adapted in human art, but also hinting that those natural forms are themselves signs of artistry. As Waterhouse represents it, nature's divine artistry shows a distinctive taste for particular recurrent patterns. In one of the most delicate carvings in Oxford, John O'Shea plays with visual resemblances between wholly different organisms in nature. The lines of the bird that stalks among the pickerel (fig. 121) beautifully vary the theme of the plant's own leaves and buds. The resulting depiction is at once richly ecological and a masterly study in form. Like O'Shea, Waterhouse is struck by the resemblance between utterly divergent types, but Owen's science and the plan and structure of the Natural History Museum lead him to explore this not ecologically by placing them among one another but rather through parallelism. Where ancient ungulates vary the same blueprint as the rams, it is the ammonites (fig. 120b) that recall most closely the form of their horns. Here there is no chronological, ecological or taxonomic link. The rams are modern, the ammonites have been extinct since the end of the Cretaceous period; the rams live on land, the ammonites lived in the

121 John O'Shea, pickerel, 1860.
Oxford University Museum.

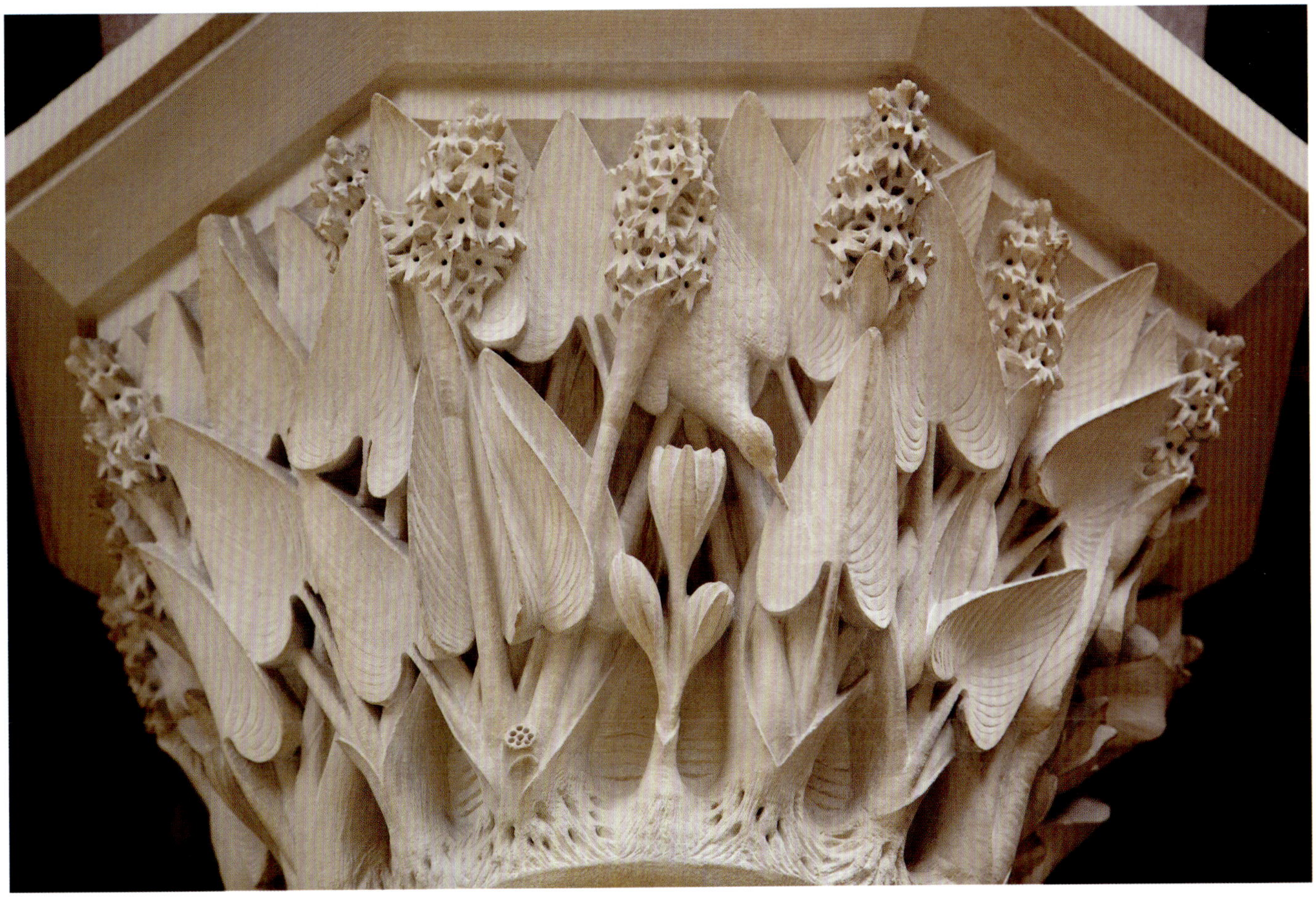

sea; the rams are vertebrates, the ammonites molluscs. Yet the resemblance between the rams' horns and the ammonites' shells is unmistakable. As if the visual resemblance itself were not enough, Waterhouse draws it out by placing the ammonites among seaweeds which themselves closely resemble the plants on either side of the rams.

The separation of distinct types and their recurrence around the Natural History Museum identify the science of natural history as the study of kinds, not individuals. At the same time, these kinds themselves suggest order within nature and a divine artistry predicated on recurrent patterns. As James O'Shea toyed with evolution in his carvings at Oxford, so Waterhouse's decorative art playfully explores the ramifications of Owen's alternative scientific conception of nature. The starkest separation between natural types inscribed in the plan of the Natural History Museum is between extinct and living forms. As we have seen, this distinction was reinforced through the museum's decoration. Both the physical distance between exhibits and the spatial symbolism of their arrangement discourage evolutionary interpretations of the relationship between species as one of descent with modification. Instead, this arrangement throws up other ways of apprehending the patterns in nature through science and art.

The arch leading from the front gallery of the east wing into the east pavilion is decorated with scrollwork alternating an archaeopteryx and a pterodactyl (fig. 122a and b). The terracotta animals on the equivalent arch in the west wing are a modern bird and a monkey (fig. 123). Like the colossal animals on the east parapet, the archaeopteryx and pterodactyl were special studies of Owen's.[60] They are both flying

122a and b M. Dujardin, archaeopteryx and pterodactyl, designed by Alfred Waterhouse. Natural History Museum, London, *c.*1878.

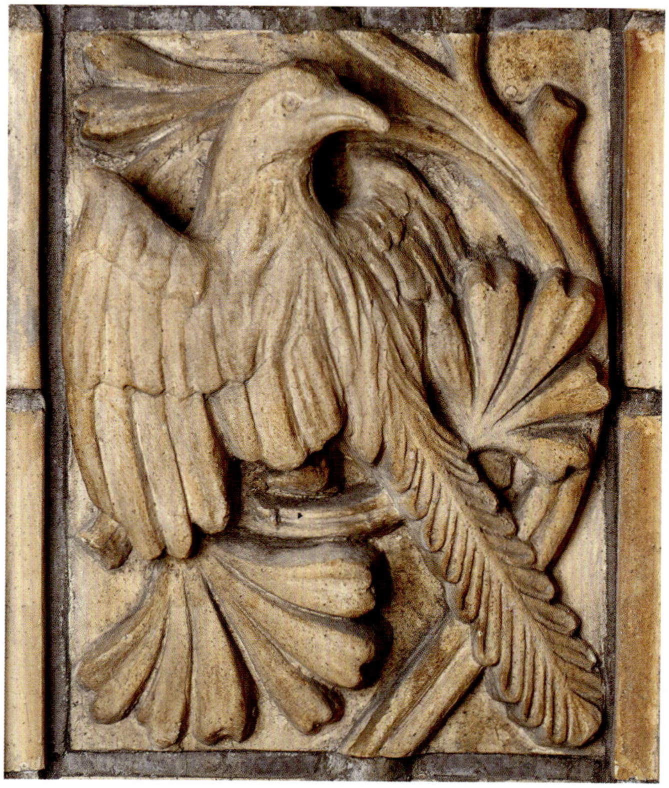

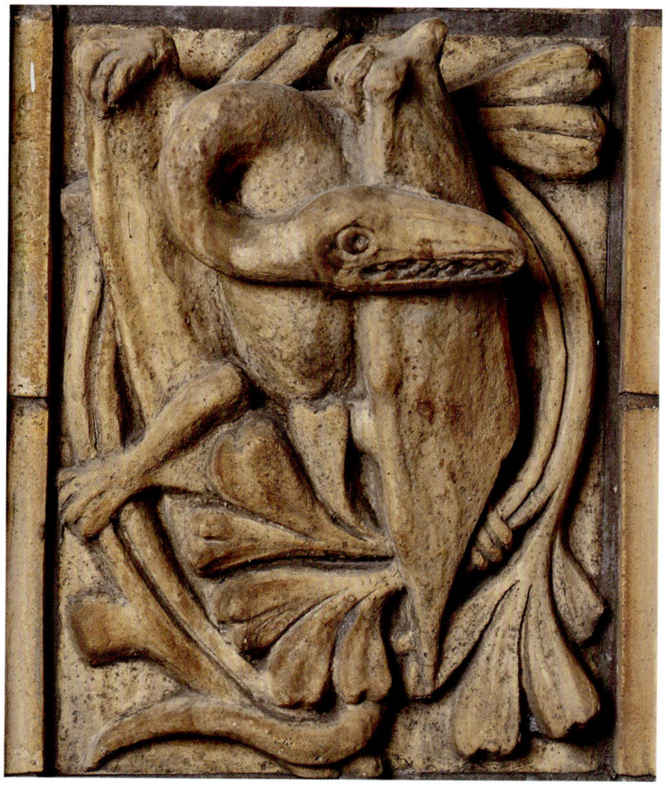

vertebrates of the late Jurassic period, discovered initially in southern Germany, so their proximity is at once chronological, geographical and ecological. The modern bird is not specific enough to be identified, but we can take it on trust that it too shares a moment in time, a location and a habitat with the monkey. But what are we to make of the equivalence of these two pairings? Like the ancient and modern fish that swim in rippling seas on the piers in each wing, their habitats are very similar. The pairing of the archaeopteryx and the bird might lead us to see the relationship as ancestral, except that the pterodactyl is clearly not the monkey's ancestor. So instead we are led to see the archaeopteryx, like the ancient and modern fishes and mammals on the piers and their capitals, as equivalent to the modern bird, not necessarily its direct forebear.

By this same reasoning, the pterodactyl and the monkey are in some sense equivalents too. Although Waterhouse's archaeopteryx is unmistakably a bird, it is clumsy and crude when compared with the delicate form of the modern bird. The modern bird's refinement is enhanced by the elegant whorls of leaves and stems that Waterhouse has created from the form of the modern plant. While the pterodactyl is unmistakably not a monkey, it does look in some respects like a grotesque anticipation of one. Both animals clutch the branch they are standing on with hand-like feet, while the curve of the pterodactyl's neck recalls that of the monkey's tail. Both have actual hands, and the pterodactyl's hideous face is prognathous like the monkey's. By setting these different creatures at equivalent sites on either side of the building, Waterhouse poses us the problem of considering what precisely that equivalence implies.

The palaeontological wing is set in the east, suggesting that what it shows to us is the dawn of life. Across the galleries, the modern animals seem more beautiful and perhaps better adapted too, hinting that life has progressed from crude early forms. The ancient fish look familiar but nightmarish, the extinct mammals on the east parapet like monstrous distortions of the noble heraldic beasts on the

123 M. Dujardin, bird and monkey, designed by Alfred Waterhouse. Natural History Museum, London, *c.*1878.

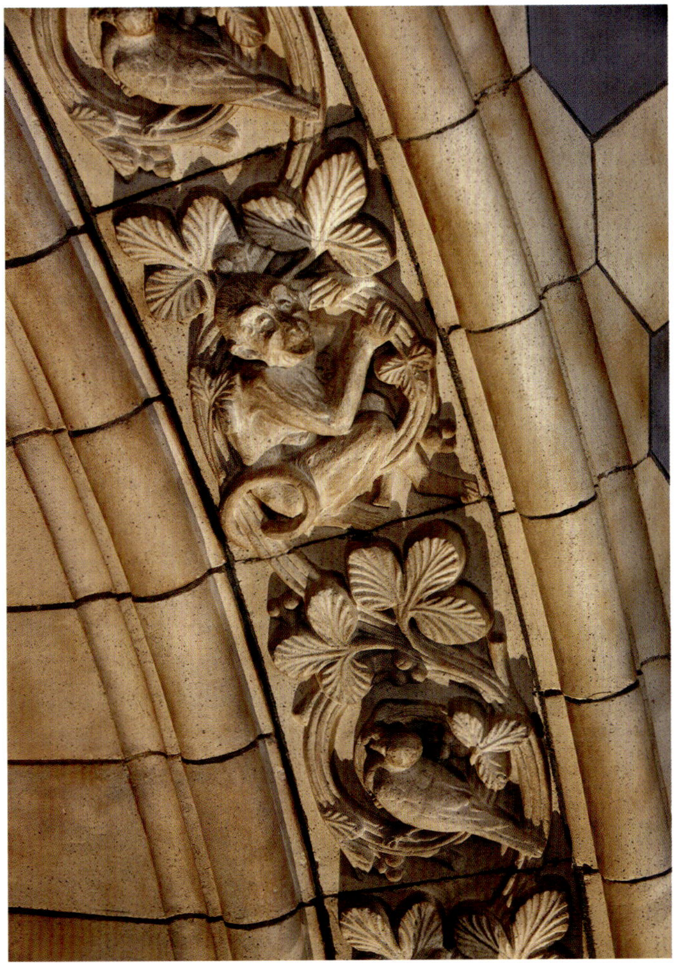

124 Alfred Waterhouse, *Great Palaeotherium*, Natural History Museum, London, 1877.

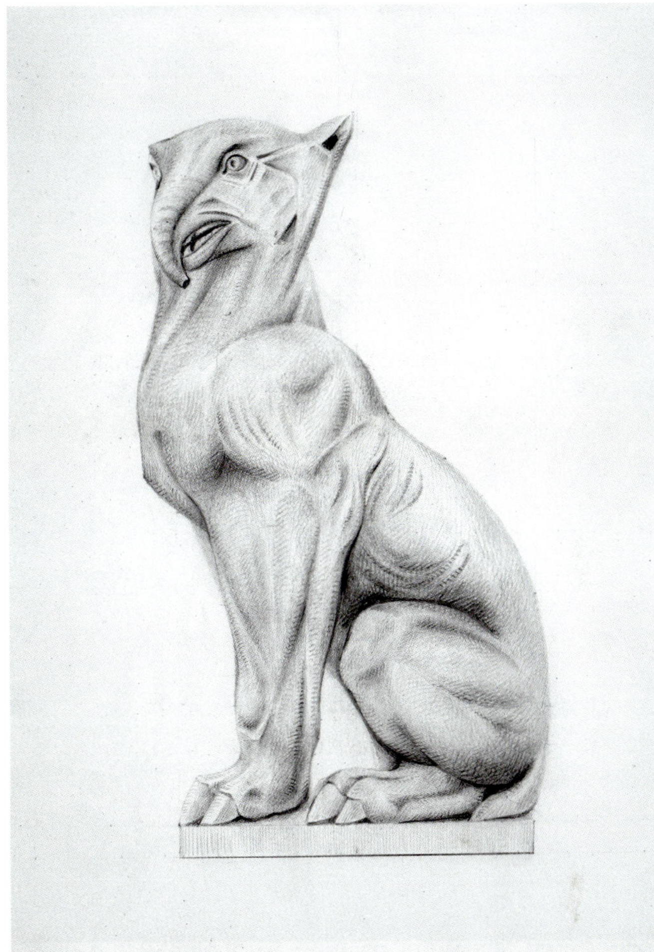

west (fig. 124). In a late lecture on 'Religion and Art' published in the *Contemporary Review*, Owen's friend and admirer Holman Hunt recapitulates Acland's idea of God as, in Hunt's words, 'the Master Artist'. The role of the artist on earth is to pay homage to this master's work:

> By his interpretations of unconsidered beauties, he teaches others to see how every order of creation progressed upwards from the roughest forms to supreme gracefulness, for when the problem of originating a new genus of life had been attained, it was often without beauty, often even a mere lump; then the heavenly artistic mind began to work out its developments; it fashioned these by slow steps into exquisite shapes; it decorated the surfaces with spots, and with delightful devices and colours, entertaining and charming first the fastidious eye of the creature's fellows . . . and then that of the far higher intelligences also to the end of time.[61]

The artistic development Hunt describes from the rough to the exquisite is reflected in the parallel positioning of the archaeopteryx and the modern bird, suggesting continuity from one creation to another, but also in the pterodactyl and the monkey, implying a more radical break between them. Matthew Rowlinson has argued that Tennyson maps the Christian concept of typology – seen in Collinson's 'The Child Jesus', for example – onto an evolutionary history of life on earth in *In Memoriam*, so that his lost friend Arthur or blessing. Above all, he represents humanity as a whole. The symbolism of the figure was captured in enraptured prose by the architect Edward Ingress Bell, who reviewed the new museum for the *Magazine of Art*: 'Glancing upwards over the whole field of its varied and orderly scheme of enrichment, the eye rests upon the consummation of the whole in the figure which terminates appropriately the highest gable. There, standing erect, is seen the 'quintessence of nature,' with outstretched arms and upward gaze directed towards a still higher power.'[72] Owen was no biblical literalist. Though he saw biology as corroborating the broad thrust of Genesis and though he remained a dedicated natural theologian, he made it clear in a lecture he gave to the Young Men's Christian Association at Exeter Hall in 1863 that it was a mistake to try to square the Bible with science where that meant rejecting the new revelations of science for crudely factual interpretations of a text that was meant as a source of moral teachings.[73] For Owen and Waterhouse, it was more accurate and compelling to see the statue of the man above the entrance to their museum as the 'quintessence of nature', at once the consummation of the natural world and above it, than as a character from what was by now largely agreed to be a symbolic myth.

POLICING PRE-RAPHAELITISM: THE NATURAL HISTORY MUSEUM REVIEWED IN *NATURE*

With its imaginative medievalism, smooth surface and stereotyped terracottas, the Natural History Museum recalls the late Pre-Raphaelitism of Burne-Jones or Morris, segueing into aestheticism on the one hand and Arts and Crafts on the other. But it was conceived as a piece of representational architecture on the same early Pre-Raphaelite principles as the Oxford University Museum, with the same commitment to accuracy and truth in its depiction of nature. The new building's architecture and decoration were held to these exacting standards by *Nature* when it reviewed the museum. Since its launch in 1869, four years after the *Fortnightly Review*, *Nature* had rapidly established itself as the leading scientific journal aimed at a well-informed readership from across the spectrum of scientific disciplines. While both journals shared a commitment to scientific naturalism, *Nature*'s remit was far more specific. It was concerned primarily with the processes and findings of science itself, engaging with the wider culture only, as in Lockyer's articles on art, where science could help in assessing culture or, as in the reviews of the Natural History Museum, where culture was impinging on the practice or communication of science.

As a whole, the architectural and artistic press gave Waterhouse's building and its decorations a warm welcome. Journals such as *The Builder*, *Building News*, *The Architect* and the *Magazine of Art* were all enthusiastic in their praise, as were many of the more generalist papers including *The Times*, the *Daily News* and the *Saturday Review*.[74] *Nature*, by contrast, was niggardly in its praise and thorough in its criticism. This reaction was partly tribal. Lockyer was broadly allied to Huxley, who was Owen's rival and nemesis. Huxley had crossed swords several times with Owen over various aspects of the proposals for the museum, including where it should be sited and what proportion of the collection should be on display.[75] At one point *Nature* had even tacitly proposed that Huxley, not Owen, should be the new museum's director.[76] There were also practical objections. *Nature* ran two editorials and a further long review on the museum between 1880 and 1882. In response to the second of these, a correspondent signing him or herself F. G. S. complained that the colour of the terracotta was too close to that of the exhibits themselves and that the impulse to make the most of the aesthetics of the architecture ran counter to the need to display the specimens themselves effectively (fig. 128).[77] *Nature*'s third notice of the museum eighteen months later raised the same complaints. It had been 'a serious mistake', the reviewer wrote, to erect 'a building with such elaborate and ornate internal decorations for museum purposes', as 'the form, position, and illumination of the cases' had been sacrificed to the 'architectural effect', while the 'peculiar tint of the terra-cotta' failed to make the museum's skeletons and casts stand out.[78]

In addition to these political and practical reservations, however, *Nature* held the museum to account for the truth, or rather the falsehood, of the vision of nature embodied in its decorations and built into its plan. For the lay press, Owen's authority and the museum's function were enough

128 The Central Hall of the Natural History Museum (now Hintze Hall) with the skeleton of a blue whale on display, since July 2017, magnificently refuting F. G. S.'s judgement on Waterhouse's choice of terracotta.

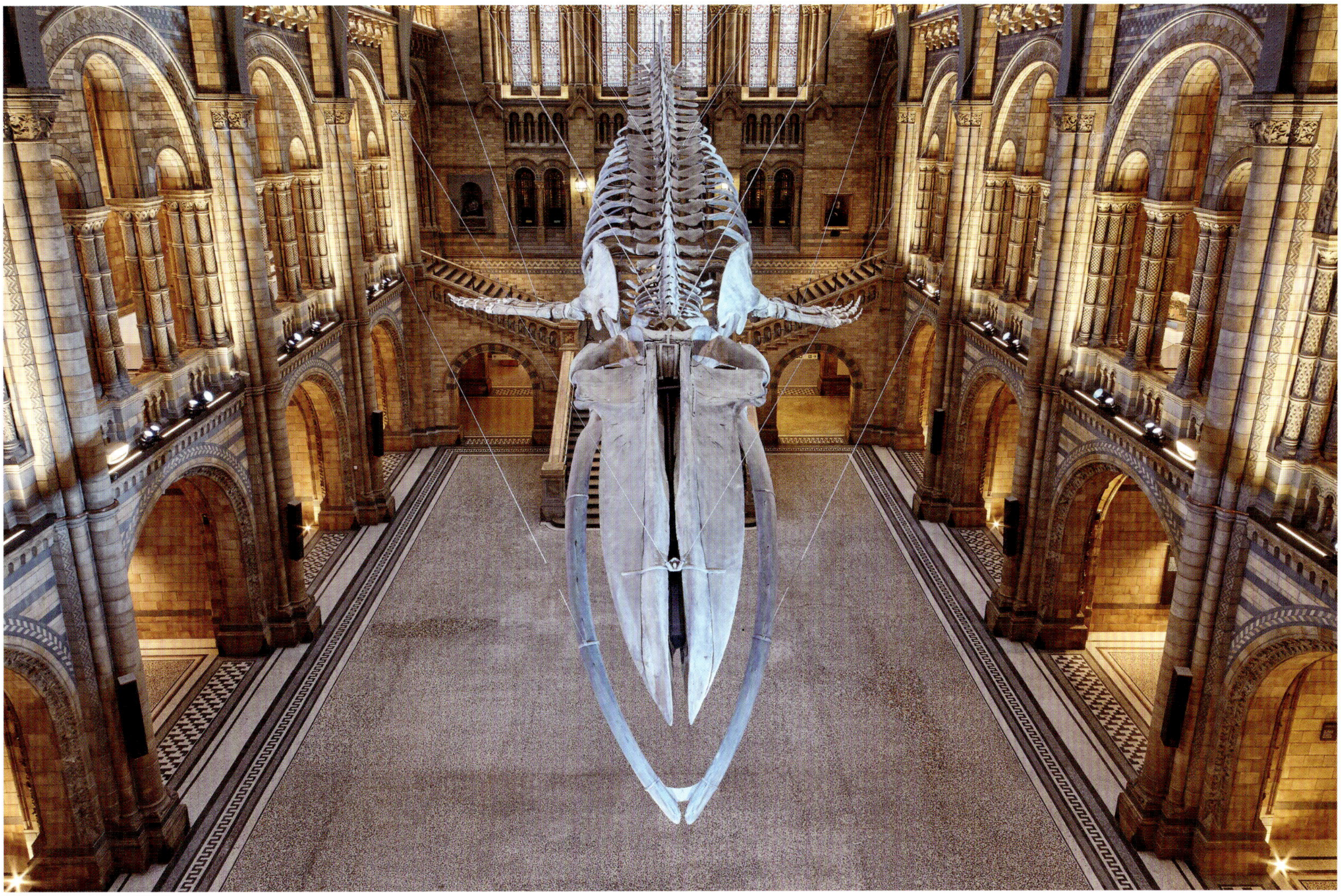

to guarantee the scientific accuracy of Waterhouse's decorations.[79] The more specialist natural history journals were less easily convinced. A reviewer for *The Field*, a magazine for naturalists, farmers and sportsmen, dismissed the terracotta animals as 'incorrect and grotesque', remarking disdainfully that the building's style was 'more adapted for a suburban tea-garden than a national museum'.[80] *Nature* applied the same criterion of accuracy. The first short editorial on the new museum was similarly flippant, remarking of the external decorations: 'It is perhaps only fair that if the inside is devoted to *natural* history the outside should be similarly devoted to unnatural history.' Architects, the reviewer added with an indulgent smile, 'must have their way'.[81]

In the second and third reviews, the same issues are discussed more seriously. Of the three, the second editorial is the most complimentary. The author was on the whole impressed with the decorations in the Central Hall, praising with a careful discrimination reminiscent of Lockyer's comments on the Royal Academy paintings three years earlier the 'nearly natural' terracottas, and remarking that the botanical paintings on the ceiling were suitably 'bold' where the trees can best be viewed from the floor and more detailed in

129 Alfred Waterhouse, Manchester Museum, opened 1887.

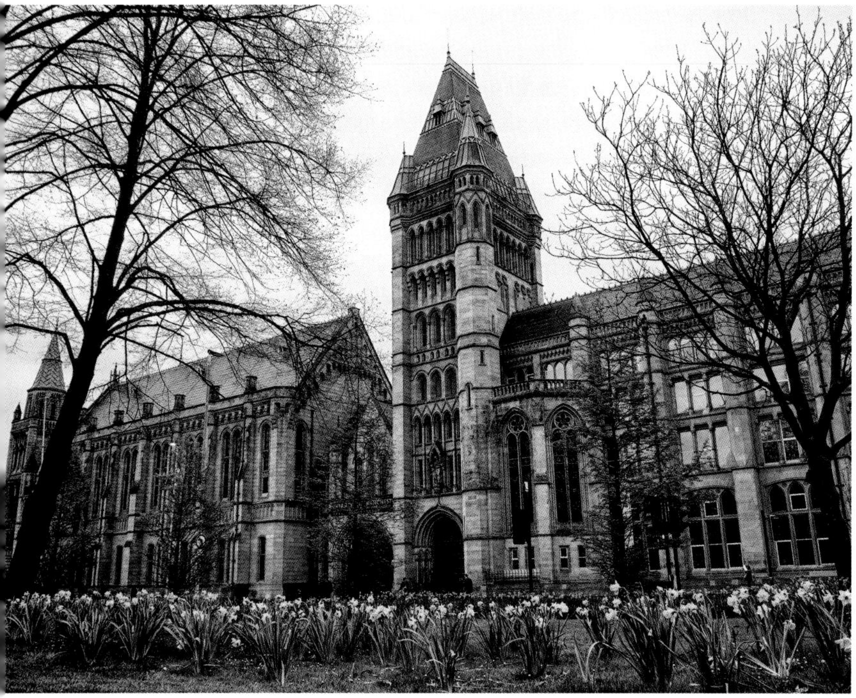

versity Museum through a direct collaboration between scientists, architects and Pre-Raphaelite artists. The Natural History Museum was built according to the same principle, followed through more systematically and on a grander scale. In both cases, the vision of nature expressed in the forms of the building was an expressly Christian one. In Oxford in the 1850s, natural theology had been in the ascendant. It could express itself with confidence, without needing to police its boundaries and eliminate dissent. By the 1870s, however, natural theology was in retreat. For most of the scientific community, Owen's conception of science was out of date even before its first brick was laid. The Natural History Museum was his last defiant act of resistance in the face of Darwinian evolution and scientific naturalism.

As Victorian scientists' conceptions of the natural world changed, so the challenge to represent science and nature through architecture changed too. While Pre-Raphaelite poetry chimed with scientific naturalism, Pre-Raphaelite architecture remained at the service of natural theology. In 1887 William Abbott Herdman, Professor of Natural History at University College, Liverpool, argued that natural history museums should ideally be built to a branching plan mapping out the phylogenetic tree of nature.[95] Had such a museum been built, it would have been both a Darwinian and a Pre-Raphaelite natural history museum. But like Owen's Natural History Museum, it also would have petrified the science of the moment, only to be given the lie as new phylogenies were traced. One lesson of the comparison between the Oxford and the London museums is that the more programmatically a building seeks to embody science, the more brittle that vision of science will be. Pollen's archway and the fall of 'Adam' notwithstanding, the Oxford University Museum is less dated than the Natural History Museum as a vision of nature. As a collaboration, it was able to combine many different perspectives on nature, both religious and secular, which are still in dialogue with one another today. As a more thoroughgoing exercise in Pre-Raphaelitism, it was predicated on the direct observation of nature above and beyond the received wisdom of science. As a Gothic building, it embodies an organic conception of architecture which embraces the growth and change Owen's plan and Waterhouse's terracottas try in vain to resist.[96]

Once Owen had retired, the Natural History Museum did its best to accommodate Darwin's science. In 1885 a marble statue by Joseph Boehm of Darwin enthroned was unveiled (fig. 130). If this was supposed to cement the victory of Darwinism over Owen, it signally failed. *Nature* praised the statue, but it could not resist expressing its usual Pre-Raphaelite scruples, regretting that – through no fault of Boehm's, as he had never met his subject – 'the hands . . . do not bear the smallest resemblance to those of Mr. Darwin, but are of a kind which, had they been possessed by him, would have rendered impossible the accomplishment of much of his work.'[97] Twelve years later Boehm's Darwin was joined by a tall bronze statue of Owen by Thomas Brock (fig. 131), which again received a careful and in this case more appreciative notice in *Nature*.[98] Initially, Darwin was on the central stairs and Owen on the floor of the Central Hall. They have been battling it out for primacy in the museum ever since, like two old wizards, one white, one black. In the 1920s, when the theory of natural selection was at its lowest ebb, Darwin was banished to the North Hall. Owen held sway over the Central Hall from the stairs, while Darwin sat

130 Joseph Boehm, *Charles Darwin*, 1885.
Natural History Museum, London.

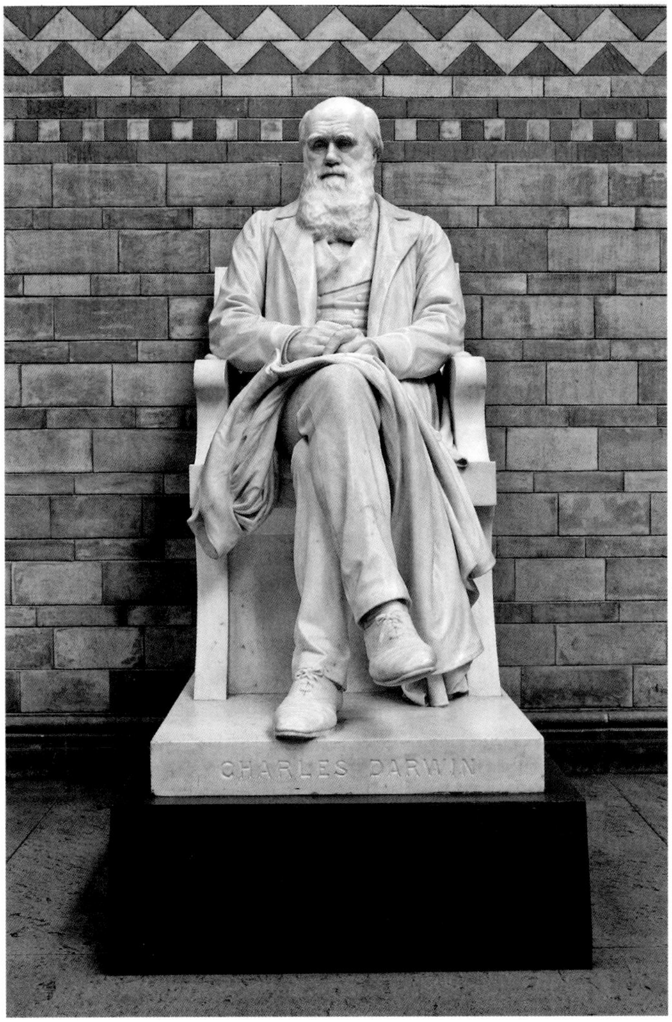

131 Thomas Brock, *Richard Owen*, 1897.
Natural History Museum, London.

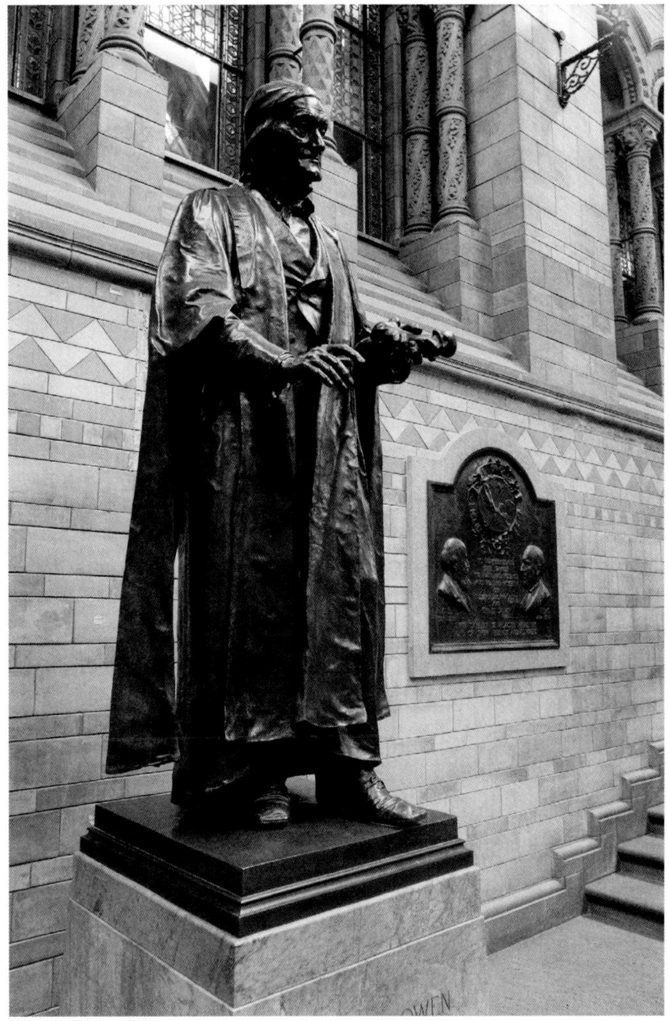

alongside his friend and ally Huxley among the diners in what became a café. Darwin was restored to primacy in time for his bicentenary in 2009, as Owen was banished in his turn to dispute with Huxley on an obscure landing on the first floor. This unseemly struggle is a fitting allegory for the divided loyalties of the museum itself, founded and built by Owen, yet committed to teaching the science of his greatest rival.

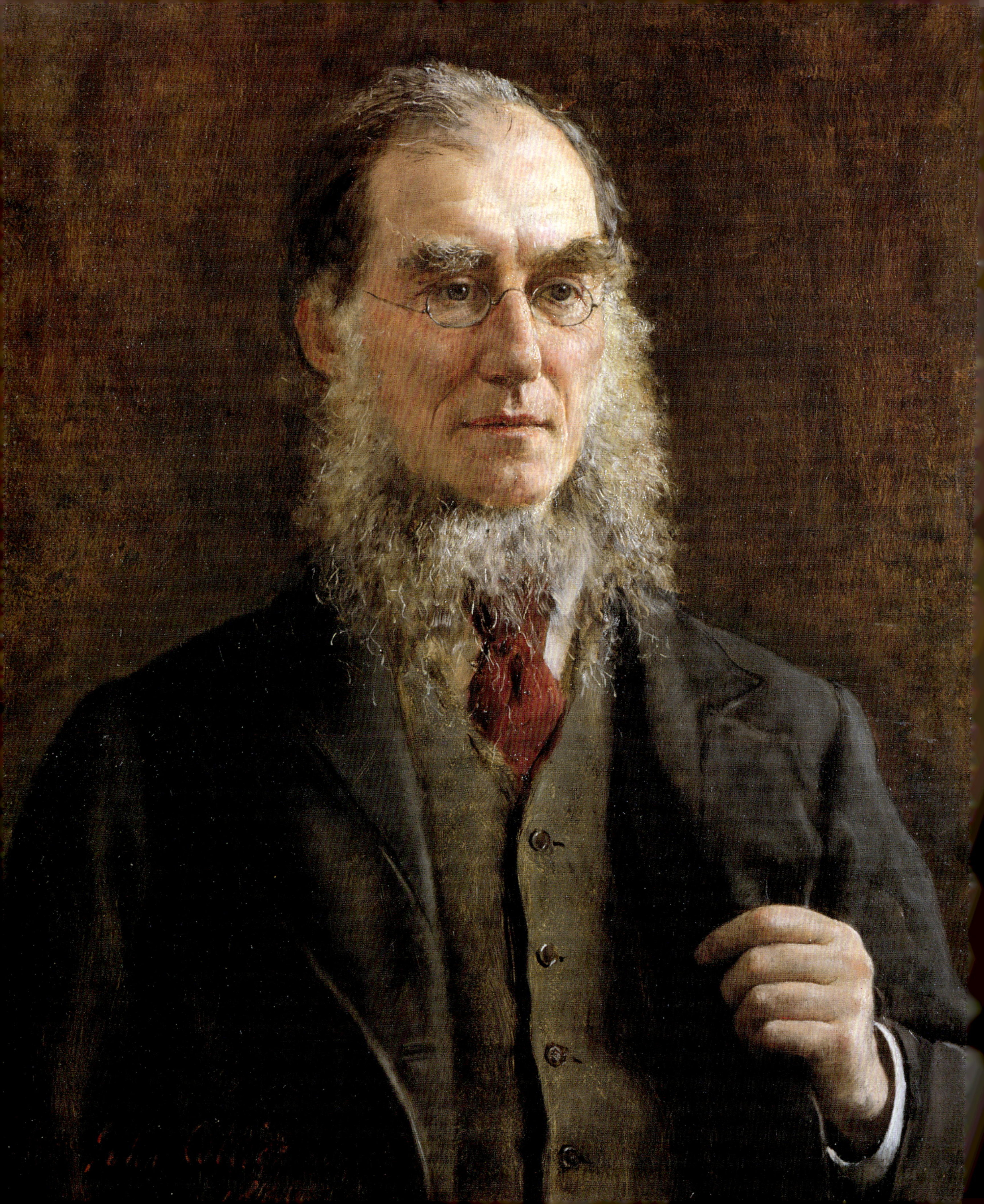

9 · Retrospect

PRE-RAPHAELITE PAINTING AND THE HISTORY OF SCIENCE

The 1860s and 1870s saw a battle for authority within the scientific community between scientific naturalists who debarred religion from science and an old guard who remained committed to the principle that what science revealed was God's creation. This struggle played out both within the institutions of science and across a wider culture that increasingly ranked science alongside religion and the liberal arts as a civilising force. Both camps took support from the Pre-Raphaelite arts. Pre-Raphaelite poetry provided a medium through which the advocates of scientific naturalism could confront the implications of materialism without abandoning the aesthetic charge of art. Pre-Raphaelite architecture, on the other hand, gave imposing form to the ordered vision of nature cherished by natural theology, while Pre-Raphaelite art theory ranged itself on both sides.

Pre-Raphaelite painting had won the approval of scientists such as Owen and Acland in the 1850s, offering them a pictorial aesthetic in which they could recognise the rigour and clarity of science. Within the new scientific culture of the 1860s and after, however, Pre-Raphaelite painting was more peripheral than either poetry or architecture. The essays of Morley, Pater and Colvin suggest that second-generation Pre-Raphaelite paintings could function as evocative mysteries, yielding aesthetic pleasure and symbolic resonance without requiring the viewer to vest belief in them as sacred art. There is a complementarity with science here, but it works by exclusion, with art and science assigned separate spheres and opposing aesthetics, both seeking in their own ways to fill a gap left by the dwindling authority of religion. The 'scientific men' who subscribed to *Nature* and the

Facing page Detail of fig. 135.

Fortnightly Review, and who read *The Earthly Paradise* with enthusiasm, left behind little evidence that they saw these late Pre-Raphaelite paintings as contributing directly to the culture of science.

Late Pre-Raphaelite painting did contribute to the formation and expression of a scientific culture in one important way, however. Back in the 1840s the Pre-Raphaelites had set out to revolutionise history painting on the model of science. In the 1880s the three major painters surviving from the original Pre-Raphaelite circle took parting glances at science in works which call attention to science's own history. On 29 December 1880 Hunt wrote to his old friend and advocate Richard Owen to ask if he might paint his portrait in order to 'register some phase of your character not before rendered for the interest of scientific and philosophic humanity'.[1] Owen replied the following day that he considered it 'so priceless a boon' to be painted by Hunt that he would be grateful to sit for him whenever he liked.[2] The portrait was exhibited at the Grosvenor Gallery the next summer. Around the same time as Hunt began his portrait of Owen, Millais paid a visit to his son John Guille Millais's friend, the elderly naturalist John Gould, who would die a few weeks later.[3] In 1885, four years after Gould's death, Millais exhibited *The Ornithologist; or, The Ruling Passion*, a painting inspired by this visit, at the Royal Academy. While Hunt and Millais were painting these canvases, Brown was at work on his last and greatest commission: twelve murals illustrating the history of Manchester in the city's new town hall. Manchester Town Hall was designed by Waterhouse and built at the same time as the Natural History Museum. It was completed in 1877, and Brown was commissioned to paint the murals the following year. They took him fifteen years to complete. Two of the twelve murals show scientists at work: *Crabtree watching the Transit of Venus A.D. 1639*, painted 1882 to 1883, and *Dalton collecting Marsh-Fire Gas*, painted 1886 to 1887.

Early Pre-Raphaelite art theory saw science as the highway to truth, which art also needed to travel if it was to offer valid insights into nature or morality. In their late retrospective paintings, Hunt, Millais and Brown cast science in a different light. Though they remain in sympathy with science, they no longer hold it up as a single, clearly defined set of methodological principles, or a single body of knowledge, ever accumulating and becoming more precise. Instead, they view it as the product of particular acts of research by individual scientists, each working within his own time, constrained by his place in society, and driven by his own convictions and those of his colleagues and rivals. The conception of science which underpinned the original Pre-Raphaelite project was decidedly ahistorical, a Baconian ideal of empiricism. In their late paintings of scientists, Hunt, Millais and Brown give science back its history, but at the cost of some of its authority – the authority on which their own project had rested.

PORTRAITURE AS CONTEMPORARY HISTORY: PAINTING THE SCIENTIST IN THE 1880S

Hunt's portrait of Owen and Millais's tribute to Gould reflect on a significant transition in the history of British science. From the early 1860s the scientific naturalists had been increasingly in the ascendant at the expense of the natural theologians. The cabal at the heart of this campaign to redefine science as a secular enterprise was the X Club, an exclusive dining society formed by nine men of science in 1864, led by Huxley, Hooker and Tyndall. Over the 1860s and 1870s the X Club took control of most of London's major scientific institutions. Huxley was already Professor of Natural History at the Royal School of Mines. He was joined there in 1865 by another of the Xs, Edward Frankland, who took on the chair in chemistry. The same year Hooker took over from his father as Director of the Royal Botanic Gardens at Kew, and in 1867 Tyndall took over from Michael Faraday as Superintendent of the Royal Institution. Two more Xs, the anthropologist John Lubbock and the mathematician Thomas Archer Hirst, added London University and the Royal Naval College at Greenwich to the club's portfolio. Lubbock was Vice-Chancellor of London University from 1872 until 1880 when he became its Member of Parliament. Hirst had been its Registrar from 1870 to 1873, at which point he became Director at Greenwich. When the Normal School of Science was established at South Kensington in 1881, Huxley was appointed Dean. From their several institutional power bases,

132 Thomas Woolner, portrait bust of Adam Sedgwick, 1860. Trinity College, Cambridge.

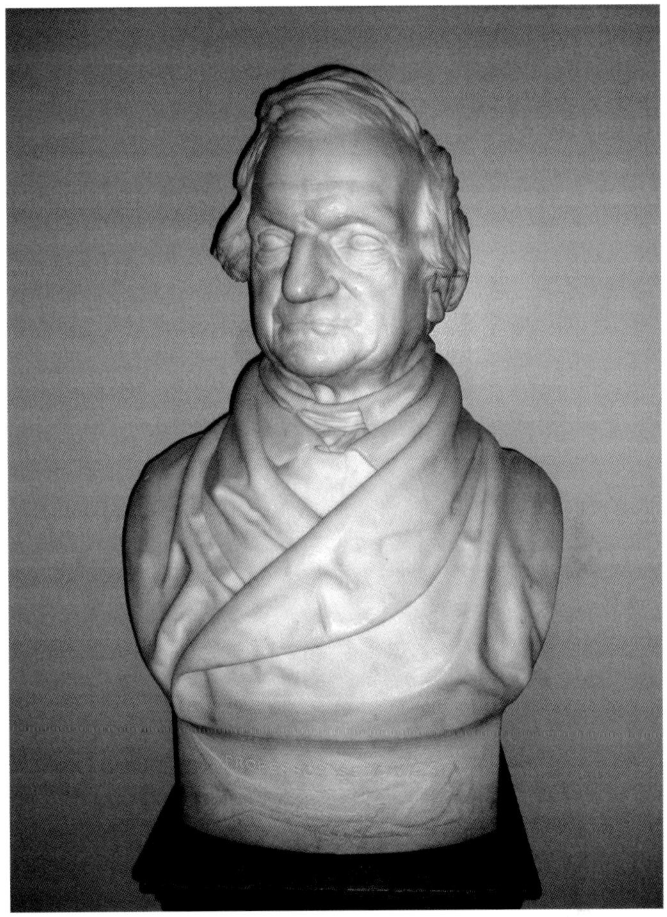

the Xs took control of the Royal Society itself. In 1871 one of their number, William Spottiswoode, became Treasurer, followed by Huxley as one of the two Secretaries in 1872 and Hooker as President in 1873. An X Club member would hold the presidency through until 1885, with Spottiswoode succeeding Hooker and Huxley succeeding him.[4]

Spottiswoode died in 1883. Hirst retired the same year, followed by Huxley, Hooker and Frankland in 1885 and Tyndall in 1887. Their collective impact in transforming the place of science within education and the wider culture had been immense. All the same, they had not secured a final victory for scientific naturalism. Both the next two Presidents of the Royal Society, George Stokes and Lord Kelvin, were committed to the principle that Christianity could be brought to bear on and substantiated through science. But the Xs had pushed these arguments increasingly towards the margins, at least within science itself. Within biology, they had championed Darwin and evolutionism in alliance with Christian evolutionists such as Acland, Rolleston and Flower, all of whom came to reject the a priori assumptions of natural theology in favour of the more comprehensive explanation of the history of life on earth proposed by Darwin. Over the same period the geologists and natural historians who had resisted Darwin's model of science and rejected its findings gradually died off. At Oxford, Daubeny died in 1867, Phillips in 1874. Adam Sedgwick from Cambridge went in 1873. Two more prominent natural theologians – Buckland's son Frank, a popular writer on natural history, and Thomas Rymer Jones, sometime Professor of Comparative Anatomy at King's College, London – followed in 1880. By the early 1880s, when Hunt and Millais painted their pictures of ageing natural theologians, it looked as though it was not just the scientists themselves who were dying, but their very idea of what science was.

Woolner had registered this impending change within science as early as 1860 in a masterly portrait bust of Sedgwick, the Woodwardian Professor of Geology at Cambridge (fig. 132). In the months when he was sitting for Woolner, Sedgwick had been dwelling on his outrage at Darwin's new book. To his mind it was 'a dish of rank materialism cleverly cooked and served up'.[5] Sedgwick attacked Darwin for abandoning Baconian induction and rejecting the evidence for final causes – the teleology of design – in nature.[6] Sedgwick was in his mid-seventies when Woolner's bust was carved. Woolner records his massive Roman nose, the heavy bags under his eyes, the baggy skin of his neck, the veins in his temples, even the careless brushing of his hair. But the features that convey his state of mind most clearly are his set frown and his thin, pursed lips. Sedgwick had been known for his charm and good humour as well as his pugnacity, but though his eyes are ringed with laugh lines he is not laughing now.[7] Instead, Woolner portrays Sedgwick as an old but implacable man, steadfast in his defiance of what he sees as dangerous error in science. Yet there is just a hint of resignation in his face when viewed sidelong.

If Woolner's bust of Darwin is a portrait of the scientist as seer (see fig. 98), his Sedgwick is the very model of the

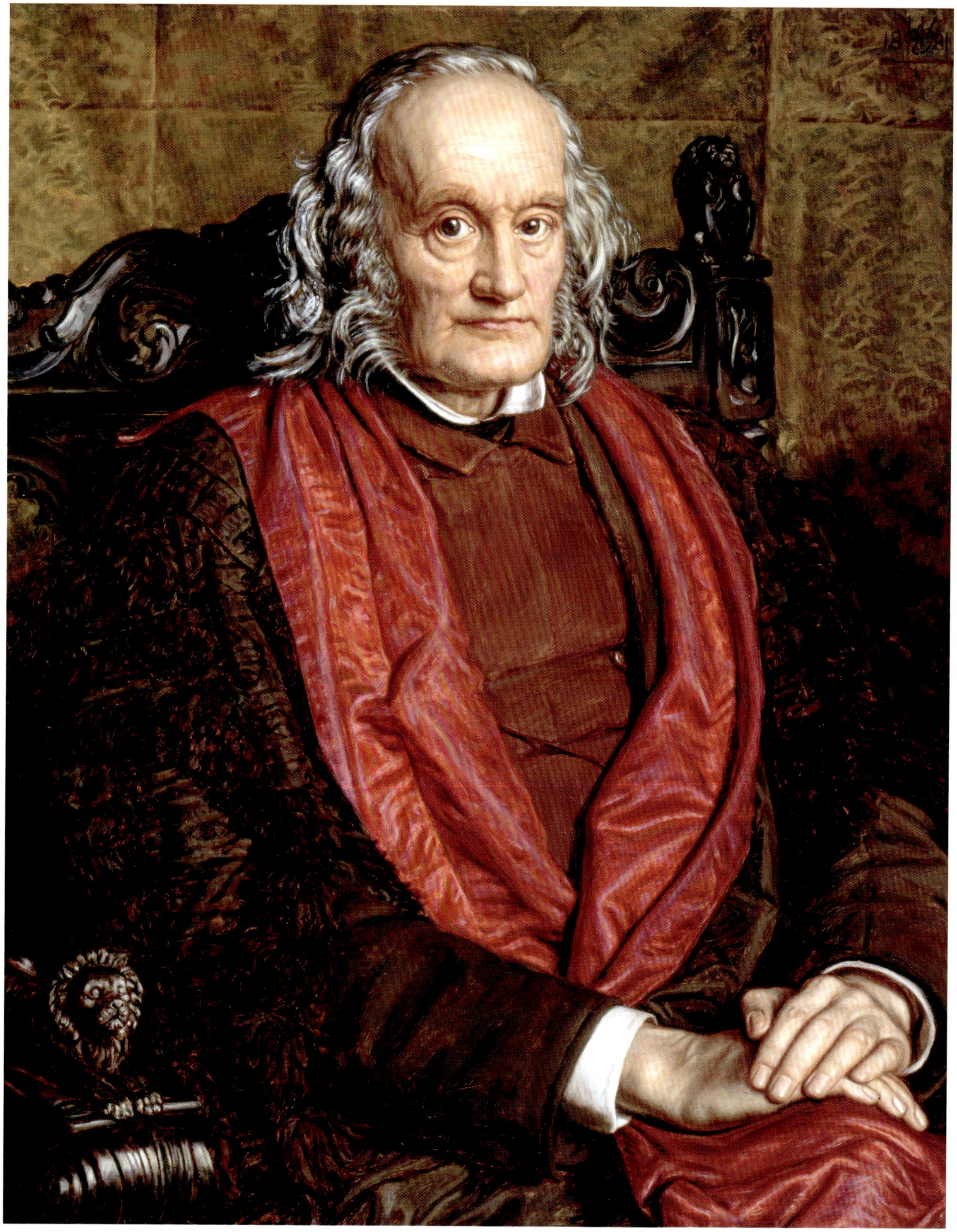

133 William Holman Hunt, *Sir Richard Owen*, 1881. Natural History Museum, London.

determined Baconian, aged and tenacious, but beneath it all somehow aware that he is increasingly obsolete. This same sense of obsolescence haunts Hunt's portrait and Millais's deathbed scene. By now in their seventies, Owen and Gould had both been influential in shaping the discipline of natural history in the mid-century, one as a palaeontologist, the other as an ornithologist. Both were skilled taxonomists, though Owen worked in museums while Gould was increasingly a field naturalist. Both popularised their own science, Owen through lectures and exhibitions, Gould through collections of illustrations by himself and others including Edward Lear. Above all, both saw the beauty and order in nature as signs of divine artistry and purpose. By the time Hunt and Millais came to paint them, however, this view of science and nature had been successfully challenged and largely supplanted among British biologists. As Darwin had predicted in 1859, it was 'young and rising naturalists' who had embraced and championed his evolutionary naturalism and methodological materialism against the older generation.[8] The rapid rise of evolutionism was a victory for Darwin's science, but it was also a consequence of the passing of the generation whose science he had challenged.

In her study of portraits of scientists, Ludmilla Jordanova draws attention to the collaborative relationship between the artist and the sitter in portraiture.[9] Hunt's portrait of Owen (fig. 133) is a collaboration between an artist and a scientist, each of whom respected and identified with the other's work. Owen shared Hunt's commitment to scrupulous attention to detail and an informed yet critical gaze. Hunt shared Owen's conviction that science was the study of the 'work of creation', as he himself put it in his late lecture on 'Religion and Art'.[10] For them both, art and science together were means of discovering and paying homage to God's divine purpose. In painting his portrait of Owen, Hunt pays tribute to this world-view which they share. As Rupke remarks, Hunt's painting 'gave expression, with the purple of Owen's outfit, the sanctity of his facial expression, and the pious pose of his hands, to his priestly, if not episcopal, aspirations' – aspirations which complemented Hunt's own increasing perception of himself as a prophet more or less in the wilderness, deserted by Rossetti and his other Pre-Raphaelite brethren.[11]

Hunt's Owen is enthroned in a heavy, dark wooden chair decorated with heraldic lions. This throne, his rich professorial robes and his age all underscore his public authority. The painting was exhibited when the new Natural History Museum had just opened. Owen's standing with the public was at a fresh high, for all that his credit with his fellow scientists had been damaged by years of imperious and not always honest dealings, by the opposition of Huxley, Darwin and their allies, and by his own refusal to accept their often compelling scientific arguments. For Hunt, though, Owen remains a consummate and exemplary scientist. His professional skill is suggested by his long-fingered hands and large, piercing eyes. His hands are as deft as those of an artist. They are surprisingly youthful, suggesting that his abilities as a comparative anatomist are unimpaired by age. At the same time, the circle formed by his strong right eyebrow and the heavy bag beneath it – signs of age, not youth – marks his eye out as an optical instrument of preternatural acuity.

Hunt pays tribute to Owen in his portrait, but he registers too the limits of his power. Like Woolner's bust of Sedgwick, Hunt's painting of Owen shows in unsparing detail the physical toll age has taken on his face. His skin is taut over his forehead but baggy on his cheeks and under his eyes. His wispy hair has receded and now frames his face in such a way that, as T. S. Eliot said of Webster, we can see the skull beneath the skin. Both his hair and robes echo the mane of the lion on the arm of his chair, but Owen is a lion in winter. There is an alertness and irascibility to his face, a suspicion in his eye, which suggests he might still bite if provoked, yet his cheeks are sunken, his mouth narrow, perhaps toothless. As a man, he is old, and cold too, buttoned up against the chill of age in two collars exactly as Woolner's Sedgwick is. Like Sedgwick, he is still intellectually alert, but where Woolner's marble gives Sedgwick the mass and solidity of a man who can still sustain a fight against Darwinism, Hunt's paint, for all its vividness and texture, shows us an Owen who seems frail and increasingly insubstantial under his robes and collars.

Hunt's Owen is a reluctant King Lear, on the brink of resigning his kingdom to impious and squabbling successors who will destroy his legacy and trample his reputation into

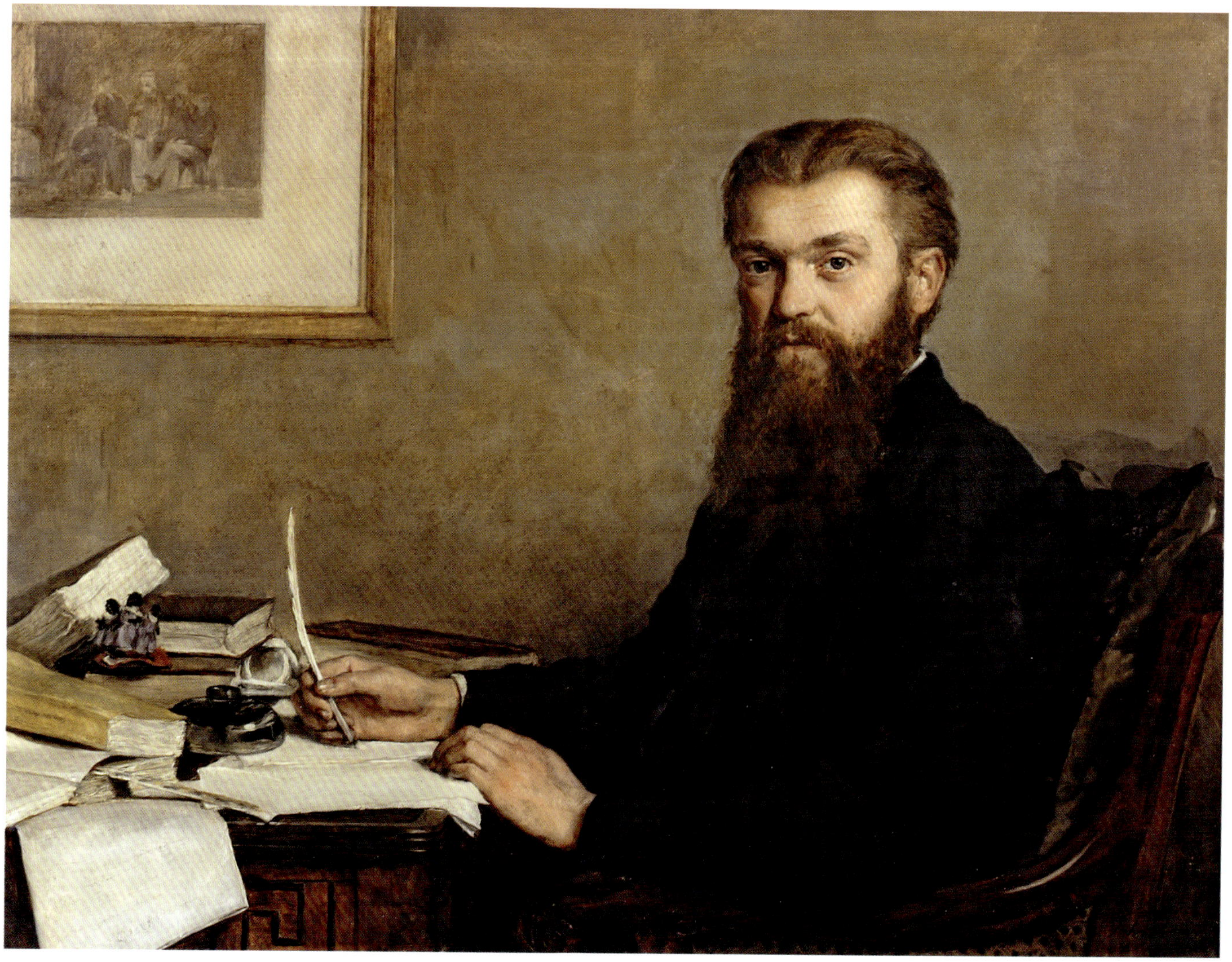

dust. When the portrait was painted, Owen was still one of England's senior men of science, presiding over the national natural history collection with the authority of the British Museum behind him. But he had long been under siege. The X Club and their allies were surrounding him and challenging his authority from their own individual fiefdoms and within the Royal Society. Huxley was breathing down his neck at the Normal School in South Kensington; Hooker had never accepted the British Museum's suzerainty over Kew; Lockyer and *Nature* were sniping at his new Natural History Museum. His old post as Hunterian Professor at the Royal College of Surgeons had been claimed by two of his staunchest opponents in turn, first Huxley, then Flower, with Flower also taking over the college's Hunterian Museum. When Owen retired as Superintendent of the Natural History Museum, two and a half years after Hunt painted him, Flower would take this over as well.

Hunt's portrait of Owen is put into relief when it is compared with the series of portraits of the scientific naturalists painted at much the same moment by John Collier. From the late 1870s to the early 1890s, several members of the X Club sat for Collier, including Hooker, Lubbock and Huxley. He also completed a posthumous portrait of Spottiswoode, and painted the X's close friend and protégé, William Kingdon Clifford, a year before he died of lung disease aged only 33 in 1879 (fig. 134).[12] Collier was personally very close to Huxley and his circle. He married Huxley's daughter, the painter Mady Huxley, in 1879, and then, in 1889, two years after her death, he married her sister Ethel in Norway to evade the ban on marrying a deceased wife's *(facing page)* 134 John Collier, *William Kingdon Clifford*, 1878. Royal Society, London.

135 John Collier, *Joseph Hooker*, 1880. Royal Society, London.

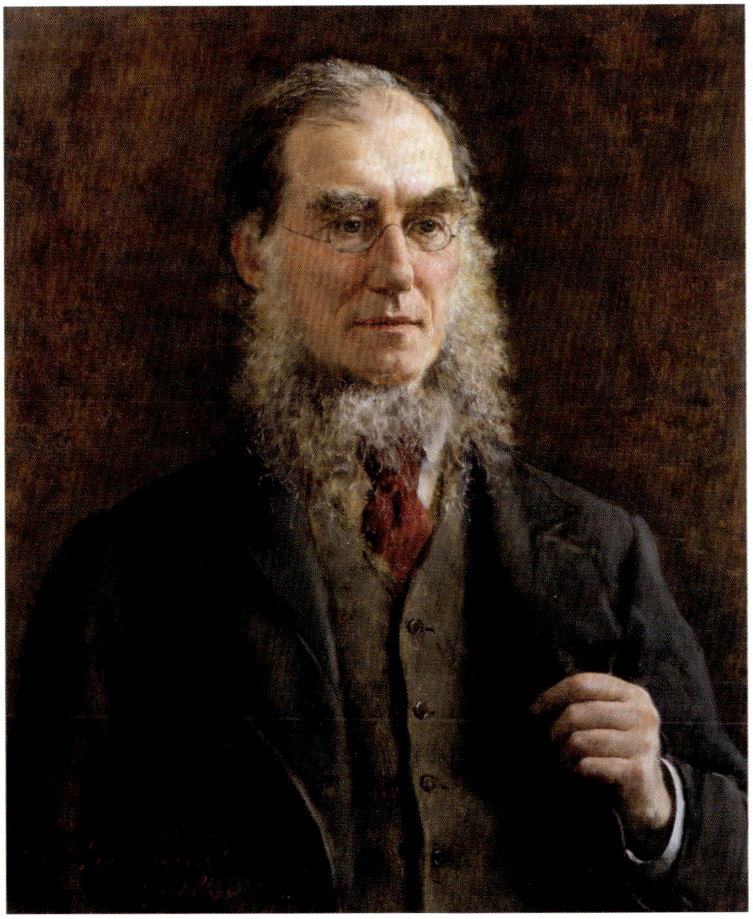

sister in England (a law Hunt also fell foul of). He was also Clifford's mainstay during the months before he died. The easy terms Collier enjoyed with his sitters is reflected in the portraits. Hooker (fig. 135) looks dignified, Lubbock genial, Clifford slightly quizzical, but they all look relaxed, without Owen's edginess. There is a touch of informality to each portrait – Hooker thumbing his lapel, Huxley with his hand in his pocket (fig. 136), Clifford sat at an overcrowded desk – befitting their friendship with Collier himself, but also suggesting their brisk self-assurance. Where Hunt's Owen insists on the status accorded him by his robe and his throne, Collier's scientific naturalists do not have to stand on ceremony, nor do they expect others to stand on ceremony for them. Instead, they are confident in their own authority as modern men of science.

Collier's portraits of Huxley and Clifford contrast with Hunt's Owen in another way too. The marker of Owen's scientific eminence in Hunt's portrait is his professorial robe and his dark wood throne. These are signs of achievements accomplished and formally recognised. Huxley and Clifford are shown rather as contributing to science as thinkers and writers here and now. Clifford sits at his desk. He is clearly posed and not in the act of composition, but the pose nonetheless identifies him with his writings. This is a conventional motif that Collier would repeat for his second, somewhat pompous portrait of Huxley, painted in 1890. Where Huxley's writing desk is orderly, Clifford's is a mess, suggesting his refusal to curb his intellectual energy. A more striking and original depiction of the scientist as thinker comes in Collier's earlier portrait of Huxley (fig. 136) from 1883, the year he took over as President of the Royal Society. Here Huxley leans on a stack of books, holding a skull in one hand, gesturing towards his work on human anatomy and evolution. If Hunt's Owen is King Lear, Collier's Huxley is Hamlet, meditating on the implications of mortality and calling into doubt the authority of words, words, words. Yet unlike Hamlet, Huxley as we see him here is not distressed or paralysed by his doubts. Instead, he has taken them in his stride. He can hold the skull – western art's most immediately vivid memento mori – without flinching, either at the finality of death or at the lessons it tells us about our own animal origins, whether as a synecdoche for the skeleton or as a metonymy for the brain. Indeed, he has taken the lead in confronting us with precisely these facts about man's place in nature. It is Hunt's Owen, nervous and diminished by time, who has not faced up to them.

Hunt's Owen is weakened by age but he remains magisterial, even in his own way heroic. The battle for natural theology may be lost in practice, Hunt implies, but the cause was right and it was worth fighting for. Millais's ornithologist, by contrast, is decrepit. *The Ruling Passion* shows him bedridden, surrounded by his grandchildren who are dutifully attentive but far from rapt (fig. 137). This version of the painting's title seems to mock this pathetic figure for his obsession with birds and their plumage, as does the clear boredom on the face of the girl sitting by the foot of his bed and the grotesque parody of taxonomy suggested by

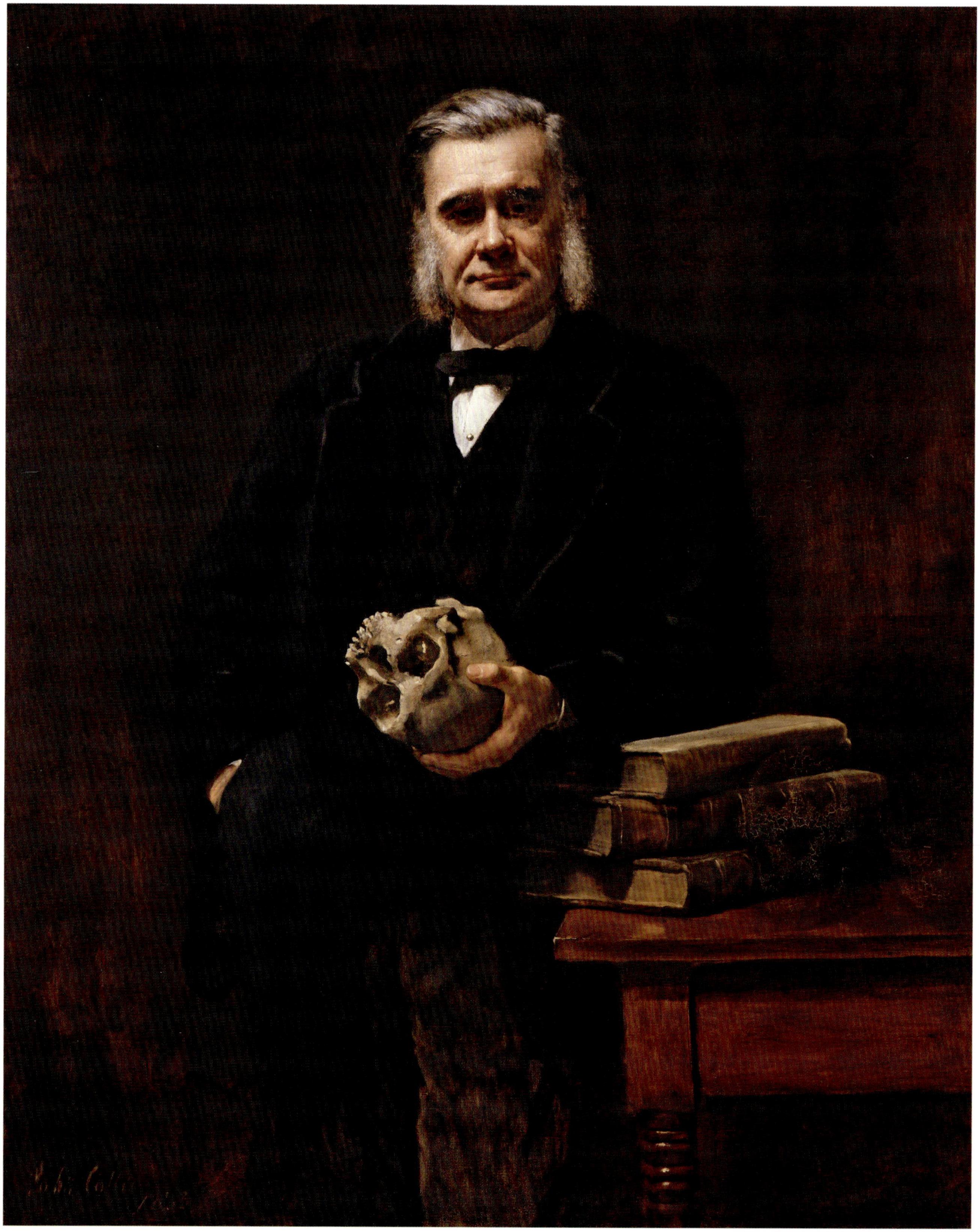

(*facing page*) 136 John Collier, *Thomas Henry Huxley*, 1883. National Portrait Gallery, London.

137 John Everett Millais, *The Ornithologist; or, The Ruling Passion*, 1885. Kelvingrove Art Gallery and Museum, Glasgow.

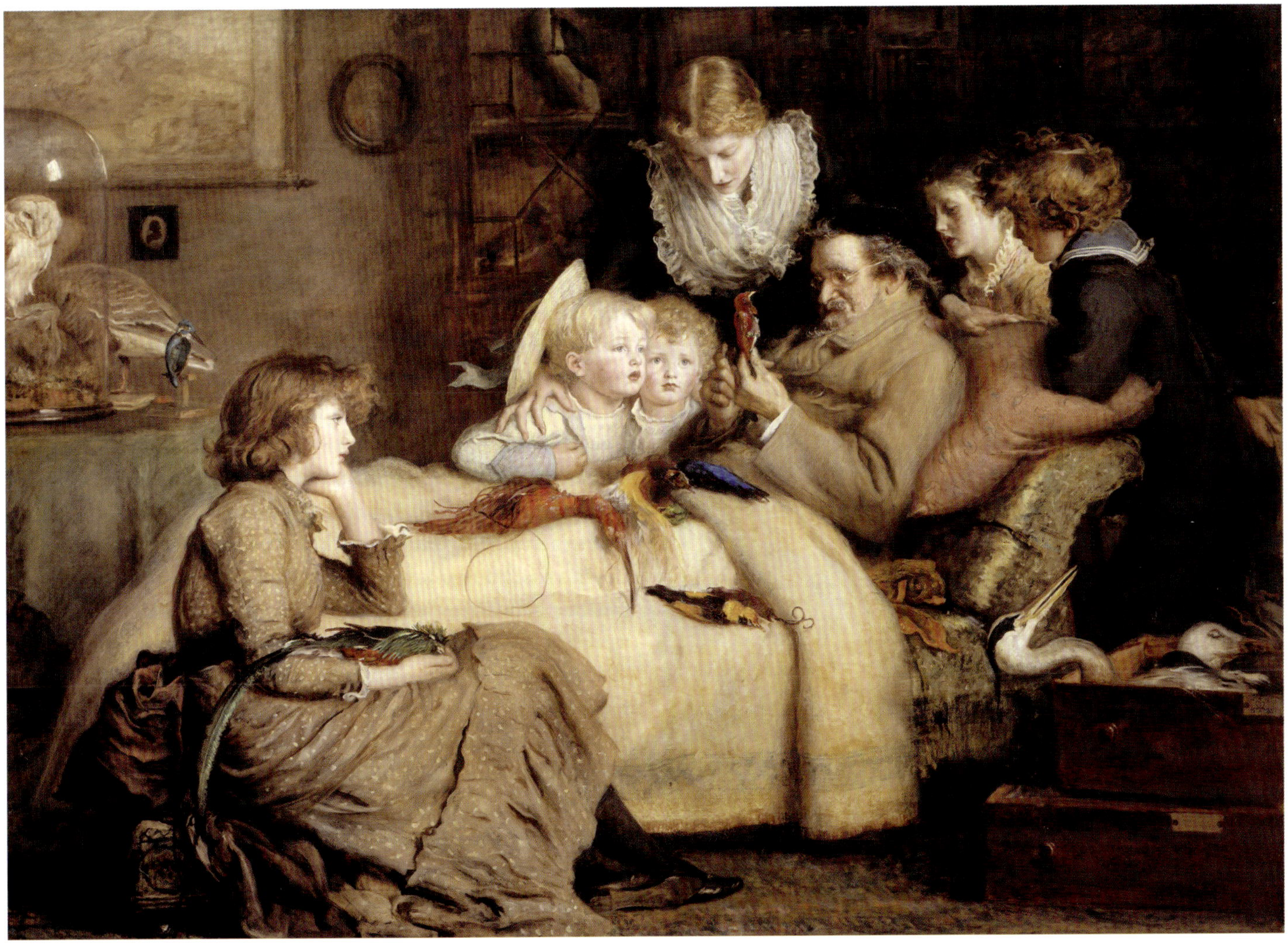

the stuffed birds kept in drawers left lying about on the floor. This element of mockery seems to align Millais's painting with Henry Stacy Marks's paintings of eighteenth-century ornithologists *Science is Measurement* (1879), and *A Treatise on Parrots*, painted in 1885, the same year as *The Ruling Passion*. Marks's paintings are calculatedly ludicrous. His ornithologists are distanced from the viewer by their eighteenth-century dress. Each is made to resemble the birds he is studying. The expert on parrots is dressed in the same colours as the parrots themselves, while the stiff pose of the man measuring the stork and the thin line which his spectacles trace from his periwig to his beaky nose recall the skeleton before him (fig. 138). As the scientists themselves are somewhat ridiculous, particularly in the earlier painting, so is their science. *Science is Measurement* precisely repudiates the claim made by its own title, while the sheets from the treatise on parrots scattered across the floor, each one a meticulous copy of a specimen, call into question the usefulness, let alone the practicability, of this kind of taxonomical cataloguing.

But while Millais's painting seems to join in this mockery of the outmoded taxonomist, its tone is very different from that of Marks's witty but ultimately facile paintings. For one thing, Millais's palette is very different from Marks's. Where Marks's paintings are characterised by an even and clear light

140 Ford Madox Brown, *John Kay, Inventor of the Fly Shuttle, AD 1753*, 1888–90. Manchester Town Hall.

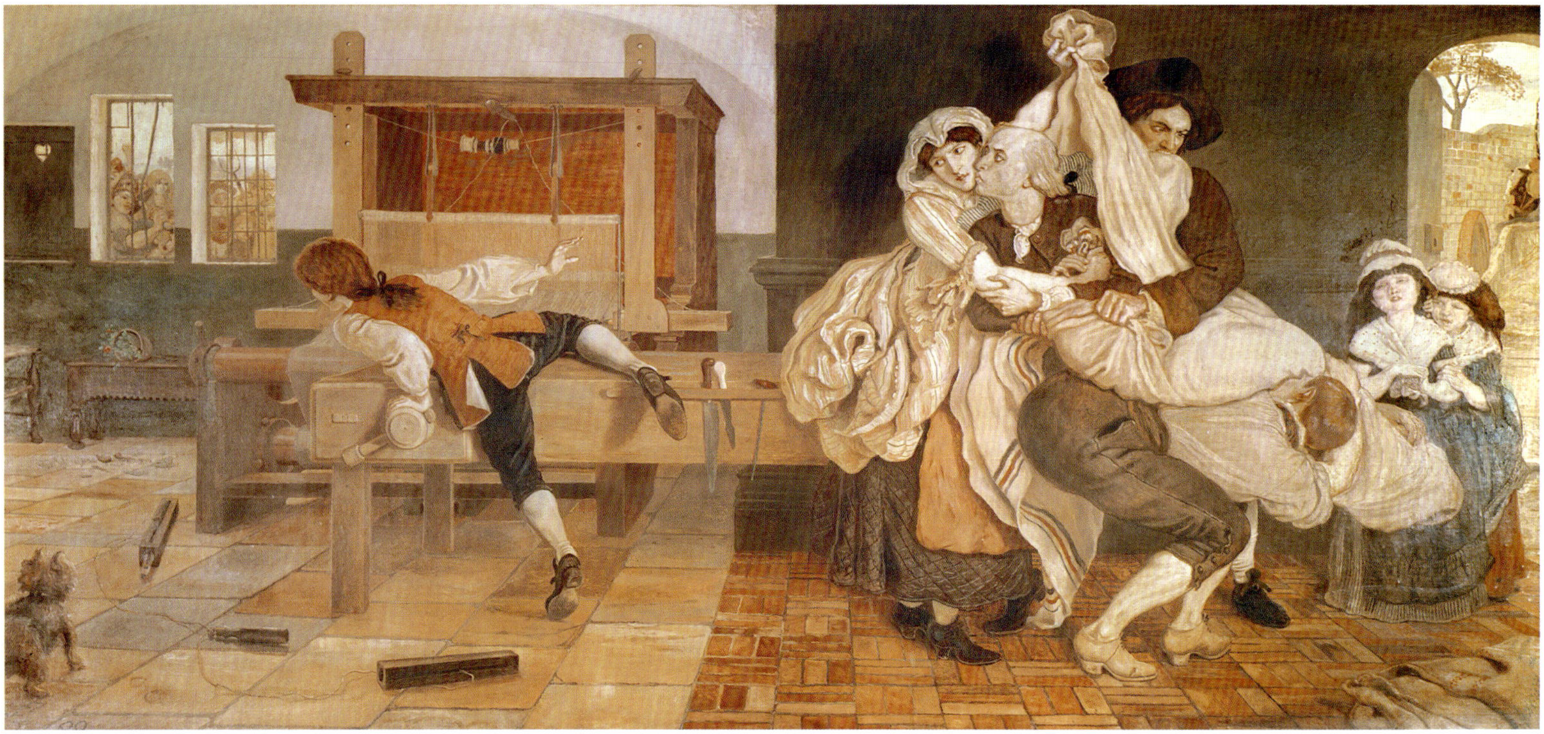

one councillor as describing it as 'one of the most painful things he has seen in the way of art for many a long year'.[29] The composition as a whole marginalises the canal's contribution to Manchester's industrial achievement in supplying the city with coal by showing only the front of one small coal barge, outshone by the celebrations of the duke's visit, with the barge boy himself cheering him on.

Kay's invention and the Bridgewater Canal stand in for the advances in technology and industry which underpin modern Manchester's prosperity, for all that Brown views them with scepticism. How does pure science fare in Brown's representation of Manchester's history by comparison, and what function does it play? Brown had depicted scientists as men of historic importance before in a series of windows commissioned in 1870 by Peterhouse, Cambridge, from Morris, Marshall, Faulkner & Co. The contrast between Brown's two historic cycles at Peterhouse and Manchester is striking. The windows on the north and south sides of Peterhouse dining hall form a pantheon of eminent students, fellows, masters and patrons of the college, including several churchmen, poets and one scientist, the eighteenth-century chemist Henry Cavendish. Three of these windows were designed by Burne-Jones, the rest by Brown.[30] Brown also drew the cartoons for the series of seven smaller portraits included in the oriel window at the end of the hall. The central figure is Hugo de Balsham, the medieval bishop who founded the college – the earliest Cambridge college. The other six have no direct connection with the college, but represent instead a collective assertion of its ideals of learning and knowledge. To the left stand three classical authors, Homer, Aristotle and Cicero; to the right, three British scientists, Roger Bacon, Francis Bacon and Newton (fig. 141).

This second, less parochial pantheon designed by Brown for Peterhouse overlaps with the pantheon of statues of scientists at the Oxford University Museum. Peterhouse's decision to identify the ancients with literature and the moderns with science is less stark than it looks, not because modern literature is accommodated through Francis Bacon but because, as at Oxford, Aristotle is cast not as a philosopher per se but as a natural philosopher specifically, making four

141 Peterhouse Hall, Cambridge, showing the oriel window designed by Ford Madox Brown, 1870.

Details of the windows depicting (142) Homer and Aristotle; (143) Hugo de Balsham and Roger Bacon; (144) Francis Bacon and Isaac Newton

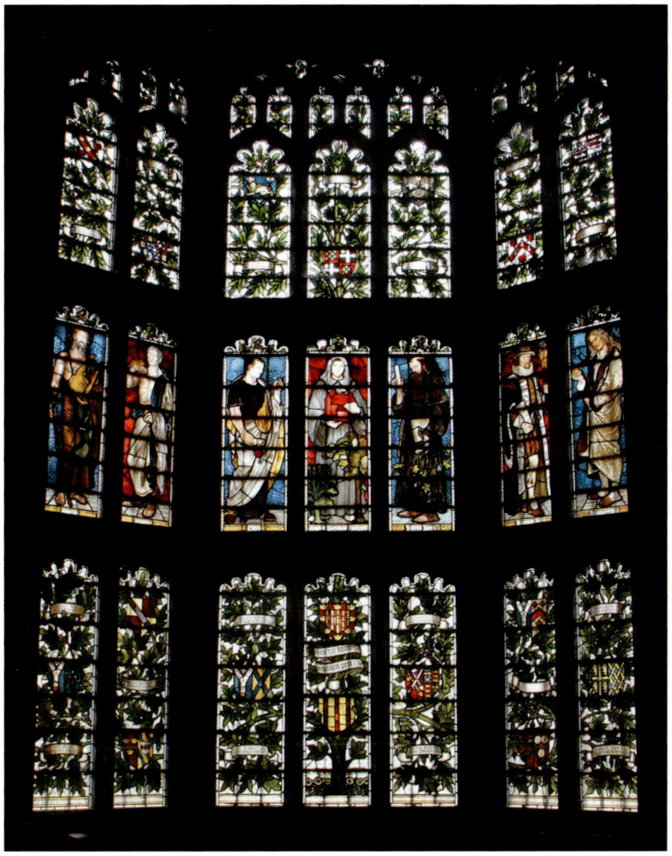

of the six scientists. The series as a whole traces a progressive tradition of empirical observation. Aristotle is shown examining a living chameleon perched on his hand (fig. 142). The chameleon was one of the animals described in most detail in Aristotle's *History of Animals*.[31] Aristotle's interest in natural history is picked up on in the portrait of Roger Bacon, who is shown holding a branch of a fig tree in his left hand (fig. 143). According to *The Cambridge History of Science*, Bacon picked up on the discussion of grafting fig trees in the ancient Greek treatise *On Plants*, then attributed to Aristotle, although his own interests centred on the nature of the plant's soul or intrinsic identity.[32] Bacon holds a copy of his *Opus Majus* under his arm, looking forward to the next window, where Francis Bacon is shown holding the *Novum Organum* under his right arm and a burning torch in his left hand, symbolising the light of science (fig. 144). As its title signifies, the *Novum Organum* is both a homage to and a revision of Aristotle's *Organon*. Where Aristotle is studying a chameleon, Roger Bacon is looking through a prism, alluding to his work on optics and looking forward to Newton, who is, predictably, looking equally intently at an apple (fig. 144). There is an appealing hint here of the old principle of organising natural history into animals, vegetables and minerals, for all that only Aristotle is concerned

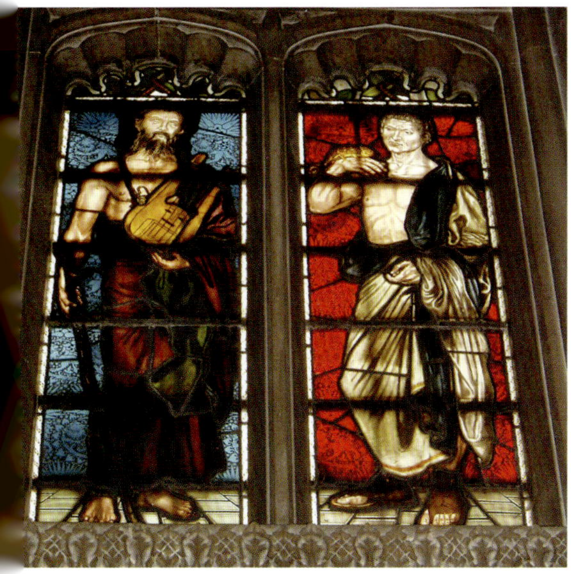

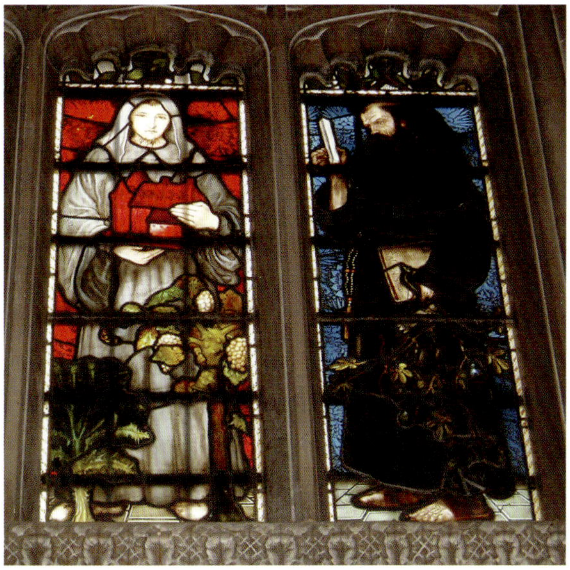

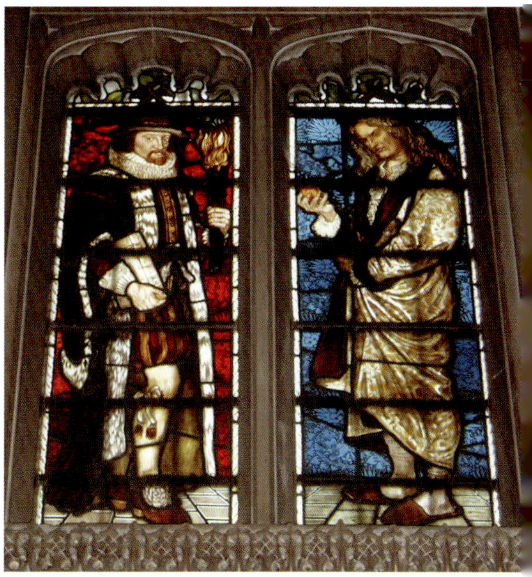

145 Ford Madox Brown, *Crabtree Watching the Transit of Venus, AD 1639*, 1882–3. Manchester Town Hall.

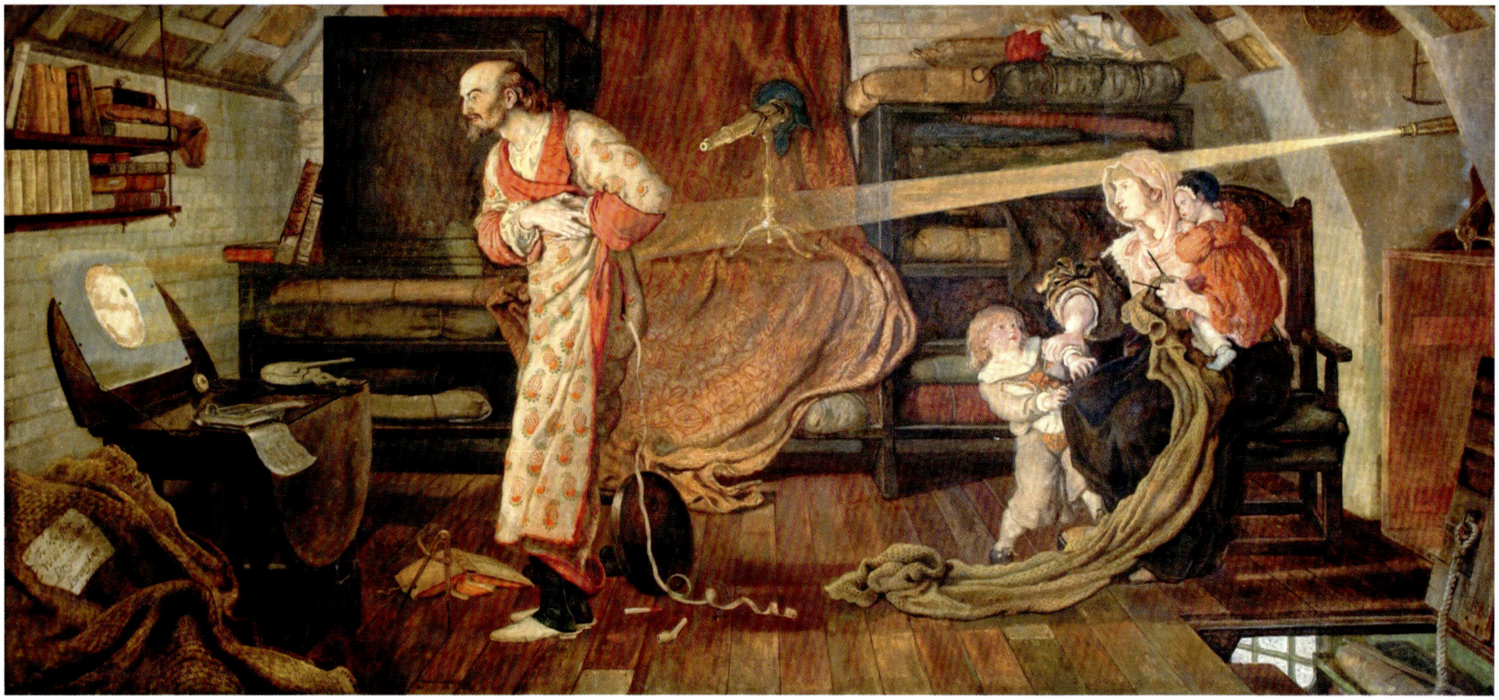

with the natural history of his object of study. There is a further narrative moving from fiction to fact, poetry to science, blindness to perception, as the windows are read across in order from Homer to Newton.

Like the Oxford statues, Brown's windows at Peterhouse establish a heroic lineage of empirical science. Where Woolner, Tupper and Munro represent the scientist as equivalent to a poet, gifted with a remarkable imagination, in the Peterhouse windows Brown restores the original Pre-Raphaelite conception of the scientist as first and foremost a dedicated observer. Aristotle, Roger Bacon and Newton are all shown in the act of scrutiny, their brows furrowed in thought as they look. Partly for this reason, and partly because of the flatness of the form, the windows lack the animation, immediacy or detail of the statues, though their framing by Morris in gem-like backgrounds of deep blue or red glass helps to intensify the impression they give of the depth and brilliance of human insight. In 1878 Brown doubled this canon of great men by adding six new cartoons, drawn for Charles Rowley and later included by Rowley in his autobiography, *Fifty Years of Work without Wages*.[33] This new series does not add any further scientists, though it does establish new parallels and associations. Shakespeare's posture and costume mirror Francis Bacon's closely, for example, establishing them as equivalent giants of the English Renaissance, and perhaps alluding to the controversy over whether 'Shakespeare' was in fact Bacon. Alfred the Great stands like Homer revitalised, his medieval harp, crown and sword picking up on Homer's lyre, laurel wreath and stick. Again, science, like kingship and the arts, is represented as the achievement of unique and determined individuals who should be honoured as great men.

Treuherz suggests that this series of twelve cartoons was intended to establish Brown's credentials as a historical painter for the town hall commission.[34] Rowley describes Brown's cartoons as 'unrivalled by any modern in dignity, insight, and invention'.[35] If this is what Waterhouse and the Manchester burghers expected from Brown, then it is no surprise if they were disappointed. His murals are insightful and inventive, but far from dignified. This applies as much to his treatment of science as it does to his depiction of other aspects of Manchester's civic history. Indeed, in their casual

146 Ford Madox Brown, *Dalton Collecting Marsh-Fire Gas*, 1886–7. Manchester Town Hall.

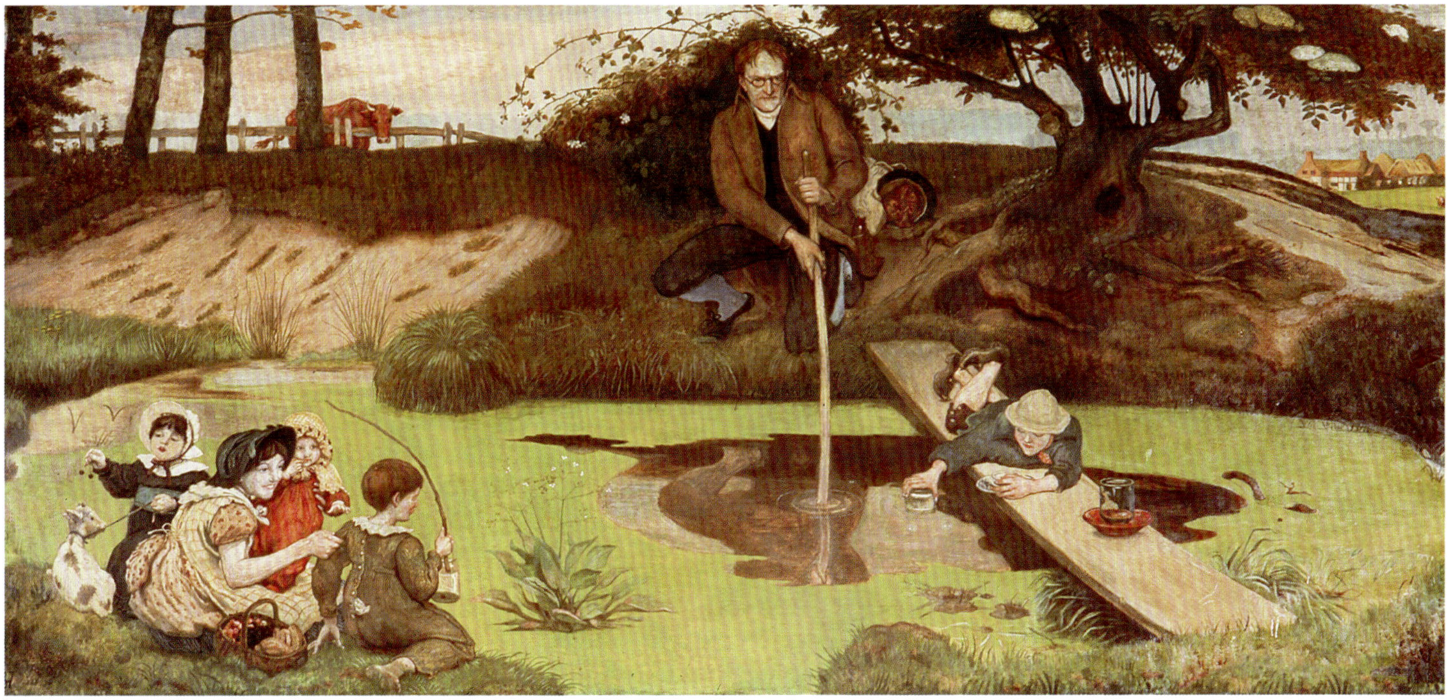

indignity, Brown's paintings of Crabtree and Dalton offer us a very different history of science from that implied by the heroic image of the scientist and the progressive image of science itself at both Oxford and Peterhouse.

The domesticity that characterises Millais's presentation of the old ornithologist in *The Ruling Passion* recurs in both of Brown's depictions of scientists in the Manchester Town Hall. The seventeenth-century draper and self-taught astronomer William Crabtree is shown watching the transit of Venus across the Sun projected down a telescope into his attic (fig. 145). He has thrown on a dressing-gown and slippers – the one continuity from the Peterhouse windows, where Newton is similarly hastily dressed – and his wife, child and baby are in the room with him. John Dalton, the inventor of modern atomic theory, is shown taking time off from his day job as a Manchester schoolmaster to collect gas from a stagnant pond (fig. 146). Like Crabtree and the ornithologist, he is surrounded by children, one of whom is helping him collect his samples, while a lugubrious cow looks over a fence in the background at them. For both Crabtree and Dalton as Brown imagines them, science is a hobby, something that happens in intervals between work. But it is also a vocation, as each is fully absorbed in attending to the phenomena in front of them.

Considering Brown's paintings of Crabtree and Dalton in the context of the series as a whole, the gentle comedy of their domesticity shows itself to be markedly less satirical than the broader, more burlesque comedy of many of the other murals. Where the applied science of technology and industry is lampooned and implicitly denigrated by Brown, science for the sake of knowledge itself is indulged and enjoyed. Crabtree may be in his dressing-gown, trailing his tape-measure on the ground and forgetting to take detailed measurements of what he sees. But this failure is precisely because he is caught up in the moment of observation – the first time anyone had seen a transit of Venus. In 1874 and again in December 1882, when Brown was painting his mural, the Victorians were themselves caught up in the excitement of witnessing the transit of Venus.[36] Crabtree's rapt gaze on the projected image of the planet's disc moving across the sun was an apt and topical image for the fascination which science held for a wider public as well as for scientists themselves. Crabtree was after all a draper by trade who only happened to be interested in astronomy, and the transit has the same hold over his wife and children as it has over him. His wife and their baby stare with the same fixed gaze along the beam of light as Crabtree himself, as she restrains the older child from disturbing his father by going to look more closely.

Brown's Manchester murals show little love for science when it is set on advancing material progress at the expense of people's existing lives and livelihoods. Instead, they offer us an alternative ideal of science which is at once democratic and anti-utilitarian. But where the Peterhouse windows sustain the idea of the scientist as a rare, heroic genius, the Manchester murals imply that anyone – a draper, a schoolteacher, any one of the poor boys educated in the surprisingly relaxed environment of Chetham's School – can become a scientist. Furthermore, they suggest that science can be undertaken by ordinary people in their usual surroundings, that it is not the preserve of public institutions like universities or of an intellectual aristocracy. Brown's Crabtree and Dalton are not lone geniuses, but everyday men of their time and place, set on the task of close observation and investigation with help from those around them who share their enthusiasm. Not overly conscious of their place in history, their humility and lack of self-regard is all the more appreciable when contrasted with the pride of the town hall itself.

In his Manchester murals Brown gives us glimpses of a social history of science, in place of the heroic history presented at Oxford and Peterhouse. At the same time, he revives Herschel's and Spedding's Baconian ideal of science as a process of discovery we can all contribute to – the same ideal that Brown and the other Pre-Raphaelites took inspiration from in their early art in the 1840s and 1850s. Yet his manner of doing so suggests that, like Hunt and Millais in their paintings of aged natural theologians, he is aware that this conception of science is increasingly old-fashioned. Like the warm reception *The Earthly Paradise* received from 'scientific men', Brown's depictions of Crabtree and Dalton reveal a nostalgia for a time when science could be conceived in Stephens's terms as 'adherence to fact, to experiment and not theory', and when the discovery of new facts that might advance scientific progress seemed to be within the reach and capabilities of ordinary people. There was still a place for the citizen-as-scientist in late Victorian England.[37] But in the 1880s the significant research into the transit of Venus was undertaken by worldwide expeditions of professional scientists, not by a draper alone in his attic with his family. Crabtree, his small brass telescope and his trailing tape-measure are part of Manchester's history – that is, its past. Dalton is too, albeit more recently. Here Brown's treatment of his subject pushes him further into the past. When Brown was first asked to paint Dalton, he remarked: 'Dalton inventing atomic theory could never be anything, pictorially, but one man thinking – unless one might represent him shaking up his atoms, like grated Parmesan, in a bottle.'[38] He solved the conundrum of how to show Dalton and his science in a picture, but only by taking him out of the laboratory implied in this comic image of a bottle of atoms and putting him back into everyday life and the outdoor landscape of natural history. In the process, he effaced the radical modernity both of Dalton's ideas and of Manchester itself as an industrial city.

Brown's murals offer a demotic critique of the grandiose historical pretensions of Waterhouse's town hall, a counter-narrative to contest the myth of Manchester's illustrious past that the building as a whole creates. Although Brown does not take war or politics, commerce or industry as seriously as the Manchester burghers might have expected him to, he values the pursuit of knowledge for its own sake and holds scientists themselves in affectionate regard. Yet where Waterhouse boldly affirms his historical myth as a foundation for the next stage of Manchester's civic greatness, it is less clear whether Brown sees the Baconian science of Crabtree or Dalton as recoverable. In his Manchester murals Brown revisits the model of science which he and his fellow Pre-Raphaelites had embraced the best part of forty years before. But he finds this ideal in the past, and though it has resonances in the present, as the topicality of Crabtree's astronomy reveals, it is not necessarily workable into the future. Certainly, the X Club would have repudiated its amateurism. On the other hand, while democratic Baconianism may no longer have been a plausible model of science itself, it offered, Brown suggests, a much better and happier model of citizenship than militarism, capitalism or deference to established authority.

In their late paintings of scientists, Brown, Millais and Hunt take a last look back at the science that inspired the Pre-Raphaelite project. They retain their admiration for the old Baconians and natural theologians, even as they accept that their conceptions of what science entailed and what it could tell us no longer hold sway. There is an integrity to Hunt's Owen even in defeat, and a love born from discovery and borne out in the sharing of knowledge in Brown's Crabtree, his Dalton and even Millais's ornithologist. In 1850 Stephens wrote that imitating science could advance the moral purposes of the arts. In the 1880s the surviving Pre-Raphaelite painters used their art to trace the positive moral legacy of mid-Victorian science.

At the same time, Hunt's and Brown's paintings belatedly reaffirm the original Pre-Raphaelite project. Hunt endorses natural theology, Brown Baconianism; Hunt maintains the hyperreal clarity and unflinching gaze of early Pre-Raphaelitism, Brown its commitment to scrupulous research and independent investigation. As the Baconians and natural theologians had passed away, so too had many of the Pre-Raphaelites. Of those who had been closely involved in the project in the late 1840s and 1850s, Munro died in 1871, Collins in 1873, Tupper in 1879. Between Hunt beginning his portrait of Owen and Brown starting to paint Crabtree on the wall of Manchester Town Hall, Collinson and Rossetti both died too. As they lost their old comrades, the surviving Pre-Raphaelites began to take a long look back over the movement, tracing its origins and development. These paintings contribute to this reflexive analysis, alongside memoirs, lectures and essays by Hunt, William Rossetti, Ruskin and Stephens.

Like these written recollections, Hunt's and Brown's reaffirmations of Pre-Raphaelitism in their late paintings cannot escape their own associations with the past. In painting Owen beleaguered and in decline, Hunt is also painting an implied self-portrait as the last Pre-Raphaelite, staunchly but fatalistically battling on for a scientific aesthetic that Rossetti, Millais and the others had long abandoned. In resigning Baconian science to history, Brown hints that his own insistence on Pre-Raphaelite practices in art may be a form of Luddism, a refusal to comply with the demands and expectations of modernity in art, science or civic life – the aesthetic equivalent of the machine-breaking that tried but ultimately failed to retard Manchester's industrial progress. Like Millais's ornithologist, by the 1880s the old Pre-Raphaelites were in terminal decline. The ruling passion of Pre-Raphaelitism itself – its commitment to facts, freshly discovered, clearly outlined and captured in brilliant colour, like the feathers of the stuffed birds laid out on the dying naturalist's blanket – had come to seem as outdated as the science that had inspired it.

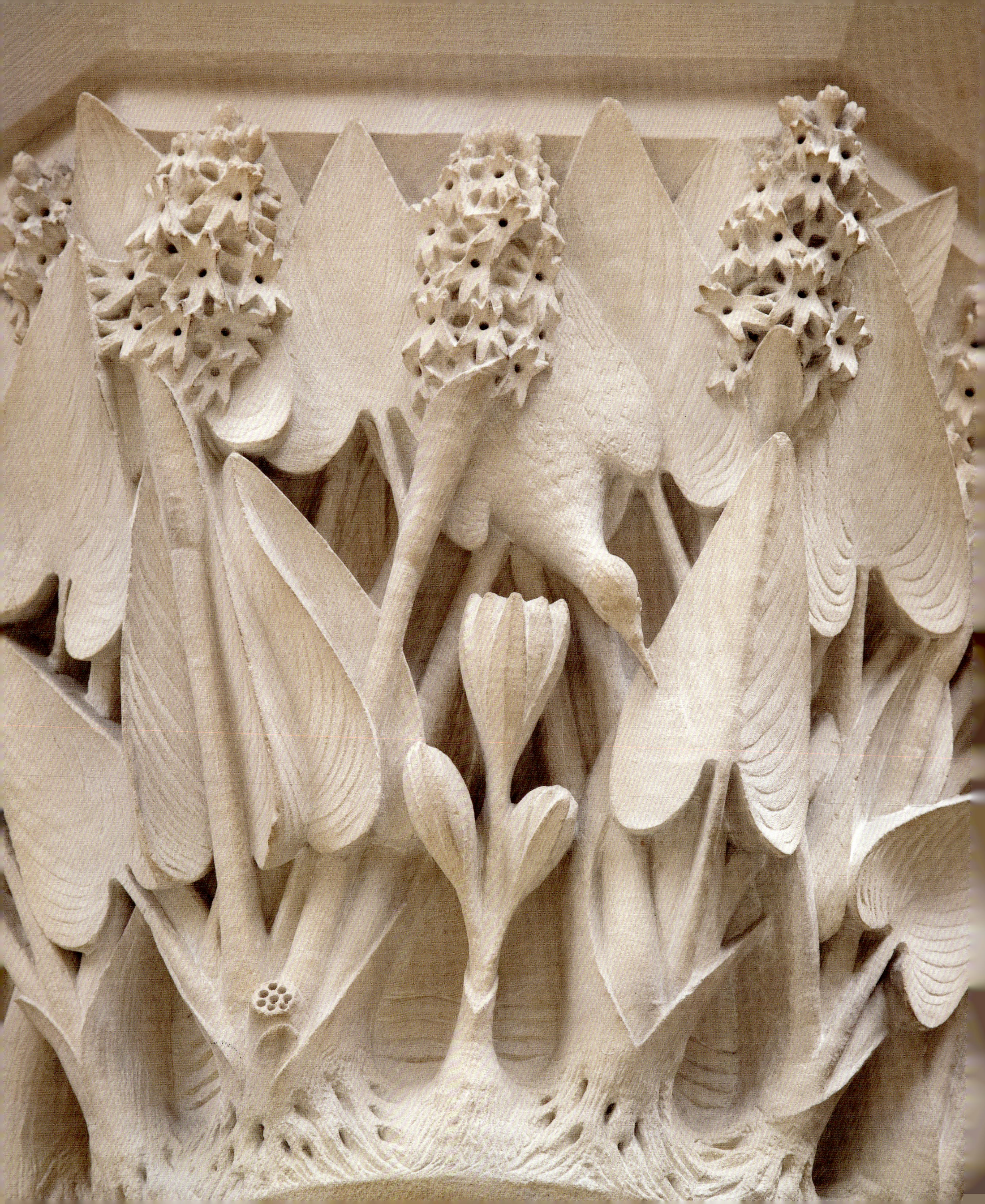

CONCLUSION

The Pre-Raphaelites and Science Today

As *Nature*'s response to the Natural History Museum suggests, and as Hunt's, Millais's and Brown's late paintings of scientists tacitly acknowledge, by the 1880s the original Pre-Raphaelite project was a spent force, like the Baconian science that had inspired it and the natural theology that had embraced it. Through their campaign to remake the arts on the model of science, the Pre-Raphaelites had brought a new rigour to painting and sculpture. They had pioneered the principle that a work of art should be an investigation, scrutinising the natural and the social worlds, and challenging its viewers' preconceptions. Their intensity of gaze and determination in research had led to fresh insights into both psychology and natural history. Their poetry had the same clarity and concentration, the same refusal to fall back into received wisdom. It too became a catalyst for fresh thought, a means to comprehend oneself and one's experiences, and to hold science itself to account.

Pre-Raphaelitism not only reinvigorated the arts. It helped to form the physical, educational and cultural landscape of Victorian science. The observational rigour of Pre-Raphaelite art set an example and a standard for the public communication of science through architecture. In so doing, it shaped public perceptions of what science revealed about nature. It even provided a model for how scientific observation itself should be practised and taught. It is impossible to quantify the impact of Pre-Raphaelitism on Victorian science and perceptions of it. We do not know how many science students were taught drawing on Pre-Raphaelite principles at Oxford, for example, or which of Lockyer's observations and discoveries took account of his consultations with artists, or

Facing page Detail of fig. 121.

how many of *Nature*'s readers took his and Brett's advice to enter into such collaborations themselves. Nor can we say precisely what impression was left by the architecture of the natural history museums in Oxford and London on the world-views of the scientists who studied or worked in them, or of the everyday Victorians who visited them. What is certain is that the Pre-Raphaelite project to ground art in science was fundamental to the design and reception of these major museums which underscored and defended natural theology through public art. Equally clearly, Pre-Raphaelitism in its various forms was an integral component of periodicals such as *Nature* and the *Fortnightly Review* which made the opposing case for scientific naturalism.

Through its symbolism and aestheticism, second-generation Pre-Raphaelitism remained a potent influence on poetry and painting well into the twentieth century. But its relationship to science had always been, and continued to be, more ambivalent and uncertain. After Rossetti's death, his friends Thomas Gordon Hake and Theodore Watts sought to construct new imaginative syntheses between his art and modern science. Hake memorialises their visits to Rossetti at Kelmscott Manor in the early 1870s in the opening sonnets of his 1890 sequence, *The New Day*, but Rossetti's real legacy for him is the form itself. Although Hake's poetic language is very different from his friend's, he uses the sonnet sequence as a structure to work through the existential challenges of scientific materialism, as Rossetti had done in the sixteen sonnets published by Morley in the *Fortnightly Review*. In this sense, the 'science-poet' that Hake prophesies in his poem is indeed a fresh synthesis of Pre-Raphaelite aesthetics and modern science.[1] Hake identified Watts as exemplifying this new ideal of the scientifically literate poet. Watts, who changed his name to Watts-Dunton in 1897, works through his own dialectic between Pre-Raphaelitism and science in his 1899 novel, *Aylwin*. The novel's narrator, Henry Aylwin, is a determined materialist in revolt against the esoteric mysticism of his father. These two world-views come together at the end in the alliance of the painter and self-declared mystic D'Arcy, based on Rossetti, and the alienist Dr Mivart, who bears some resemblance to Hake. Together they use magnetism to save Henry's sweetheart, Winifred Wynne, from hysteria induced by trauma. Their cure is represented as at once scientific in its method and spiritual in its effects. As D'Arcy advises Henry in the novel, it is only possible to gain clarity if one accepts that one is 'entangled . . . in a net of conflicting evidence – the evidence of the spiritual and the evidence of the natural world'.[2]

Around the turn of the century, Rossetti himself was cited as evidence for the direction of human evolution. For Max Nordau, a self-styled disciple of the Italian criminologist Cesare Lombroso, the Pre-Raphaelites were degenerates symptomatic of a biological as well as the moral collapse within western civilisation. Rossetti's sensuality and vague transcendentalism are characteristic, Nordau writes in his treatise *Degeneration*, of 'the brain-work of weak degenerate minds'.[3] If Rossetti is an imbecile, Swinburne's 'mysticism partakes more of the depraved and the criminal'.[4] Even the 'conscientious distinctness' employed without differentiation in the foreground and background of early Pre-Raphaelite paintings is a sign of the Pre-Raphaelites' 'incapacity for attention' equivalent to 'the disconnected speech of a weak mind'.[5] The spiritualist and poet George Barlow also held Rossetti to be 'organically different from others'. For Barlow, however, Rossetti represented a higher, not a lower, evolutionary type. As he explains in his pamphlet on *The Higher Love*, Rossetti's love poetry reveals that he 'possessed senses which others do not possess'. Love in his poems was 'peculiarly ethereal and refined', Barlow asserted, because within his biological make-up 'actual atomic changes had taken place by which physical atoms had been converted into what we may call soul-atoms.'[6]

Aylwin and *Degeneration* were both bestsellers, while *The Higher Love* was reprinted from essays published in the widely read *Contemporary Review*. For all their distribution, however, these texts remained very much on the margins of science. What is more, they all cast Rossetti as a mystic, positioning him implicitly in opposition to modern science and setting aside the deep vein of scepticism which had characterised the poems he published in the *Fortnightly Review*. Only Hake built on the affinity between Rossetti's own world-view and that of scientific materialism which Morley had recognised. *The New Day* was one of several sonnet sequences written from the 1870s through to the First World War which took a lead from Rossetti in seeking to work through what it might mean to live in a world deprived of the

147 Caryatides by Rudolf Weyr, room X, Naturhistorische Museum, Vienna, *c.*1884–5.

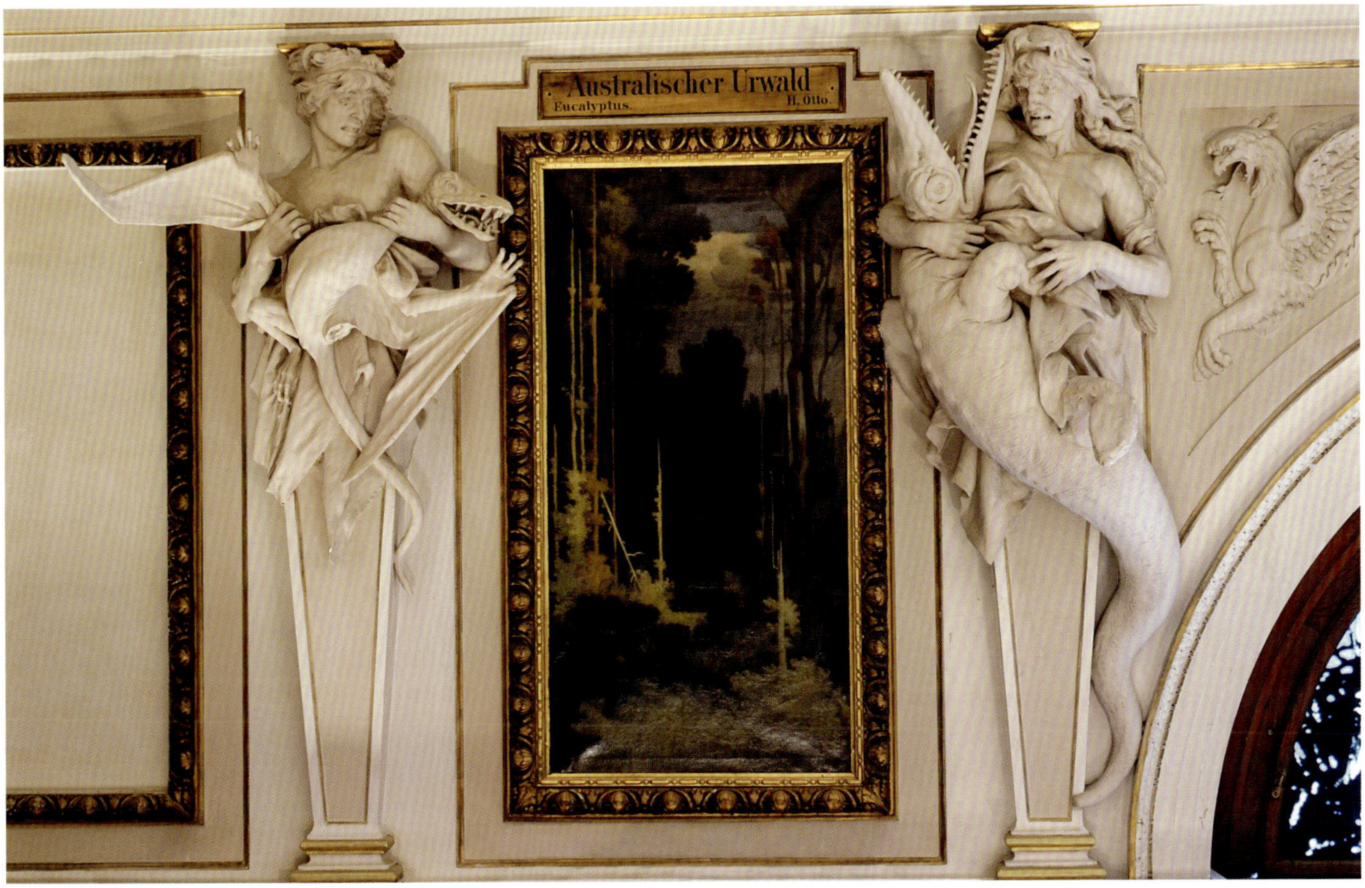

security of conventional beliefs.[7] Many of these poems engaged more explicitly than Rossetti himself had with science as a theme, but *The New Day* was unique in placing science at the centre of its own belief system.

Ultimately, the richest legacy of the Pre-Raphaelite engagement with science came not in poetry or painting but in architecture. The Oxford University Museum and the Natural History Museum became key points of reference for architects and scientists wanting to interpret the world through the fabric of their buildings. The Naturhistorische Museum in Vienna and the Royal Ontario Museum in Toronto are just two of the major museums that extended the programme of representational architecture on what had been Pre-Raphaelite principles to encompass radically different styles and different conceptions of science. As Stefanie Jovanović-Kruspel has explained, like Oxford, the Naturhistorische Museum (fig. 147) was born of a collaboration between the architects, the scientists who would be curating the collections, and the artists who were commissioned to execute its sculptures and paintings. The museums in Oxford and London were built in Gothic and Romanesque styles to convey different inflections of natural theology. In Vienna, a Darwinian view of nature was expressed through a baroque style which could accommodate a hybrid schema combining naturalistic painting, rococo humour and allegorical sculpture.[8] In Toronto, when a second wing was added to the Royal Ontario Museum in the early 1930s, the chosen style was Art Deco, albeit incorporating the round Romanesque arches that are characteristic of the late nineteenth- and early twentieth-century architecture of the city.[9] The museum's brief was to cover geology, mineralogy, palaeontology, zoology and archaeology. Its motto, 'The Record of Nature through Countless Ages' and 'The Arts of Man through All the Years', placed archaeology on its own as equal to the other four combined. The decorative schema cemented this hierarchy. The vast majority of the relief sculptures on the museum's façade and of the designs within the mosaic ceiling in the rotunda represent different human cultures or their symbols, while only a few depict natural forms, and those far from naturalistically.

On the face of it, the architecture of the Royal Ontario Museum is more remote from the Pre-Raphaelite project to represent the natural world truthfully than that of the Naturhistorische Museum, for all that stylistically it retains more traces of the Gothic Revival. But there is a deeper continuity here. The Egyptologist Charles Currelly was the museum's principal advocate and its director for over thirty years. Currelly was yet another influential scientist with close connections to the Pre-Raphaelites. As a young archaeologist he apprenticed himself to Flinders Petrie, widely recognised as the founder of archaeology as a scientific discipline, but his closest personal relationship was with Edith and Holman Hunt. The roots of Currelly's interest in archaeology were Biblical, grounded in an intense Protestantism combined with an intellectual openness to the higher criticism. He shared Hunt's conviction, given pictorial form in *The Finding of the Saviour in the Temple* and *The Shadow of Death*, that to apprehend an ancient culture requires an intimate knowledge of its material forms and conditions. The young archaeologist and the old painter quickly formed a rapport. The bond between them became so firm that it was Currelly who arranged Hunt's cremation when he died and who carried the urn containing his ashes up the steps of St Paul's Cathedral at his funeral.[10]

In his introduction to Currelly's autobiography, the influential Canadian literary critic Northrop Frye comments:

> Now that archaeology is one of the world's major sciences, it is hard to realize how new it is. We can see in the book how the idea of 'evolution' was transformed, in archaeology, into the idea of continuity. In excavation this replaced the mere digging for buried treasure with the careful observing and comparing of strata. In collecting it replaced the notion of 'great periods' with the conception of all periods as being of equal primary interest.[11]

Frye's gloss on Currelly's story tallies with the decoration of his museum. As John Hungerford Pollen's entranceway to the Oxford University Museum unites cell theory and the Bible in an affirmation of natural theology, so the decorations at the Royal Ontario Museum affirm that archaeology is now indeed a major science. The museum's motto – itself carved in stone on two plaques on either side of the main entrance – brings the discoveries of archaeology into an evolutionary continuum with the 'Record of Nature'. A second motto, 'That all men may know His work', taken from the Book of Job (37. 7) and incorporated into the rotunda mosaic, quietly reaffirms the earlier Pre-Raphaelite museums' commitment to natural theology, now incorporating all the 'Arts of Man' as, ultimately, works of God. This is underscored by a further equivalence across the decorative schema, not of archaeology and the natural historical sciences, but within archaeology itself. In refusing to elevate particular civilisations or geographies over others, the reliefs confirm Frye's view that, in Currelly's science, evolution does not translate into racial or cultural hierarchies. Far from repudiating the model of the Pre-Raphaelite natural history museum, then, the Royal Ontario Museum continues the project of using architectural design to convey a particular vision of science and (our place in) nature. Its transcultural egalitarianism too is indebted to the Pre-Raphaelite refusal to prejudge the results of any artistic investigation, via Pater's insistence that, to the aesthetic critic, 'all periods, types, schools of taste, are in themselves equal,' and that the 'essence of humanism', embodied in the life and work of Pico della Mirandola, is the conviction that 'nothing which has ever interested living men and women can wholly lose its vitality.'[12] This embrace of human culture in its rich diversity is reflected in the choice of a human-headed, winged bull from Assyria rather than a Greek sculpture as the primary symbol of the arts and the old world in the carved arch above the museum's entrance, balancing the Canadian bison chosen as the emblem of the natural and new worlds (fig. 148). Eighty years before, this self-same sculpture had prompted Dante Gabriel Rossetti to undertake his own interrogation of the implications of archaeology in his poem 'The Burden of Nineveh' at the height of the Pre-Raphaelite experiment in scientific art.

148 Chapman and Oxley, Queen's Park Building, Royal Ontario Museum, 1929–33, detail showing the mix of natural history and cultural traditions within the decorative schema.

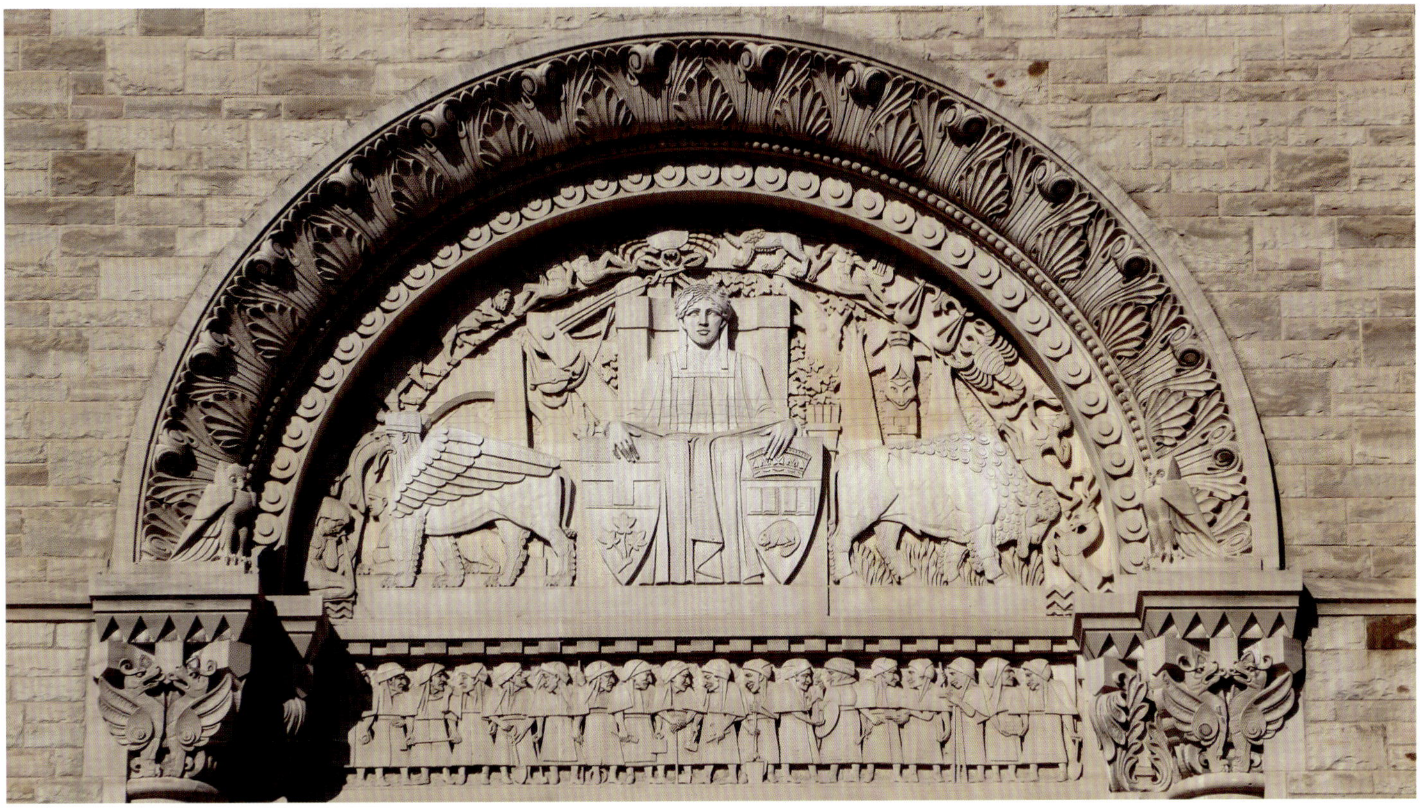

Natural history museums form a tangible and durable legacy. In 2015 the Oxford University Museum of Natural History received over 640,000 visitors, the Natural History Museum over five million.[13] Again, it is impossible to discriminate exactly how the buildings will have shaped their visitors' conceptions of science or the natural world. At the least, they will have established new associations in their minds, and perhaps new intimations of the value and meanings that we ascribe to nature. The directors and curators who work at these museums are more and more aware that their own premises are a rich resource for engaging the public and for stimulating artistic responses to nature, precisely because the buildings were built on Pre-Raphaelite principles to be representational models of science and nature. In 2016 the Oxford museum hosted a programme of events and exhibitions under the title 'Visions of Nature', including two original art exhibitions and a residency for three poets, charged with responding to the museum's collection, its architecture and its prior meanings.[14] One of the exhibitions in particular – 'Microsculpture' by the photographer Levon Biss – combined a hyper-Pre-Raphaelite attention to detail with a direct response to the museum's collection and an allusion to its sculptural heritage through a series of immense and immensely detailed photographs of insect specimens (fig. 149).[15] After over a century of battling against the architectural legacy of Owen's outdated science, the Natural History Museum has recently come to see its own architecture as an asset to be worked with, not against, both in displays and in new architectural projects. The landscape architect Kim Wilkie, for example, who began work on the grounds in front of the museum in 2014, has a brief to represent evolutionary history in the gardens in front of the east wing of the Waterhouse Building, where extinct animals dominate the decorations, and to use the grounds in front of the west wing, which is decorated with living forms, to represent ecology.[16] Wilkie's designs (fig. 150), informed by the museum's

149 Levon Biss, *Jewel Longhorn Beetle*, included the 'Microsculpture' exhibition, Oxford University Museum of Natural History, 2016.

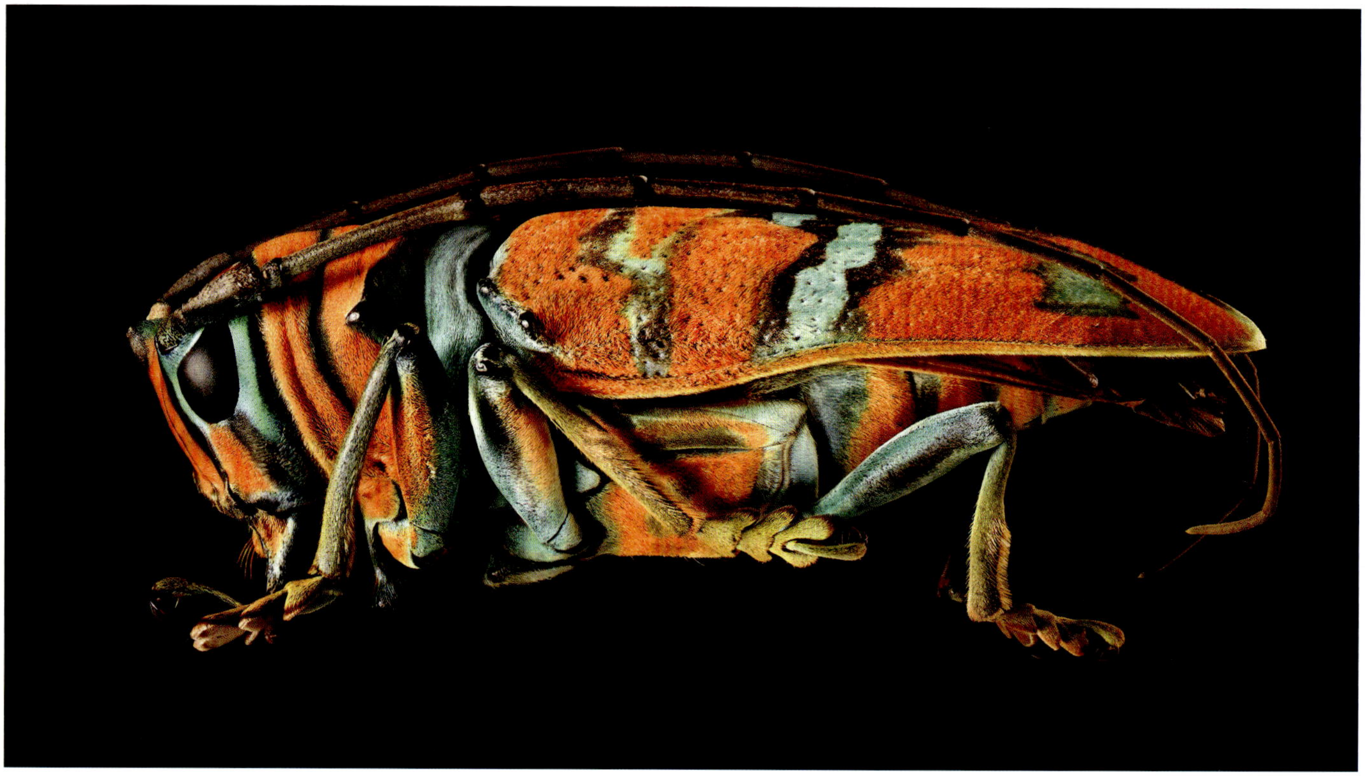

programme of science communication, are a further iteration of the Pre-Raphaelite project to represent the truth of science through the forms of art.

The Pre-Raphaelite project to ground art in science has recently been revived within the natural history museums that still embody it. Beyond them, there is no longer a direct lineage that can be traced back from contemporary painting and poetry to early Pre-Raphaelite art. What Pre-Raphaelitism offers artists and poets today is not principally a heritage but rather a model for how the arts can engage with science, and what they can gain by doing so. The Pre-Raphaelites did not respond to science as a series of findings but as a method. Science set a standard for impartial accuracy guaranteed by precise observation and for procedural rigour in research and experimentation. By matching this standard, the Pre-Raphaelites set out to regenerate what they felt to be the declining integrity and intellectual authority of the arts. Once their own rigour had been established, they were able to collaborate directly with 'scientific men' in interpreting the world that science described, while continuing to contribute their own independent observations and critiques to the programme of science itself. Today the sciences have considerably more authority within public discourse than the arts. The Pre-Raphaelite project suggests one way for the arts to re-establish their own authority by reasserting their rigour and relevance. At the same time, the arts have roles to play both in holding science to account and in awakening and accommodating an apprehensive or even dismissive public to its inconvenient truths. Again, the Pre-Raphaelites' interrogations of science on its own terms and their collaborations with scientists and 'scientific men' are exemplary.

Whether or not artists and poets today find any mileage in the model of the Pre-Raphaelites' engagement with science, it remains significant for the study of the history of science and its relations with other cultural forms. Scholarship on art and literature that engages with science has typically emphasised more or less direct allusions to

150 Kim Wilkie, projected view of the Natural History Museum, London, including Jurassic garden, 2014.

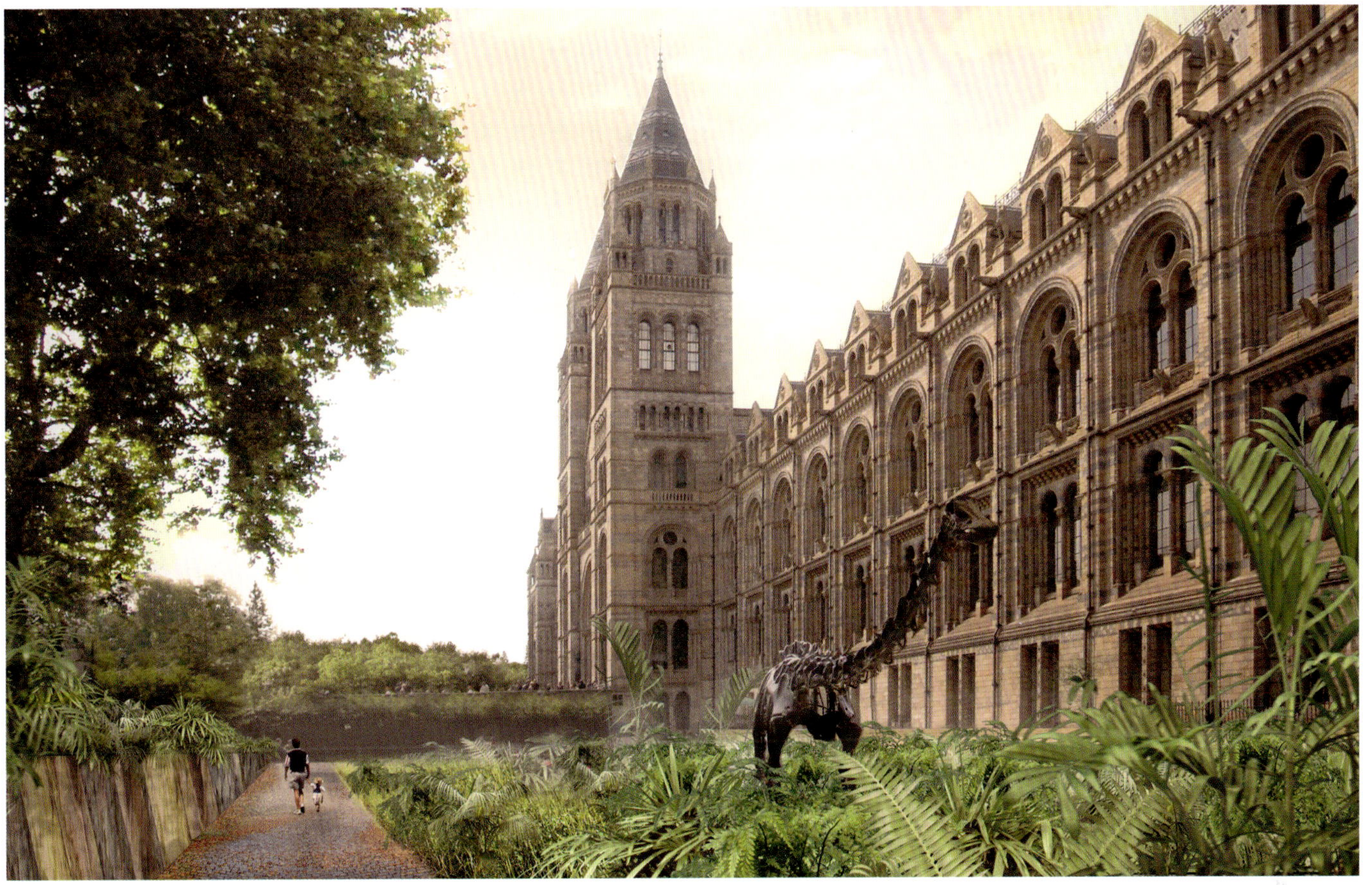

scientific theories or data within the work of art. Dyce's *Pegwell Bay* is one example of this; Tennyson's *In Memoriam* is another. The Pre-Raphaelite project shows how art can work with science and even as science without making explicit reference to it. It is possible to get a sense of the relationship between Pre-Raphaelite art and science by reading their own art theory and the critical responses of Ruskin, Acland, Morley and others who recognised this quality in their art. But the only way to trace how they put their ambition to raise art to the level of science into practice is to study their paintings, poems, sculptures and buildings with the same care and precision that they brought to creating them, to share in their investigations and follow them through to their findings. If we want to comprehend the complex interactions between science and the arts, we cannot let ourselves be limited to direct allusions or by the record that happens to survive. Any given culture's engagement with science, and science's role within that culture, will be far more pervasive and subtle than this evidence alone would ever enable us to recover. Instead, we need to make full use of the methods of literary and art criticism – in essence, close reading and interpretation of the works themselves, informed by a detailed knowledge of the wider intellectual, material and cultural context. In doing so, we can extend our knowledge not just of the arts but of the history and the meanings of science itself. In the case of the Pre-Raphaelites, we can see how central the idea of science was to the most influential movement in Victorian art, and at the same time how the arts gave expression to the key struggle within Victorian science, between Baconianism and natural theology, on the one hand, and scientific naturalism and evolutionary materialism, on the other.

NOTES

INTRODUCTION

1 John Ruskin, *The Art of England* (Orpington: George Allen, 1884), p. 7.

2 Ibid., p. 52.

3 Sidney Colvin, 'English Painters and Painting in 1867', *Fortnightly Review*, N.S. 2 (1867), 464–76 (pp. 470–71).

4 Elizabeth Prettejohn, *The Art of the Pre-Raphaelites* (London: Tate, 2000), p. 251.

5 Tim Barringer, *The Pre-Raphaelites: Reading the Image* (London: Weidenfeld and Nicolson, 1998), p. 16.

6 In addition to Prettejohn, p. 251, and Barringer, p. 15, this passage is cited as the key statement of Pre-Raphaelite attitudes to science by Stephanie Grilli in 'Pre-Raphaelitism and Phrenology', in Leslie Parris (ed.), *Pre-Raphaelite Papers* (London: Tate, 1984), 44–60 (p. 45); Lynn L. Merrill in *The Romance of Victorian Natural History* (Oxford: Oxford University Press, 1989), p. 172; and both Alison Smith in '"The Enfranchised Eye"' and Allen Staley in 'Rejecting Nothing, Selecting Nothing', in Allen Staley et al., *Pre-Raphaelite Vision: Truth to Nature* (London: Tate, 2004), 11–21 (p. 16), and 25–31 (p. 30).

7 Prettejohn, *Art of the Pre-Raphaelites*, p. 251.

8 Barringer, *The Pre-Raphaelites*, p. 16. The most compelling case for viewing the Pre-Raphaelites as the first of the modern avant-garde art movements is made by Prettejohn in *The Art of the Pre-Raphaelites*, especially pp. 63–5. Prettejohn's work was underscored by Joyce H. Townsend, Jacqueline Ridge and Stephen Hackney (eds), *Pre-Raphaelite Painting Techniques: 1848–56* (London: Tate, 2004). This interpretation of the Pre-Raphaelites as an avant-garde was realised on a grand scale in the retrospective exhibition, curated by Barringer together with Jason Rosenfeld and Alison Smith, which opened at Tate Britain in London in September 2012 and went on to the National Gallery of Art in Washington, DC and the State Pushkin Museum of Fine Arts in Moscow the following year. See Tim Barringer, Jason Rosenfeld and Alison Smith, *Pre-Raphaelites: Victorian Avant-Garde* (London: Tate, 2012), with contributions by Elizabeth Prettejohn and Diane Waggoner. For the record, it should be added that, while this interpretation has become increasingly widely accepted among scholars of Pre-Raphaelitism, it is not always given credence by reviewers. See, for example, Laura Cumming, 'Pre-Raphaelites: Victorian Avant-Garde – review', published in *The Observer*, 16 September 2012 (www.guardian.co.uk/artandde-

sign/2012/sep/16/pre-raphaelites-victorian-avant-review), accessed 29 January 2013.

9 Prettejohn, *Art of the Pre-Raphaelites*, p. 251. Prettejohn's own discussion of scientific and technological contexts for Pre-Raphaelitism (pp. 251–7) is rich and suggestive, but necessarily brief.

10 See Staley et al., *Pre-Raphaelite Vision*, passim, in particular Christopher Newall, 'Understanding the Landscape', 133–43; and Jason M. Rosenfeld, 'New Languages of Nature in Victorian England: The Pre-Raphaelite Landscape, Natural History and Modern Architecture in the 1850s' (unpublished doctoral dissertation, New York University, 1999).

11 Individual chapters and articles have been published on Pre-Raphaelitism and natural history, phrenology and ethnography, or on particular painters, such as John Everett Millais and John Brett. See respectively Merrill, *Romance of Victorian Natural History*, pp. 163–89; Grilli, 'Pre-Raphaelitism and Phrenology'; Marcia Pointon, 'The Artist as Ethnographer: Holman Hunt and the Holy Land', in Marcia Pointon (ed.), *Pre-Raphaelites Re-viewed* (Manchester: Manchester University Press, 1989), 22–44; Julie F. Codell, 'Empiricism, Naturalism and Science in Millais's Paintings', in Debra N. Mancoff (ed.), *John Everett Millais: Beyond the Pre-Raphaelite Brotherhood* (New Haven and London: Yale University Press, 2001), 119–47; and Christiana Payne, *John Brett: Pre-Raphaelite Landscape Painter* (New Haven and London: Yale University Press, 2010), pp. 104–23. For a recent preliminary exploration of how early Pre-Raphaelite art might be seen as unifying Romantic symbolism with scientific realism, see Lea Felicitas Döding, 'Symbolic Realism: Science and the Poetic in Early Pre-Raphaelite Painting', *Pre-Raphaelite Society Review*, 24.3 (Autumn 2016), 44–54.

12 D. G. Rossetti, *The Correspondence of Dante Gabriel Rossetti*, ed. William E. Fredeman, 9 vols (Cambridge: D. S. Brewer, 2002–10), I, pp. 64–7.

13 On the Cyclographic Society, see Appendix 3 of W. M. Rossetti, *The P.R.B. Journal: William Michael Rossetti's Diary of the Pre-Raphaelite Brotherhood 1849–1853*, ed. William E. Fredeman (Oxford: Oxford University Press, 1975), pp. 108–12.

14 W. M. Rossetti, *P.R.B. Journal*, pp. 71–2, 78, 241–2.

15 Bodleian Library, Oxford, MS Facs. d. 277, fols 148–9.

16 On Tupper's career as a sculptor, see Joanna Barnes and Alexander Kader, 'The Sculpture of John Lucas Tupper: "The Extremest Edge of P.R.Bism"', in Benedict Read and Joanna Barnes (eds), *Pre-Raphaelite Sculpture: Nature and Imagination in British Sculpture 1848–1914* (London: Henry Moore Foundation, 1991), 66–70.

17 George P. Landow, 'John Lucas Tupper (1824?–1879)', www.victorianweb.org/sculpture/tupper/index.html (accessed 13 January 2013).

18 John Lucas Tupper, 'Extracts from the Diary of an Artist', *The Crayon*, 2 (1855), 159–60, 271–2, 319–21, 336–7, 368–9, 400–1; 3 (1856), 43–5, 75–6, 102–5, 234–6; 4 (1857), 103–7, 145–9, 200–5. There are some anomalies within the diary as published which call its authenticity into question. Both the first entry, dated 8 December 1836, and the entry for 10 December 1840 (1856, p. 75) begin 'Admitted to-day a student of the Royal Academy'. According to Barnes and Kader, 'The Sculpture of John Lucas Tupper', p. 70, neither of these dates corresponds with the Royal Academy's own records, which place his admittance in 1838. There is also the question of Tupper's age. According to Landow, 'When Was Tupper Born?', www.victorianweb.org/sculpture/tupper/birthdate.html (accessed 13 January 2013), public records suggest that he was born in 1824, he himself told Hunt that he was forty in October 1863, while William Rossetti placed his birth in or around 1826. On 10 January 1841 (1856, p. 103), the supposed diarist notes that he is twenty, making him two to five years older than Tupper himself by these estimates. A further layer of complexity is generated by the editorial interventions into the supposed source material. From the second instalment on, these interventions are typically signed 'J.T.', seemingly identifying the editor with the artist. For his own part, the editor refers to himself rather as 'the collater of the manuscripts' (1855, p. 320). At one point, he even raises directly the question of the relation between himself and their author: 'These, and any subsequent criticisms which the compiler may think fit to extract from the Diary, will be understood *distinctly as not emanating from him as their author*, and this, whatever the reader may opine of the personal reality of the Diarist. He (compiler), is not a reviewer, and has here no intention of reviewing either the past or present works of living artists; indeed he would have suppressed these and similar memoranda, were it not that their omission would interpose a gap in *that faithful registry of art experience and artist life*' (1857, p. 104). This passage, again signed 'J.T.', can be read equally as identifying the compiler as the diarist (he was the author of these criticisms, but they do not emanate from him, in that he no longer endorses them) and as denying that very identification (on an alternative reading of the same syntax, he is not their author after all). Similarly, it raises the possibility that the diarist may not even exist (at least, the reader may have come to that conclusion), yet it also insists that the diary is genuine (it is a faithful record which the compiler is unwilling to tamper with). It is possible to reconcile these contradictions into a single account, taking Tupper's editorial role as merely a mannered way of representing the distance between himself in the mid-1850s and the young artist he was when he wrote the diary, and reading his challenge to the reader as an affirmation of the diarist's reality. But even then he is at least hinting at another way of reading his text – as fiction – while his self-conscious performance of the role of editor suggests too a playfulness in its evasion of a straightforward affirmation that these are his own diaries. Against this evidence for doubting the diary's authenticity is the accuracy of its topical references. On 14 April 1841, Tupper records, 'Saw the Oxford and Cambridge match – Cantabs won', which they did, and on that day. Two weeks before, he notes that he heard Michael Balfe's opera *Keolanthe* at the English Opera, where it was indeed playing. Two weeks earlier again, he describes watching the railway station at Nine Elms burn down, which it did (1856, p. 235). Details like these seem too arbitrary to have been written into the text merely for verisimilitude, suggesting that a core of the text does indeed comprise Tupper's diary from this time. At the very least its publication in the mid-1850s affirms Tupper's active endorsement of the principles of Pre-Raphaelitism at that point, regardless of whether or not he did indeed anticipate them. In this study I will assume that the diaries are genuine, while recognising that, as I am treating

them as contributions to the Pre-Raphaelite project and not making arguments about priority, it does not ultimately matter for these purposes whether his biographical account of himself is factually correct or partly fictionalised.

19 John Rylands Library, Manchester, Ryl. Eng. MS 1214, fol. 13.

20 William Michael Rossetti (ed.), *Præraphaelite Diaries and Letters* (London: Hurst and Blackett, 1900), pp. 206–7.

21 W. M. Rossetti, *P.R.B. Journal*, p. 3.

22 Ibid., p. 96. Tupper's bas-relief appears no longer to survive. See Barnes and Kader, 'The Sculpture of John Lucas Tupper', p. 68.

23 W. M. Rossetti, *Præraphaelite Diaries and Letters*, p. 207.

24 For referencing purposes, all notes on articles and poems from this magazine use the original title. There was, in effect, only one volume of *The Germ*; the pagination across its four issues is therefore continuous, with issue 1 running to p. 48, issue 2 to p. 96, issue 3 to p. 144 and issue 4 to p. 192.

25 William Holman Hunt, 'The Pre-Raphaelite Brotherhood: A Fight for Art', *Contemporary Review*, 49 (1886), 471–88, 737–50, 820–33 (p. 739).

26 Barringer, *The Pre-Raphaelites*, p. 14.

27 These letters, from 13 and 30 May 1851, are reprinted by Stephen Wildman in his very useful collection of Ruskin's writings on the Pre-Raphaelites, *Ruskin and the Pre-Raphaelites* (London: Pallas Athene, 2012), pp. 51–6.

28 Marcia Werner, *Pre-Raphaelite Painting and Nineteenth-Century Realism* (Cambridge: CUP, 2005), pp. 18–57.

29 See George P. Landow, *'Your Good Influence on Me': The Correspondence of John Ruskin and William Holman Hunt* (Manchester: John Rylands University Library, 1977), p. 43.

30 W. M. Rossetti, *Dante Gabriel Rossetti: His Family-Letters with a Memoir*, 2 vols (London: Ellis and Elvey, 1895), I, p. 135.

31 John Guille Millais, *The Life and Letters of Sir John Everett Millais*, 2 vols (London: Methuen, 1899), I, pp. 61–2.

32 Amy Woolner, *Thomas Woolner, R.A., Sculptor and Poet: His Life in Letters* (London: Chapman and Hall, 1917), pp. 86–8.

33 Wildman, *Ruskin and the Pre-Raphaelites*, pp. 65, 77.

34 Jan Marsh and Pamela Gerrish Nunn, *Women Artists and the Pre-Raphaelite Movement* (London: Virago, 1989) and *Pre-Raphaelite Women Artists* (Manchester: Manchester City Art Galleries, 1997).

35 Prettejohn, *Art of the Pre-Raphaelites*, pp. 66–85 (quotation p. 69).

36 W. M. Rossetti, *Some Reminiscences*, 2 vols (New York: Scribner's, 1906), I, p. 74.

37 Ibid., I, pp. 74–5.

38 See, respectively, Christina Rossetti (as Ellen Alleyn), 'Dream Land', *The Germ*, 20; 'An End' and 'Repining', *The Germ*, 48 and 111–17; and 'A Testimony' and 'Sweet Death', *The Germ*, 73–5 and 117.

39 C. Rossetti, *Annus Domini: A Prayer for Each Day of the Year, Founded on a Text of Holy Scripture* (Oxford: James Parker, 1874), p. 234.

40 On Tupper's career, see Sushma Kapoor, 'John L. Tupper, to 1863: "King of the Cadaverals"', *Journal of Pre-Raphaelite Studies*, 4.2 (1984), 70–86, and 'John Lucas Tupper, 1865–79: The Rise of "Outis"', *JPRS*, 6.2 (1986), 37–44; George P. Landow, 'An Annotated Checklist of J. L. Tupper's Works', *JPRS*, 7.1 (1986), 63–8; and James H. Coombs et al. (eds), *A Pre-Raphaelite Friendship: The Correspondence of William Holman Hunt and John Lucas Tupper* (Ann Arbor, Mich.: UMI Research Press, 1986).

41 Hunt, 'The Pre-Raphaelite Brotherhood', pp. 739–40, 826.

42 W. M. Rossetti, *Some Reminiscences*, I, p. 31.

43 Millais, *Life and Letters*, I, p. 2; Woolner, *Thomas Woolner*, p. 2.

44 James A. Secord, *Visions of Science: Books and Readers at the Dawn of the Victorian Age* (Oxford: OUP, 2014) and *Victorian Sensation: The Extraordinary Publication, Reception, and Secret Authorship of 'Vestiges of the Natural History of Creation'* (Chicago: University of Chicago Press, 2000).

45 Arnold first made this claim in the second lecture of his lecture series, 'On Translating Homer', in 1860. He reiterated it forcefully at the beginning of his 1864 article, 'The Function of Criticism at the Present Time'; see Matthew Arnold, *Complete Prose Works*, ed. R. H. Super, 11 vols (Ann Arbor: University of Michigan Press, 1960–77), I, p. 140, and III, p. 258.

46 Quoted in Peter Addinall, *Philosophy and Biblical Interpretation: A Study in Nineteenth-Century Conflict* (Cambridge: Cambridge University Press, 1991), p. 87.

47 Addinall, *Philosophy and Biblical Interpretation*, pp. 86–171.

48 See W. M. Rossetti, *Some Reminiscences*, I, p. 122, for his account of his loss of faith.

49 John Tyndall, *Fragments of Science: A Series of Detached Essays, Addresses and Reviews*, 6th edn, 2 vols (London: Longmans, 1879), II, p. 199.

50 Luke Herrmann, *Nineteenth Century British Painting* (London: Giles de la Mare, 2000), p. 241.

1 ART ON THE MODEL OF SCIENCE

1 Marysa Demoor, 'In the Beginning, There Was *The Germ*: The Pre-Raphaelites and "Little Magazines"', in Peter Brooker and Andrew Thacker (eds), *The Oxford Critical and Cultural History of Modernist Magazines, Volume 1: Britain and Ireland 1880–1955* (Oxford: Oxford University Press, 2009), 51–65; Paola Spinozzi and Elisa Bizzotto, *The Germ: Origins and Progenies of Pre-Raphaelite Interart Aesthetics* (Bern: Peter Lang, 2012), pp. 175–219.

2 For a comprehensive discussion of science in the critical prose of *The Germ* specifically, see John Holmes, 'Pre-Raphaelitism, Science, and the Arts in *The Germ*', *Victorian Literature and Culture*, 43 (2015), 689–703. Marcia Werner includes an insightful discussion of the aesthetic theory of *The Germ* in *Pre-Raphaelite Painting and Nineteenth-Century Realism* (Cambridge: Cambridge University Press, 2005), pp. 58–73.

3 Stephanie Grilli, 'Pre-Raphaelitism and Phrenology', in Leslie Parris (ed.), *Pre-Raphaelite Papers* (London: Tate, 1984), 44–60 (p. 45).

4 Frederic George Stephens [as John Seward], 'The Purpose and Tendency of Early Italian Art', *The Germ*, 58–64 (p. 61).

5 Gowan Dawson and Bernard Lightman (eds), *Victorian Science and Literature*, 8 vols (London: Pickering and Chatto, 2011–12), I, p. ix.

6 Samuel Maunder, *The Treasury of Knowledge and Library of Reference*, 11th edn, 2 vols (London: Longman, 1839), I, 'A New and Enlarged Dictionary

of the English Language: Science'. The first edition of this compendium was published in 1830.

7 C. R. Leslie, *Memoirs of the Life of John Constable, Esq. R.A.* (London: Longman, 1845), p. 333.

8 Ibid., p. 355.

9 Ford Madox Brown, 'On the Mechanism of a Historical Picture, Part I: The Design', *The Germ*, 70–73 (p. 73).

10 F. G. Stephens, 'Modern Giants', *The Germ*, 169–73 (p. 173).

11 John Ruskin, *Modern Painters*, 6 vols (1843–60; London: George Allen, 1897), I, xxxiv (Preface to 2nd edn, § 27).

12 Stephen Wildman (ed.), *Ruskin and the Pre-Raphaelites* (London: Pallas Athene, 2012), pp. 72–3, 96 (original emphasis).

13 Benedict Read and Joanna Barnes (eds), *Pre-Raphaelite Sculpture: Nature and Imagination in British Sculpture 1848–1914* (London: Henry Moore Foundation, 1991), pp. 141–2, 160–61; Paul Barlow, 'Grotesque Obscenities: Thomas Woolner's *Civilization* and its Discontents', in Colin Trodd, Paul Barlow and David Amigoni (eds), *Victorian Culture and the Idea of the Grotesque* (Aldershot: Ashgate, 1999), 97–113 (p. 108); Charles Darwin, *The Descent of Man, and Selection in Relation to Sex*, ed. James Moore and Adrian Desmond (London: Penguin, 2004), p. 32. Darwin's letter to Woolner is in the Bodleian Library, Oxford, MS Eng. Lett. d. 292, fol. 77.

14 Darwin, *Descent of Man*, p. 32.

15 William Michael Rossetti, 'Pre-Raphaelitism', *The Spectator*, 24 (1851), 955–7 (pp. 955–6). For an extensive discussion of Rossetti's art theory and criticism over the course of his career, see Julie L'Enfant, *William Rossetti's Art Criticism: The Search for Truth in Victorian Art* (Lanham, Md: University Press of America, 1999).

16 Stephens, 'Modern Giants', p. 171.

17 Charles Daubeny, *An Introduction to the Atomic Theory* (London: Murray, 1831), pp. 2–3.

18 John Frederick William Herschel, *A Preliminary Discourse on the Study of Natural Philosophy* (Cambridge: Cambridge University Press, 2009; reprinted from the 1831 edition), p. 64.

19 [Henry Holland,] 'Article II – Modern Chemistry', *Quarterly Review*, 83 (1848), 37–70 (p. 38).

20 John Lucas Tupper, 'The Subject in Art', *The Germ*, 11–18, 118–25 (pp. 13–14).

21 W. M. Rossetti, 'Pre-Raphaelitism', p. 956; [F. G. Stephens], 'The Two Pre-Raphaelitisms', *The Crayon*, 3 (1856), 225–8, 289–92, 321–4, 353–6 (p. 322); John Lucas Tupper, 'Extracts from the Diary of an Artist', *The Crayon*, 4 (1857), p. 148; William Holman Hunt, *Pre-Raphaelitism and the Pre-Raphaelite Brotherhood*, 2 vols (London: Macmillan, 1905), I, p. 87.

22 Robert Hunt, 'On the Applications of Science to the Fine and Useful Arts: The Geometrical Principles of Beauty', *The Art-Journal* (1850), 357–9 (p. 357).

23 Tupper, 'Subject in Art', p. 14.

24 Alice Jenkins, *Space and the 'March of Mind': Literature and the Physical Sciences in Britain, 1815–1850* (Oxford: Oxford University Press, 2007), pp. 158–60, 172–4.

25 See ibid., pp. 173–4.

26 William Leithead, *Electricity: Its Nature, Operation, and Importance in the Phenomena of the Universe* (London: Longman, 1837), 275. This is the first citation given for 'electrophysiological' in the *Oxford English Dictionary*.

27 Tupper refers to Smee in *Hiatus: The Void in Modern Education; Its Cause, and Antidote* (London: Macmillan, 1869), p. 185, published under the pseudonym Outis. The full title of Smee's book is *Elements of Electro-Biology, or, The Voltaic Mechanism of Man; Of Electro-Pathology, Especially of the Nervous System; and of Electro-Therapeutics* (London: Longman, 1849).

28 Tupper, 'Subject in Art', p. 119.

29 James A. Secord, *Visions of Science: Books and Readers at the Dawn of the Victorian Age* (Oxford: Oxford University Press, 2014), pp. 81, 87.

30 Ibid., pp. 91–2, quoting Herschel, *Preliminary Discourse*, p. 8.

31 Herschel, *Preliminary Discourse*, p. 76.

32 Timothy Hilton, *The Pre-Raphaelites* (London: Thames and Hudson, 1970), p. 72; Lynn L. Merrill, *The Romance of Victorian Natural History* (Oxford: Oxford University Press, 1989), pp. 173–5; Jason Rosenfeld in Tim Barringer, Jason Rosenfeld and Alison Smith, *Pre-Raphaelites: Victorian Avant-Garde* (London: Tate, 2012), p. 96. Alison Smith in Allen Staley et al., *Pre-Raphaelite Vision: Truth of Nature* (London: Tate, 2004), pp. 33–4.

33 For discussions of the Pre-Raphaelites' early experiments in painting outdoors, see Allen Staley, *The Pre-Raphaelite Landscape* (Oxford: Oxford University Press, 1973), pp. 1–30, and Elizabeth Prettejohn, *The Art of the Pre-Raphaelites* (London: Tate, 2000), pp. 152–63. See also Barringer et al., *Pre-Raphaelites*, pp. 44, 55–6, 92, 96–9.

34 [Stephens], 'The Two Pre-Raphaelitisms' (1856), p. 353. John Ruskin makes a very similar claim about Hunt's next painting, *Our English Coasts*, in *The Art of England* (Orpington: George Allen, 1884), p. 11, where he remarks: 'It showed to us, for the first time in the history of art, the absolutely faithful balances of colour and shade by which actual sunshine might be transposed into a key in which the harmonies possible with material pigments should yet produce the same impressions upon the mind which were caused by the light itself'.

35 F. G. Stephens, *William Holman Hunt and his Works: A Memoir of the Artist's Life, with Description of his Pictures* (London: Nisbet, 1860), p. 20.

36 [Stephens], 'The Two Pre-Raphaelitisms' (1856), p. 354; Stephens, *William Holman Hunt*, p. 19.

37 For the list of Immortals, see Hunt, *Pre-Raphaelitism*, I, p. 159.

38 Helen Glanville discusses how Hunt's paintings in particular correlate with contemporary theories of optics as developed by Brewster and others in 'Contemporary Colour Theory' in Joyce H. Townsend, Jacqueline Ridge and Stephen Hackney (eds), *Pre-Raphaelite Painting Techniques: 1848–56* (London: Tate, 2004), 29–37.

39 [Stephens], 'The Two Pre-Raphaelitisms' (1856), p. 354.

40 [F. G. Stephens], 'The Two Pre-Raphaelitisms', *The Crayon*, 4 (1857), 261–5, 298–302, 325–9, 361–3 (p. 325).

41 Colin Cruise discusses this sketch in *Pre-Raphaelite Drawing* (London: Thames and Hudson, 2011), pp. 58–9.

42 Prettejohn, *Art of the Pre-Raphaelites*, p. 172.

43 Richard Yeo, 'An Idol of the Market-Place: Baconianism in Nineteenth-Century Britain', *History of Science*, 23 (1985), 251–98, reprinted in Yeo, *Science in the Public Sphere: Natural Knowledge*

in British Culture 1800–1860, Variorum Collected Studies Series (Aldershot: Ashgate, 2001), p. 285.

44 Ibid., p. 267.

45 Herschel, *Preliminary Discourse*, pp. 104–5, 113–14.

46 [Holland], 'Modern Chemistry', p. 39; Daubeny, *Introduction to the Atomic Theory*, p. 2, and 2nd edn (Oxford: Oxford University Press, 1850), p. 3.

47 Secord, *Visions of Science*, p. 153.

48 H. E. Strickland, *On Geology in Relation to the Studies of the University of Oxford* (Oxford: Vincent, 1852), pp. 12–13.

49 John Phillips, *A Guide to Geology* (London: Longman, 1834), p. 4, and 4th edn (London: Longman, 1854), p. 5.

50 Hunt, *Pre-Raphaelitism*, I, p. 18.

51 Amy Woolner, *Thomas Woolner, R.A., Sculptor and Poet: His Life in Letters* (London: Chapman and Hall, 1917), p. 93.

52 [Stephens], 'The Two Pre-Raphaelitisms' (1857), p. 362.

53 As Uglow suggests, the list of Immortals is 'an act of identification' which '*shows* an identity it does not *tell*'. See Nathan Uglow, 'The Relation of the "List of Immortals" to the Pre-Raphaelite Movement', in Paul Hardwick and Martin Hewitt (eds), *The Pre-Raphaelite Ideal*, Leeds Centre Working Papers in Victorian Studies, vol. 7 (Leeds: Trinity and All Saints and University of Leeds, 2004), 9–15 (p. 10).

54 Jonathan Smith, *Fact and Feeling: Baconian Science and the Nineteenth-Century Literary Imagination* (Madison: University of Wisconsin Press, 1994), p. 13.

55 Ibid., 12.

56 Aside from Leonardo, the only other Immortal with a clear connection to science on the list is Goethe. The list as printed has no defined order, but assuming it bears some similarity to the order in which Hunt and Rossetti jotted down the names, Bacon and Newton are paired together, while Goethe heads a series of Romantic heroes including the Polish patriot Kosciusko and the poets Byron, Wordsworth, Keats and Shelley. For Rossetti, at least, Goethe was principally the poet of *Faust*; see W. M. Rossetti, *Dante Gabriel Rossetti: His Family-Letters with a Memoir*, 2 vols (London: Ellis and Elvey, 1895), I, p. 59. Little evidence has come to light that the Pre-Raphaelites engaged with his work on botany, although Hunt certainly did make use of Goethe's *Theory of Colours* later in his career when researching pigments; see Carol Jacobi, *William Holman Hunt: Painter, Painting, Paint* (Manchester: Manchester University Press, 2006), p. 119. Glanville, 'Contemporary Colour Theory', notes that Charles Eastlake translated the *Theory of Colours* in 1840, and traces parallels between Goethe's observations and Hunt's practice as a colourist. On the other hand, Stephens noticeably does not mention Goethe as one of the experts on optics whose theories Hunt's discoveries corroborate in his account of *The Hireling Shepherd*.

57 Huntington Library, San Marino, Calif., HM 12981, fol. 103r. The poem also exists in a typescript transcript of this MS, HM 12982, with a transcription error ('bleak' for 'blank' in line 11). It is quoted from the typescript by Eben E. Bass in *Dante Gabriel Rossetti: Poet and Painter* (New York: Peter Laing, 1990), pp. 65–6, and by Werner, *Pre-Raphaelite Painting and Nineteenth-Century Realism*, p. 106, quoting Bass. Mary Lutyens cites the poem from the manuscript in 'Walter Howell Deverell (1827–1854)', in Leslie Parris (ed.), *Pre-Raphaelite Papers* (London: Tate, 1984), 76–96 (p. 84), but transcribes 'Swimming' as Swimmingly' in line 4, and 'That' as 'The' at the beginning of line 14.

58 Bass, *Dante Gabriel Rossetti*, p. 66.

59 Thomas Woolner, 'Emblems', *The Germ*, 127–8 (p. 128).

60 Dante Gabriel Rossetti, 'From the Cliffs: Noon', *The Germ*, 129.

61 D. G. Rossetti, 'The Blessed Damozel', *The Germ*, 80–83 (p. 80).

62 John Lucas Tupper, 'An Incident in the Siege of Troy, seen from a modern Observatory', *The Germ*, 131–5. For a brief account of Knox as a transcendental anatomist, see Philip F. Rehbock, 'Transcendental Anatomy', in Andrew Cunningham and Nicholas Jardine (eds), *Romanticism and the Sciences* (Cambridge: Cambridge University Press, 1990), 144–60 (pp. 155–6); for his controversy with Charles Bell over the use of anatomy in art, see A. Smith, *The Victorian Nude: Sexuality, Morality and Art* (Manchester: Manchester University Press, 1996), p. 24.

63 Stephens, 'Modern Giants', p. 171.

64 W. M. Rossetti, *Fine Art, Chiefly Contemporary: Notices Reprinted, with Revisions* (London: Macmillan, 1867), p. 147; see also pp. 99, 142, and 'Pre-Raphaelitism', p. 956.

65 William Dyce, *Theory of the Fine Arts: An Introductory Lecture* (1844), extracted in Inga Bryden (ed.), *The Pre-Raphaelites: Writings and Sources*, 4 vols (London: Routledge/Thoemmes, 1998), III, 103–8 (pp. 106–7).

66 See Hilton, *Pre-Raphaelites*, p. 129; Staley, *Pre-Raphaelite Landscape*, pp. 167–8; Julian Treuherz, *Victorian Painting* (London: Thames and Hudson, 1993), pp. 94–5; Tim Barringer, *The Pre-Raphaelites: Reading the Image* (London: Weidenfeld and Nicolson, 1998), pp. 79–81; Staley et al., *Pre-Raphaelite Vision*, pp. 188–9; Barringer et al., *Pre-Raphaelites*, pp. 110–11.

67 Barringer et al, *Pre-Raphaelites*, p. 110.

68 Hilton, *Pre-Raphaelites*, p. 129.

69 W. M. Rossetti, 'Introduction' to *The Germ: Thoughts towards Nature in Poetry, Literature and Art, being a Facsimile reprint of the Literary Organ of the Pre-Raphaelite Brotherhood, published in 1850* (London: Elliot Stock, 1901), p. 6.

70 Marcia Pointon, 'The Representation of Time in Painting: A Study of William Dyce's *Pegwell Bay: A Recollection of October 5th, 1858*', *Art History*, 1 (1978), 99–103.

71 Staley et al., *Pre-Raphaelite Vision*, p. 188.

72 Barringer, *Pre-Raphaelites*, pp. 79–81; Rebecca Bedell, 'The History of the Earth: Darwin, Geology and Landscape Art', in Diana Donald and Jane Munro (eds), *Endless Forms: Charles Darwin, Natural Science and the Visual Arts* (New Haven and London: Yale University Press, 2009), 48–79 (pp. 62–4).

73 Ibid., p. 110.

74 Ibid., p. 110; Hilton, *Pre-Raphaelites*, p. 129; Staley et al., *Pre-Raphaelite Vision*, p. 188.

75 Bedell refers to Darwin and Mantell ('History of the Earth', pp. 63–4), Smith to these two periodicals (Staley et al., *Pre-Raphaelite Vision*, p. 188).

76 W. M. Rossetti, 'Introduction' to *The Germ*, p. 6.

77 Gillian Beer, *Open Fields: Science in Cultural Encounter* (Oxford: Oxford University Press, 1996), p. 149.

78 William Bell Scott, *Autobiographical Notes of the Life of William Bell Scott*, ed. W. Minto, 2 vols (London: Osgood, McIlvaine and Co., 1892), I, p. 251. This passage is quoted by Lindsay Smith in *Victorian Photography, Painting and Poetry: The Enigma of Visibility in Ruskin, Morris and the Pre-Raphaelites* (Cambridge: Cambridge University Press, 1995), p. 99; by Diane Waggoner, Tim Barringer and Britt Salvesen in Diane Waggoner et al., *The Pre-Raphaelite Lens: British Photography and Painting, 1848–1875* (Washington, DC: National Gallery of Art, 2010), pp. 6, 28, 180; and by Waggoner again in Barringer et al., *Pre-Raphaelites*, p. 90.

79 Sidney Colvin, *Memories and Notes of Persons and Places 1852 to 1912* (London: Edward Arnold, 1921), p. 74.

80 Waggoner et al., *The Pre-Raphaelite Lens*, p. 6.

81 Wildman, *Ruskin and the Pre-Raphaelites*, p. 126.

82 [Stephens], 'The Two Pre-Raphaelitisms' (1856), p. 354.

83 Staley, *Pre-Raphaelite Landscape*, p. 167.

84 Waggoner et al., *Pre-Raphaelite Lens*, p. 24.

85 See Prettejohn, *Art of the Pre-Raphaelites*, pp. 166–71, for a comparison of Millais's painting with Ruskin's drawing and a modern photograph of the same spot, along with Ruskin's description of the flow of water from *Modern Painters* I. For discussions of Millais's painting and Ruskin's drawing in relation to geology, and of Ruskin's more determined interest in it than Millais's, see Staley et al., *Pre-Raphaelite Vision*, pp. 134–6, 145–7.

86 Stephens, 'Modern Giants', p. 170.

87 See Barringer, *Pre-Raphaelites*, pp. 105–6; Prettejohn, *Art of the Pre-Raphaelites*, pp. 254–5.

88 Robert Mitchell, *Experimental Life: Vitalism in Romantic Science and Literature* (Baltimore, Md: John's Hopkins University Press, 2013), p. 4.

89 D. G. Rossetti, *The Correspondence of Dante Gabriel Rossetti*, ed. William E. Fredeman, 9 vols (Cambridge: D. S. Brewer, 2002–10), I, pp. 102–3.

90 Werner, *Pre-Raphaelite Painting and Nineteenth-Century Realism*, p. 164.

91 D. G. Rossetti, *Correspondence*, I, p. 104.

92 D. G. Rossetti, 'Antwerp to Ghent', in *The Works of Dante Gabriel Rossetti*, ed. W. M. Rossetti (London: Ellis, 1911), pp. 186–7. The original manuscript source for this poem is no longer extant.

93 Tupper, 'Diary of an Artist' (1857), p. 204.

94 W. M. Rossetti, 'Fancies at Leisure', *The Germ*, 76–8 (pp. 77–8).

95 Rossetti included a second series of five poems under the same overall title in the third issue of *The Germ*.

96 The date of 'Idle Blessedness' is given in W. M. Rossetti, *The P.R.B. Journal: William Michael Rossetti's Diary of the Pre-Raphaelite Brotherhood 1849–1853*, ed. William E. Fredeman (Oxford: Oxford University Press, 1975), p. 11. Tupper's poem is printed as part of his diary entry for 9 May 1841 in 'Diary of an Artist' (1857), p. 105.

97 W. M. Rossetti, 'Introduction' to *The Germ*, p. 22.

98 W. M. Rossetti, '*Christmas Eve and Easter Day*: by Robert Browning', *The Germ*, 187–92; '*The Bothie of Toper-na-fuosich: A Long-vacation Pastoral*. By Arthur Hugh Clough', *The Germ*, 34–46.

99 W. M. Rossetti, '*The Bothie of Toper-na-fuosich*', pp. 36, 45 (original emphasis).

100 W. M. Rossetti, '*The Strayed Reveller; and other Poems*. By A.', *The Germ*, 84–96 (p. 96).

101 Herschel, *Preliminary Discourse*, p. 14.

102 Ibid., pp. 120–21.

103 Ibid., p. 16.

104 Yeo, 'Idol of the Market-Place', p. 276.

105 W. M. Rossetti, *Fine Art*, p. 242.

106 Wildman, *Ruskin and the Pre-Raphaelites*, p. 61.

107 W. M. Rossetti, *Fine Art*, p. 241.

108 Wildman, *Ruskin and the Pre-Raphaelites*, p. 60.

109 Stephens, 'The Two Pre-Raphaelitisms' (1856), pp. 354–5.

110 Alan Bowness et al., *The Pre-Raphaelites* (London: Tate, 1984), p. 94; John 10:11–14. For a helpful summary of Hunt's engagement with religious debate in the painting, see Bowness et al., *The Pre-Raphaelites*, pp. 94–6. One prominent suggestion is that the painting is allegorical, with the shepherd representing negligent pastors in the ecclesiastical sense. Hunt's own comments on this question are ambiguous, however. When the painting was acquired by the Manchester Art Gallery in 1897, Hunt wrote a letter drawing an analogy between the shepherd in the painting and negligent clergy or pastors. The same analogy was used by Ruskin and Dyce in pamphlets published in 1851, and critics have typically followed this through and interpreted the painting accordingly. On the other hand, in a manuscript fragment in the John Rylands Library in Manchester, Hunt explicitly states that the painting 'was not painted with any distinctly symbolic or allegorical view' (Ryl. Eng. MS 1216, fol. 31). Furthermore, he does not mention this interpretation in his memoir *Pre-Raphaelitism*, which is peculiar given his keenness to identify his own project with Ruskin's by this stage. As Staley has pointed out, none of the contemporary critics offers this interpretation either, including Stephens and W. M. Rossetti, who were Hunt's close artistic allies and in his confidence. Staley points out that Hunt only offered this interpretation forty-five years after the painting was exhibited, when he himself was increasingly casting himself as a Christian prophet. On balance, it seems more likely to have been an afterthought, long after the fact, than part of the original meaning of the painting. See Staley, *Pre-Raphaelite Landscape*, pp. 20–21.

111 See Hunt, *Pre-Raphaelitism*, I, pp. 353–4. Hunt quotes a letter from Jane Carlyle in which she quotes her husband as describing one of Hunt's pictures as 'The greatest Picture that he has seen painted by any modern man!' but it is unclear from Hunt's account whether this refers to *The Hireling Shepherd* or *Our English Coasts*.

112 Barringer et al., *Pre-Raphaelites*, p. 98.

113 Carol Jacobi, 'William Holman Hunt (1827–1910)', in Elizabeth Prettejohn (ed.), *The Cambridge Companion to the Pre-Raphaelites* (Cambridge: Cambridge University Press, 2012), 116–32 (p. 122).

114 W. M. Rossetti, *Fine Art*, p. 236; see also [Stephens], 'The Two Pre-Raphaelitisms' (1856), p. 353.

115 Sophia Andres, *The Pre-Raphaelite Art of the Victorian Novel: Narrative Challenges to Visual Gendered Boundaries* (Columbus: Ohio State University Press, 2005), p. 26. Andres herself cites Prettejohn, *Art of the Pre-Raphaelites*, p. 214, making a similar point.

116 Prettejohn, *Art of the Pre-Raphaelites*, p. 190.

117 George Eliot, *Selected Critical Writings*, ed. Rosemary Ashton (Oxford: Oxford University Press, 1992), pp. 261, 264.

2 THE ECOLOGICAL VISION OF PRE-RAPHAELITE ART

1 W. M. Rossetti (ed.), *Præraphaelite Diaries and Letters* (London: Hurst and Blackett, 1900), pp. 206–7.

2 W. M. Rossetti, *The P.R.B. Journal: William Michael Rossetti's Diary of the Pre-Raphaelite Brotherhood 1849–1853*, ed. William E. Fredeman (Oxford: Oxford University Press, 1975), pp. 6, 17.

3 Amy Woolner, *Thomas Woolner, R.A., Sculptor and Poet: His Life in Letters* (London: Chapman and Hall, 1917), p. 2; J. G. Millais, *The Life and Letters of Sir John Everett Millais*, 2 vols (London: Methuen, 1899), I, p. 2.

4 D. G. Rossetti, *The Correspondence of Dante Gabriel Rossetti*, ed. William E. Fredeman, 9 vols (Cambridge: D. S. Brewer, 2002–10), I, p. 106.

5 W. M. Rossetti, 'Fancies at Leisure', *The Germ*, p. 77 (my emphasis).

6 W. M. Rossetti, 'Introduction' to *The Germ: Thoughts towards Nature in Poetry*... (London: Elliot Stock, 1901), pp. 20, 22.

7 W. M. Rossetti, 'Fancies at Leisure', pp. 76–7.

8 Lynn L. Merrill, *The Romance of Victorian Natural History* (Oxford: Oxford University Press, 1989), pp. 182, 184.

9 John Ruskin, *Modern Painters*, 6 vols (1843–60; London: George Allen, 1897), I, p. xxxiv (Preface to 2nd edn, § 27).

10 Ibid., I, pp. xxxiv–v (Preface to 2nd edn, § 27–8).

11 F. G. Stephens [as John Seward], 'The Purpose and Tendency of Early Italian Art', *The Germ*, p. 64.

12 Ibid.

13 Stephen Wildman (ed.), *Ruskin and the Pre-Raphaelites* (London: Pallas Athene, 2012), p. 52.

14 Ruskin, *Modern Painters*, I, pp. xxxvi–vii (Preface to 2nd edn, § 30).

15 John Lucas Tupper, 'Extracts from the Diary of an Artist', *The Crayon*, 4 (1857), p. 105.

16 Elizabeth Prettejohn, *The Art of the Pre-Raphaelites* (London: Tate, 2000), p. 172. For a suggestive selection of depictions of flowers in Pre-Raphaelite art, viewed both as botanical illustration and through the language of floral symbolisms, see Debra N. Mancoff, *The Pre-Raphaelite Language of Flowers* (Munich: Prestel, 2012).

17 W. H. Deverell, 'A Modern Idyll', *The Germ*, 177–9 (p. 179).

18 George Eliot, *Selected Critical Writings*, ed. Rosemary Ashton (Oxford: Oxford University Press, 1992), p. 263.

19 Prettejohn, *Art of the Pre-Raphaelites*, p. 156.

20 Wildman, *Ruskin and the Pre-Raphaelites*, p. 127.

21 For the catalogue of an exhibition which made a compelling case for Pre-Raphaelite landscape art in 2004 to 2005, see Allen Staley et al., *Pre-Raphaelite Vision: Truth to Nature* (London: Tate, 2004).

22 Christiana Payne, *John Brett: Pre-Raphaelite Landscape Painter* (New Haven and London: Yale University Press, 2010), p. 17.

23 Ibid., p. 25.

24 Ibid., p. 18.

25 On Rosa Brett's career as an artist, see Jan Marsh and Pamela Gerrish Nunn, *Women Artists and the Pre-Raphaelite Movement* (London: Virago, 1989), pp. 52–8.

26 Staley et al., *Pre-Raphaelite Vision*, p. 51.

27 Wildman, *Ruskin and the Pre-Raphaelites*, p. 82.

28 Ibid., p. 204.

29 Ibid., p. 194.

30 Ibid., p. 206.

31 Payne, *John Brett*, p. 45.

32 Quoted in Robert P. McIntosh, *The Background of Ecology: Concept and Theory* (Cambridge: Cambridge University Press, 1985), pp. 7–8.

33 Ibid., pp. 7–8; Donald Worster, *Nature's Economy: A History of Ecological Ideas*, 2nd edn (Cambridge: Cambridge University Press, 1994), p. 114; Frank Egerton, *Roots of Ecology: Antiquity to Haeckel* (Berkeley: University of California Press, 2012), p. 165.

34 Quoted in Egerton, *Roots of Ecology*, p. 82.

35 Egerton, *Roots of Ecology*, pp. 120–25, 130–35, 143–6, 165–75. On Humboldt, see also Andrea Wulf, *The Invention of Nature: The Adventures of Alexander von Humboldt, the Lost Hero of Science* (London: John Murray, 2015).

36 Karl Kroeber, *Ecological Literary Criticism: Romantic Imagining and the Biology of the Mind* (New York: Columbia University Press, 1994); James C. McKusick, *Green Writing: Romanticism and Ecology*, 2nd edn (New York: Palgrave Macmillan, 2010); Onno Oerlemans, *Romanticism and the Materiality of Nature* (Toronto: University of Toronto Press, 2002).

37 For persuasive readings of Clare's poetry as ecological, see McKusick, *Green Writing*, pp. 77–94; Oerlemans, *Romanticism and the Materiality of Nature*, pp. 137–43; and M. M. Mahood, *The Poet as Botanist* (Cambridge: Cambridge University Press, 2008), pp. 112–46. For a discussion of his scepticism towards the terms and strictures of Linnaean taxonomy, see Francesca Cuojati, 'John Clare: The Poetics and Politics of Taxonomy', in Caroline Patey and Laura Scuriatti (eds), *The Exhibit in the Text: The Museological Practices of Literature* (Bern: Peter Lang, 2009), 29–48.

38 Diana Donald, *Picturing Animals in Britain 1750–1850* (New Haven and London: Yale University Press, 2007), pp. 28–100; Diana Donald, 'The "Struggle for Existence" in Nature and Human Society' in Diana Donald and Jane Munro (eds), *Endless Forms: Charles Darwin, Natural Science and the Visual Arts* (New Haven and London: Yale University Press, 2009), 81–99.

39 Donald, 'The "Struggle for Existence" in Nature and Human Society', p. 94. See also Donald, *Picturing Animals in Britain*, pp. 89–100.

40 A. H. Palmer, *The Life of Joseph Wolf, Animal Painter* (London: Longman, 1895), p. 67.

41 See Bodleian Library, Oxford, MS Facs. d. 278, fol. 83; W. M. Rossetti, 'Animal-Design and Landscape: Aspects of their Contemporary Treatment', *Macmillan's Magazine*, 8 (1863), 116–23 (pp. 120–21). The manuscript source is a note by Rossetti glossing a letter to him by Palmer asking him about Wolf. See also Palmer, *Life of Joseph Wolf*, p. 67.

42 Dennis T. Lanigan, 'The First Pre-Raphaelite Group Exhibition', *Journal of Pre-Raphaelite Studies*, 17 (2008), 9–19.

43 Paul Barlow, 'John Everett Millais (1829–1896)', in Elizabeth Prettejohn (ed.), *The Cambridge Companion to the Pre-Raphaelites* (Cambridge: Cambridge University Press, 2012), 133–47 (pp. 141, 143).

44 Paul Barlow, *Time Present and Time Past: The Art of John Everett Millais* (Aldershot: Ashgate, 2005), p. 36.

45 Ibid., p. 36.

46 Ibid., p. 21.

47 See Coventry Patmore, 'A Pre-Raphaelite Exhibition', *Saturday Review*, 4 (1857), 11–12, and 'Walls and Wall Painting at Oxford', *Saturday Review*, 4 (1857), 583–4.

48 W. M. Rossetti, *P.R.B. Journal*, pp. 3–7.

49 Coventry Patmore, *Poems* (London: Bell, 1915), p. 412.

50 F. G. Stephens, *Exhibition of the Works of Sir John E. Millais, Bart., R.A., with Notes* (London: Grosvenor Gallery, 1886), p. 68.

51 Ibid., p. 67.

52 Ibid., p. 68.

53 For a discussion of oak ecology and its relation to human ecology, see Paul Hatcher and Nick Battey, *Biological Diversity: Exploiters and Exploited* (Chichester: Wiley, 2011), pp. 183–204.

54 See the *Oxford English Dictionary* online, 'shaw, *n.1*'.

55 For discussions of the relationship between geology and Pre-Raphaelite aesthetics in *The Glacier of Rosenlaui*, see Kenneth Bendiner, *An Introduction to Victorian Painting* (New Haven and London: Yale University Press, 1985), pp. 47–63; Payne, *John Brett*, pp. 31–3; Staley et al., *Pre-Raphaelite Vision*, pp. 152–3.

56 Recent studies of differing temperaments among domestic sheep include Angela M. Sibbald and Russell J. Hooper, 'Sociability and the Willingness of Individual Sheep to move away from their Companions in order to graze', *Applied Animal Behaviour Science*, 86 (2004), 51–62; Ngaio J. Beausoleil et al., 'Exploring the Basis of Divergent Selection for "Temperament" in Domestic Sheep', *Applied Animal Behaviour Science*, 109 (2008), 261–74; Amandine Ramseyer et al., 'Decision Making in Group Departures of Sheep is a Continuous Process', *Animal Behaviour*, 78 (2009), 71–8; and Ngaio J. Beausoleil et al., 'Selection for Temperament in Sheep: Domain-General and Context-Specific Traits', *Applied Animal Behaviour Science*, 139 (2012), 74–85.

57 [F. G. Stephens], 'The Two Pre-Raphaelitisms', *The Crayon*, 3 (1856), p. 354. D. J. Snook has recently pointed out the inaccuracies in Stephens's memory of this painting, along with other grounds for doubting these interpretations of Hunt's painting, in 'His Romancing Humour', *Pre-Raphaelite Society Review*, 21.1 (Spring 2013), 2–11.

58 William Holman Hunt, 'The Pre-Raphaelite Brotherhood: A Fight for Art', *Contemporary Review*, 49 (1886), pp. 750, 820.

59 Ibid., p. 828.

60 W. M. Rossetti, *Fine Art, Chiefly Contemporary: Notices Reprinted, with Revisions* (London: Macmillan, 1867), p. 237.

61 Ibid., p. 238.

62 J. G. Millais, *The Life and Letters of Sir John Everett Millais*, 2 vols (London: Methuen, 1899), I, p. 130.

63 Anne Clark Amor, *William Holman Hunt: The True Pre-Raphaelite* (London: Constable, 1989), p. 89.

64 W. M. Rossetti, *Fine Art*, p. 237. For an excellent discussion of Landseer's dogs, see Donald, *Picturing Animals*, pp. 126–58. Donald sees Landseer's paintings of wild animals as ecological in a similar way to Wolf's; see pp. 89–100.

65 Charles Darwin, *The Expression of the Emotions in Man and Animals* (London: Murray, 1872), p. 131.

66 [Stephens], 'The Two Pre-Raphaelitisms' (1856), p. 354.

67 Darwin, *Expression of the Emotions*, p. 85.

68 Elizabeth E. Shillito, 'A Comparison of the Role of Vision and Hearing in Lambs finding their own Dams', *Applied Animal Ethology*, 1 (1975), 369–77. See also G. Alexander, 'Role of Auditory and Visual Cues in Mutual Recognition between Ewes and Lambs in Merino Sheep', *Applied Animal Ethology*, 3 (1977), 65–81; G. Alexander and Elizabeth E. Shillito, 'Importance of Visual Cues from Various Body Regions in Maternal Recognition of the Young in Merino Sheep (*Ovis aries*)', *Applied Animal Ethology*, 3 (1977), 137–43; and Elizabeth E. Shillito Walser, 'A Comparison of the Role of Vision and Hearing in Ewes finding their own Lambs', *Applied Animal Ethology*, 4 (1978), 71–9.

69 [Stephens], 'The Two Pre-Raphaelitisms' (1856), p. 354.

70 See D. G. Rossetti, *Correspondence*, I, pp. 64–7, 73–4.

71 James Collinson, 'The Child Jesus', *The Germ*, 49–57 (pp. 50–51).

72 Ibid., p. 51.

73 Luke 22:44; Hebrews 5:7. The two passages are cross-referenced in the King James Bible.

74 Luke 22:42. The equivalent passages, with variant wording, are Matthew 26:42 and Mark 14:36.

75 Matthew 26:41. See also Mark 14:38.

76 Collinson, 'The Child Jesus', p. 53.

77 Ibid., p. 54.

78 See John Holmes, *Darwin's Bards: British and American Poetry in the Age of Evolution* (Edinburgh: Edinburgh University Press, 2009), pp. 159–64.

79 William Paley, *Natural Theology or Evidence of the Existence and Attributes of the Deity, collected from the Appearances of Nature*, ed. Matthew D. Eddy and David Knight (Oxford: Oxford University Press, 2006), p. 238.

80 John Lucas Tupper, 'The Subject in Art', *The Germ*, p. 122.

81 Julian Treuherz, *Ford Madox Brown: Pre-Raphaelite Pioneer* (Manchester: Manchester Art Gallery, 2011), p. 164. Staley points out that the window was in fact his landlady's, not his own. See Staley et al., *Pre-Raphaelite Vision*, p. 71.

82 Staley et al., *Pre-Raphaelite Vision*, p. 79.

83 Staley suggests that Brown may have taken a lead from Collins's painting in Staley et al., *Pre-Raphaelite Vision*, p. 79. For an insightful discussion of Collins's painting itself by Alison Smith, see Staley et al., *Pre-Raphaelite Vision*, p. 69.

84 On Boyce's career as an artist, see Christopher Newall and Judy Egerton, *George Price Boyce* (London: Tate, 1987), and Staley, *Pre-Raphaelite Landscape*, pp. 107–10.

85 Newall and Egerton, *George Price Boyce*, p. 57.

86 Newall, the leading authority on Boyce's painting, discusses his interest in vernacular architecture in Newall and Egerton, *George Price Boyce*,

pp. 21–5, and Staley et al., *Pre-Raphaelite Vision*, p. 91.

87 Treuherz, *Ford Madox Brown*, pp. 164–5.

88 Tim Barringer, 'The Effects of Industry: Ford Madox Brown and Artistic Identities in Victorian Britain', in Tessa Sidey (ed.), *Ford Madox Brown: The Unofficial Pre-Raphaelite* (London: D. Giles, 2008), 16–31 (p. 27).

89 Alastair Ian Wright, 'Suburban Prospects: Vision and Possession in Ford Madox Brown's *An English Autumn Afternoon*', in Margaretta Frederick Watson (ed.), *Collecting the Pre-Raphaelites: The Anglo-American Enchantment* (Aldershot: Ashgate, 1997), 185–97 (pp. 189–90); Staley et al., *Pre-Raphaelite Vision*, p. 71.

90 Wildman, *Ruskin and the Pre-Raphaelites*, p. 72.

91 Jonathan Bate, *Romantic Ecology: Wordsworth and the Environmental Tradition* (London: Routledge, 1991), p. 40.

92 Ford Madox Brown, *The Exhibition of Work and other Paintings* (London: M'Corquodale & Co., 1865), quoted in Treuherz, *Ford Madox Brown*, p. 165.

93 Quoted in Tim Barringer, Jason Rosenfeld and Alison Smith, *Pre-Raphaelites: Victorian Avant-Garde* (London: Tate, 2012), p. 68.

94 T. Woolner, 'Of my Lady. In Death', *The Germ*, 5–10 (p. 6).

95 Ibid., p. 6.

96 William Wordsworth, *The Poetical Works*, ed. Thomas Hutchinson and Ernest de Selincourt (London: Oxford University Press, 1950), p. 380 ('A Poet's Epitaph', ll. 19–20).

97 D. G. Rossetti, *Correspondence*, I, p. 96.

98 T. Woolner, 'Of my Lady. In Death', pp. 6–7.

3 EMBODIED PSYCHOLOGY IN PRE-RAPHAELITE PAINTING AND POETRY

1 F. G. Stephens [as John Seward], 'The Purpose and Tendency of Early Italian Art', *The Germ*, p. 58. The other major exponents of Pre-Raphaelite landscape painting alongside the Bretts and Boyce were John William Inchbold and Thomas Seddon. See Christopher Newall, *John William Inchbold: Pre-Raphaelite Landscape Artist* (Leeds: Leeds City Art Galleries, 1993) and Allen Staley, *The Pre-Raphaelite Landscape* (Oxford: Oxford University Press, 1973), pp. 96–106, 111–23. For a valuable discussion of how Pre-Raphaelite aesthetics built on developments in earlier nineteenth-century landscape painting, see W. F. Axton, 'Victorian Landscape Painting: A Change in Outlook', in U. C. Knoepflmacher and G. B. Tennyson (eds), *Nature and the Victorian Imagination* (Berkeley: University of California Press, 1977), 281–308.

2 W. M. Rossetti, *Fine Art, Chiefly Contemporary: Notices Reprinted, with Revisions* (London: Macmillan, 1867), p. 33.

3 Sally Shuttleworth, *The Mind of the Child: Child Development in Literature, Science, and Medicine, 1840–1900* (Oxford: Oxford University Press, 2010), p. 28.

4 On the early history of psychology as a discipline in Britain, see L. S. Hearnshaw, *A Short History of British Psychology 1840–1940* (London: Methuen, 1964), pp. 1–24; Rick Rylance, *Victorian Psychology and British Culture 1850–1880* (Oxford: Oxford University Press, 2000), pp. 70–143; and Alan Collins, 'England', in David B. Baker (ed.), *The Oxford Handbook of the History of Psychology: Global Perspectives* (Oxford: Oxford University Press, 2012), 182–210 (pp. 182–8).

5 Shuttleworth, *Mind of the Child*, p. 33.

6 F. G. Stephens, 'Modern Giants', *The Germ*, p. 171.

7 Ibid., p. 172. This paragraph was reprinted without the reference to American poetry in 'The Two Pre-Raphaelitisms', *The Crayon*, 3 (1856), p. 292.

8 Gregory Tate, *The Poet's Mind: The Psychology of Victorian Poetry 1830–1870* (Oxford: Oxford University Press, 2012). Tate's work builds on the earlier study by Ekbert Faas, *Retreat into the Mind: Victorian Poetry and the Rise of Psychiatry* (Princeton: Princeton University Press, 1988).

9 Tate, *The Poet's Mind*, p. 9.

10 For a study of the heart in Victorian poetry, see Kirstie Blair, *Victorian Poetry and the Culture of the Heart* (Oxford: Oxford University Press, 2006). In remarking that 'Browning could be described as a poet more interested in the movements of the mind than of the heart' (p. 20), Blair anticipates Tate's argument rather than rehearsing Stephens's.

11 Matthew Arnold, *The Poems of Matthew Arnold*, ed. Kenneth and Miriam Allott, 2nd edn (London: Longman, 1979), p. 654.

12 See J. B. Bullen, *The Pre-Raphaelite Body: Fear and Desire in Painting, Poetry, and Criticism* (Oxford: Oxford University Press, 1998).

13 Lucy Hartley, *Physiognomy and the Meaning of Expression in Nineteenth-Century Culture* (Cambridge: Cambridge University Press, 2001), p. 62. For Hartley's full discussion of Bell's theories of anatomy and art, see pp. 44–79.

14 John Lucas Tupper, 'Extracts from the Diary of an Artist' *The Crayon*, 3 (1856), p. 44.

15 William Holman Hunt, *Pre-Raphaelitism and the Pre-Raphaelite Brotherhood* (London: Macmillan, 1905), 2 vols, I, pp. 100–1.

16 W. M. Rossetti, *Dante Gabriel Rossetti: His Family Letters with a Memoir* (London: Ellis and Elvey, 1895), 2 vols, I, p. 89.

17 Hartley, *Physiognomy and the Meaning of Expression*, p. 74.

18 Ibid., p. 69.

19 Julie F. Codell, 'Expression over Beauty: Facial Expression, Body Language, and Circumstantiality in the Paintings of the Pre-Raphaelite Brotherhood', *Victorian Studies*, 29 (1986), 255–90 (pp. 272–4); Julie F. Codell, 'Empiricism, Naturalism and Science in Millais's Paintings', in Debra N. Mancoff (ed.), *John Everett Millais: Beyond the Pre-Raphaelite Brotherhood* (New Haven and London: Yale University Press, 2001), pp. 130–32; Stephanie Grilli, 'Pre-Raphaelitism and Phrenology', in Leslie Parris (ed.), *Pre-Raphaelite Papers* (London: Tate, 1984), pp. 56–60.

20 W. M. Rossetti, *The P.R.B. Journal: William Michael Rossetti's Diary of the Pre-Raphaelite Brotherhood 1849–1853*, ed. William E. Fredeman (Oxford: Oxford University Press, 1975), pp. 22, 24.

21 Hunt, *Pre-Raphaelitism*, I, pp. 257–61. Rossetti too records Donovan associating poetry with scrofula in *P.R.B. Journal*, p. 21.

22 Tupper, 'Diary of an Artist' (1855), p. 336. For 'Suspiria de Profundis', see Thomas De Quincey, *Confessions of an English Opium-Eater and Other Writings*, ed. Grevel Lindop (Oxford: Oxford University Press, 1985), pp. 87–181.

23 Hearnshaw, *A Short History of British Psychology*, p. 22.

24 Codell, 'Empiricism, Naturalism and Science', p. 131.

25 Elizabeth Prettejohn, *The Art of the Pre-Raphaelites* (London: Tate, 2000), p. 256.

26 Barbara Larson, 'Evolution and Victorian Art', in Bernard Lightman and Bennett Zon (eds), *Evolution and Victorian Culture* (Cambridge: Cambridge University Press, 2014), 121–48 (p. 136).

27 Tupper, 'Diary of an Artist' (1857), p. 148.

28 W. M. Rossetti, 'Pre-Raphaelitism', *The Spectator*, 24 (1851), p. 955.

29 Tupper, 'Diary of an Artist' (1855), p. 368; (1856), p. 44.

30 Elizabeth Prettejohn, *The Modernity of Ancient Sculpture: Greek Sculpture and Modern Art from Winckelmann to Picasso* (London: Tauris, 2012), pp. 60–73.

31 B. R. Haydon, *Lectures on Painting and Design*, 2 vols (London: Longman, 1844–6), II, p. 219.

32 Tupper, 'Diary of an Artist' (1855), p. 400.

33 Ibid., p. 368.

34 Ibid., p. 369.

35 Haydon, *Lectures on Painting and Design*, II, p. 220; Tupper, 'Diary of an Artist' (1855), p. 368.

36 Tupper, 'Diary of an Artist' (1855), p. 369.

37 Ibid., p. 400 (original emphasis).

38 Prettejohn, *The Modernity of Ancient Sculpture*, p. 63; Haydon, *Lectures on Painting and Design*, II, p. 220.

39 Ford Madox Brown, 'On the Mechanism of a Historical Picture, Part I: The Design', *The Germ*, p. 71.

40 Oscar Wilde, *The Complete Works of Oscar Wilde*, 5th edn (London: Harper Collins, 2003), p. 17.

41 For glimpses of the history of the composition of this poem between September 1849 and May 1850, see D. G. Rossetti, *The Correspondence of Dante Gabriel Rossetti*, ed. William E. Fredeman, 9 vols (Cambridge: D. S. Brewer, 2002–10), I, pp. 93, 96, 144; W. M. Rossetti, *P.R.B. Journal*, pp. 24, 30, 32, 37, 40, 48–9.

42 D. G. Rossetti, *The Works of Dante Gabriel Rossetti*, ed. W. M. Rossetti (London: Ellis, 1911), p. 24.

43 Ibid., p. 20.

44 Ibid., p. 25.

45 Carole Silver, 'Visions and Revisions', in Katharine Lochnan and Carol Jacobi (eds), *Holman Hunt and the Pre-Raphaelite Vision* (Toronto: Art Gallery of Ontario, 2008), 15–30 (p. 20).

46 William Shakespeare, *The Complete Works: Compact Edition*, ed. Stanley Wells and Gary Taylor (Oxford: Oxford University Press, 1988), p. 23 (*The Two Gentlemen of Verona*, Act 5, Scene 4, ll. 82–3).

47 Ibid., ll. 73–4).

48 [F. G. Stephens], 'The Two Pre-Raphaelitisms', *The Crayon*, 3 (1856), p. 355.

49 For this text of this sonnet, see http://www.rossettiarchive.org/docs/n8640.a8.rad.html#p18 [accessed 20 September 2013].

50 D. G. Rossetti, 'Sonnets for Pictures', *The Germ*, 180–82 (p. 180).

51 D. G. Rossetti, *Correspondence*, I, p. 128. For the attribution of this painting, see http://www.rossettiarchive.org/docs/41-1849.raw.html [accessed 23 September 2013].

52 Rossetti's itinerary in Belgium took him from Brussels to Antwerp, on to Ghent, then Bruges, then back to Ghent. See D. G. Rossetti, *Correspondence*, I, pp. 119–31, and *Works*, pp. 182–7.

53 D. G. Rossetti, *Correspondence*, I, p. 128.

54 W. M. Rossetti, *P.R.B. Journal*, p. 29.

55 Elizabeth K. Helsinger, *Poetry and the Pre-Raphaelite Arts: Dante Gabriel Rossetti and William Morris* (New Haven and London: Yale University Press, 2008), p. 44.

56 Luke, 1:26–38.

57 Tim Barringer, Jason Rosenfeld and Alison Smith, *Pre-Raphaelites: Victorian Avant-Garde* (London: Tate, 2012), p. 120.

58 Ibid. See also Helsinger, *Poetry and the Pre-Raphaelite Arts*, pp. 44–5.

59 J. R. Seeley, *The Expansion of England: Two Courses of Lectures* (London: Macmillan, 1883), pp. 8, 10–11, and passim.

60 W. M. Rossetti, *P.R.B. Journal*, p. 4.

61 On Smith, see Juliet Peers, 'Bernard Smith: "The Missing Brother"', in Benedict Read and Joanna Barnes (eds), *Pre-Raphaelite Sculpture: Nature and Imagination in British Sculpture 1848–1914* (London: Henry Moore Foundation, 1991), 12–20.

62 David F. Mitch, *The Rise of Popular Literacy in Victorian England: The Influence of Private Choice and Public Policy* (Philadelphia: University of Pennsylvania Press, 1992), p. 1.

63 Ibid., p. xvii.

64 W. M. Rossetti, *P.R.B. Journal*, p. 18.

65 On the well-established relationship between Collinson's art and the tradition of genre painting, see Barringer et al., *Pre-Raphaelites*, pp. 36–7; Julian Treuherz, *Victorian Painting* (London: Thames and Hudson, 1993), p. 84, and *Pre-Raphaelite Paintings from Manchester City Art Galleries*, 2nd edn (Manchester: Manchester City Art Galleries, 1993), p. 22.

66 D. G. Rossetti, *Works*, p. 24.

67 'A Last Confession' was included in the collection of poems buried with Elizabeth Siddal in 1862 and exhumed in 1869. It was published after some revision and expansion in D. G. Rossetti's *Poems* in 1870. For the textual history of the poem, insofar as it is recoverable, see http://www.rossettiarchive.org/docs/1-1849.raw.html [accessed 6 September 2013]. 'Mrs. Holmes Grey', originally entitled 'An Exchange of News' and then 'A Plain Story of Life', was written while William Rossetti was on holiday on the Isle of Wight in September 1849. It was intended for *The Germ*, but it had already outgrown this format even had the magazine not folded. For this early textual history, see W. M. Rossetti, *P.R.B. Journal*, pp. 14–20, 24, 49, 222. It was eventually published in *The Broadway* in 1868, 'not indeed without some revision,' as Rossetti admitted in an accompanying note, 'but without the least alteration in its general character and point of view' (quoted in *P.R.B. Journal*, p. 130). Fredeman's edition of the *P.R.B. Journal* includes a reprint of the full poem (pp. 131–54), which is the text followed here.

68 Faas, *Retreat into the Mind*, p. 158.

69 W. M. Rossetti, *Selected Letters of William Michael Rossetti*, ed. Roger W. Peattie (University Park: Pennsylvania State University Press, 1990), pp. 11–12; W. M. Rossetti, '*Christmas Eve and Easter Day*: by Robert Browning', *The Germ*, p. 191.

70 D. G. Rossetti, *Works*, p. 47.

71 Ibid., pp. 47–8.

72 D. G. Rossetti, 'A Last Confession (fair copy manuscript with corrections, Fitzwilliam Museum)',

http://www.rossettiarchive.org/docs/1-1849.fizms.rad.html [accessed 6 September 2013].

73 D. G. Rossetti, *Works*, p. 53.

74 Ibid.

75 Ibid., p. 44.

76 W. M. Rossetti, *P.R.B. Journal*, p. 141, ll. 380, 384.

77 Ibid., p. 154, ll. 809–10.

78 Ibid., p. 151, ll. 731–2.

79 Ibid., pp. 142–3, ll. 411, 422–6.

80 Ibid., p. 145, ll. 516–18.

81 Ibid., p. 148, ll. 600–2, 611–14.

82 Ibid., p. 148, ll. 614–17.

83 Ibid., p. 147, ll. 580–83.

84 Ibid., p. 151, ll. 711–16.

85 Ibid., p. 151, ll. 720–23.

86 Helen Small, *Love's Madness: Medicine, the Novel, and Female Insanity 1800–1865* (Oxford: Oxford University Press, 1996), p. 33.

87 D. G. Rossetti, *Correspondence*, I, p. 112.

88 MacKenzie Bell, *Christina Rossetti: A Biographical and Critical Study* (London: Thomas Burleigh, 1898), p. 20.

89 W. M. Rossetti, *P.R.B. Journal*, pp. 20, 32.

90 D. M. R. Bentley, '"A Very Clever and Finished Piece of Writing": William Michael Rossetti's "Mrs. Holmes Grey"', *Journal of Pre-Raphaelite Studies*, N.S. 20.1 (2011), 5–25 (p. 22).

91 W. M. Rossetti, *P.R.B. Journal*, p. 151, ll. 725–6, 728.

92 W. M. Rossetti, 'Pre-Raphaelitism', p. 955.

4 SACRED ART ON SCIENTIFIC PRINCIPLES

1 F. G. Stephens [as John Seward], 'The Purpose and Tendency of Early Italian Art', *The Germ*, p. 61.

2 For a commentary on Rossetti's early, unpublished collection of poems 'Songs of the Art Catholic', see http://www.rossettiarchive.org/docs/11-1847.raw.html [accessed 18 September 2013].

3 William Holman Hunt, *Pre-Raphaelitism and the Pre-Raphaelite Brotherhood*, 2 vols (London: Macmillan, 1905), I, p. 355.

4 Stephens, 'Early Italian Art', p. 64 (original emphasis).

5 W. M. Rossetti, 'The Externals of Sacred Art', *The Crayon*, 5 (1858), 333–6 (p. 333).

6 John Ruskin, *The Art of England* (Orpington: George Allen, 1884), p. 40 (original emphasis).

7 W. M. Rossetti, 'Externals of Sacred Art', p. 333.

8 Ibid., p. 334.

9 Anna Jameson, *Memoirs of the Early Italian Painters, and of the Progress of Painting in Italy*, 2 vols (London: Knight, 1845), I, p. 5.

10 D. G. Rossetti, 'Hand and Soul', *The Germ*, pp. 26–7.

11 D. G. Rossetti, 'Sonnets for Pictures', *The Germ*, p. 180.

12 W. M. Rossetti, 'Externals of Sacred Art', p. 334.

13 J. G. Millais, *The Life and Letters of Sir John Everett Millais*, 2 vols (London: Methuen, 1899), I, p. 78.

14 Stephanie Grilli, 'Pre-Raphaelitism and Phrenology', in Leslie Parris (ed.), *Pre-Raphaelite Papers* (London: Tate, 1984), p. 56.

15 W. M. Rossetti, 'The Abstract and Naturalism in Art', *The Crayon*, 6 (1859), 65–8 (p. 66).

16 J. G. Millais, *Life and Letters*, I, p. 78.

17 For a detailed discussion of *Christ in the House of his Parents* as 'a gesture of defiance' towards the traditions of religious art, see Michaela Giebelhausen, *Painting the Bible: Representation and Belief in Mid-Victorian Britain* (Aldershot: Ashgate, 2006), pp. 111–26 (p. 111).

18 Colin Cruise, *Pre-Raphaelite Drawing* (London: Thames and Hudson, 2011), p. 17.

19 W. M. Rossetti, 'Externals of Sacred Art', p. 334.

20 Ibid., p. 334.

21 Ibid., p. 335.

22 Ibid., p. 335.

23 F. G. Stephens, *Exhibition of the Works of Sir John E. Millais, Bart., R.A., with Notes* (London: Grosvenor Gallery, 1886), p. 13.

24 Collinson, 'The Child Jesus', *The Germ*, pp. 50, 55.

25 W. M. Rossetti, 'Introduction' to *The Germ: Thoughts towards Nature in Poetry*... (London: Elliot Stock, 1901), pp. 20–21.

26 W. M. Rossetti, *The P.R.B. Journal: William Michael Rossetti's Diary of the Pre-Raphaelite Brotherhood 1849–1853*, ed. William E. Fredeman (Oxford: Oxford University Press, 1975), p. 73.

27 Stephen Wildman (ed.), *Ruskin and the Pre-Raphaelites* (London: Pallas Athene, 2012), p. 80.

28 Ford Madox Brown, 'On the Mechanism of a Historical Picture, Part I: The Design', *The Germ*, p. 70.

29 William Holman Hunt, 'The Pre-Raphaelite Brotherhood: A Fight for Art', *Contemporary Review*, 49 (1886), p. 826.

30 For a more detailed discussion of the relationship between Hunt's project and the higher criticism, see Giebelhausen, *Painting the Bible*, pp. 128–33, 159–87.

31 Hunt, 'Pre-Raphaelite Brotherhood', p. 829.

32 Ibid.

33 For an excellent discussion of Hunt's sense of his own religious mission, and how this plays out in his paintings, see Carol Jacobi, *William Holman Hunt: Painter, Painting, Paint* (Manchester: Manchester University Press, 2006), pp. 61–87.

34 William Holman Hunt, 'Painting "The Scapegoat"', *Contemporary Review*, 52 (1887), 21–38, 206–20; Hunt, *Pre-Raphaelitism*, I, pp. 446–512; II, pp. 1–16. For a comparison of Hunt's published accounts and his journals for this trip, see Judith Bronkhurst, '"An interesting series of adventures to look back upon": William Holman Hunt's visit to the Dead Sea in November 1854', in Parris, *Pre-Raphaelite Papers*, 111–25.

35 Hunt, *Pre-Raphaelitism*, I, p. 458.

36 John Rylands Library, Manchester, Ryl. Eng. MS 1213, fol. 12 (punctuation and capitalisation as in the original).

37 John Rylands Library, Manchester, Ryl. Eng. MS 1210, fol. 4 (punctuation and capitalisation as in the original).

38 Hunt, *Pre-Raphaelitism*, I, p. 477.

39 Alexis Drahos, 'Imagining Geological Time: *The Scapegoat* by William Holman Hunt', *Review of the Pre-Raphaelite Society*, 19.2 (2011), 40–45; Tim Barringer, Jason Rosenfeld and Alison Smith, *Pre-Raphaelites: Victorian Avant-Garde* (London: Tate, 2012), p. 140.

40 John Rylands Library, Manchester, Ryl. Eng. MS 1210, fol. 7. For a succinct account of the three

goats, see Mary Bennett, *Artists of the Pre-Raphaelite Circle – The First Generation: Catalogue of Works in the Walker Art Gallery, Lady Lever Art Gallery and Sudley Art Gallery* (London: Lund Humphries, 1988), p. 72.

41 [F. G. Stephens], 'Two Pre-Raphaelitisms', *The Crayon*, 3 (1856), p. 355.

42 Wildman, *Ruskin and the Pre-Raphaelites*, pp. 163–4.

43 John Rylands Library, Manchester, Ryl. Eng. MS 1213, fol. 12 (punctuation and capitalisation as in the original).

44 Drahos, 'Imagining Geological Time', p. 43; Nicholas Tromans, 'Palestine: Picture of Prophecy', in Katharine Lochnan and Carol Jacobi (eds), *Holman Hunt and the Pre-Raphaelite Vision* (Toronto: Art Gallery of Ontario, 2008), 135–58 (p. 139).

45 F. G. Stephens, *William Holman Hunt and his Works: A Memoir of the Artist's Life, with Description of his Pictures* (London: Nisbet, 1860), pp. 57–8.

46 Ibid., pp. 47–50 (original emphasis).

47 Hunt, *Pre-Raphaelitism*, II, pp. 25, 27.

48 Stephens, *William Holman Hunt and his Works*, p. 58.

49 Ibid., p. 58.

50 In *The Stones of Venice*, Ruskin refutes 'the strange and vain supposition, that the original conception of Gothic architecture had been derived from vegetation', arguing that it is given the lie by the forms of the earlier architecture from which Gothic in fact derives. Instead, he proposes that 'the Gothic did not arise out of, but developed itself into, a resemblance to vegetation.' See John Ruskin, *The Stones of Venice*, 3 vols (London: Smith, Elder, 1851–3), II, p. 201 (Ch. VI, § 70).

51 Stephens, *William Holman Hunt and his Works*, p. 58.

52 Ibid., p. 59, references 2 Chronicles 3:7 as Hunt's source here.

53 Ibid., p. 60.

54 For a largely sceptical response to Stephens's account of *The Finding of the Saviour in the Temple*, see Kenneth Bendiner, *An Introduction to Victorian Painting* (New Haven and London: Yale University Press, 1985), pp. 73–7.

55 Alan Bowness et al., *The Pre-Raphaelites* (London: Tate, 1984), pp. 158–9.

56 Stephens, *William Holman Hunt and his Works*, pp. 58–9.

57 Anon., 'Binocular Vision and the Painter's Art', *Saturday Review*, 14 (1862), 163–4 (p. 164).

58 John Lucas Tupper, 'Addendum: How We Write on Art', in *Hiatus: The Void in Modern Education; Its Cause, and Antidote* (London: Macmillan, 1869), 239–51; this essay was written in 1862 but not published until it was appended to Tupper's book in 1869 (p. 204).

59 [J. W. Jackson,] 'Art in Relation to Comparative Anatomy', *Anthropological Review*, 5 (1867), 28–46 (p. 30). This article is identified as by Jackson in the endpaper to his book *Man Contemplated Physically, Morally, Intellectually and Spiritually* (London: James Burns, 1871). Mary Cowling discusses this article in *The Artist as Anthropologist: The Representation of Type and Character in Victorian Art* (Cambridge: Cambridge University Press, 1989), pp. 109–11.

60 Elizabeth Prettejohn, *The Art of the Pre-Raphaelites* (London: Tate, 2000), p. 248.

61 John Rylands Library, Manchester, Ryl. Eng. MS 1210, fol. 4; Bronkhurst, '"An interesting series of adventures"', p. 123.

62 Drahos remarks that 'the presence of animal skeletons in a wild landscape surrounded by mountains may have paleontological connotations' ('Imagining Geological Time', p. 43).

63 Alfred Tennyson, *The Poems of Tennyson*, ed. Christopher Ricks, 2nd edn, 3 vols (Harlow: Longman, 1987), II, pp. 373–4 (*In Memoriam*, LVI, ll. 15, 22).

64 John Rylands Library, Manchester, Ryl. Eng. MS 1210, fols 4–5. For an insightful discussion of the complexity of Hunt's physical and spiritual response to this landscape, see Jacobi, *William Holman Hunt*, pp. 50–58.

65 Leviticus 16:22.

66 John Rylands Library, Manchester, Ryl. Eng. MS 1213, fol. 12.

67 Hunt, *Pre-Raphaelitism*, I, p. 458.

68 Prettejohn, *Art of the Pre-Raphaelites*, p. 249.

69 The catalogue entry is quoted in full in Bennett, *Artists of the Pre-Raphaelite Circle*, p. 70.

70 George P. Landow and Ruth M. Landow (eds), 'The Exhibition Pamphlet for W. Holman Hunt's *The Shadow of Death*', *Journal of Pre-Raphaelite Studies*, 16.1 (2007), 9–29 (p. 16).

71 For the classic analysis of this prejudice, see Edward W. Said, *Orientalism* (London: Penguin, 2003).

72 Jacobi, *William Holman Hunt*, p. 93.

73 Carol Jacobi, 'William Holman Hunt (1827–1910)', in Elizabeth Prettejohn (ed.), *The Cambridge Companion to the Pre-Raphaelites* (Cambridge: Cambridge University Press, 2012), p. 130.

74 Landow and Landow, 'Exhibition Pamphlet', p. 14.

75 Ibid., pp. 23–4.

76 Barringer et al., *Pre-Raphaelites*, p. 152.

77 Hunt, 'Pre-Raphaelite Brotherhood', p. 829.

78 Landow and Landow, 'Exhibition Pamphlet', p. 15.

79 See Marcia Pointon, 'The Artist as Ethnographer: Holman Hunt and the Holy Land', in Marcia Pointon (ed.), *Pre-Raphaelites Re-viewed* (Manchester: Manchester University Press, 1989), 22–44; Tim Barringer, *The Pre-Raphaelites: Reading the Image* (London: Weidenfeld and Nicolson, 1998), pp. 119–28; Tromans, 'Palestine', pp. 147–8.

80 Barringer, *The Pre-Raphaelites*, p. 121.

81 Ibid., p. 125.

82 This key is quoted in full in Bennett, *Artists of the Pre-Raphaelite Circle*, pp. 81–3.

83 Barringer, *The Pre-Raphaelites*, p. 125.

84 Stephens, *William Holman Hunt and his Works*, p. 63.

5 PRE-RAPHAELITISM AND SCIENCE AT THE OXFORD MUSEUM

1 John Lucas Tupper, 'Extracts from the Diary of an Artist', *The Crayon*, 3 (1856), p. 43.

2 There are several detailed studies of the architecture of the Oxford University Museum. The most comprehensive is in Frederick O'Dwyer, *The Architecture of Deane and Woodward* (Cork: Cork University Press, 1997), pp. 152–283. For briefer interpretative accounts, see Eve Blau, *Ruskinian Gothic: The Architecture of Deane and Woodward 1845–61* (Princeton: Princeton University Press, 1982), pp. 48–81; Michael W. Brooks, *John Ruskin*

and Victorian Architecture (London: Thames and Hudson, 1989), pp. 112–42; Howard Colvin, *Unbuilt Oxford* (New Haven and London: Yale University Press, 1983), pp. 125–34; Trevor Garnham, *Oxford Museum: Deane and Woodward (Architecture in Detail)* (London: Phaidon, 1992); Birkin Haward, *Oxford University Museum: Its Architecture and Art* (Oxford: Oxford University Museum, 1991); Peter Howell, '"As beautiful as anything I know in civil Gothic', or 'a very shabby bit of work of mine': Ruskin and the Oxford Museum', in Rebecca Daniels and Geoff Brandwood (eds), *Ruskin and Architecture* (Reading: Spire Books, 2003), 56–85; John Illingworth, 'Ruskin and Tradition: The Case of Museums', in Michael Wheeler and Nigel Whiteley (eds), *The Lamp of Memory: Ruskin, Tradition and Architecture* (Manchester: Manchester University Press, 1992), 39–53; Caroline van Eck, *Organicism in Nineteenth-Century Architecture: An Inquiry into its Theoretical and Philosophical Background* (Amsterdam: Architectura & Natura Press, 1994), pp. 203–12; and Carla Yanni, *Nature's Museums: Victorian Science and the Architecture of Display* (New York: Princeton Architectural Press, 2005). For specific discussions of the Oxford Museum in relation to the Pre-Raphaelites, see O'Dwyer, *Architecture of Deane and Woodward*, pp. 209–22, and Jason M. Rosenfeld, 'New Languages of Nature in Victorian England: The Pre-Raphaelite Landscape, Natural History and Modern Architecture in the 1850s' (unpublished doctoral dissertation, New York University, 1999), pp. 238–305. On the significance of the museum to the history of science at Oxford, see Robert Fox, 'The University Museum and Oxford Science, 1850–1880', in M. G. Brock and M. C. Curthoys (eds), *The History of the University of Oxford*, vol. VI: *Nineteenth-Century Oxford, Part 1* (Oxford: Oxford University Press, 1997), 641–91. For a discussion of the museum in the wider general context of nineteenth-century displays of natural history and prehistory, see Yanni, *Nature's Museums*, and Michael Freeman, *Victorians and the Prehistoric: Tracks to a Lost World* (New Haven and London: Yale University Press, 2004), pp. 222–57. In addition to primary material and these secondary sources, my account of the establishment and building of the museum draws on its own archives, including a collection of papers prepared by J. M. Edmonds, Curator of the Geological Collection from 1955 to 1976, for a history of the museum which he never completed.

3 Anon., 'The Oxford Museum', *Saturday Review*, 10 November 1855, 31–2 (p. 31); Oxford University Museum, Edmonds, J 7.

4 Fox, 'The University Museum and Oxford Science', pp. 645–6.

5 Ibid., pp. 649–50.

6 O'Dwyer, *Architecture of Deane and Woodward*, pp. 166–73.

7 A Member of Convocation, 'The New Museum' (privately printed at Oxford, 1855); Fox, 'The University Museum and Oxford Science', p. 654.

8 Fox, 'The University Museum and Oxford Science', p. 641; O'Dwyer, *Architecture of Deane and Woodward*, p. 186.

9 Richard Greswell, *Memorial on the (Proposed) Oxford University Lectures Rooms, Library, Museums, etc.* (privately printed at Oxford, 1853).

10 George Edmund Street, *An Urgent Plea for the Revival of True Principles of Architecture in the Public Buildings of the University of Oxford* (Oxford: James Parker, 1853).

11 Anon. [J. H. Parker?], *The Old English Style of Architecture, as applicable to Modern Requirements, or Suggestions for the New Museum at Oxford* (privately printed, [1853]). Colvin attributes this pamphlet to Parker in *Unbuilt Oxford*, pp. 131–2. See also Yanni, *Nature's Museums*, p. 75.

12 *Oxford Architectural Society: Reports of Meetings from July 1853, to May 31, 1856* [no publication details], p. 70.

13 Yanni, *Nature's Museums*, pp. 74–5.

14 Anne Clark Amor, *William Holman Hunt: The True Pre-Raphaelite* (London: Constable, 1989), p. 92.

15 J. B. Atlay, *Sir Henry Wentworth Acland, Bart, K.C.B., F.R.S., Regius Professor of Medicine in the University of Oxford: A Memoir* (London: Smith, Elder, 1903), pp. 173–4, 392; Timothy Hilton, *John Ruskin* (New Haven and London: Yale University Press, 2002), pp. 189–90. Acland had had lessons in painting from the watercolour artist John Varley, and his own style tended to be much looser than Ruskin's or the Pre-Raphaelites'. See Atlay, *Sir Henry Wentworth Acland*, pp. 467–9.

16 Bodleian Library, Oxford, MS Acland d. 72, fol. 43.

17 Oxford University Museum, Edmonds, J 3.

18 Ibid.; O'Dwyer, *Architecture of Deane and Woodward*, p. 209.

19 Hilton, *John Ruskin*, pp. 220–21.

20 Ibid., p. 222.

21 O'Dwyer, *Architecture of Deane and Woodward*, pp. 211–12; Howell, 'Ruskin and the Oxford Museum', pp. 68–9; Oxford University Museum, HBM 01, fol. 99. Rossetti was certainly asked to undertake decorative carvings and a statue for the museum. There is a persistent scholarly rumour that he was asked to paint a mural of Newton gathering pebbles on the shore of the Ocean of Truth at the museum – see Blau, *Ruskinian Gothic*, p. 75; Brooks, *John Ruskin and Victorian Architecture*, p. 125; Rosalie Mander, 'Rossetti and the Oxford Murals, 1857', in Leslie Parris (ed.), *Pre-Raphaelite Papers* (London: Tate, 1984), 170–83 (p. 171); O'Dwyer, *Architecture of Deane and Woodward*, p. 212 – but none of these sources for this claim cites a primary source themselves, and I have not been able to corroborate it.

22 D. G. Rossetti, *The Correspondence of Dante Gabriel Rossetti*, ed. William E. Fredeman, 9 vols (Cambridge: D. S. Brewer, 2002–10), II, pp. 44, 109–11, 114–15.

23 Katharine Macdonald, 'Alexander Munro: Pre-Raphaelite Associate', in Benedict Read and Joanna Barnes (eds), *Pre-Raphaelite Sculpture: Nature and Imagination in British Sculpture 1848–1914* (London: Henry Moore Foundation, 1991), 46–8, 57–65 (pp. 46–7, 58–9).

24 Benedict Read, 'Introduction', in Read and Barnes, *Pre-Raphaelite Sculpture*, 7–11 (p. 9).

25 D. G. Rossetti, *Correspondence*, II, pp. 107–8. Given that Woolner advised Armstead on his composition, it is also possible that he recommended him as Hunt was a close mutual friend. See Bodleian Library, Oxford, MS Acland d. 95, fol. 63.

26 Bodleian Library, Oxford, MS Acland d. 72, fols 68, 92. Ten of these designs are now in the Ashmolean Museum in Oxford and one each in the Birmingham Museums and Art Gallery and

the Ruskin Library and Research Centre at Lancaster University. For a discussion of Ruskin's designs and their relation to the finished carvings on the façade of the museum, see John Holmes, 'Ruskin's Windows at the Oxford Museum', *Ruskin Review and Bulletin*, 9.2 (2013), 23–33.

27 D. G. Rossetti, *Correspondence*, II, p. 175; O'Dwyer, *Architecture of Deane and Woodward*, pp. 220–21.

28 O'Dwyer, *Architecture of Deane and Woodward*, p. 292.

29 Ibid., pp. 292–3, 306; Brooks, *John Ruskin and Victorian Architecture*, p. 137; [Rossetti, quoted in Alexander Gilchrist,] 'The Late Mr. Woodward', *The Critic*, 23 (1861), 181–2 (p. 182).

30 Anne Pollen, *John Hungerford Pollen 1820–1902* (London: Murray, 1912), pp. 268–71.

31 On Trevelyan's and Scott's contribution to the museum, see O'Dwyer, *Architecture of Deane and Woodward*, pp. 228–30, and Blair J. Gilbert, 'Puncturing an Oxford Myth: The Truth about the "Infamous" O'Sheas and the Oxford University Museum', *Oxoniensia*, 74 (2009), 87–112 (pp. 95, 108). On Tyrwhitt's, see Rosenfeld, 'New Languages of Nature', pp. 279–96. Siddal prepared designs for decorative work on Woodward's request, but it is not clear whether these were for the Oxford museum or the Museum Building at Trinity College Dublin, which Woodward's firm was working on at the same time. Either way, they appear not to have been carved – see O'Dwyer, pp. 146–8, 210–12. Blau, *Ruskinian Gothic*, p. 75, repeats a rumour, which she traces to Georgiana Burne-Jones, that Rossetti and Morris each carved one capital, but there appears to be no evidence for this claim. Blau does not supply a reference, and there is no mention of these capitals in Georgiana Burne-Jones, *Memorials of Edward Burne-Jones*, 2 vols (London: Macmillan, 1906). As Gilbert points out, 'Given the personalities of both Rossetti and Morris, it is likely that if they had carved capitals for the museum it would have been common knowledge' (p. 91).

32 Henry W. Acland and John Ruskin, *The Oxford Museum*, new [5th] edn (London: George Allen, 1893), p. 57. This edition has been collated against the earlier editions. The 1st edn appeared in 1859, followed by a moderately revised 2nd edn in 1860. The 3rd edn, printed in 1866, omits Ruskin's letters but incorporates a detailed description of the collections on display. This text was reprinted with new pagination as a 4th edn in 1867. The 1893 edition reprints the 2nd edn with a new preface and appendices. Subsequent references are to this final edition unless otherwise specified. Where the text quoted is specific to the 2nd/5th edn, and not the 1st edn, this is noted in the body of my own text.

33 On Swan's work at the museum, see O'Dwyer, *Architecture of Deane and Woodward*, pp. 213–14. Both J. W. Mackail in *The Life of William Morris*, 2 vols (London: Longman, 1899), I, pp. 126–7, and Burne-Jones, *Memorials of Edward Burne-Jones*, I, p. 189, record that Swan worked with Morris on the ceiling of the Oxford Union Society debating chamber (now the library) in October to November 1857. In 1859 he was working on the museum. None of these sources gives Swan's full name, but he is most likely the Henry Swan who studied under Ruskin at the Working Men's College and went on to be the curator of the museum of the Guild of St George at Walkley in Sheffield in the 1870s and 1880s. On Henry Swan, see James S. Deardon, *John Ruskin: A Life in Pictures* (Sheffield: Sheffield Academic Press, 1999), p. 48. I am grateful to Paul Smith, Director of the Oxford University Museum of Natural History, for this reference.

34 J. Mordaunt Crook, 'Ruskinian Gothic', in John Dixon Hunt and Faith M. Holland (eds), *The Ruskin Polygon: Essays on the Imagination of John Ruskin* (Manchester: Manchester University Press, 1982), 65–93 (p. 75).

35 See Acland and Ruskin, *The Oxford Museum*, pp. 42, 104–9; Howell, 'Ruskin and the Oxford Museum', p. 81.

36 Frederick O'Dwyer, 'O'Shea Brothers', in Paula Murphy (ed.), *The Art and Architecture of Ireland*, vol. III: *Sculpture 1600–2000* (New Haven and London: Yale University Press), 282–4 (p. 284).

37 [Gilchrist,] 'The Late Mr. Woodward', p. 182.

38 F. G. Stephens, 'The Oxford University Museum', *Macmillan's Magazine*, 5 (1862), 525–33 (pp. 527–31).

39 Brooks, *John Ruskin and Victorian Architecture*, pp. 174–5.

40 Amy Woolner, *Thomas Woolner, R.A., Sculptor and Poet: His Life in Letters* (London: Chapman and Hall, 1917), p. 213.

41 On Phillips's career as a scientist, see Jack Morrell, *John Phillips and the Business of Victorian Science* (Aldershot: Ashgate, 2005).

42 Yanni, *Nature's Museums*, pp. 54, 76, 83; Patrick Wyse Jackson, 'A Victorian Landmark: Trinity College's Museum Building', *Irish Arts Review Yearbook*, 11 (1995), 149–54 (pp. 152–3). Phillips had been Professor of Geology and Mineralogy at Trinity College briefly from 1844 to 1845, but he left well before the Museum Building was commissioned. See Gordon L. Herries Davies, 'Geology', in C. H. Holland (ed.), *Trinity College Dublin and the Idea of a University* (Dublin: Trinity College Dublin Press, 1991), 294–311 (pp. 297–8); Patrick Wyse Jackson, 'A History of Geology in Trinity College Dublin', in Patrick Wyse Jackson (ed.), *In Marble Halls: Geology in Trinity College Dublin* (Dublin: Department of Geology, Trinity College Dublin, 1994), 11–78 (pp. 19–20).

43 Acland and Ruskin, *The Oxford Museum*, pp. 91–100.

44 O'Dwyer, *Architecture of Deane and Woodward*, p. 230. Phillips's original papers connected with the museum remain in its archive (Oxford University Museum, HBM 02, folder 6).

45 Blau, *Ruskinian Gothic*, p. 49.

46 William Holman Hunt, *Pre-Raphaelitism and the Pre-Raphaelite Brotherhood*, 2 vols (London: Macmillan, 1905), I, p. 137.

47 Michael J. Lewis, *The Gothic Revival* (London: Thames and Hudson, 2002), p. 108.

48 Lewis, *Gothic Revival*, pp. 122, 160–61.

49 Deborah Cherry, 'The Hogarth Club: 1858–1861', *Burlington Magazine*, 122 (1980), 236–44 (p. 243). Michael Hall discusses the affinity between Pre-Raphaelitism and the Gothic Revival, including their connection through the Hogarth Club, in his recent study *George Frederick Bodley and the Later Gothic Revival in Britain and America* (New Haven and London: Yale University Press, 2014), pp. 30–49.

50 W. M. Rossetti (ed.), *Ruskin: Rossetti: Preraphaelitism Papers 1854 to 1862* (London: George Allen, 1899), pp. 216–17.

51 G. E. Street, 'On the Future of Art in England', *Ecclesiologist*, 19 (1858), 232–40 (pp. 234,

240). On Street's view of the Pre-Raphaelite movement more generally, see Arthur Edmund Street, *Memoir of George Edmund Street, R.A. 1824–1881* (London: Murray, 1888), p. 38.

52 W. M. Rossetti, *Fine Art, Chiefly Contemporary: Notices Reprinted, with Revisions* (London: Macmillan, 1867), p. 142.

53 Street, *Urgent Plea*, p. 4.

54 F. G. Stephens [as John Seward], 'The Purpose and Tendency of Early Italian Art', *The Germ*, p. 61.

55 *Oxford Architectural Society: Reports*, pp. 69–71.

56 Ibid., p. 69.

57 O'Dwyer, *Architecture of Deane and Woodward*, p. 263; for a full account of the roofing and reroofing of the museum, see pp. 257–71.

58 John Ruskin, *The Two Paths: Being Lectures on Art, and its Application to Decoration and Manufacture, Delivered in 1858–9* (New York: Wiley, 1859), pp. 151–97. Skidmore made the same claims for the decorative potential of wrought iron in a lecture in 1854 – see Blau, *Ruskinian Gothic*, p. 61.

59 Acland and Ruskin, *The Oxford Museum*, p. 21.

60 Ibid., p. 21.

61 Anon., 'The New Museum at Oxford', *The Athenaeum*, 1411 (11 November 1854), 1372–3.

62 James Fergusson, *History of the Modern Styles of Architecture* (London: Murray, 1862), pp. 327–8. On the chemistry laboratory at the museum, see Maurice Crosland, 'Difficult Beginnings in Experimental Science at Oxford: The Gothic Chemistry Laboratory', *Annals of Science*, 60 (2003), 399–421.

63 Yanni, *Nature's Museums*, p. 74.

64 Acland and Ruskin, *The Oxford Museum*, pp. 50–51.

65 Street, *Urgent Plea*, p. 12.

66 John Ruskin, *The Stones of Venice*, 3 vols (London: Smith, Elder), II, p. 201 (Ch. VI, § 69).

67 Street, *Urgent Plea*, p. 17.

68 Lewis, *Gothic Revival*, pp. 152, 154–5.

69 Blau, *Ruskinian Gothic*, pp. 35–47, makes a cogent case for seeing the Trinity Museum Building as a concerted attempt at building along the lines set out by Ruskin in *The Seven Lamps of Architecture*. For the O'Sheas' carving from living plants in Dublin, see Blau, *Ruskinian Gothic*, pp. 42, 175.

70 Anon., 'The Museums, Trinity College, Dublin', *The Builder*, 12 (1854), 467.

71 Wyse Jackson, 'A Victorian Landmark', p. 152.

72 Quondam, 'Ireland – Its Progress in Architecture', *The Builder*, 14 (1856), 143–4.

73 Fox, 'The University Museum and Oxford Science', p. 660; Acland and Ruskin, *The Oxford Museum*, p. xiii.

74 Rosenfeld, 'New Languages of Nature', p. 260.

75 Charles Daubeny, *Miscellanies: Being a Collection of Memoirs and Essays on Scientific and Literary Subjects*, 4 parts in 2 vols (Oxford: James Parker, 1867), III, p. 172.

76 J. S. Duncan, cited in *A Catalogue of the Ashmolean Museum, Descriptive of the Zoological Specimens, Antiquities, Coins and Miscellaneous Curiosities* (Oxford: Collingwood, 1836) [attributed to P. B. Duncan], p. vii.

77 H. E. Strickland, *On Geology in Relation to the Studies of the University of Oxford* (Oxford: Vincent, 1852), p. 5; Greswell, *Memorial*, p. 7.

78 John Phillips, *Life on the Earth: Its Origin and Succession* (Cambridge: Macmillan, 1860), p. 3.

79 Henry Wentworth Acland, *The Harveian Oration 1865* (London: Macmillan, 1865), p. 13.

80 Acland and Ruskin, *The Oxford Museum*, pp. 8, 56–7.

81 See O'Dwyer, *Architecture of Deane and Woodward*, pp. 187–9.

82 Ibid., p. 192; Wyse Jackson, 'A Victorian Landmark', pp. 151–2.

83 Bodleian Library, Oxford, MS Acland d. 95, fol. 62; O'Dwyer, *Architecture of Deane and Woodward*, p. 220.

84 On the wooden board on which the Oxford copy is mounted Woolner has written: 'Design for the tympanum for the Grand Entrance to the New Museum by T. W. Woolner June 1860'. The same copy includes a small slip of paper on the back which reads, 'The is a design by Woolner for the spandrels over the principal entrance to the New Museum at Oxford. Woolner gave it to me in 1860. It was not adopted.' The note is signed with a monogram comprised of the letters T and C.

85 O'Dwyer, *Architecture of Deane and Woodward*, pp. 220–21.

86 William Swainson, *A Preliminary Discourse on the Study of Natural History* (London: Longman, 1834), p. 5.

87 Richard Owen, 'Lecture on the Comparison of the Apes or Anthropoid Quadrumana with Man', *Medical Times*, no. 203 (20 May 1854), 509–14 (p. 514).

88 O'Dwyer, *Architecture of Deane and Woodward*, p. 221.

89 Ibid., p. 221.

90 Marquis of Salisbury and Henry W. Acland, *The Unveiling of the Statue of Sydenham in the Oxford Museum* (Oxford: University of Oxford, 1894), p. 36.

91 Salisbury and Acland, *Statue of Sydenham*, pp. 34–6.

92 On the significance of early cell theory, see Laura Otis, *Membranes: Metaphors of Invasion in Nineteenth-Century Literature, Science and Politics* (Baltimore: Johns Hopkins University Press, 1999), pp. 13–15.

93 See his pamphlet *Christianity and Rationalism in their Relations to Natural Science* (1867), reprinted in Daubeny, *Miscellanies*, IV, pp. 1–24.

94 Strickland, *On Geology*, pp. 14–16.

95 While Strickland argued that geology supplied 'unmistakeable proofs of frequent suspension of the Laws of Nature, and of the special interpositions of Creative power' (*On Geology*, p. 15), Daubeny argued in his pamphlet that this mode of explanation failed to recognise that miracles were entirely beyond the comprehension of science.

96 O'Dwyer, *Architecture of Deane and Woodward*, p. 236. O'Dwyer proposes that this was probably James O'Shea, though there is no conclusive evidence either way.

97 Garnham, *Oxford Museum* (pages unnumbered).

98 Stephens, 'Oxford University Museum', p. 528.

99 Acland and Ruskin, *The Oxford Museum*, p. 47.

100 Crook, 'Ruskinian Gothic', p. 75.

101 I follow Stephens's contemporary account (see n. 38) and Gilbert's careful scholarship (see n. 31) in the attributions of individual carvings to specific members of the O'Shea family.

102 Street, *Urgent Plea*, p. 12.

103 Stephens, 'Oxford University Museum', p. 530.

104 Gilbert, 'Puncturing an Oxford Myth', p. 107.

105 On truth-to-nature as a principle in natural history illustration, by contrast with the later concept of objectivity, see Lorraine Daston and Peter Galison, *Objectivity*, 2nd edn (New York: Zone Books, 2010), pp. 55–113.

106 Acland and Ruskin, *The Oxford Museum*, pp. 25, 99.

107 Stephens, 'Oxford University Museum', pp. 528, 530.

108 Gilbert, 'Puncturing an Oxford Myth', p. 99.

109 D. G. Rossetti, *The Works of Dante Gabriel Rossetti*, ed. W. M. Rossetti (London: Ellis, 1911), pp. xxviii, 205.

110 See http://www.rossettiarchive.org/docs/1-1870.penkb.trox.rad.html [accessed 17 March 2014].

111 Acland and Ruskin, *The Oxford Museum*, p. 99.

112 Gilbert, 'Puncturing an Oxford Myth', p. 108, points out that the subjects for these three capitals were initially based on Trevelyan's sketches.

113 Stephens, 'Oxford University Museum', p. 528.

114 W. Tuckwell, *Reminiscences of Oxford*, 2nd edn (London: Smith, Elder, 1907), p. 52.

115 Oxford University Museum, Edmonds, J 7; O'Dwyer, *Architecture of Deane and Woodward*, p. 247.

116 Gilbert, 'Puncturing an Oxford Myth', p. 99.

117 William Paley, *Natural Theology or Evidence of the Existence and Attributes of the Deity, Collected from the Appearances of Nature*, ed. Matthew D. Eddy and David Knight (Oxford: Oxford University Press, 2006), p. 238.

118 Charles Darwin, *Autobiographies*, ed. Michael Neve and Sharon Messenger (London: Penguin, 2002), p. 52.

119 Charles Darwin, *On the Origin of Species by Means of Natural Selection*, ed. Joseph Carroll (Peterborough, Ontario: Broadview, 2003), p. 397. For succinct but suggestive accounts of how Darwin's thinking contributes to ideas of ecology, see James Justus, 'Darwin's Evolutionary Ecology' and David Steffes, 'Darwin and the Environment', in Michael Ruse (ed.), *The Cambridge Encyclopedia of Darwin and Evolutionary Thought* (Cambridge and New York: Cambridge University Press, 2013), 383–90 and 391–6 respectively.

120 Blau, *Ruskinian Gothic*, p. 68, notes the similarity of these two capitals.

121 O'Shea's sketch is in the Bodleian Library, Oxford, MS Acland d. 95, fol. 42. Acland tells this story in his 1893 reissue of his pamphlet on the museum. See Acland and Ruskin, *Oxford Museum*, pp. 106–9. Acland told the story to Hunt too (see Bodleian Library, Oxford, MS Acland d. 71, fols 19–20), who garbles it in *Pre-Raphaelitism*, II, pp. 158–9, reversing the transformation so that cats become monkeys. For an in-depth analysis of the story, its inconsistencies and the likely truth of the matter, see O'Dwyer, *Architecture of Deane and Woodward*, pp. 231–42. O'Dwyer's account is thorough and largely persuasive, although it is mistaken in its claim that the parrots and owls which Acland says O'Shea began to carve around the doorway as caricatures of the Oxford University Congregation were not carved, as their basic forms can still clearly be seen on the right-hand side of the arch.

122 For a full account of this debate, see Ian Hesketh, *Of Apes and Ancestors: Evolution, Christianity, and the Oxford Debate* (Toronto: University of Toronto Press, 2009); for a close critical analysis of the different accounts of who said what, see Frank A. J. L. James, 'An "Open Clash between Science and the Church"?: Wilberforce, Huxley and Hooker on Darwin at the British Association, Oxford, 1860', in David M. Knight and Matthew D. Eddy (eds), *Science and Beliefs: From Natural Philosophy to Natural Science, 1700–1900* (Aldershot: Ashgate, 2005), 171–93; see pp. 192–3 for a list of the people known to have been in attendance at the meeting. For a discussion of the meeting in the context of the building of the Oxford museum, see Yanni, *Nature's Museums*, pp. 88–90.

123 James, 'An "Open Clash between Science and the Church"?', p. 178.

124 Daubeny, *Miscellanies*, III, pp. 107.

125 O'Dwyer, *Architecture of Deane and Woodward*, p. 245.

126 Hilton, *John Ruskin*, p. 296; O'Dwyer, *Architecture of Deane and Woodward*, p. 243.

127 Darwin, *Origin of Species*, p. 210.

128 See Paul Barlow, 'Grotesque Obscenities: Thomas Woolner's *Civilization* and its Discontents', in Colin Trodd, Paul Barlow and David Amigoni (eds), *Victorian Culture and the Idea of the Grotesque* (Aldershot: Ashgate, 1999), pp. 107–10, for a discussion of the relationship between the grotesque and evolutionary ideas in Woolner's *Puck* and other sculptures.

129 Charles Daubeny (ed.), *Fugitive Poems Connected with Natural History and Physical Science* (Oxford: James Parker, 1869), pp. 139–40, 146–50.

130 Darwin, *Origin of Species*, p. 398.

131 Phillips, *Life on the Earth*, pp. 213–16.

132 Daubeny, *Miscellanies*, III, pp. 102, 107.

133 Atlay, *Sir Henry Wentworth Acland*, p. 307.

134 Acland, *Harveian Oration*, pp. 24–6. Darwin reported Herschel's remark, which had been reported to him, in a letter to Lyell on 10 December 1859. See the Darwin Correspondence Project, Letter 2575, http://www.darwinproject.ac.uk/letter/DCP-LETT-2575.xml [accessed 20 July 2017].

135 Henry Wentworth Acland and John Phillips, *The Oxford Museum*, 3rd edn (Oxford: James Parker, 1866), p. 63.

136 Acland and Phillips, *The Oxford Museum*, 3rd edn, pp. 60, 65.

137 Ibid., pp. 51, 63–4.

138 Salisbury and Acland, *Statue of Sydenham*, p. 36.

139 Henry Wentworth Acland, *Remarks on the Extension of Education at the University of Oxford* (Oxford: James Parker, 1848), p. 39.

140 Oxford University Museum, HBM 02, fol. 144: 'Oxford University Museum' (printed at Oxford, 1 June 1855).

141 'Oxford Museum. – Balliol Chapel. – Blenheim.', *The Builder*, 17 (1859), 401–2 (p. 401).

142 Acland and Ruskin, *The Oxford Museum*, p. 25.

143 Martina Droth, Jason Edwards and Michael Hatt (eds), *Sculpture Victorious: Art in an Age of Inven-*

tion, 1837–1901 (New Haven and London: Yale University Press, 2014), pp. 25–7.

144 Benedict Read, 'Thomas Woolner: PRB, RA', in Read and Barnes, *Pre-Raphaelite Sculpture*, 21–33 (p. 26).

145 Macdonald, 'Alexander Munro', p. 47; Joanna Barnes and Alexander Kader, 'The Sculpture of John Lucas Tupper', in Read and Barnes, *Pre-Raphaelite Sculpture*, p. 68.

146 [F. G. Stephens], 'The Two Pre-Raphaelitisms', *The Crayon* (1857), p. 362.

147 John Lucas Tupper, 'English Artists of the Present Day. XXVII. – Thomas Woolner, A.R.A.', *The Portfolio*, 2 (1871), 97–101 (p. 98).

148 Elizabeth Prettejohn, *The Art of the Pre-Raphaelites* (London: Tate, 2000), p. 252.

149 W. M. Rossetti, *Some Reminiscences*, 2 vols (New York: Scribner's, 1906), I, p. 160; John Lucas Tupper, *Poems*, ed. W. M. Rossetti (London: Longman, 1897), p. ix.

150 Ruskin, *The Two Paths*, pp. 166–8.

151 Read, 'Thomas Woolner', p. 26.

152 Ford Madox Hueffer, *Ford Madox Brown: A Record of his Life and Work* (London: Longman, 1896), pp. 134–5.

153 Oxford University Museum, HBM 02, fols 136, 139–40. A plaster cast of a statue of Oersted was eventually sent over from Denmark to the museum in 1885 – see Oxford University Museum, Edmonds, J 7.

154 W. M. Rossetti, *Fine Art*, pp. 62, 75–6.

155 F. G. Stephens, 'Costume in Sculpture', *Fraser's Magazine*, 75 (1867), 382–94 (p. 390).

156 For an image of this bust, see the National Trust Collections website http://www.nationaltrustcollections.org.uk/object/922459 [accessed 1 November 2016].

157 Stephens, 'Two Pre-Raphaelitisms' (1857), p. 362.

158 A. Woolner, *Thomas Woolner*, pp. 117, 128–9, 144.

159 Acland and Ruskin, *The Oxford Museum*, p. 25.

160 Quoted in Richard Yeo, 'Genius, Method and Morality: Images of Newton in Britain, 1760–1860', *Science in Context*, 2 (1988), 257–84 (p. 276), reprinted in Yeo, *Science in the Public Sphere: Natural Knowledge in British Culture 1800–1860*, Variorum Collected Studies Series (Aldershot: Ashgate, 2001).

161 Bodleian Library, Oxford, MS Acland d. 95, fol. 63.

162 Oxford University Museum, Edmonds, J 7. According to further documents held at the museum (HBM 02, fol. 137), Tupper had to return to the museum to repair the statue in 1862, so it is possible that this broke off within two years of it being erected.

163 Tupper, *Poems*, pp. 21–2.

164 Hall Caine, *Recollections of Rossetti* (London: Cassell, 1928), p. 112.

165 Stephen Gill, *Wordsworth and the Victorians* (Oxford: Oxford University Press, 2001), p. 2; Acland and Ruskin, *The Oxford Museum*, pp. xviii–xix. For a discussion of how *The Excursion* contributed to nineteenth-century debates on the relationship between science and culture, see David Amigoni, *Colonies, Cults and Evolution: Literature, Science and Culture in Nineteenth-Century Writing* (Cambridge: Cambridge University Press, 2007), pp. 57–83, and Robert M. Ryan, *Charles Darwin and the Church of Wordsworth* (Oxford: Oxford University Press, 2016), pp. 13–15, 83–90, and passim.

166 William Wordsworth, *The Poetical Works*, ed. Thomas Hutchinson and Ernest de Selincourt (London: Oxford University Press, 1950), p. 594 (*The Excursion*, I, ll. 270–79).

167 Ibid. (*The Excursion*, Preface, ll. 28–30).

168 Ibid. (*The Excursion*, IV, l. 1252).

169 Ibid. (*The Excursion*, IV, ll. 332–45).

170 Stephens, 'Oxford University Museum', p. 527.

171 Tupper, *Poems*, p. 19.

172 'Twelfth Annual Report of the Delegates of the University Museum (For 1899)', p. 4.

173 On Tylor's grounding in Darwinism, see George W. Stocking, Jr, *Victorian Anthropology* (New York: Macmillan, 1987), pp. 156–64.

174 On Pitt Rivers's Darwinian anthropology and his collection, see Stocking, *Victorian Anthropology*, pp. 180–81, 264–5; and Peter Rivière, 'General Pitt-Rivers and the Evolutionist Anthropologists', *Museum History Journal*, 7:2 (2014), 155–67. Pitt Rivers was originally Augustus Lane Fox, but changed his name in 1880 according to the terms of a very large inheritance.

175 Robert P. McIntosh, *The Background of Ecology: Concept and Theory* (Cambridge: Cambridge University Press, 1985), p. 29.

176 Acland and Ruskin, *Oxford Museum*, p. 28.

177 Oxford University Museum, HM Box 1, 1.5.

178 Richard Owen, letter to Horatio Walpole, 5 November 1882, quoted in Nicolaas Rupke, *Richard Owen: Biology without Darwin*, 2nd edn (Chicago: University of Chicago Press, 2009), p. 179.

6 PRE-RAPHAELITISM AND SCIENCE AFTER 1859

1 Alan Bowness et al., *The Pre-Raphaelites* (London: Tate, 1984), p. 159.

2 Amy Woolner, *Thomas Woolner, R.A., Sculptor and Poet: His Life in Letters* (London: Chapman and Hall, 1917), p. 194.

3 D. G. Rossetti, *The Correspondence of Dante Gabriel Rossetti*, ed. William E. Fredeman, 9 vols (Cambridge, D. S. Brewer, 2002–10), II, p. 276. This letter is quoted in the Rossetti Archive entry for *Bocca Baciata*, http://www.rossettiarchive.org/docs/s114.rap.html [accessed 29 August 2016].

4 Deborah Cherry, 'Hogarth Club: 1858–1861', *Burlington Magazine*, 122 (1980), 236–44 (p. 241).

5 Theodore Watts, 'The Truth about Rossetti', *Nineteenth Century*, 13 (1883), 404–23 (p. 408).

6 For a rich and full critical discussion of the aesthetic movement in art, see Elizabeth Prettejohn, *Art for Art's Sake: Aestheticism in Victorian Painting* (New Haven and London: Yale University Press, 2007). The fullest exploration of symbolism in British art is Andrew Wilton and Robert Upstone (eds), *The Age of Rossetti, Burne-Jones and Watts: Symbolism in Britain 1860–1910* (London: Tate, 1997).

7 Elizabeth Prettejohn's edited collection *After the Pre-Raphaelites: Art and Aestheticism in Victorian England* (Manchester: Manchester University Press, 1999) includes a section on science, comprising three case studies: Robyn Asleson, 'Nature and Abstraction in the Aesthetic Development of Albert Moore', 115–34; Caroline Arscott, 'Poynter and the

Arty', 135–51; and Kate Flint, 'Edward Burne-Jones's *The Mirror of Venus*: Surface and Subjectivity in the Art Criticism of the 1870s', 152–64. Other studies of science and aesthetic visual art include Caroline Arscott, *William Morris and Edward Burne-Jones: Interlacings* (New Haven and London: Yale University Press, 2008), which links Morris and Burne-Jones to Victorian discussions of physiology, anthropology, armaments and horticulture; and James Mussell, 'Specular Reflections: John Brett and the Mirror of Venus', *19: Interdisciplinary Studies in the Long Nineteenth Century*, 17 (2013), http://www.19.bbk.ac.uk/articles/10.16995/ntn.670/ [accessed 12 February 2014].

8 On the history of evolutionary theories before Darwin, see Peter J. Bowler, *Evolution: The History of an Idea*, 3rd edn (Berkeley: University of California Press, 2003), pp. 48–140; Michael Ruse, *Monad to Man: The Concept of Progress in Evolutionary Biology* (Cambridge, Mass.: Harvard University Press, 1996), pp. 42–135; and Rebecca Stott, *Darwin's Ghosts: In Search of the First Evolutionists* (London: Bloomsbury, 2012). For an exhaustive discussion of the place of *Vestiges* in Victorian Culture, see James A. Secord, *Victorian Sensation: The Extraordinary Publication, Reception, and Secret Authorship of 'Vestiges of the Nature History of Creation'* (Chicago: University of Chicago Press, 2000).

9 For a brief summary of developments in prehistoric archaeology and the comparative study of human and ape anatomy independent of Darwin, see Peter J. Bowler, *Darwin Deleted: Imagining a World without Darwin* (Chicago: University of Chicago Press, 2013), pp. 101–5.

10 For brief introductions to the immediate reception of Darwin's theory, see Bowler, *Evolution*, pp. 177–223, and M. J. S. Hodge, 'England', in T. F. Glick (ed.), *The Comparative Reception of Darwinism* (Chicago: University of Chicago Press, 1988), 3–31, reprinted in Hodge, *Darwin Studies: A Theorist and his Theories in their Context*, Variorum Collected Studies Series (Farnham: Ashgate, 2009). For accounts of the competing evolutionary theories current in the late nineteenth and early twentieth centuries, see Bowler, *Evolution*, pp. 224–324, *The Eclipse of Darwinism: Anti-Darwinian Evolutionary Theories in the Decades around 1900* (Baltimore: Johns Hopkins University Press, 1983) and *The Non-Darwinian Revolution: Reinterpreting a Historical Myth*, 2nd edn (Baltimore: Johns Hopkins University Press, 1992). As I have argued elsewhere (John Holmes, *Darwin's Bards: British and American Poetry in the Age of Evolution* (Edinburgh: University of Edinburgh Press, 2009), pp. 6–19), although Bowler's work on non-Darwinian evolutionary theories and research has been immensely valuable, in my view he overstates the extent to which natural selection itself was rejected by the scientific community as a whole in this period.

11 Charles Darwin, *Autobiographies*, ed. Michael Neve and Sharon Messenger (London: Penguin, 2002), p. 72.

12 For a stimulating account of the importance Darwin's researches on barnacles in his life and work, see Rebecca Stott, *Darwin and the Barnacle* (London: Faber, 2003).

13 James Moore, *The Post-Darwinian Controversies: A Study of the Protestant Struggle to Come to Terms with Darwin in Great Britain and America, 1870–1900* (Cambridge: Cambridge University Press, 1979), pp. 194–6.

14 On Lankester's career, see Joseph Lester, *E. Ray Lankester and the Making of Modern British Biology*, ed. Peter J. Bowler (Oxford: British Society for the History of Science, 1995).

15 E. Ray Lankester, *Degeneration: A Chapter in Darwinism* (London: Macmillan, 1880), pp. 2, 5, 9–10.

16 Robert Chambers, *Vestiges of the Natural History of Creation and Other Evolutionary Writings*, ed. James A. Secord (Chicago: University of Chicago Press, 1994), pp. 26, 153, 348.

17 Charles Darwin, *On the Origin of Species by Means of Natural Selection*, ed. Joseph Carroll (Peterborough, Ontario: Broadview, 2003), p. 394.

18 For a careful exposition of Huxley, Tyndall and Victorian agnosticism, see Bernard Lightman, *The Origins of Agnosticism: Victorian Unbelief and the Limits of Knowledge* (Baltimore: Johns Hopkins University Press, 1987). For a recent series of essays on the same cluster of scientists as scientific naturalists, see Gowan Dawson and Bernard Lightman (eds), *Victorian Scientific Naturalism: Community, Identity, Continuity* (Chicago: University of Chicago Press, 2014).

19 Stephen Jay Gould expounds his conception of 'NOMA' or the 'non-overlapping magisteria' of science in *Rocks of Ages: Science and Religion in the Fullness of Life* (New York: Ballantine, 1999).

20 Darwin, *Origin of Species*, p. 397.

21 Chambers, *Vestiges*, pp. 347–8.

22 John Brett, 'Natural Science at the Royal Academy', *Nature*, 2 (1870), 157–8 (p. 158).

23 Henry Wentworth Acland and John Phillips, *The Oxford Museum*, 3rd edn (Oxford: James Parker, 1866), p. 25.

24 James H. Coombs et al. (eds), *A Pre-Raphaelite Friendship: The Correspondence of William Holman Hunt and John Lucas Tupper* (Ann Arbor, Mich.: UMI Research Press, 1986), p. 165.

25 John Lucas Tupper, *Hiatus: The Void in Modern Education; Its Cause, and Antidote* (London: Macmillan, 1869), p. 45.

26 Coombs et al., *A Pre-Raphaelite Friendship*, p. 164.

27 Tupper, *Hiatus*, p. 36.

28 Ibid., p. 45.

29 Ibid., pp. 44–5.

30 Charles Daubeny, 'Address to the Members of the Devonshire Association for the Advancement of Science, Literature, and Art' (1865), p. 14, reprinted in *Miscellanies: Being a Collection of Memoirs and Essays on Scientific and Literary Subjects*, 4 parts in 2 vols (Oxford: James Parker, 1867), III, 181–214.

31 Tupper, *Hiatus*, p. 48.

32 Ibid., p. 48.

33 Christiana Payne, *John Brett: Pre-Raphaelite Landscape Painter* (New Haven and London, Yale University Press, 2010), p. 107.

34 On Brett's astronomy and its relation to his art, see Payne, *John Brett*, pp. 111–13; and Mussell, 'Specular Reflections'. On Brett's contributions to *Nature*, see also Roy M. MacLeod, 'The Social Framework of *Nature* in its First Fifty Years', *Nature*, 224 (1969), 441–66, reprinted in MacLeod, *The 'Creed of Science' in Victorian England*, Variorum Collected Studies Series (Aldershot: Ashgate, 2000), ch. XI, pp. 10–12, 15.

35 Ruth Barton, 'Scientific Authority and Scientific Controversy in *Nature*: North Britain against the X Club', in Louise Henson et al. (eds),

Culture and Science in the Nineteenth-Century Media (Aldershot: Ashgate, 2004), 223–35.

36 George P. Landow, 'A Check List of the Writings of John Lucas Tupper, Friend of the Pre-Raphaelites', *Journal of Pre-Raphaelite Studies*, 7.1 (1986), 63–8 (pp. 66–7).

37 Coombs et al., *A Pre-Raphaelite Friendship*, pp. 171, 174–5.

38 Melinda Baldwin, 'The Successors to the X Club? Late Victorian Naturalists and *Nature*, 1869–1900', in Dawson and Lightman (eds), *Victorian Scientific Naturalism*, 288–308 (p. 301).

39 T. Mary Lockyer and Winifred L. Lockyer, *Life and Work of Sir Norman Lockyer* (London: Macmillan, 1928), pp. 15–21, 211; A. J. Meadows, *Science and Controversy: A Biography of Sir Norman Lockyer* (London: Macmillan, 1972), pp. 16–22, 211, 234–5.

40 J. Norman Lockyer, 'Physical Science for Artists', *Nature*, 18 (1878), 29–31, 58–61, 87–9, 122–6, 154–7, 223–4; 'Science for Artists', *Nature*, 36 (1887), 199–201; 'Physical Science for Artists', *Nature*, 44 (1891), 175–8, 227–9.

41 Brett, 'Natural Science at the Royal Academy', p. 157.

42 Ibid., p. 158.

43 Ibid.

44 Lockyer, 'Physical Science for Artists' (1878), p. 29.

45 Ibid., pp. 30–31.

46 Ibid., p. 30.

47 Meadows, *Science and Controversy*, p. 235.

48 Lockyer, 'Physical Science for Artists' (1878), p. 30.

49 John Ruskin, *The Art of England* (Orpington: George Allen, 1884), p. 19.

50 Mary Bennett, *Artists of the Pre-Raphaelite Circle – The First Generation: Catalogue of Works in the Walker Art Gallery, Lady Lever Art Gallery and Sudley Art Gallery* (London: Lund Humphries, 1988), p. 88.

51 Quoted in ibid., pp. 91–2.

52 Ruskin, *Art of England*, p. 19.

53 Ibid., pp. 8–9.

54 On Ruskin's responses to late Victorian science, see Robert Hewison, '"Paradise Lost": Ruskin and Science', in Michael Wheeler (ed.), *Time and Tide: Ruskin and Science* (London: Pilkington Press, 1996), 29–44; George Levine, *Realism, Ethics and Secularism: Essays on Victorian Literature and Science* (Cambridge: Cambridge University Press, 2008), pp. 75–99; Francis O'Gorman, '"The Eagle and the Whale?": Ruskin's Argument with John Tyndall', in Wheeler, *Time and Tide*, 45–64; Phillip Prodger, 'Ugly Disagreements: Darwin and Ruskin Discuss Sex and Beauty', in Barbara Larson and Fae Brauer (eds), *The Art of Evolution: Darwin, Darwinisms, and Visual Culture* (Hanover, N.H.: Dartmouth College Press, 2009), 40–58; Jonathan Smith, *Charles Darwin and Victorian Visual Culture* (Cambridge: Cambridge University Press, 2006), pp. 24–32, 125–36, 165–78; and Clive Wilmer, '"No Such Thing as a Flower . . . No Such Thing as a Man": John Ruskin's Response to Darwin', in Valerie Purton (ed.), *Darwin, Tennyson and their Readers* (New York: Anthem Press, 2013), 97–108.

55 John Ruskin, *The Eagle's Nest: Ten Lectures on the Relation of Natural Science to Art* (Orpington: Allen, 1880), p. 42.

56 F. G. Stephens, 'The Triumph of the Innocents', *The Portfolio*, 16 (1885), 80–82 (pp. 80–81).

57 Ibid., p. 81.

58 Ibid., pp. 81–2.

59 Ibid., p. 82.

60 [F. G. Stephens], 'The Two Pre-Raphaelitisms', *The Crayon*, 3 (1856), pp. 291–2, 323, 353, 355.

61 [F. G. Stephens], 'The Two Pre-Raphaelitisms', *The Crayon*, 4 (1857), p. 302.

62 D. G. Rossetti, *The Works of Dante Gabriel Rossetti* (London: Ellis, 1911), p. 270.

63 W. M. Rossetti, *Dante Gabriel Rossetti: His Family Letters with a Memoir*, 2 vols (London: Ellis and Elvey, 1895), I, p. 102; William Holman Hunt, 'The Pre-Raphaelite Brotherhood: A Fight for Art', *Contemporary Review*, 49 (1886), pp. 739–40.

64 D. G. Rossetti, *Works*, p. 649.

65 W. M. Rossetti, *Dante Gabriel Rossetti*, I, p. 149; William Holman Hunt, *Pre-Raphaelitism and the Pre-Raphaelite Brotherhood*, 2 vols (London: Macmillan, 1905), I, p. 208.

66 W. M. Rossetti, *The P.R.B. Journal: William Michael Rossetti's Diary of the Pre-Raphaelite Brotherhood 1849–1853*, ed. William E. Fredeman (Oxford: Oxford University Press, 1975), p. 15; Hunt, *Pre-Raphaelitism*, I, p. 135.

67 John Orchard, 'A Dialogue on Art', *The Germ*, 146–67 (p. 154). Rossetti wrote the introduction to Orchard's posthumous essay, p. 146. For a discussion of how Orchard's arguments contest the scientific ideal of art developed by the Pre-Raphaelites across the critical essays in *The Germ*, and how Stephens in turn replies to Orchard's position in 'Modern Giants', see John Holmes, 'Pre-Raphaelitism, Science and the Arts in *The Germ*', *Victorian Literature and Culture*, 43 (2015), pp. 697–700.

68 Hunt, 'Pre-Raphaelite Brotherhood', p. 740; see also, Hunt, *Pre-Raphaelitism*, I, p. 149.

69 D. G. Rossetti, 'The Blessed Damozel', *The Germ*, pp. 80–81.

70 For influential early discussions of Siddal's painting and poetry, see Jan Marsh and Pamela Gerrish Nunn, *Women Artists and the Pre-Raphaelite Movement* (London: Virago, 1989), pp. 65–73, and Jan Marsh, *The Legend of Elizabeth Siddal* (London: Quartet, 1992), pp. 173–213; for an insightful analysis of the creative relationship between Rossetti and Siddal, see Elizabeth Prettejohn, *The Art of the Pre-Raphaelites* (London: Tate, 2000), pp. 74–7.

71 On the building and decoration of the Oxford Union Society library, see Coventry Patmore, 'Walls and Wall Painting at Oxford', *Saturday Review*, 4 (1857), 583–4; Eve Blau, *Ruskinian Gothic: The Architecture of Deane and Woodward 1845–61* (Princeton: Princeton University Press, 1982), pp. 82–92; Frederick O'Dwyer, *The Architecture of Deane and Woodward* (Cork: Cork University Press, 1997), pp. 302–11; John D. Renton, *The Oxford Union Murals* (privately printed, 1996); and Clare A. P. Willsdon, *Mural Painting in Britain 1840–1940: Image and Meaning* (Oxford: Oxford University Press, 2000), pp. 258–9.

72 Robert Buchanan [as Thomas Maitland], 'The Fleshly School of Poetry: Mr D. G. Rossetti', *Contemporary Review*, 18 (1871), 334–50. Buchanan expanded his essay into a pamphlet under the title *The Fleshly School of Poetry and Other Phenomena of the Day* (London: Strahan, 1872).

73 John Masefield paid tribute to Rossetti's formative influence on his poetry in his late book *Thanks Before Going: Notes on Some of the Original Poems of Dante Gabriel Rossetti* (London: Heine-

mann, 1946). For a brief discussion of how Masefield's sonnets respond to Rossetti's poems, see John Holmes, *Dante Gabriel Rossetti and the Late Victorian Sonnet Sequence: Sexuality, Belief and the Self* (Aldershot: Ashgate, 2005), pp. 157–67.

74 Fiona MacCarthy, *The Last Pre-Raphaelite: Edward Burne-Jones and the Victorian Imagination* (London: Faber, 2011), pp. 38, 402. Caroline Arscott puts forward a suggestive reading of this late series of paintings, exploring their symbolism through tracing the associations of the wild rose in the horticultural literature of the period, mapping them onto writings about physiology, and teasing out the interaction of plants, skin and armour in the paintings themselves; see Arscott, *William Morris and Edward Burne-Jones*, pp. 104–25.

75 Prettejohn, *Art for Art's Sake*, pp. 48–50.

76 D. G. Rossetti, *Works*, p. 619 (emphasis in original).

77 William Allingham, *William Allingham's Diary*, ed. H. Allingham and D. Radford, intro. Geoffrey Grigson (Fontwell: Centaur Press, 1967), p. 162.

78 See the commentaries on these paintings in, respectively, Julian Treuherz, Elizabeth Prettejohn and Edwin Becker, *Dante Gabriel Rossetti* (Zwolle: Waanders, 2003), pp. 189–90, and the Rossetti Archive, http://www.rossettiarchive.org/docs/s232.raw.html [accessed 1 April 2014]. Allingham, *Diary*, p. 160, also describes Rossetti bringing a picture he was working on with him on his visit to Lymington in 1867 and refusing the offer of some beautiful roses for it by one Captain Barton because they were not the correct kind.

79 W. M. Rossetti, *Dante Gabriel Rossetti as Designer and Writer* (London: Cassell, 1889), p. 87.

80 Prettejohn, *Art for Art's Sake*, pp. 21–3.

81 Ibid., p. 39.

82 Jonathan Smith, 'Evolutionary Aesthetics and Victorian Visual Culture', and Jane Munro, '"More Like a Work of Art than of Nature": Darwin, Beauty and Sexual Selection', in Diana Donald and Jane Munro (eds), *Endless Forms: Charles Darwin, Natural Science and the Visual Arts* (New Haven and London: Yale University Press, 2009), 237–51 and 253–91; Laurence Shafe, 'Why is the peacock's tail so beautiful?', in Barbara Larson and Sabine Flack (eds), *Darwin and Theories of Aesthetics and Cultural History* (Farnham: Ashgate, 2013), 37–51.

83 Edward J. Poynter, 'Beauty and Realism, in Construction and Decoration', *Fortnightly Review*, N.S. 9 (1871), 709–30 (pp. 720–21).

84 See the Rossetti Archive, http://www.rossettiarchive.org/docs/s173.r-2.rap.html [accessed 1 April 2014].

85 Jordan Kistler, *Arthur O'Shaughnessy: A Pre-Raphaelite Poet in the British Museum* (Abingdon: Routledge, 2016), pp. 23–60; John Holmes, 'Algernon Swinburne, Anthropologist', *Journal of Literature and Science*, 9.1 (2016), 16–39.

86 Stephen F. Eisenman (ed.), *Design in the Age of Darwin: From William Morris to Frank Lloyd Wright* (Evanston, Ill.: Northwestern University Press, 2008), pp. 1–25; Eisenmann, 'Communism in Furs: A Dream of Pre-History in William Morris's *John Ball*', *Art Bulletin*, 87 (2005), 92–110; Piers J. Hale, 'Labor and the Human Relationship with Nature: The Naturalization of Politics in the Work of Thomas Henry Huxley, Herbert George Wells, and William Morris', *Journal of the History of Biology*, 36 (2003), 249–84; Hale, 'Of Mice and Men: Evolution and the Socialist Utopia; William Morris, H. G. Wells, George Bernard Shaw', *Journal of the History of Biology*, 43 (2010), 17–66; Hale, 'William Morris, Human Nature and the Biology of Utopia', in Phillippa Bennett and Rosie Miles (eds), *William Morris in the Twenty-First Century* (Bern: Peter Lang, 2010), 107–27; Jessica Kuskey, 'Bodily Beauty, Socialist Evolution, and William Morris's *News from Nowhere*', *Nineteenth-Century Prose*, 38.1 (2011), 147–82; and Jed Mayer, 'William Morris and the Greening of Science', *Journal of Pre-Raphaelite Studies*, 17 (2008), 56–76.

87 Ruskin, *Art of England*, p. 52.

88 For a further discussion of Pater and science, see Kanarakis Yannis, 'The Aesthete as a Scientist: Walter Pater and Nineteenth Century Science', *Victorian Network*, 2 (2010), 88–105. See also Sara Lyons, *Algernon Swinburne and Walter Pater: Victorian Aestheticism, Doubt, and Secularisation* (Abingdon: MHRA and Routledge, 2015).

89 Walter Pater, *Studies in the History of the Renaissance*, ed. Matthew Beaumont (Oxford: Oxford University Press, 2010), p. 4.

90 Ibid., p. 3.

91 Ibid., p. 61.

92 Ibid., pp. 30–31.

93 Ibid., p. 118.

94 George Meredith, *The Poems of George Meredith*, ed. Phyllis B. Bartlett, 2 vols (New Haven and London: Yale University Press, 1978), I, p. 199.

7 SCIENTIFIC NATURALISM AND PRE-RAPHAELITE POETRY

1 Sidney Colvin, 'English Painters of the Present Day: Edward Burne-Jones', *The Portfolio*, 1 (1870), 17–22 (pp. 18–19).

2 Algernon Swinburne, *The Swinburne Letters*, ed. Cecil Y. Lang, 6 vols (New Haven: Yale University Press, 1959–62), II, p. 335.

3 See Gowan Dawson, *Darwin, Literature and Victorian Respectability* (Cambridge: Cambridge University Press, 2007).

4 Edwin Mallard Everett, *The Party of Humanity: The Fortnightly Review and its Contributors 1865–74* (Chapel Hill: University of North Carolina Press, 1939; reissued New York: Russell and Russell, 1971), pp. 18–19, 42–3; for a full account of Lewes's editorship, see pp. 28–73.

5 John Morley, 'France in the Seventeenth Century', *Fortnightly Review*, N.S. 1 (1867), 1–18 (pp. 15–18), and 'Mr. Froude on the Science of History, *Fortnightly Review*, N.S. 2 (1867), 226–37; George Henry Lewes, 'The Reign of Law', *Fortnightly Review*, N.S. 2 (1867), 96–111; Francis T. Palgrave, 'On the Scientific Study of Poetry', *Fortnightly Review*, N.S. 6 (1869), 163–78.

6 T. H. Huxley, 'On the Physical Basis of Life', *Fortnightly Review*, N.S. 5 (1869), 129–45; John Tyndall, 'Climbing in Search of the Sky', *Fortnightly Review*, N.S. 7 (1870), 1–15.

7 Robert Buchanan, 'Immorality in Authorship', *Fortnightly Review*, O.S. 6 (1866), 289–300.

8 John Morley, 'Causeries', *Fortnightly Review*, N.S. 1 (1867), 100–3 (pp. 101–2).

9 For Morley's and Buchanan's reviews of *Poems and Ballads*, see Clyde K. Hyder (ed.), *Algernon Swinburne: The Critical Heritage* (London: Routledge, 1970), pp. 22–34.

10 Swinburne, *Letters*, I, p. 149, and *The Uncollected Letters of Algernon Charles Swinburne*, ed. Terry L. Meyers, 3 vols (London: Pickering and Chatto, 2004–5), I, p. 57.

11 Swinburne, *Letters*, I, p. 301; D. G. Rossetti, *The Correspondence of Dante Gabriel Rossetti*, ed. William E. Fredeman, 9 vols (Cambridge: D. S. Brewer, 2002–10), IV, pp. 123, 127.

12 Ioan Williams (ed.), *George Meredith: The Critical Heritage* (London: Routledge, 1971), pp. 32–5, 97–9.

13 On evolution in Meredith's poetry, see John Holmes, *Darwin's Bards: British and American Poetry in the Age of Evolution* (Edinburgh: Edinburgh University Press, 2009), pp. 54–62, 107–16, 170–75, 199–212, 245–59; Georg Roppen, *Evolution and Poetic Belief: A Study in Some Victorian and Modern Writers* (Oslo: Oslo University Press, 1956), pp. 209–82; Lionel Stevenson, *Darwin among the Poets* (Chicago: University of Chicago Press, 1932), pp. 183–236.

14 The only other poet who published original poetry in the *Fortnightly* in this period was Robert Lytton, whose 'Lyrical Fables' series was published in four instalments in the latter half of 1871.

15 Swinburne, *Uncollected Letters*, I, p. 183.

16 Everett, *Party of Humanity*, p. 266.

17 Ibid., pp. 265, 268–76.

18 Sidney Colvin, 'English Painters and Painting in 1867', *Fortnightly Review*, N.S. 2 (1867), 470–72.

19 Walter Pater, *Studies in the History of the Renaissance*, ed. Matthew Beaumont (Oxford: Oxford University Press, 2010), pp. 146, 149, 152, 156. For a discussion of Pater's contributions to the *Fortnightly Review*, see Gowan Dawson, 'Science in the Periodical Press', in John Holmes and Sharon Ruston (eds), *The Routledge Research Companion to Nineteenth-Century British Literature and Science* (London: Routledge, 2017), 172–86 (pp. 176–9).

20 John Morley, 'Mr. Pater's Essays', *Fortnightly Review*, N.S. 13 (1873), 469–77 (pp. 476–7).

21 On positivism in Britain, see T. R. Wright, *The Religion of Humanity: The Impact of Comtean Positivism on Victorian Britain* (Cambridge: Cambridge University Press, 1986). For a brief discussion of positivism in the *Fortnightly Review* which builds on Wright's work, see Mark W. Turner, *Trollope and the Magazines: Gendered Issues in Mid-Victorian Britain* (Basingstoke: Macmillan, 2000), pp. 96–102.

22 James Cotter Morison, 'The Significance of Ritualism', *Fortnightly Review*, N.S. 1 (1867), 68–77 (pp. 75, 77).

23 George Meredith, 'In the Woods', *Fortnightly Review*, N.S. 8 (1870), 179–83 (pp. 181, 183).

24 For a detailed analysis of this sequence as a whole, see Holmes, *Darwin's Bards*, pp. 107–16.

25 Morison, 'The Significance of Ritualism', p. 75.

26 Algernon Swinburne, 'Child's Song in Winter', *Fortnightly Review*, N.S. 1 (1867), 19–26 (p. 21).

27 Ibid., pp. 23–4.

28 Ibid., p. 24.

29 Ibid., p. 26.

30 J. F. McLennan, 'The Worship of Animals and Plants', *Fortnightly Review*, N.S. 6 (1869), 407–27, 562–82 (p. 422), and *Fortnightly Review*, N.S. 7 (1870), 194–216.

31 Swinburne, 'Child's Song in Winter', pp. 25–6.

32 John Morley, 'Valedictory', *Fortnightly Review*, N.S. 32 (1882), 511–21 (p. 519), and 'Mr. Pater's Essays', p. 476.

33 Algernon Swinburne, 'Ode on the Insurrection in Candia', *Fortnightly Review*, N.S. 1 (1867), 284–9 (pp. 288–9).

34 Algernon Swinburne, 'The Halt before Rome', *Fortnightly Review*, N.S. 2 (1867), 539–46 (p. 542).

35 William Bell Scott, *Poems* (London: Longman, 1875), p. vii.

36 William Bell Scott, *Autobiographical Notes of the Life of William Bell Scott*, ed. W. Minto, 2 vols (London: Osgood, McIlvaine and Co., 1892), II, p. 233.

37 See John Tyndall, 'Miracles and Special Providences', *Fortnightly Review*, N.S. 1 (1867), 645–60; Frances Power Cobbe, 'The Devil', *Fortnightly Review*, N.S. 10 (1871), 180–91; Francis Galton, 'Statistical Inquiry into the Efficacy of Prayer', *Fortnightly Review*, N.S. 12 (1872), 125–35.

38 Scott, *Autobiographical Notes*, II, p. 233.

39 Cobbe, 'The Devil', p. 190.

40 William Bell Scott, 'Anthony', *Fortnightly Review*, N.S. 4 (1868), 81–9 (p. 89).

41 Frances Power Cobbe, 'What is Progress, and Are We Progressing?', *Fortnightly Review*, N.S. 1 (1867), 357–70.

42 John Morley, 'On "The Ring and the Book"', *Fortnightly Review*, N.S. 5 (1869), 331–43 (p. 338).

43 Ibid., p. 333.

44 Ibid., p. 331.

45 John Morley, 'The Earthly Paradise: A Poem, by William Morris', *Fortnightly Review*, N.S. 3 (1868), 713–15 (p. 714).

46 Sidney Colvin, *Memories and Notes of Persons and Places 1852 to 1912* (London: Edward Arnold, 1921), p. 66.

47 Quoted in Alfred Tennyson, *The Poems of Tennyson*, ed. Christopher Ricks, 2nd edn, 3 vols (Harlow: Longman, 1987), II, p. 312.

48 Ibid., pp. 415, 459 (*In Memoriam*, XCVI, ll. 11–12, and Epilogue, ll. 141–4).

49 Lewes, 'The Reign of Law', p. 97.

50 On Tennyson's reaffirmation of theistic evolution in his poetry after Darwin, see Holmes, *Darwin's Bards*, pp. 62–73.

51 D. G. Rossetti, 'Of Life, Love, and Death: Sixteen Sonnets', *Fortnightly Review*, N.S. 5 (1869), 266–73 (p. 270).

52 Ibid., p. 269.

53 Ibid., p. 272.

54 Ibid., p. 271.

55 Frederic Harrison, 'The Subjective Synthesis', *Fortnightly Review*, N.S. 8 (1870), 184–97 (p. 194).

56 I. A. Richards, *Poetries and Science* (New York: Norton, 1970), pp. 68–9.

57 Harry Quilter, 'The Art of Rossetti', *Contemporary Review*, 43 (1883), 190–203 (p. 200).

58 William Morris, *The Earthly Paradise: A Poem* (London: Ellis, 1868), p. 3. This is the first edition of what became parts I and II of Morris's poem.

59 For a summary of each tale, together with critical remarks and notes on sources, see http://morrisedition.lib.uiowa.edu/Poetry/EarthlyParadise/eptexts.html [accessed 27 July 2017].

60 In June 1868 Eliot wrote to John Blackwood: 'We take Morris's poem into the woods with us and read it aloud, greedily, looking to see how much *more* there is in store for us. *If ever* you have an idle afternoon, bestow it on the *Earthly*

Paradise' (quoted in Peter Faulkner (ed.), *William Morris: The Critical Heritage* (London: Routledge, 1973), p. 10).

61 Dawson, *Darwin, Literature and Victorian Respectability*, pp. 204–5.

62 Faulkner, *William Morris*, p. 150.

63 Ibid.

64 Morley, 'The Earthly Paradise', p. 714.

65 Faulkner, *William Morris*, p. 68.

66 Ibid., p. 122.

67 Morley, 'The Earthly Paradise', p. 714; Faulkner, *William Morris*, p. 172.

68 Faulkner, *William Morris*, p. 168.

69 Morris, *Earthly Paradise*, I/II, pp. 394–5.

70 Ibid., p. 400.

71 Ibid., p. 401.

72 Faulkner, *William Morris*, p. 151.

73 Ibid., p. 103.

74 William Morris, *The Earthly Paradise: A Poem*, Part III (London: Ellis, 1870), pp. 398–9.

75 Ibid., III, p. 369.

76 Faulkner, *William Morris*, p. 76.

77 Ibid., pp. 112–14.

78 Ibid., pp. 171–2.

79 Ibid., p. 105.

80 Ibid., pp. 102–5.

81 Morris, *Earthly Paradise*, I/II, p. 407.

82 William Morris, *The Earthly Paradise: A Poem*, Part IV (London: Ellis, 1870), pp. 200–1.

83 Ibid., IV, p. 389.

84 Faulkner, *William Morris*, p. 173.

85 Morris, *Earthly Paradise*, I/II, p. 615.

86 Ibid., IV, p. 365.

87 Faulkner, *William Morris*, p. 89.

88 Ibid., p. 91.

89 Ibid.

90 Ibid., p. 92.

91 Ibid., pp. 89–90.

92 Pater, *Studies in the History of the Renaissance*, pp. 118, 177–8.

93 Walter Pater, *Plato and Platonism: A Series of Lectures* (London: Macmillan, 1893), p. 1.

94 Ibid., p. 14.

95 Ibid., pp. 14–15.

96 Ibid., p. 14.

97 Faulkner, *William Morris*, p. 91.

98 Elizabeth Prettejohn, *Art for Art's Sake: Aestheticism in Victorian Painting* (New Haven and London: Yale University Press, 2007), p. 270.

99 Jeffrey Skoblow, *Paradise Dislocated: Morris, Politics, Art* (Charlottesville: University Press of Virginia, 1993), p. 3.

100 Faulkner, *William Morris*, p. 88.

101 William Morris, *News from Nowhere and Other Writings*, ed. Clive Wilmer (London: Penguin, 1993), p. 252.

102 D. G. Rossetti, 'The Cloud Confines', *Fortnightly Review*, N.S. 11 (1872), 14–15.

103 Ibid., p. 15.

104 D. G. Rossetti, *Correspondence*, VII, p. 23.

105 The text of 'The Question' is transcribed in the Rossetti Archive, http://www.rossettiarchive.org/docs/1-1882.s241.oup.rad.html [accessed 17 April 2014].

106 A. Smith (ed.), *Exposed: The Victorian Nude* (London: Tate, 2001), p. 153.

107 D. G. Rossetti, 'The Cloud Confines', p. 15.

108 Ibid., p. 14.

109 Ibid.

8 THE NATURAL HISTORY MUSEUM VERSUS *NATURE*

1 Gowan Dawson, *Darwin, Literature and Victorian Respectability* (Cambridge: Cambridge University Press, 2007), pp. 169–71, 205; M. J. Walhouse, 'Some Vestiges of Girl Sacrifices, Jar Burial, and Contracted Interments in India and the East', *Journal of the Anthropological Institute*, 11 (1882), 415–23 (p. 419).

2 On the design and building of the Natural History Museum, see J. B. Bullen, 'Alfred Waterhouse's Romanesque "Temple of Nature": The Natural History Museum, London', *Architectural History*, 49 (2006), 257–85; Colin Cunningham, *The Terracotta Designs of Alfred Waterhouse* (London: Natural History Museum, 2001); Mark Girouard, *Alfred Waterhouse and the Natural History Museum*, 2nd edn (London: Natural History Museum, 1999); F. H. W. Shepherd (ed.), *Survey of London*, 38: *The Museums Area of South Kensington and Westminster* (London: Athlone Press, 1975), pp. 201–16; William T. Stearn, *The Natural History Museum at South Kensington: A History of the Museum 1753–1980*, 2nd edn (London: Natural History Museum, 1998), pp. 41–63; John Thackeray and Bob Press, *The Natural History Museum: Nature's Treasurehouse* (London: Natural History Museum, 2001), pp. 55–65; and Carla Yanni, *Nature's Museums: Victorian Science and the Architecture of Display* (New York: Princeton Architectural Press, 2005), pp. 111–46. The fullest and most authoritative studies of Owen and Waterhouse are Nicolaas Rupke, *Richard Owen: Biology without Darwin*, 2nd edn (Chicago: University of Chicago Press, 2009), and Colin Cunningham and Prudence Waterhouse, *Alfred Waterhouse 1830–1905: Biography of a Practice* (Oxford: Oxford University Press, 1992). See also Stuart Allen Smith, 'Alfred Waterhouse: Civic Grandeur', in Jane Fawcett (ed.), *Seven Victorian Architects* (London: Thames and Hudson, 1976), 102–21; and Sally Maltby, Sally MacDonald and Colin Cunningham, *Alfred Waterhouse 1830–1905* (London: Royal Institute of British Architects, 1983).

3 'Owen Memorial', p. 3, in the Owen Collection, Natural History Museum Archives, Box 38.

4 William Holman Hunt, 'The Pre-Raphaelite Brotherhood: A Fight for Art', *Contemporary Review*, 49 (1886), p. 750. For a fuller account of their first meeting, perhaps embellished, implying Owen's endorsement of Pre-Raphaelitism, see Hunt, *Pre-Raphaelitism and the Pre-Raphaelite Brotherhood*, 2 vols (London: Macmillan, 1905), I, pp. 321–3.

5 Amy Woolner, *Thomas Woolner R.A., Sculptor and Poet: His Life in Letters* (London: Chapman and Hall, 1917), p. 165.

6 J. G. Millais, *The Life and Letters of Sir John Everett Millais*, 2 vols (London: Methuen, 1899), I, pp. 435–6.

7 On J. G. Millais and his son Raoul as naturalists, artists and sportsmen, see J. N. P. Watson, *Millais: Three Generations in Nature, Art and Sport* (London: Sportsman's Press, 1988).

8 Bodleian Library, Oxford, MS Eng. Lett. d. 308, fol. 138 and MS Don. e. 78, fols 76–9.

9 Millais, *Life and Letters*, I, p. 435.

10 Bodleian Library, Oxford, MS Don. e. 78, fol. 78v.

11 Cunningham and Waterhouse, *Alfred Waterhouse*, pp. 16–17, 34–5.

12 On Dujardin's contribution to the museum, see Cunningham, *Terracotta Designs*, pp. 62–79. Cunningham's book includes reproductions of all the designs in the first portfolio. The second, bound and known as the Dujardin album, includes several extra designs that have not been reproduced.

13 Royal Institute of British Architects, London, PA 1525, 1938, 1939, 1950 and 1952.

14 See Yanni, *Nature's Museums*, pp. 115–19, for a discussion of these early plans.

15 Richard Owen, *York 1881: Address to the Biological Section of the British Association* (London: Spottiswoode, 1881), p. 8.

16 Anon., 'New Natural History Museum at South Kensington', *Building News*, 25 January 1878, 76–7 (p. 77) (press cutting in Natural History Museum Archives, DF5014/2).

17 Report by Owen to the Trustees of the British Museum, 20 November 1873, in Natural History Museum Archives, DF930/1, fol. 23.

18 Letter from Waterhouse to Owen, 12 October 1874, quoted in Cunningham, *Terracotta Designs*, pp. 59–60.

19 Yanni, *Nature's Museums*, pp. 132–9.

20 Richard Owen, 'Description of some Fossil Remains of *Chaeropotamus*, *Palaeotherium*, *Anoplotherium*, and *Dichobunes* from the Eocene *Formation*, Isle of Wight', *Transactions of the Geological Society of London*, N.S. 6 (1842), 41–5; Owen, *Description of the Skeleton of an Extinct Gigantic Sloth* (London: Van Voorst, 1842); Owen, *A History of British Fossil Mammals* (London: Van Voorst, 1846), pp. 179–83, 316–24.

21 Richard Owen, *Memoir on the Dodo* (London: Taylor and Francis, 1866). Charlotte Sleigh includes the first plate from this *Memoir* in her lively and beautiful historical survey of illustrations of animals, *The Paper Zoo: 500 Years of Animals in Art* (London: British Library, 2016), p. 45.

22 Cunningham, *Terracotta Designs*, p. 23.

23 Ibid., pp. 161–2.

24 Richard Owen, *Palaeontology, or A Systematic Summary of Extinct Animals and their Geological Relations*, 2nd edn (Edinburgh: Adam & Charles Black, 1861), pp. 153–4, 164, 170–71. The copy consulted is Owen's own copy, held at the Natural History Museum.

25 A. C. Ramsay, *The Physical Geology and Geography of Great Britain: A Manual of British Geology*, 5th edn (London: Stanford, 1878), p. 114.

26 H. Alleyne Nicholson, *The Ancient Life-History of the Earth: A Comprehensive Outline of the Principles and Leading Facts of Palaeontological Science* (Edinburgh: Blackwood, 1877), pp. 153–4.

27 See the designs reproduced in Cunningham, *Terracotta Designs*, pp. 179–80.

28 Stearn, *Natural History Museum*, p. 49, states: 'The terracotta Seychelles Gecko (*Phelsuma sundbergi*) was based on a specimen in alcohol of what was then a new species.' Cunningham, *Terracotta Designs*, p. 53, suggests that Waterhouse's roundel of a rhea on the balcony of the Central Hall may be based on a plate from the *Zoology of the Voyage of HMS Beagle*.

29 Greswell consulted Owen on his proposals for the Oxford museum; see Richard Greswell, *Memorial on the (Proposed) Oxford University Lecture Rooms, Library, Museums, etc.* (privately printed at Oxford, 1853), pp. 7–8, 13, 19. Acland studied comparative anatomy under Owen at the Royal College of Physicians; see J. B. Atlay, *Sir Henry Wentworth Acland, Bart, K.C.B., F.R.S., Regius Professor of Medicine in the University of Oxford: A Memoir* (London: Smith Elder, 1903), pp. 98–9.

30 Owen, *York 1881*, p. 10.

31 Letter from Owen to Acland, 23 March 1855, quoted in Rupke, *Richard Owen*, p. 253.

32 Natural History Museum Archives, DF2000/3/20, fol. 91; Owen, *York 1881*, p. 10.

33 Natural History Museum Archives, DF2000/6, fol. 4.

34 Girouard, *Waterhouse and the Natural History Museum*, pp. 26–7, 42. In 'Alfred Waterhouse's Romanesque "Temple of Nature"', Bullen argues that Romanesque also embodied a second, complementary metaphor of secular prosperity, through allusion to the arcades of Venice and more recent commercial architecture derived from them.

35 Untitled article, *The Times*, 18 April 1881, p. 9 (press cutting in Natural History Museum Archives, DF5014/1, fol. 4).

36 Letter to J. Winter Jones, 25 August 1873, in Natural History Museum Archives, DF930/1, fol. 11.

37 Anon., 'New Natural History Museum at South Kensington', *Building News*, 1878, p. 77.

38 Anon., 'The New Museum of Natural History', *Nature*, 23 (1881), 549–52 (p. 549).

39 The fullest account of the ceiling paintings is Sandra Knapp and Bob Press, *The Gilded Canopy: Botanical Ceiling Paintings of the Natural History Museum* (London: Natural History Museum, 2005). On Lea's role as painter, see Cunningham, *Terracotta Designs*, pp. 45–7; Knapp and Press, *Gilded Canopy*, p. 20. On Carruthers's likely involvement, see Cunningham, *Terracotta Designs*, p. 58; Knapp and Press, *Gilded Canopy*, p. 19; Stearn, *Natural History Museum*, p. 52.

40 Girouard, *Waterhouse and the Natural History Museum*, p. 36.

41 Ibid., p. 52.

42 Ibid.; Yanni, *Nature's Museums*, p. 132.

43 The foundational essay for this kind of approach is Carol Duncan and Alan Wallach, 'The Universal Survey Museum', *Art History*, 3 (1980), 448–69. For more recent examples of this mode of analysis with reference to natural history museums, see Sophie Forgan, 'The Architecture of Display: Museums, Universities and Objects in Nineteenth-Century Britain', *History of Science*, 32 (1994), 139–62; Fabienne Galangau-Quérat, 'La Grande Galerie de l'Evolution: An Alternative Cognitive Experience', in Suzanne MacLeod (ed.), *Reshaping Museum Space: Architecture, Designs, Exhibitions* (London: Routledge, 2005), 95–107; Sophia Psarra, 'Spatial Culture, Way-finding and the Educational Message: The Impact of Layout on the Spatial, Social and Educational Experiences of Visitors to Museums and Galleries', in MacLeod (ed.), *Reshaping Museum Space*, 78–94; and Yanni, *Nature's Museums*.

44 Richard Owen, *On the Extent and Aims of a National Museum of Natural History*, 2nd edn (London: Saunders, Otley, 1862), p. 41.

45 Owen, *Memoir on the Dodo*, pp. 49–50.

46 On Owen's engagement with evolutionism, see Peter J. Bowler, *Evolution: The History of an Idea*, 3rd edn (Berkeley: University of California Press, 2003), pp. 124–6, 204; Rupke, *Richard Owen*, pp. 141–81.

47 Richard Owen, *On the Anatomy of the Vertebrates*, 3 vols (London: Longman, 1866–8), III, p. 808.

48 Ibid. The lines are quoted from Tennyson's *Idylls of the King*, 'Lancelot and Elaine', ll. 1365–6, originally published as 'Elaine' in 1859. In the poem, the lines read: 'To doubt her fairness were to want an eye, / To doubt her pureness were to want a heart.' Alfred Tennyson, *The Poems of Tennyson*, ed. Christopher Ricks, 2nd edn, 3 vols (Harlow: Longman, 1987), III, p. 461.

49 Owen, *Extent and Aims*, p. 36.

50 Stearn, *Natural History Museum*, p. 57.

51 Owen, *Extent and Aims*, p. 41.

52 F. G. Stephens, 'The Oxford University Museum', *Macmillan's Magazine*, 5 (1862), p. 530.

53 Cunningham and Waterhouse, *Alfred Waterhouse*, p. 17.

54 Quoted ibid., p. 191.

55 Girouard, *Waterhouse and the Natural History Museum*, pp. 53, 61.

56 Cunningham, *Terracotta Designs*, p. 57.

57 On the vertebrate archetype, see Rupke, *Richard Owen*, pp. 90–140, and Adrian Desmond, *Archetypes and Ancestors: Palaeontology in Victorian London 1850–1875* (Chicago: University of Chicago Press, 1982).

58 Peter J. Bowler, *Darwin Deleted: Imagining a World without Darwin* (Chicago: University of Chicago Press, 2013), pp. 139, 146, 162.

59 A drawing by Waterhouse, dated 2 March 1874, which includes a sketched design for these snakes, is reproduced by Cunningham in *Terracotta Designs*, p. 11.

60 Richard Owen, 'On the *Archaeopteryx* of *von Meyer*, with a description of the Fossil Remains of a Long-tailed Species, from the Lithographic Stone of Solenhofen', *Philosophical Transactions*, 153 (1863), 33–47; Owen, *A Monograph of the Fossil Reptilia of the Liassic Formations*, 3 parts (London: Palaeontographical Society, 1861–81), III, pp. 41–80.

61 William Holman Hunt, 'Religion and Art', *Contemporary Review*, 71 (1897), 41–52 (p. 43).

62 Matthew Rowlinson, 'History, Materiality and Type in Tennyson's *In Memoriam*', in Valerie Purton (ed.), *Darwin, Tennyson and their Readers: Explorations in Victorian Literature and Science* (London: Anthem Press, 2013), 35–48.

63 Natural History Museum Archives, DF2000/7, fols 20–21. See also Owen, *Extent and Aims*, p. 66.

64 Owen, *Extent and Aims*, p. 68.

65 Letter from Waterhouse to Sir Edward Bond, Principal Librarian of the British Museum, 20 March 1880, in Natural History Museum Archives, DF930/2, fol. 430. Waterhouse's watercolour drawing is reproduced by Yanni in *Nature's Museums*, p. 133. See also an elevation drawing of the main staircase including a statue in classical style on a stair post at the north end of the Central Hall (Royal Institute of British Architects, London, PA1950, WAT A [71] 96).

66 Royal Institute of British Architects, London, PA1950, WAT A [71] 7.

67 Royal Institute of British Architects, PA1950, WAT A [71] 13–15.

68 Natural History Museum Archives, DF2001/6, no. 6.

69 Royal Institute of British Architects, PA1950, WAT A [71] 40.

70 Natural History Museum Archives, DF2001/6, nos 7, 12, 13; Royal Institute of British Architects, PA1950, WAT A [71] 39.

71 Royal Institute of British Architects, PA1950, WAT A [71] 16. This drawing also includes a pair of figures, one male, one female, neither clearly identifiable, between the windows of the second floor above the main arch.

72 E. Ingress Bell, 'The New Natural History Museum', *Magazine of Art*, 4 (1881), 358–62, 463–5.

73 Richard Owen, *Instances of the Power of God as Manifested in his Animal Creation* (London: Nisbet, 1864).

74 For discussions of the initial responses to the museum in the press, see Bullen, 'Alfred Waterhouse's Romanesque "Temple of Nature"', pp. 280–81; and Yanni, *Nature's Museums*, pp. 139–44. See also the file of press cuttings in Natural History Museum Archives, DF5014.

75 On Huxley's disagreements with Owen with regard to the museum, see Yanni, *Nature's Museums*, pp. 112–15, 128–31.

76 Anon., 'Our Natural History Collections', *Nature*, 18 (1878), 513–14 (p. 513).

77 F. G. S., 'The New Museum of Natural History', *Nature*, 23 (1881), 577. There are no specific grounds to believe that this correspondent was Stephens, though it may have been.

78 Anon., 'The New Natural History Museum', *Nature*, 27 (1882), 54–6 (pp. 54–5).

79 Anon., 'New Natural History Museum at South Kensington', p. 77; Bell, 'The New Natural History Museum', p. 362; Yanni, *Nature's Museums*, p. 142.

80 Anon., 'Natural History at South Kensington', *The Field*, 57 (1881), 545.

81 Anon., 'The New Natural History Museum', *Nature*, 22 (1880), 237.

82 Anon., 'The New Museum of Natural History', *Nature*, 23, p. 549.

83 Anon., 'The New Natural History Museum', *Nature*, 27, p. 55.

84 Anon., 'Sir Richard Owen', *Nature*, 51 (1894), 169–71 (p. 171).

85 Anon., 'The New Natural History Museum', *Nature*, 27, p. 55.

86 Ibid., p. 56.

87 W. H. Flower, 'The New Natural History Museum', *Nature*, 2 (1870), 61.

88 W. H. Flower, *Essays on Museums and other Subjects Connected with Natural History* (London: Macmillan, 1898), p. 11.

89 W. H. Flower, 'The Oxford University Museum', report reprinted in *Nature*, 44 (1891), 619–21 (p. 621).

90 Flower, *Essays*, p. 29.

91 Bowler, *Darwin Deleted*, p. 138.

92 Flower, 'The New Natural History Museum', p. 61; Stearn, *Natural History Museum*, p. 96.

93 W. Boyd Dawkins, 'The Organisation of Natural History Museums', *Nature*, 16 (1877), 137–8 (p. 137).

94 Samuel J. M. M. Alberti, *Nature and Culture: Objects, Disciplines and the Manchester Museum* (Manchester: Manchester University Press, 2009), pp. 32–8.

95 W. A. Herdman, 'An Ideal Natural History Museum', in Thomas Greenwood, *Museums and Art Galleries* (London: Simpkin, 1888), 180–95.

96 See Caroline van Eck, *Organicism in Nineteenth-Century Architecture: An Inquiry into its Theoretical and Philosophical Background* (Amsterdam: Architecture & Natura Press, 1994), pp. 203–12.

97 Anon., 'The Darwin Memorial', *Nature*, 32 (1885), 121–2 (p. 121).

98 Anon., 'The Memorial Statue of Sir Richard Owen', *Nature*, 55 (1897), 561–2.

9 PRE-RAPHAELITE PAINTING AND THE HISTORY OF SCIENCE

1 John Rylands Library, Manchester, Ryl. Eng. MS 1268, fol. 11.

2 Jacob W. Gruber, Calendar of Owen Letters, Natural History Museum archives, Owen Collection, OC 87 (2), p. 619.

3 J. G. Millais, *The Life and Letters of Sir John Everett Millais*, 2 vols (London: Methuen, 1899), II, pp. 169–70. J. G. Millais specifies that this visit took place in winter when Gould was an invalid. As Gould died in February 1881, it is likely that this visit took place at much the same time as Hunt wrote to Owen.

4 For scholarship on the X Club, see Ruth Barton, '"An Influential Set of Chaps": The X-Club and Royal Society Politics 1864–85', *British Journal for the History of Science*, 23 (1990), 53–81; Barton, '"Huxley, Lubbock, and Half a Dozen Others": Professionals and Gentlemen in the Formation of the X Club, 1851–1864', *Isis*, 89 (1998), 410–44; and Barton, 'Scientific Authority and Scientific Controversy in *Nature*: North Britain against the X Club', in Louise Henson et al. (eds), *Culture and Science in the Nineteenth-Century Media* (Aldershot: Ashgate, 2004); see also Adrian Desmond, *Huxley: From Devil's Disciple to Evolution's High Priest* (Harmondsworth: Penguin, 1998), pp. 327–32, 520–21, 573–4; John Holmes, 'The X Club: Romanticism and Victorian Science', in Amanda Mordavsky Caleb (ed.), *(Re)creating Science in Nineteenth-Century Britain* (Newcastle: Cambridge Scholars Publishing, 2007), 12–31; J. Vernon Jensen, 'The X Club: Fraternity of Victorian Scientists', *British Journal for the History of Science*, 5 (1970), 63–72; Jensen, 'Interrelationships within the Victorian "X Club"', *Dalhousie Review*, 51 (1971–2), 539–52; Roy M. MacLeod, 'The X-Club: A Social Network of Science in Late-Victorian England', *Notes and Records of the Royal Society of London*, 24 (1970), 305–22, reprinted in *The 'Creed of Science' in Victorian England*, ch. I; and Colin A. Russell, *Edward Frankland: Chemistry, Controversy and Conspiracy in Victorian England* (Cambridge: Cambridge University Press, 1996), pp. 317–41.

5 John Willis Clark and Thomas McKenny Hughes, *The Life and Letters of the Reverend Adam Sedgwick*, 2 vols (Cambridge: Cambridge University Press, 1890), II, p. 360.

6 Richard Yeo, 'Scientific Method and the Rhetoric of Science in Britain, 1830–1917', in John Schuster and Richard Yeo (eds), *The Politics and Rhetoric of Scientific Method: Historical Studies* (Dordrecht: Kluwer, 1986), 259–97, reprinted in Yeo, *Science in the Public Sphere: Natural Knowledge in British Culture 1800–1860*, Variorum Collected Studies Series (Aldershot: Ashgate, 2001), p. 276; John Willis Clark and Thomas McKenny Hughes, *The Life and Letters of the Reverend Adam Sedgwick*, 2 vols (Cambridge: Cambridge University Press, 1890), II, pp. 356–62.

7 For examples of Sedgwick's humour, see Adelene Buckland, *Novel Science: Fiction and the Invention of Nineteenth-Century Geology* (Chicago: University of Chicago Press, 2013), pp. 59, 71–3.

8 Charles Darwin, *On the Origin of Species by Means of Natural Selection*, ed. Joseph Carroll (Peterborough, Ontario: Broadview, 2003), p. 393.

9 Ludmilla Jordanova, *Defining Features: Scientific and Medical Portraits 1660–2000* (London: Reaktion, 2000), pp. 140–42.

10 William Holman Hunt, 'Religion and Art', *Contemporary Review*, 71 (1897), p. 43.

11 Nicolaas Rupke, *Richard Owen: Biology without Darwin*, 2nd edn (Chicago: University of Chicago Press, 2009), p. 253.

12 Collier's portraits of Clifford, Hooker and Spottiswoode are at the Royal Society, his portrait of Lubbock at the Guildhall Gallery, and his portraits of Huxley at the National Portrait Gallery and the Bolton Museum and Art Gallery. They are all reproduced on https://artuk.org/.

13 Peter J. Bowler, *Evolution: The History of an Idea*, 3rd edn (Berkeley: University of California Press, 2003), pp. 154–5; Adrian Desmond and James Moore, *Darwin* (Harmondsworth: Penguin, 1991), pp. 209–10.

14 John Gould, *An Introduction to the Trochilidae, or Family of Humming-Birds* (London: privately printed, 1861), p. 5; see also J. Smith, *Charles Darwin and Victorian Visual Culture*, p. 96.

15 Jonathan Smith, *Charles Darwin and Victorian Visual Culture* (Cambridge: Cambridge University Press, 2006), p. 94; see pp. 99–114 for a full discussion of *The Birds of Great Britain* as a reply to Darwinism.

16 Ibid., p. 97.

17 Julie F. Codell, 'Empiricism, Naturalism and Science in Millais's Paintings', in Debra N. Mancoff (ed.), *John Everett Millais: Beyond the Pre-Raphaelite Brotherhood* (New Haven and London: Yale University Press, 2001), pp. 137–8; Paul Barlow, *Time Present and Time Past: The Art of John Everett Millais* (Aldershot: Ashgate, 2005), pp. 163–8. On Ruskin's response to Darwin in *Love's Meinie*, including his use of Gould, see J. Smith, *Charles Darwin and Victorian Visual Culture*, pp. 125–36.

18 Gould, *Introduction to the Trochilidae*, p. 18.

19 Duke of Argyll, *The Reign of Law* (London: Strahan, 1867), extracted in Gowan Dawson and Bernard Lightman (eds), *Victorian Science and Literature*, 8 vols (London: Pickering and Chatto, 2011–12) III, 197–212 (p. 207).

20 Charles Darwin, *The Descent of Man, and Selection in Relation to Sex*, ed. James Moore and Adrian Desmond (London: Penguin, 2004), p. 538.

21 See, for example, Darwin, *Descent of Man*, pp. 434–6, 446, 511–14, 538, 543.

22 Barlow, *Time Present and Time Past*, p. 166.

23 Darwin, *Origin of Species*, pp. 392–3.

24 Roger Dixon and Stefan Muthesius, *Victorian Architecture*, revised 2nd edn (London: Thames and Hudson, 2008), pp. 164, 168.

25 Julian Treuherz, *Ford Madox Brown: Pre-Raphaelite Pioneer* (Manchester: Manchester Art Gallery, 2011), p. 49.

26 Ibid., p. 50.

27 Julian Treuherz, 'Ford Madox Brown and the Manchester Murals', in John H. G. Archer (ed.), *Art and Architecture in Victorian Manchester: Ten Illustrations of Patronage and Practice* (Manchester: Manchester University Press, 1985), 162–207 (p. 169).

28 Treuherz, *Ford Madox Brown*, p. 50. On the Manchester Town Hall and the Manchester murals, see John H. G. Archer, 'A Classic of its Age', in Archer (ed.), *Art and Architecture in Victorian Manchester*, 127–61; Colin Cunningham and Prudence

Waterhouse, *Alfred Waterhouse 1830–1905: Biography of a Practice* (Oxford: Oxford University Press, 1992), pp. 51–62; Dixon and Muthesius, *Victorian Architecture*, pp. 166–8; Treuherz, 'Ford Madox Brown and the Manchester Murals' and *Ford Madox Brown*, pp. 47–59, 283–309.

29 Treuherz, *Ford Madox Brown*, p. 307.

30 Personal communication from Dr Roger Lovatt, Emeritus Fellow and College Archivist, Peterhouse, Cambridge, 20 February 2013.

31 Aristotle, *History of Animals*, trans. Richard Cresswell (London: Bohn, 1862), pp. 34–5.

32 Karen Meier Reeds and Tomoni Kinukawa, 'Medieval Natural History', in David C. Lindberg and Michael H. Shank (eds), *The Cambridge History of Science, Volume 2: Medieval Science* (Cambridge and New York: Cambridge University Press, 2013), 569–89 (p. 573).

33 Charles Rowley, *Fifty Years of Work without Wages* (London: Hodder and Stoughton, 1911), between pp. 28 and 29.

34 Treuherz, *Ford Madox Brown*, p. 274.

35 Rowley, *Fifty Years of Work without Wages*, p. 28.

36 See Jessica Ratcliff, *The Transit of Venus Enterprise in Victorian Britain* (London: Pickering and Chatto, 2008). On the Pre-Raphaelite painter and astronomer John Brett's involvement in controversy over observations of Venus, see James Mussell, 'Specular Reflections: John Brett and the Mirror of Venus', *19: Interdisciplinary Studies in the Long Nineteenth Century*, 17 (2013), http://www.19.bbk.ac.uk/articles/10.16995/ntn.670/ [accessed 12 February 2014].

37 See the ongoing research of the project 'Constructing Scientific Communities: Citizen Science in the 19th and 21st Centuries', based at the University of Oxford: https://conscicom.org [accessed 8 September 2016].

38 Quoted by Treuherz, *Ford Madox Brown*, p. 308.

CONCLUSION

1 Thomas Gordon Hake, *The New Day: Sonnets*, ed. W. Earl Hodgson (London: Remington, 1890), p. 10. For a fuller discussion of Hake's engagement with science in this poem, see John Holmes, '*The New Day*: Dr Hake and the Poetry of Science', *Journal of Victorian Culture*, 9 (2004), 68–89.

2 Theodore Watts-Dunton, *Aylwin* (London: Hurst and Blackett, 1899), p. 454. For a detailed critical discussion of *Aylwin*, see Catherine Maxwell, *Second Sight: The Visionary Imagination in Late Victorian Literature* (Manchester: Manchester University Press, 2008), pp. 166–96.

3 Max Nordau, *Degeneration*, 5th edn (New York: Appleton, 1895), p. 91.

4 Ibid., p. 94.

5 Ibid., pp. 83–4.

6 George Barlow, *The Higher Love: A Plea for a Nobler Conception of Human Love* (London: Fifield, 1905), pp. 10–11.

7 See John Holmes, *Dante Gabriel Rossetti and the Late Victorian Sonnet Sequence: Sexuality, Belief and the Self* (Aldershot: Ashgate, 2005), for a full discussion of the legacy of Rossetti's sonnets in later poets' explorations of their beliefs and identities.

8 Stefanie Jovanović-Kruspel and Alice Schumacher, *The Natural History Museum: The History of the Construction, its Conception and Architecture* (Vienna: Naturhistorischen Museums, 2017).

9 The fullest account of the architecture of the Royal Ontario Museum is Kelvin Browne, *Bold Visions: The Architecture of the Royal Ontario Museum* (Toronto: Royal Ontario Museum, 2008).

10 Charles Trick Currelly, *I Brought the Ages Home* (Toronto: Ryerson Press, 1956; reprinted Royal Ontario Museum, 1976), pp. 8, 29, 65–6, 127, 134, 159, 178, 185, 191–2; see also Katharine Lochnan, 'The Canadian Diaspora: Last Rights' in Lochnan and Jacobi (eds), *Holman Hunt and the Pre-Raphaelite Vision*, 191–206 (pp. 196–206).

11 Northrop Frye, 'Editor's Introduction' to Currelly, *I Brought the Ages Home*, vii–x (p. ix).

12 Walter Pater, *Studies in the History of the Renaissance*, ed. Matthew Beaumont (Oxford: Oxford University Press, 2010), pp. 4, 28.

13 Association of Leading Visitor Attractions, 2015 Visitor Figures, http://www.alva.org.uk/details.cfm?p=606 [accessed 12 September 2016].

14 The exhibitions were by Kurt Jackson ('Bees (and the odd wasp) in my Bonnet') and Levon Biss ('Microsculpture'). The poets in residence were John Barnie, Steven Matthews and Kelley Swain. Their poems in response to their time at the museum were published, together with a selection of Victorian poetry associated with the museum, as *Guests of Time: Poetry from the Oxford University Museum of Natural History* (Scarborough: Valley Press, 2016), ed. John Holmes with photographs by Scott Billings.

15 The catalogue of this exhibition was published as *Microsculpture: The Insect Photography of Levon Biss* (Oxford: Oxford University Museum of Natural History, 2016).

16 For a snapshot of Wilkie's plans for the Natural History Museum, see his website, http://www.kimwilkie.com/london/natural-history-museum/ [accessed 17 September 2016].

PHOTOGRAPH CREDITS

© National Portrait Gallery, London, 2017: 2, 3, 4, 136; © Tate Images 2017: 5, 7, 11, 12, 15, 23, 24, 27, 32, 33, 34, 40, 42, 46, 47, 48, 99; © *The Complete Works of Charles Darwin Online* 2017: 6; © Bridgeman Images: 8, 9, 13, 14, 17, 19, 21, 22, 25, 26, 29, 30, 31, 36, 38, 49, 51, 55, 101, 102, 104, 105, 140, 145, 146; © Manchester Art Gallery 2017: 16, 44, 54; Samuel and Mary R. Bancroft Memorial / Bridgeman Images: 28; © Trustees of the British Museum 2017: 35, 39, 67, 89; © Fitzwilliam Museum, University of Cambridge 2017: 37; © Groeninge Museum, Bruges 2017: 41; © Birmingham Museums & Art Gallery 2017: 43, 56; © Oxford University Museum of Natural History 2017: 57, 59, 60, 61, 62, 66, 69, 70, 71, 72, 73, 74, 75, 76, 77, 78a, 78b, 78c, 79, 80, 81, 82, 83, 84a, 84b, 85a, 85b, 87, 88, 91, 92, 93, 94, 95, 96, 109, 115, 116, 121; © Ruskin Library and Research Centre 2017: 58; © Anthony Majanlahti: 63; © the author: 64, 65, 86, 98, 119, 122a, 122b, 123, 132, 142, 143, 144, 148; Photo © Tarker / Bridgeman Images: 68; © University of Amsterdam 2017: 90; Gift of James Lawrence / Bridgeman Images: 97; Photo © Christie's Images / Bridgeman Images: 100; © Bequest of Grenville L. Winthrop / Bridgeman Images: 103; © The Trustees of the Natural History Museum, London 2017: 106, 107, 108, 111, 112, 113, 114, 117, 118a, 118b, 118c, 120a, 120b, 124, 125, 126, 127, 128, 131, 133; © Dr Phoenix: 129; © Robert Freidus and the *Victorian Web* 2011: 130; © The Royal Society 2017: 134, 135; © CSG CIC Glasgow Museums Collection / Bridgeman Images: 137; © Look and Learn / Illustrated Papers Collection / Bridgeman Images: 138; © Mark Andrew: 139; © Jonathan Goddard / DarkHarte Travel 2017: 141; © Naturhistorische Museum, Vienna 2017, photograph by Alice Schumacher: 147; © Levon Biss 2017: 149; © Kim Wilkie 2017: 150

INDEX

Page references in italics refer to images, including the subjects of portraits and illustrations from printed texts, and to poems quoted in full. Endnotes are indexed where they contain additional substantive information or discussion.